Universitext

For other titles published in this series, go to
www.springer.com/series/223

Eberhard Freitag

Complex Analysis 2

Riemann Surfaces, Several Complex Variables,
Abelian Functions, Higher Modular Functions

 Springer

Eberhard Freitag
Universität Heidelberg
Mathematisches Institut
Im Neuenheimer Feld 288
69120 Heidelberg
Germany
freitag@mathi.uni-heidelberg.de

ISSN 0172-5939   ISSN 2191-6675 (eBook)
ISBN 978-3-642-20553-8   ISBN 978-3-642-20554-5 (eBook)
DOI 10.1007/978-3-642-20554-5
Springer Heidelberg Dordrecht London New York

Library of Congress Control Number: 2011930840

Mathematics Subject Classification (2010): 30, 31, 32

Cover design: deblik, Berlin

Printed on acid-free paper

Springer is part of Springer Science+Business Media (www.springer.com)

for Hella

Contents

Introduction

This book is intended for readers who are familiar with the basics of elementary complex function theory, i.e. the topics that are usually covered in an introductory course on the theory of complex functions. Additionally, it would be useful for the reader to be familiar with the theory of elliptic functions and the theory of elliptic modular functions. These theories were treated in detail in the textbook Complex Analysis by Rolf Busam and the present author ([FB] in the reference list of the present book). There will be many cross-references to that book.

The goal of this book is to outline the new epoch of *classical complex analysis,* which was shaped decisively by Riemann. More than a half of this volume, Chaps. I–IV, is devoted to the theory of *Riemann surfaces.*

The theory of Riemann surfaces provides a new foundation for complex analysis on a higher level. As in elementary complex analysis, the subject matter is analytic functions. But the notion of an analytic function will have now a broader meaning. The domains of definition are not exclusively open parts of the complex plane or the Riemann sphere, but more general surfaces. Such functions automatically come up when one wants to describe an a priori multivalued function such as $f(z) = \sqrt{z^4 + 1}$ completely by a single-valued function. The natural domain of definition of this function f will turn out to be a twofold covering of the sphere which has the shape of a torus.

This example shows in outline that, in the theory of Riemann surfaces, we have to struggle with topological problems. The notion of "topology" here has a double meaning.

First, in the present-day mathematical world, topology is a *universal linguistic* tool for addressing questions of convergence in a context that is as broad as possible. This purpose is served by the notion of a topological space and derived notions such as "open set", "closed set", "neighborhood", "continuity", "convergence", and "compactness", just to give a few important examples, similarly to *set theory,* which is also a universally valid linguistic tool in mathematics. Readers of the first volume of our book have probably gained more mathematical experience in the meantime, so we can assume that they are familiar with the language of topological spaces. For the sake of completeness, we nevertheless introduce the fundamental terms of this language in an introductory section (I.0). This contains, very briefly, all of what we need. Most of the simple proofs will be skipped.

A second aspect of topology is that it is a mathematical discipline for investigating nontrivial *geometric problems*. For example, it is an important geometric fact that every compact, orientable surface is homeomorphic to a sphere with p handles. The number p is a topological invariant of the surface which determines its topological type. Topological theorems of significant mathematical substance will be derived completely in this book. Besides the topological classification of compact oriented surfaces, we shall also give an outline of covering theory. In particular, the universal covering and its relation to the fundamental group will be treated.

By the way, the development of topology was related to the fact that it is advantageous in the theory of Riemann surfaces, as well as providing a linguistic tool and means to attack serious geometric problems in the theory.

One of the main achievements of the theory of Riemann surfaces was that it enabled a proof of the *Jacobi inversion theorem* and, moreover, opened a deep understanding of it. We shall give a complete proof of the inversion theorem in this volume.

In a similar way to that in which meromorphic functions with two independent periods, called elliptic functions, arise in the inversion of elliptic integrals, we shall be led to meromorphic functions of several complex variables $z_1, \ldots z_p$ with $2p$ independent periods. Such functions are called *abelian functions*.

The inversion theorem is the prelude to a new mathematical development. It is necessary now to fix the notion of a meromorphic function of several complex variables. So, we are forced to establish a theory of complex functions of several variables. We can then introduce the notion of an *abelian function* and develop a theory of them which generalizes the theory of elliptic functions. One of the main results of this theory is that the field of abelian functions is finitely generated. It is an algebraic function field of transcendental degree $m \leq p$. Unlike the case $p = 1$, we can have $m = p$ in the case $p > 1$ only under very restrictive conditions. The Riemann period relations must hold. These relations are satisfied for the abelian functions which arise from the inversion of abelian integrals. It is not only for this reason that the case $m = p$ is the most interesting one.

By studying the *manifold* of all lattices $L \subset \mathbb{C}$, we are led to the elliptic modular functions. In the same manner, abelian functions lead us to a theory of *modular functions of several complex variables*. In the last chapter of this book, we give an introduction to this theory, which has been kept as simple as possible but nevertheless leads to quite deep results.

Therefore this book is a continuation of [FB] on a higher level. The usual Cauchy–Weierstrass theory of complex functions corresponds to the theory of Riemann surfaces and to the foundation of some basics of the theory of complex functions of several complex variables. The theory of elliptic functions is replaced by the theory of abelian functions, and the theory of elliptic modular functions by the theory of Siegel's modular functions.

We have tried to proceed *in as elementary a way as possible,* to give complete proofs and to develop all that is needed. Even small excursions into algebra to develop the necessary algebraic tools are included.

It is a great pleasure for me to thank the co-author of the first volume, Rolf Busam, for his help with the figures and with the general foundations of the theory.

I. Riemann Surfaces

The first four chapters are devoted to the theory of Riemann surfaces. It can be assumed that readers are already acquainted with several Riemann surfaces even if they are not familiar with the notion of a Riemann surface. In the book *Complex Analysis* [FB], the following examples occurred:

1) The torus $\mathbb{C}/L$, $L \subset \mathbb{C}$ a lattice ([FB], Chap. V).

2) Modular spaces $\mathbb{H}/\Gamma$, $\Gamma \subset \mathrm{SL}(2, \mathbb{Z})$ a congruence subgroup ([FB], Chap. VI).

3) Some plane affine or projective algebraic curves ([FB] Appendix to Sect. V.3).

A central aim of our description of the theory of Riemann surfaces is to treat these examples from a higher point of view and to deepen them. We shall be led to new insights. For example, we shall obtain dimension formulae for spaces of elliptic modular forms, which cannot be obtained by means of the elementary methods of [FB]. But the theory of Riemann surfaces is not at all exhausted by these examples.

In Chap. I, we shall treat the elementary theory of Riemann surfaces. This contains the basic definitions, i.e. the language of Riemann surfaces will be developed and important examples will be treated.

The second chapter is devoted to central *constructive problems*. It turns out to be useful to consider *harmonic* functions instead of analytic ones. The real parts of analytic functions are harmonic, and each harmonic function is at least locally the real part of an analytic function. We have to investigate *boundary value problems* and *singularity problems* for harmonic functions. Our main tool for their construction will be the alternating method of Schwarz.

The subject of Chap. III is the theory of *uniformization*. At its center stands the *uniformization theorem*, which states that a simply connected Riemann surface is conformally equivalent to the Riemann sphere, the complex plane, or the unit disk.

The big Chap. IV is devoted the theory of *compact Riemann surfaces*. It turns out that this theory is equivalent to the theory of *algebraic functions* of one variable. Historically, it was a big problem to generalize the theory of elliptic integrals to the theory of integrals of arbitrary algebraic functions. Riemann surfaces turned out to be a suitable instrument for solving this problem. At the center of the theory there are prominent theorems such as the *Riemann–Roch theorem, Abel's theorem,* and *Jacobi's inversion theorem.*

We shall now give a more detailed description of the present chapter. We start (Sect. 0) with a collection of some basic notions in topology. This concerns only topology as a linguistic device, as nowadays it is used in most mathematical disciplines. We can assume that the reader is more or less acquainted with these notions. To have a safe foundation we have collected together the necessary definitions and properties, but the mostly simple proofs have been skipped.

In Sect. 1, we introduce as quickly as possible the notion of a Riemann surface and the notion of an analytic map between Riemann surfaces. The simplest examples of Riemann surfaces are the Riemann sphere and tori.

In Sect. 2 we introduce an example of great historical significance, the *analytisches Gebilde*. *Gebilde* means something like "shaped object". This is a Riemann surface which arises in a natural way if an analytic function is analytically continued along paths. The point is that the continuation may depend on the choice of the path. When one considers all possible analytic continuations one obtains something like a multivalued function, such as $\sqrt{z}$, which can be considered as a two-valued function. For such a multivalued function a Riemann surface can be constructed, which covers the complex plane and is such that the originally multivalued function appears as a single-valued function on this covering. The *analytisches Gebilde* is an example of an abstract topological construction. The language of topological spaces finds a justification here. Nevertheless, one should not overrate this example. We shall make no use of it in the following theory and its applications. Hence it can be skipped by any reader who wants to proceed as quickly as possible.

In Sect. 3, an example of fundamental importance occurs, the Riemann surface of an *algebraic function*. More precisely, we shall associate a *compact connected Riemann surface* with an algebraic function of one variable. In Chap. IV we shall see that one can obtain all compact connected Riemann surfaces in this way. A priori, algebraic functions are multivalued. Hence it is natural to use the *analytisches Gebilde* for the construction of the surface. This is possible, and we describe this approach in Sect. 3. Independently, one can use a different approach, where we consider the *algebraic curve* which is associated with the algebraic function. To our mind, this a more elegant way. In both cases, in the first instance we obtain a surface which is not yet compact. It is more difficult to *compactify* it. In the literature, the compactification usually is managed by adding so-called Puiseux elements. This approach is concrete and explicit. But the pure topological background remains hidden. It rests on a pure topological proposition, a special case of covering theory:

Let $f : X \to \mathbb{E}^{\cdot}$ be a proper and locally topological map of a nonempty connected Hausdorff space X into the punctured disk. Then there exist a topological map $\sigma : X \to \mathbb{E}^{\cdot}$ and a natural number n such that f corresponds to

$$\mathbb{E}^{\cdot} \longrightarrow \mathbb{E}^{\cdot}, \quad q \longmapsto q^{n},$$

i.e. $f(x) = \sigma(x)^{n}$.

Hence the abstract space X has a hole. It is natural to extend the space X by adding an additional point, $\tilde{X} = X \cup \{a\}$, and to extend the maps f and σ to maps $\tilde{f} : \tilde{X} \to \mathbb{E}$ and $\tilde{\sigma} : \tilde{X} \to \mathbb{E}$ by $\tilde{f}(a) = 0$ and $\tilde{\sigma}(a) = 0$. One can topologize $\tilde{X}$ in such a way that $\tilde{\sigma}$ gets a topological map. The map $\tilde{f}$ is still proper but is no longer locally topological.

It seemed worthwhile to us to work out this special case of covering theory in connection with the construction of the compact Riemann surface of an algebraic function. Covering theory will not be treated in full generality until Chap. III.

0. Basic Topological Notions

A *topology* $\mathcal{T}$ on a set X is a system of subsets with the following properties:

1) $\emptyset$, $X \in \mathcal{T}$.
2) The intersection of finitely many sets from $\mathcal{T}$ belongs to $\mathcal{T}$.
3) The union of arbitrarily many sets from $\mathcal{T}$ belongs to $\mathcal{T}$.

A *topological space* is a pair $(X, \mathcal{T})$ consisting of a set X and a topology $\mathcal{T}$ on X. Since it is usually clear from the context which topology is being considered on the given set X at any moment, one usually writes X instead of $(X, \mathcal{T})$,

$$"X = (X, \mathcal{T})".$$

The elements of $\mathcal{T}$ are called the *open* sets of X.

We now describe some important construction principles for a topology.

I Metric Spaces and Their Topology

A metric d on a set X is a map

$$d : X \times X \longrightarrow \mathbb{R}_{\geq 0}$$

with the properties

a) $d(a, b) = 0 \iff a = b$;
b) $d(a, b) = d(b, a)$;
c) $d(a, c) \leq d(a, b) + d(b, c)$ $(a, b, c \in X)$.

We associate the metric space (X, d) with the "usual topology". A subset $U \subset X$ is called *open* if for every $a \in U$ there exists $\varepsilon > 0$ such that

$$U_\varepsilon(a) \subset U \quad (U_\varepsilon(a) := \{x \in X;\ d(a, x) < \varepsilon\}).$$

Example. The real line, the complex plane $\mathbb{C}$, or, more generally, $\mathbb{R}^n$ can be equipped with the Euclidean metric and henceforth with a structure in the form of a topological space.

II The Induced Topology

Let Y be a subset of a topological space $X = (X, \mathcal{T})$. Then Y can be equipped with a topology $\mathcal{T}|Y$, which is called the *induced topology* or *subspace topology*. A subset $V \subset Y$ belongs to $\mathcal{T}|Y$ iff there exists a subset $U \subset X$, $U \in \mathcal{T}$, such that

$$V = U \cap Y.$$

If Y already is an open subset of X, this simply means $V \in \mathcal{T}$.

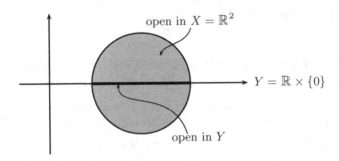

open in $X = \mathbb{R}^2$

$Y = \mathbb{R} \times \{0\}$

open in Y

III The Quotient Topology

Let X be a topological space and let $f : X \to Y$ be a map onto a set Y. Then Y is equipped with the *quotient topology*. A subset $V \subset Y$ is called *open* if its preimage $U := f^{-1}(V)$ is open in X.

Special case. Let "$\sim$" be an equivalence relation on X and Y, the set of all equivalence classes, and let $f : X \to Y$ be the canonical projection. Then Y is called the *quotient space* of X with respect to the given equivalence relation.

Examples.
a) The torus $X = \mathbb{C}/L$ ($L \subset \mathbb{C}$ a lattice).
b) The "modular space" $\mathbb{H}/\operatorname{SL}(2, \mathbb{Z})$.

IV The Product Topology

Let $X_1, \ldots, X_n$ be finitely many topological spaces. The Cartesian product

$$X = X_1 \times \cdots \times X_n$$

carries the *product topology*.

    A subset $U \subset X$ is called *open* if, for every point $a \in U$, there exist open subsets $U_1 \subset X_1, \ldots, U_n \subset X_n$ such that $a \in U_1 \times \cdots \times U_n \subset U$:

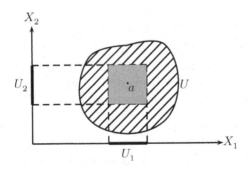

When $\mathbb{R}^n$ is equip]
comes from the Euclid
product topology of n
follows from the well-k
metric and the maxim
lent (see Exercise 1).

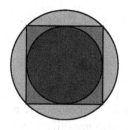

Derived Topological Notions

a) For subsets of a topological space X, we can say the following:

1) A subset $A \subset X$ is called *closed* if its complement $X - A$ is open.

2) A subset $M \subset X$ is called a *neighborhood* of a
point $a \in X$ if there exists an open subset $U \subset X$
with $a \in U \subset M$.

3) A point $a \in X$ is called a *boundary point* of $M \subset X$ if, in any neighborhood of a, one can find points
of M and points in the complement $X - M$.

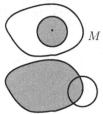

Notation.

$$\partial M := \text{set of boundary points,}$$
$$\bar{M} := M \cup \partial M.$$

We can show that $\bar{M}$ is the smallest closed subset of X which contains M, i.e.

$$\bar{M} = \bigcap_{\substack{M \subset A \subset X, \\ A \text{ closed}}} A.$$

Furthermore, we have

$$M \text{ closed} \iff M = \bar{M}.$$

We call $\bar{M}$ the *closure* of M.

b) For mappings $f : X \to Y$ between topological spaces, the map f is called
continuous at a point $a \in X$ if the preimage $f^{-1}(V(b))$ of any neighborhood
of $b := f(a)$ is a neighborhood of a (in X). We call f *continuous* when f is
continuous at all points.

The following properties are equivalent:

1) f is continuous,
2) the preimage of an arbitrary open set in Y is open (in X),
3) the preimage of an arbitrary closed set in Y is closed (in X).

The composition of two continuous maps

$$X \xrightarrow{f} Y \xrightarrow{g} Z$$

is continuous.

Universal Properties of the Constructed Topologies

The Induced Topology

Let Y be a subset of a topological space X which has been equipped with the induced topology. A map $f : Z \to Y$ from a third topological space Z into Y is continuous iff its composition with the natural inclusion i,

$$i \circ f : Z \longrightarrow X \qquad (i : Y \hookrightarrow X, \ i(y) = y),$$

is continuous. In particular, i is continuous:

$$
\begin{array}{ccc}
Z & \xrightarrow{\ f\ } & Y \\
 & {\scriptstyle i \circ f}\searrow & \downarrow{\scriptstyle i} \\
 & & X
\end{array}
$$

The Quotient Topology

Let $f : X \to Y$ be a surjective map of topological spaces where Y carries the quotient topology. A map $h : Y \to Z$ into a third topological space Z is continuous if and only if the composition

$$h \circ f : X \longrightarrow Z$$

is continuous:

$$
\begin{array}{ccc}
X & \xrightarrow{\ f\ } & Y \\
 & {\scriptstyle h \circ f}\searrow & \downarrow{\scriptstyle h} \\
 & & Z
\end{array}
$$

(In particular, f is continuous.)

The Product Topology

Let $X_1, \ldots, X_n$ be topological spaces and let

$$f : Y \longrightarrow X_1 \times \cdots \times X_n$$

be a map from another topological space Y into the Cartesian product, which has been equipped with the product topology. The map f is continuous iff all its components

$$f_j = p_j \circ f : Y \longrightarrow X_j,$$
$$p_j : X_1 \times \cdots \times X_n \longrightarrow X_j \quad j\,\text{th projection,}$$

are continuous:

$$Y \xrightarrow{\;f\;} X_1 \times \cdots \times X_n$$

with $p_\nu \circ f$ and p_ν to X_ν.

(In particular, the projections p_j are continuous.)

Topological Mappings

A map $f : X \to Y$ of topological spaces is called *topological* if it is bijective and if f and f^{-1} are both continuous. Two topological spaces X, Y are called *topologically equivalent* (or *homeomorphic)* if there exists a topological map $f : X \to Y$.

Examples
(The following two examples will be treated in more detail in Sect. 1.)

1) The 2-sphere

$$S^2 = \{x \in \mathbb{R}^3; \quad x_1^2 + x_2^2 + x_3^2 = 1\}$$

and the Riemann sphere are homeomorphic,

$$S^2 \simeq \bar{\mathbb{C}} = \mathbb{C} \cup \{\infty\}.$$

This can be shown, for example, by means of the stereographic projection (see [FB], Chap. III, in the appendix to Sects. 4 and 5 after Theorem A.8):

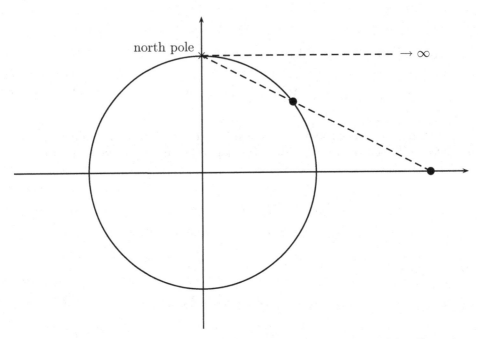

2) When $L \subset \mathbb{C}$ is a lattice, the torus $\mathbb{C}/L$ is homeomorphic to the Cartesian product of two circles,

$$\boxed{\mathbb{C}/L = \text{torus} \simeq S^1 \times S^1.}$$

Some Properties of Topological Spaces

1) A topological space X is called a *Hausdorff* space if for any two distinct points $a, b \in X$ there exist disjoint neighborhoods $U(a)$ and $U(b)$ (such that $U(a) \cap U(b) = \emptyset$).

2) A topological space X is called *compact* if it is Hausdorff and if it possesses the Heine–Borel covering property: when

$$X = \bigcup_{j \in I} U_j$$

is an arbitrary covering of X by open subsets, there exists a *finite* subset $J \subset I$ such that

$$X = \bigcup_{j \in J} U_j.$$

A subset Y of a topological space X is called *compact* if it is a compact topological space after it has equipped with the induced topology.

Some Properties of Compact Spaces

a) Compact subsets are closed.
b) A closed subset of a compact space is compact.
c) When $f : X \to Y$ is a continuous map between Hausdorff spaces, the image $f(K)$ of a compact subset $K \subset X$ is compact.
d) Let X be compact, let Y a Hausdorff space, and let $f : X \to Y$ be bijective and continuous. Then f is topological.
e) A Cartesian product $X_1 \times \cdots \times X_n$ of compact spaces is compact.

Locally Compact Spaces and Proper Mappings

A topological space X is called *locally compact* if it is Hausdorff and if each point admits a compact neighborhood.

A continuous map

$$f : X \longrightarrow Y$$

of locally compact spaces X and Y is called *proper* if the preimage $f^{-1}(K)$ of a compact set $K \subset Y$ is compact.

We now formulate two important facts for proper maps.

0.1 Remark. *Let $f : X \to Y$ be a proper map. Then the image $f(A)$ of a closed subset $A \subset X$ is closed.*

Proof. In a locally compact space, a set is closed iff its intersection with any compact subset is compact. So, let $K \subset Y$ be compact. Obviously, $K \cap f(A)$ is the image of the compact set $A \cap f^{-1}(K)$ and hence is compact. □

0.2 Remark. *Let $f : X \to Y$ be a proper map. Let $K \subset Y$ be a compact set and let $U \subset X$ be an open subset which contains the preimage of K,*

$$U \supset f^{-1}(K).$$

Then there exists an open subset

$$V \subset Y, \quad K \subset V,$$

with the property

$$f^{-1}(V) \subset U.$$

Proof. The set $X - U$ is closed. Since f is proper, $f(X - U)$ is closed in Y. We can take $V = Y - f(X - U)$. □

Convergent Sequences

A sequence (a_n) in a Hausdorff space X converges to $a \in X$ if for each neighborhood $U(a)$ there exists a number $N \in \mathbb{N}$ with

$$a_n \in U \text{ for } n \geq N.$$

We write, for brevity,
$$a_n \longrightarrow a \text{ for } n \longrightarrow \infty.$$

The limit a is uniquely determined (because of the Hausdorff property).

A map $f : X \to Y$ between Hausdorff spaces is called *sequence continuous* if

$$a_n \longrightarrow a \Longrightarrow f(a_n) \longrightarrow f(a).$$

A subset $A \subset X$ of a topological Hausdorff space is called *sequence closed* if for every sequence

$$[a_n \longrightarrow a, \quad a_n \in A \text{ for all } n] \Longrightarrow a \in A,$$

and it is called *sequence compact* if any sequence in A admits a cumulation point in A. (A point a is called a cumulation point of a sequence (a_n) if there exists a subsequence, which converges to a.)

We can show that

$$\begin{aligned}
\text{continuous} &\Longrightarrow \text{sequence continuous,} \\
\text{closed} &\Longrightarrow \text{sequence closed,} \\
\text{compact} &\Longrightarrow \text{sequence compact.}
\end{aligned}$$

The inverse direction is true for Hausdorff spaces with a *countable basis of the topology*.

This means the following.

There exists a sequence $U_1, U_2, U_3, \ldots$ of open subsets such that every open subset U can be written as the union of certain sets U_n. Every subset of X then has a countable basis of the topology as well.

Example. The spaces $\mathbb{R}^n$. Take Euclidean balls with a rational radius around centers which have a rational radius.

Connectedness

A topological space X is called *arcwise connected* if any two points of X can be joined by a curve in X. (A curve in X is a continuous map from a real interval into X.)

A topological space X is called *connected* if one the following two equivalent conditions is satisfied:

1) Every locally constant map $f : X \to M$ into any set M is constant. It is sufficient to take for M any fixed set which contains at least two elements.

2) When $X = U \cup V$ is the union of two disjoint open subsets U, V, then U or V is empty (hence $V = X$ or $U = X$).

By the mean value theorem, every real interval is connected. As a consequence, every arcwise connected space is connected. Usually, the reverse direction is false. But for manifolds, and in particular for surfaces, see below.

Arc Components

Two points of a topological space X are called equivalent iff they can be joined by a curve. The equivalence classes with respect to this equivalence relation are called arc components of X. They are arcwise connected.

A *(topological) manifold* X is a Hausdorff space such that every point admits an open neighborhood which is homeomorphic to an open subset of some $\mathbb{R}^n$. A nontrivial result states that n is unique, but we shall not make use of this. We obviously have the following result.

Let X be a manifold. Then the arc components are open in X.

Hence the arc components are manifolds themselves. The arc components of a manifold are called *connected components*.

As a special case, we can see that a manifold is connected if and only it is arcwise connected. In the theory of manifolds, it is usually sufficient to restrict ourselves to connected manifolds.

Exercises for Sect. I.0

1. Two metrics d, d' on a set X are called (strictly) *equivalent* if there exist constants c, c' with the property

$$cd(x,y) \le d'(x,y) \le c'd(x,y) \quad (x,y \in X).$$

Show that equivalent metrics define the same topology.

Example. The maximum metric and the Euclidean metric of $\mathbb{R}^n$ are equivalent; more precisely,

$$\max_{1 \le \nu \le n} |x_\nu - y_\nu| \le \sqrt{\sum_{\nu=1}^{n} (x_\nu - y_\nu)^2} \le \sqrt{n} \max_{1 \le \nu \le n} |x_\nu - y_\nu|.$$

2. Let $\mathcal{T}$ and $\mathcal{T}'$ be two topologies on a set X. We say that $\mathcal{T}$ is *finer* than $\mathcal{T}'$ or $\mathcal{T}'$ is *coarser* than $\mathcal{T}$ if $\mathcal{T}' \subset \mathcal{T}$, i.e. every open subset with respect to $\mathcal{T}'$ is open with respect to $\mathcal{T}$. Show that the following statements are true:

a) Let $X = (X, \mathcal{T})$ be a topological space $Y \subset X$ which has been equipped with the induced topology $\mathcal{T}|Y$, and let

$$j : Y \longrightarrow X, \quad y \longmapsto y,$$

be the canonical injection. Then $\mathcal{T}|Y$ is the coarsest topology on Y such that the canonical injection is continuous.

b) The product topology on a product $X = X_1 \times \ldots \times X_n$ of topological spaces X_j is the coarsest topology on X such that the projections

$$p_j : X \longrightarrow X_j, \quad (x_1, \ldots, x_n) \longmapsto x_j,$$

are continuous.

c) Let $f : X \to Y$ be a surjective mapping of a topological space onto a set Y. The quotient topology on Y is the finest topology for which f is continuous.

3. Show that
   a) every proper injective map $\mathbb{R} \to \mathbb{R}$ is surjective;
   b) every proper analytic map $\mathbb{C} \to \mathbb{C}$ is surjective.

4. Let X be a Hausdorff space with a countable basis of the topology and such that the projection

$$X \times \bar{\mathbb{C}} \longrightarrow \bar{\mathbb{C}}$$

is closed. Show that X is compact.

Hint. Argue indirectly and assume that there exists a sequence (a_n) without a cumulation point. Then the set $\{(a_n, n); \ n \in \mathbb{N}\}$ is closed. By assumption, its image in $\bar{\mathbb{C}}$ is closed.

5. From the previous exercise, deduce that the following is true.

   Let X, Y be locally compact spaces with a countable basis of the topology. A continuous map $f : X \to Y$ is proper if and only if it is universally closed. This means that the map

$$X \times Z \longrightarrow Y \times Z; \quad (x, z) \longmapsto (f(x), z),$$

is closed for every Hausdorff space Z.

1. The Notion of a Riemann Surface

Riemann surfaces are surfaces in the sense of topology with an additional structure. Surfaces are special manifolds (Sect. 0). For the sake of completeness,
we introduce again the notion of a (topological) surface. In the following, X
denotes a topological Hausdorff space.

1.1 Definition. *A (two-dimensional topological)* **chart** φ *on* X *is a topological map*

$$\varphi : U \longrightarrow V$$

of an open subset $U \subset X$ *onto an open subset* $V \subset \mathbb{C}$ *of the complex plane.*

Let

$$\varphi : U \longrightarrow V, \quad \psi : U' \longrightarrow V'$$

be two charts on X. Then we can consider the maps

$$\varphi_0 : U \cap U' \longrightarrow \varphi(U \cap U'), \quad \varphi_0(a) = \varphi(a),$$
$$\psi_0 : U \cap U' \longrightarrow \psi(U \cap U'), \quad \psi_0(a) = \psi(a).$$

In contrast to the severe set-theoretic convention, we denote the map $\psi_0 \circ \varphi_0^{-1}$
simply by $\psi \circ \varphi^{-1}$:

$$\psi \circ \varphi^{-1} : \varphi(U \cap U') \longrightarrow \psi(U \cap U').$$

Obviously, $\varphi(U \cap U')$, $\psi(U \cap U')$ are open subsets of the complex plane. The
map $\psi \circ \varphi^{-1}$ is called the *chart transformation*. It is only of interest when the
intersection $U \cap U'$ is not empty (but can be considered also in this case, since
the empty set is very patient).

By definition, a (topological) surface is a Hausdorff space such that every
point $a \in X$ admits a chart, such that a is contained in its domain of definition.

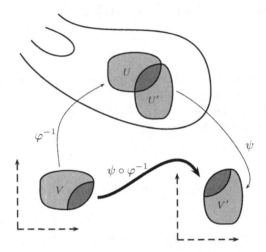

1.2 Definition. *A (two-dimensional) **atlas** $\mathcal{A}$ on a topological space X is a set of (two-dimensional) charts $\varphi : U_\varphi \to V_\varphi$ such that their domains of definition cover X:*

$$X = \bigcup_{\varphi \in \mathcal{A}} U_\varphi.$$

When there exists a two-dimensional atlas on the topological space X, then X looks locally like the complex plane $\mathbb{C} \simeq \mathbb{R}^2$.

So a (topological) surface is a Hausdorff space which admits a (two-dimensional) atlas.

1.3 Definition. *Two charts φ, ψ on a surface are called **analytically compatible** if the chart transformation*

$$\psi \circ \varphi^{-1} : \varphi(U \cap U') \longrightarrow \psi(U \cap U')$$

is biholomorphic (= conformal).

1.4 Definition. *An atlas $\mathcal{A}$ on a surface X is called **analytic** if any two charts from $\mathcal{A}$ are analytically compatible.*

Of course, a given topological surface may admit many atlases if there exists one. When $\mathcal{A}$ and $\mathcal{B}$ are analytic atlases on X, then it may happen that $\mathcal{A} \cup \mathcal{B}$ is also an analytic atlas. This means that every chart of $\mathcal{A}$ is analytically equivalent to any chart of $\mathcal{B}$. Such atlases will do the same job. Therefore we shall call them "essentially equal".

1.5 Definition.   *Two analytic atlases $\mathcal{A}$, $\mathcal{B}$ are called **essentially equal** if $\mathcal{A} \cup \mathcal{B}$ is analytic as well.*

Obviously, the relation "essentially equal" is an equivalence relation. We denote the class of all atlases which are essentially equal to $\mathcal{A}$ by $[\mathcal{A}]$.

1.6 Definition.   *A **Riemann surface** is a pair $(X, [\mathcal{A}])$ which consists of a topological surface X and a full class of essentially equal analytic atlases.*

Hence Riemann surfaces are topological surfaces with a distinguished analytic atlas. Two analytic atlases define the same Riemann surface if they are essentially equal. We shall see that on a given topological surface there can exist many analytic atlases which are essentially different and hence define different structures as a Riemann surface.

We allow the notation $(X, \mathcal{A})$ instead of $(X, [\mathcal{A}])$. But we must bear in mind that $(X, \mathcal{A})$ and $(X, \mathcal{B})$ are equal when $\mathcal{A}$ and $\mathcal{B}$ are analytically equivalent. As a rule, it will be clear from the context which analytic atlas is being considered at any moment. In this case we simply write $X = (X, [\mathcal{A}])$.

The definition of analytic compatibility has been devised such that notions from complex analysis of the complex plane, which are invariant under conformal transformation, can be transferred to Riemann surfaces. A fundamental example of this is the notion of an *analytic map*, which we shall give our attention to now.

Let $f : X \longrightarrow Y$ be a continuous map between Riemann surfaces $X = (X, \mathcal{A})$ and $Y = (Y, \mathcal{B})$. We consider two charts

$$\varphi : U_\varphi \longrightarrow V_\varphi \quad \text{from } \mathcal{A},$$
$$\psi : U_\psi \longrightarrow V_\psi \quad \text{from } \mathcal{B}.$$

Using somewhat sloppy notation, we can consider the function

$$f_{\varphi,\psi} = \psi \circ f \circ \varphi^{-1}.$$

Its domain of definition is an open part of the complex plane, namely

$$\varphi(f^{-1}(U_\psi) \cap U_\varphi) \subset V_\varphi.$$

It takes values in V_ψ:

$$f_{\varphi,\psi} : \varphi(f^{-1}(U_\psi) \cap U_\varphi) \longrightarrow V_\psi, \qquad f_{\varphi,\psi}(z) = \psi f \varphi^{-1}(z).$$

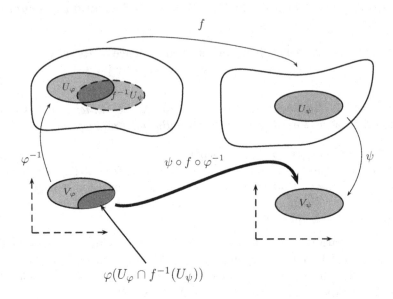

1.7 Lemma.   *Let $f : X \longrightarrow Y$ be a continuous map between topological surfaces. Assume that analytic atlases $\mathcal{A}$ on X and $\mathcal{B}$ on Y have been distinguished. The following two statements are equivalent.*

a) *There exists a pair of charts $\varphi \in \mathcal{A}$ such that $a \in U_\varphi$ and $\psi \in \mathcal{B}$ such that $b \in U_\psi$. The function $f_{\varphi,\psi}$ (which is defined on an open neighborhood of $\varphi(a)$) is analytic in an open neighborhood of $\varphi(a)$.*

b) *Condition a) holds analogously for **every** pair of charts $\varphi \in \mathcal{A}$ such that $a \in U_\varphi$ and $\psi \in \mathcal{B}$ such that $b \in U_\psi$.*

Additional remark. *Conditions a) and b) carry over from $\mathcal{A}$ and $\mathcal{B}$ to any pair of essentially equivalent atlases $\mathcal{A}'$, $\mathcal{B}'$.*

The proof is an immediate consequence of the definition of analytic compatibility.   □

1.8 Definition.   *A continuous map $f : X \to Y$ of Riemann surfaces $X = (X, \mathcal{A})$, $Y = (Y, \mathcal{B})$ is called analytic at a point $a \in X$ if the conditions a) and b) in Lemma 1.7 are satisfied.*

It is obvious that Lemma 1.7 is unavoidable for a meaningful definition of the notion of an "analytic map". The notion of analytic compatibility has been devised in such a way that Lemma 1.7 holds:

> The notion of a Riemann surface has been molded in such a way that one can define analytic mappings between Riemann surfaces in a meaningful way .

A map $f : X \to Y$ between Riemann surfaces is called *analytic* (or *holomorphic*) if it is continuous and if it is analytic at any point.

Some Simple Permanence Properties

1) The identity map
$$\mathrm{id}_X : X \longrightarrow X$$

   is analytic.

2) The composition of analytic maps between Riemann surfaces
$$X \xrightarrow{\ f\ } Y \xrightarrow{\ g\ } Z$$

   is analytic.

1.9 Definition. *A map*
$$f : X \longrightarrow Y$$

between Riemann surfaces is called **biholomorphic** *or* **conformal** *if it is topological and if f and f^{-1} are both analytic.*

As in the case of open subsets of the complex plane, the map f^{-1} is automatically analytic when f is bijective and analytic. This is a statement of a local nature, and can be reduced to the case of open subsets of $\mathbb{C}$ by means of charts (see [FB], Sect. IV.4.2). See also Exercise 4 in this section.

Two Riemann surfaces X, Y are called *biholomorphically equivalent* or *conformally equivalent* if there exists a biholomorphic map $f : X \to Y$. They are then also topologically equivalent. The reverse statement is false.

Simple Examples of Riemann Surfaces

To define a Riemann surface, one has to find an analytic atlas $\mathcal{A}$ on a topological surface X.

Let $U \subset X$ be an open subset of a Riemann surface $X = (X, \mathcal{A})$ and let $\varphi : U_\varphi \to V_\phi$ be a chart on X. We can then consider the restricted chart

$$\varphi|U : U \cap U_\varphi \xrightarrow{\ \sim\ } \varphi(U \cap U_\varphi).$$

Obviously, the set

$$\mathcal{A}|U := \{\varphi|U; \quad \varphi \in \mathcal{A}\}$$

is an analytic atlas on U and thus provides U with a structure in the form of a Riemann surface. Of course, the class $[\mathcal{A}|U]$ depends only on the class $[\mathcal{A}]$. We shall always equip an open subset with this structure.

A map $f : Y \to U$ is analytic if and only if its composition with the natural inclusion $i : U \hookrightarrow X$ is analytic.

We have, furthermore, the following result. Let $f : X \to Y$ be a map between Riemann surfaces and let

$$X = \bigcup_{i \in I} U_i$$

be a covering of X by open subsets. Then f is analytic iff all restrictions

$$f|U_i : U_i \longrightarrow Y$$

are analytic.

The Complex Plane as a Riemann Surface

We can consider the "identical chart"

$$\mathrm{id}_{\mathbb{C}} : \mathbb{C} \longrightarrow \mathbb{C}.$$

This forms an obviously analytic atlas and as such establishes $\mathbb{C}$ with a structure in the form of a Riemann surface. As a consequence, every open part $D \subset \mathbb{C}$ is equipped with such a structure. The identity $\mathrm{id}_D : D \to D$ gives an analytic atlas.

Let $X = (X, \mathcal{A})$ be an arbitrary Riemann surface. An analytic map

$$f : X \longrightarrow \mathbb{C}$$

is called an analytic function. This means that for any chart $\varphi \in \mathcal{A}$, the function

$$f_\varphi := f \circ \varphi^{-1} : V_\varphi \longrightarrow \mathbb{C}$$

is analytic in the usual sense. In the special case of an open subset $X \subset \mathbb{C}$, we of course obtain the usual notion of an analytic function.

We denote by $\mathcal{O}(X)$ the set of all analytic functions on X. Obviously, the following are true:

1) $f, g \in \mathcal{O}(X) \Longrightarrow f + g, \ f \cdot g \in \mathcal{O}(X)$.
2) The constant functions are in $\mathcal{O}(X)$.

In particular, $\mathcal{O}(X)$ is a $\mathbb{C}$-algebra.

The Riemann Sphere as Riemann Surface

Recall ([FB], Appendix to Sects. III.4 and III.5) that a subset $U \subset \bar{\mathbb{C}} = \mathbb{C} \cup \{\infty\}$ is called open iff $U \cap \mathbb{C}$ is open and if, in the case $\infty \in U$, there exists a number $C > 0$ with the property

$$z \in \mathbb{C}, \quad |z| > C \Longrightarrow z \in U.$$

Obviously, this defines a topology which equips $\bar{\mathbb{C}}$ with a structure in the form of a Hausdorff space. When this topology is induced on the (open) subset $\mathbb{C}$, we obtain the usual topology of the complex plane.

We define two charts on $\bar{\mathbb{C}}$:

1) $$\bar{\mathbb{C}} - \{\infty\} = \mathbb{C} \xrightarrow{\text{id}_{\mathbb{C}}} \mathbb{C}.$$
2) $$\bar{\mathbb{C}} - \{0\} \longrightarrow \mathbb{C}. \qquad z \longmapsto 1/z \quad (1/\infty = 0).$$

The chart transformation is

$$\mathbb{C} - \{0\} \longrightarrow \mathbb{C} - \{0\}, \qquad z \longmapsto 1/z.$$

This is a conformal map. Hence the two charts define an analytic atlas. In this way, we obtain a structure in the form of a Riemann surface on $\bar{\mathbb{C}}$.

Now we can consider holomorphic maps $f : X \to \bar{\mathbb{C}}$ from an arbitrary Riemann surface into the Riemann sphere. Such maps have already been considered in [FB] in the appendices to Sects. III.4 and III.5, where we introduced *meromorphic* functions. We reformulate Definition A1 given there.

1.10 Remark. *Let $U \subset \mathbb{C}$ be an open subset of the complex plane. For a map $f : U \to \bar{\mathbb{C}}$ from U into the Riemann sphere, the following two statements are equivalent:*

1) *f is a meromorphic function.*
2) *f is an analytic map of Riemann surfaces and the set of points $f^{-1}(\{\infty\})$ which map to ∞ is discrete (in U).*

This simple observation leads us to the following definition.

1.11 Definition. *A meromorphic function f on a Riemann surface X is an analytic map*

$$f : X \longrightarrow \bar{\mathbb{C}}$$

such that the set of points which map to ∞, $f^{-1}(\{\infty\})$, is a discrete subset of X.

Of course, the constant function $f(z) = \infty$ is an analytic map but not a meromorphic function when X is not empty.

We denote by $\mathcal{M}(X)$ the set of all meromorphic functions on X. If $f, g \in$
$\mathcal{M}(X)$ are two meromorphic functions, then we can define in an obvious way
(compare [FB], Chap. III, Appendix A) their sum and product

$$f + g, \quad f \cdot g \in \mathcal{M}(X).$$

In particular, $\mathcal{M}(X)$ is a ring. If f is a meromorphic function whose set of
zeros is discrete, one can define the meromorphic function $1/f$. The set of all
holomorphic functions $\mathcal{O}(X)$ can be embedded into $\mathcal{M}(X)$:

$$\mathcal{O}(X) \hookrightarrow \mathcal{M}(X),$$
$$f \mapsto i \circ f \quad (i : \mathbb{C} \hookrightarrow \bar{\mathbb{C}} \text{ canonical inclusion}).$$

The image consists of all meromorphic functions which do not have the value
∞. Usually, we identify f and $i \circ f$. Hence analytic functions are meromorphic
functions which do not take the value ∞.

We want to show that $\mathcal{M}(X)$ is a field when X is connected and nonempty.
For this, we need a generalization of the identity theorem (see [FB], Proposition
III.3.1).

1.12 Lemma. *Let*

$$f, g : X \longrightarrow Y$$

*be two analytic maps of a connected Riemann surface into another Riemann
surface Y. Assume that there exists a subset $S \subset X$ which has a cumulation
point[*] in X and is such that f and g coincide on S. Then $f = g$.*

Corollary. *Let $f : X \to Y$ (X connected) be a nonconstant analytic map.
Then the set $f^{-1}(b)$ is discrete in X for all $b \in Y$.*

Corollary. *Let $f : X \to \bar{\mathbb{C}}$ (X connected) be an analytic map which is not
constant ∞. Then f is a meromorphic function.*

Corollary. *The set $\mathcal{M}(X)$ of all meromorphic functions on a connected
nonempty Riemann surface is a field.*

For the proof of Lemma 1.12, we consider the set of all cumulation points of
the set coincidence set $\{x \in X; \quad f(x) = g(x)\}$. We have to show that this
set is open and closed. Since this statement is of local nature, we reduce it
by taking charts to open subsets of the plane. Then we can apply the usual
identity theorem ([FB], Theorem III.3.2).

---

[*] This is a point $a \in X$ such that every neighborhood of a contains infinitely many
elements of S.

The Torus as a Riemann Surface

Let L be a lattice in the projective plane $\mathbb{C}$. We equip the torus

$$X = \mathbb{C}/L := \mathbb{C}/\!\sim \quad (a \sim b \Longleftrightarrow a - b \in L)$$

with the quotient topology to obtain a compact connected topological space:

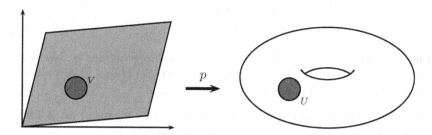

An open subset $V \subset \mathbb{C}$ is called "small" if two different points from V modulo L are never equivalent. For example, V is small when it is contained in the interior of a fundamental parallelogram.

We denote by $U \subset X$ the image of V under the canonical projection

$$p : \mathbb{C} \longrightarrow X, \qquad a \longmapsto [a] \quad (= \{b \in \mathbb{C},\ b - a \in L\}).$$

The restriction of p defines a bijective map

$$V \xrightarrow{\sim} U.$$

It is easy to see that this map is topological. Its inverse,

$$\varphi_V : U \xrightarrow{\sim} V,$$

is a chart on X. The set of all these charts is an atlas $\mathcal{A}$ on X.

Claim. The atlas $\mathcal{A}$ is analytic.

Proof. Let V and $\tilde{V}$ be two small open subsets of $\mathbb{C}$. We have to show that the chart transformation is analytic. For this, we can assume (possibly after shrinking V and $\tilde{V}$) that the images U and $\tilde{U}$ in X are equal. We can also assume that V and $\tilde{V}$ are connected. Then the chart transformation γ is a topological map

$$\gamma : V \xrightarrow{\sim} \tilde{V}.$$

For every $a \in V$, there has to exist an $\omega(a) \in L$ such that

$$\gamma(a) = a + \omega(a).$$

Since φ is continuous and L is discrete, the function $\omega(a) = \gamma(a) - a$ has to be locally constant. Since V is connected, ω must be constant. Hence the chart transformation is a translation; in particular, it is analytic. □

Hence the torus $\mathbb{C}/L$ has been equipped with a structure in the form of a Riemann surface.

Obviously, we can make the following statement.

1.13 Remark. *Let U be an open subset of the torus. A map*

$$f : U \longrightarrow Y$$

into a Riemann surface Y is analytic if and only if the composition with the projection

$$f \circ p : p^{-1}(U) \longrightarrow Y$$

is analytic.

Corollary. *(Special case $Y = \bar{\mathbb{C}}$). The meromorphic functions on the torus X are in one-to-one correspondence with the elliptic functions with respect to L (by means of $F \mapsto f = F \circ p$).*

Any two $\mathbb{R}$-bases can be transformed into each other by means of an $\mathbb{R}$-linear map. Hence two tori are always topologically equivalent:

$$\mathbb{C}/L \approx (\mathbb{R} \times \mathbb{R})/(\mathbb{Z} \times \mathbb{Z}) \approx \mathbb{R}/\mathbb{Z} \times \mathbb{R}/\mathbb{Z} \quad (\approx S^1 \times S^1).$$
$$\uparrow \qquad\qquad\qquad\qquad \uparrow$$
$$\text{topologically} \qquad\qquad \text{circle}$$

As a rule, $\mathbb{R}$-linear maps are usually not $\mathbb{C}$-linear. Hence tori need not be conformally equivalent. Actually, the following is true.

1.14 Proposition. *Two tori $\mathbb{C}/L$, $\mathbb{C}/L'$ are conformally equivalent iff L and L' can be transformed into each other by rotation and scaling (i.e. by multiplying by a complex number).*

This also shows that topologically equivalent Riemann surfaces need not be biholomorphic. On a given topological surface, there may exist essentially different structures in the form of a Riemann surface.

A proof of Proposition 1.14 follows easily from the *covering theory of Riemann surfaces*. However, since the tools for a proof are already available at this point (even if they are not so easily used), we shall sketch a proof here.

Proof of Proposition 1.14. Let

$$f : \mathbb{C}/L \longrightarrow \mathbb{C}/L'$$

be biholomorphic. Step by step, one shows the following:

1) There exists a continuous map $F : \mathbb{C} \to \mathbb{C}$ such that the diagram

$$
\begin{array}{ccc}
\mathbb{C} & \xrightarrow{\ F\ } & \mathbb{C} \\
\downarrow & & \downarrow \\
\mathbb{C}/L & \xrightarrow{\ f\ } & \mathbb{C}/L'
\end{array}
$$

commutes.

2) F is analytic.

3) dF/dz is an elliptic function with respect to L and hence is constant. This implies

$$F(z) = az + b \qquad (a \neq 0).$$

4) We have

$$aL = L'.$$

It follows inversely that the associated tori are conformally equivalent.

Only the statement 1) is not obvious.

Proof of 1). We denote by $p : \mathbb{C} \to \mathbb{C}/L$ and $p' : \mathbb{C} \to \mathbb{C}/L'$ the natural projections. First, we prove the following uniqueness statement.

Let $M \subset \mathbb{C}$ be an arcwise connected subset of $\mathbb{C}$ and let $a \in M$, $b \in \mathbb{C}$ be points with the property $f(p(a)) = p(b)$. Then there exists at most one continuous map $F : M \to \mathbb{C}$ with the properties $F(a) = b$, $f(p(z)) = p'(F(z))$ for all $z \in M$. This is trivial if there exists a small open (with respect to L') subset $U \subset \mathbb{C}$, $b \in U$ with the property $f(p(M)) \subset p'(U)$, since for reasons of connectedness $F(M)$ must be contained in U. For the general proof of the uniqueness statement, we can assume that M is the image of a curve. By dividing the parameter interval into small pieces, we reduce the statement to the first case. We call F a lifting of f.

Now, the following statement follows from the statement of uniqueness. Let M and N be two subsets of $\mathbb{C}$ whose intersection is arcwise connected, and let $F : M \to \mathbb{C}$, $G : N \to \mathbb{C}$ be two lifts of f which agree in at least one point of the intersection $M \cap N$. Then they agree on the whole intersection and glue to a continuous function on the union.

By the way, if F, G are two lifts, we may choose a lattice element ω' such that $F(z)$ and $G(z) + \omega'$ agree in a point of the intersection.

After this preparation, we prove the existence of a (continuous) lift. Because of the uniqueness statement, we can restrict ourselves to constructing the lift F for a *compact* rectangle M. We decompose the rectangle into four congruent subrectangles by halving its edges. Because of the second step, it is enough to construct lifts for each of the four subrectangles. We can divide the four subrectangles in the same way. We continue this procedure until each rectangle R is so small that $f(p(R))$ is contained in $p'(U)$ for a small open set $U \subset \mathbb{C}$. Now the existence of a lift $F : R \to \mathbb{C}$ is trivial. A simple compactness argument, which we leave to the reader, shows that after finitely many steps we obtain a subdivision with the desired property.

The Maximal Atlas of a Riemann Surface

Since open subsets (using the induced structure) of Riemann surfaces are Riemann surfaces as well, the following definition is possible.

1.15 Definition. *Let $(X, \mathcal{A})$ be a Riemann surface. An **analytic chart** on X is a biholomorphic map $\varphi : U \to V$ of an open subset $U \subset X$ onto an open subset V of the plane.*

Of course, elements of the defining atlas $\mathcal{A}$ are analytic charts, and the elements of an essentially equal atlas are also analytic charts. It is also clear that two analytic charts are analytically compatible. This means nothing more than that the set of all analytic charts is the biggest atlas which is essentially equal to $\mathcal{A}$. For this reason, we denote the set of all analytic charts by $\mathcal{A}_{\max}$. Two analytic atlases $\mathcal{A}$ and $\mathcal{B}$ on X are essentially equal iff the associated maximal atlases coincide. We can express this also as follows: in any class of essentially equal analytic atlases there exists a unique maximal atlas, and this is the set of all analytic charts in the sense of Definition 1.15.

During the introduction of the notion of a Riemann surface, we equipped topological surfaces with an equivalence class of essentially equal analytic atlases. Alternatively, we could have equipped the surface with a maximal analytic atlas and called this a Riemann surface. It is only a question of taste how to start. We feel that the notion of a maximal analytic atlas (before it is recognized as the set of all analytic charts) is somewhat artificial and unesthetic.

Usually, the maximal atlas is much bigger than the atlas $\mathcal{A}$ with which we started. For example, we have equipped $\mathbb{C}$ with the tautological atlas $\{\mathrm{id}_\mathbb{C}\}$. In this case $\mathcal{A}_{\max}$ consists of all conformal mappings $\varphi : U \to V$ between open sets U, V from the plane.

The defining atlas $\mathcal{A}$ should be considered just as a vehicle to define arbitrary analytic charts. In the literature, Riemann surfaces are usually introduced as surfaces which are equipped with a maximal analytic atlas. But this requires formal effort before one can introduce even the simplest examples of Riemann surfaces, such as the complex plane.

Some Elementary Properties of Riemann Surfaces

We now formulate some results which may be deduced without any effort from the usual type of complex analysis, as can be found for example in [FB].

1.16 Remark.

1) *A nonconstant analytic map between connected Riemann surfaces is open.*

2) *An analytic function on a connected Riemann surface whose absolute values attain a maximum is constant.*

3) *Let $f : X \to Y$ be a continuous map between Riemann surfaces which is analytic outside a discrete subset $S \subset X$. Then f is analytic everywhere.*

4) *Let $f : X \to Y$ be an injective analytic map. Then $f(X)$ is open and the induced map $X \to f(X)$ is biholomorphic.*

We skip the simple proofs (see the Exercises).

In the usual complex analysis, the local mapping behavior of analytic functions is described as follows (see the proof of Remark I.3.3 in [FB]):

An analytic function f such that $f(0) = 0$ is, in a small neighborhood of 0, either constant or the composition of a conformal map with an nth power. Here n is a natural number.

We want to formulate this result for Riemann surfaces. First, we notice that for every point $a \in X$ of a Riemann surface there exists an analytic chart $\varphi : U \to \mathbb{E}$, $a \in U$, whose image is the unit disk. We choose an arbitrary analytic chart $\psi : U' \to V'$, $a \in U'$, and then replace U' by the inverse image of a small disk around $\psi(a)$. Now φ is obtained by restricting ψ to this inverse image and composing it with a conformal map from the small disk onto the unit disk. This construction gives an arbitrarily small U in the sense that, for a given neighborhood W of a, we can find U such that $a \in U \subset W$.

1.17 Remark. *Let $f : X \to Y$ be a nonconstant analytic map of a connected Riemann surface X into a Riemann surface Y. Let $a \in X$ be a point and let $b = f(a)$ be its image. There exist analytic charts*

$$\varphi : U \longrightarrow \mathbb{E}, \ a \in U \subset X, \qquad \psi : V \longrightarrow \mathbb{E}, \ b \in V \subset Y, \qquad f(U) = V,$$

and a natural number n such that the diagram

commutes $(\psi(f(x)) = \varphi(x)^n)$.

For the proof, we can assume that X and Y are open subsets of $\mathbb{C}$ and that $a = b = 0$. Then we can use the local mapping property of the usual type of analytic functions. (see [FB], Theorem III.3.3).   □

Exercises for Sect. I.1

1. Let a be a point on a Riemann surface X. Show that any analytic function $f : X - \{a\} \to \mathbb{E}$ extends to an analytic function $X \to \mathbb{E}$.

2. Let $f : X \to Y$ be a nonconstant analytic map of a connected Riemann surface into a Riemann surface V. Show that f is open, which means that open sets are mapped onto open sets. In particular, $f(X)$ is open.

3. Any nonconstant analytic map $f : X \to Y$ of a *compact* Riemann surface into a connected Riemann surface is surjective. As polynomials can be considered as holomorphic mappings from the Riemann sphere into itself, show that one can obtain a new proof of the "fundamental theorem of algebra".

4. Let $f : X \to Y$ be a bijective and analytic map of Riemann surfaces. Show that f is biholomorphic.

5. Let $f : X \to Y$ be an injective analytic map of Riemann surfaces. Show that f induces a biholomorphic map from X onto the open (!) subset $f(X) \subset Y$.

6. Show that when $\varphi : X \to Y$ is an analytic map of Riemann surfaces, then

$$\varphi^* : \mathcal{O}(Y) \longrightarrow \mathcal{O}(X), \quad f \longmapsto f \circ \varphi$$

defines a ring homomorphism. This is injective when X and Y are connected and φ is not constant. It is an isomorphism if φ is biholomorphic.

The purpose of the following exercises is to show another way in which Riemann surfaces can be introduced. In this approach, the analytic functions themselves are introduced in an axiomatic way. The advantage is that we do not need the chart transformations. This new approach is important because it admits broad generalizations.

7. Let X be a topological space. A *sheaf* of continuous functions is a map which assigns to any open subset $U \subset$ a subring $\mathcal{O}_X(U)$ of the ring of all continuous functions $f : U \to \mathbb{C}$ such that the following conditions are satisfied:
   a) If $f \in \mathcal{O}_X(U)$ and $V \subset U$ is a further open subset, then $f|V \in \mathcal{O}_X(V)$.
   b) Let $U = \bigcup_i U_i$ be an open covering of an open subset U of X and let f be a continuous function on U such that its restrictions to U_i are contained in $\mathcal{O}_X(U_i)$. Then $f \in \mathcal{O}_X(U)$.
   A ringed space is a pair $(X, \mathcal{O}_X)$ consisting of a topological space X and a sheaf $\mathcal{O}_X$ of continuous functions.

   Let X be a Riemann surface. Show that the assignment

   $$U \longmapsto \mathcal{O}_X(U) = \text{set of all analytic functions}$$

   defines a structure of a ringed space.

8. A morphism $f : (X, \mathcal{O}_X) \to (Y, \mathcal{O}_Y)$ of ringed spaces is a continuous map between the underlying topological spaces such that the following condition is satisfied:
   If $V \subset Y$ is an open subset and $g \in \mathcal{O}_Y(V)$, then $g \circ f \in \mathcal{O}_X(U)$, $U = f^{-1}(V)$.

   Let X, Y be two Riemann surfaces which have been equipped with the sheafs of analytic functions. Show that a map $f : X \to Y$ is analytic in the sense of Riemann surfaces if and only if it defines a morphism of ringed spaces.

9. An isomorphism $f : (X, \mathcal{O}_X) \to (Y, \mathcal{O}_Y)$ of ringed spaces is a bijective map between the underlying topological spaces such that f and f^{-1} are morphisms of ringed spaces.

Let $U \subset X$ be an open subset of U. Then we can define the restricted sheaf $\mathcal{O}_X|U$. For open subsets $V \subset U$, we can define $(\mathcal{O}_X|U)(V) := \mathcal{O}_X(V)$.

Let $(X, \mathcal{O}_X)$ be a ringed space, X Haussdorff. Assume that every point $a \in X$ admits an open neighborhood $U \subset X$ and an open subset $V \subset \mathbb{C}$ of the complex plane such that the ringed spaces $(U, \mathcal{O}_X|U)$ and $(V, \mathcal{O}_V)$ are isomorphic. Here $\mathcal{O}_V$ means the sheaf of analytic functions in the usual sense. Show that there exists a unique structure of a Riemann surface on X such that $\mathcal{O}_X$ is the sheaf of analytic functions.

2. The Analytisches Gebilde

Important examples of Riemann surfaces can be obtained by use of the *analytisches Gebilde*. The *analytisches Gebilde* of a power series is obtained by gluing all of its analytic continuations to a surface. On this surface, all analytic continuations appear as a unique single-valued analytic function. The *analytisches Gebilde* is one of the historical motivations for the notion of a Riemann surface. An important example is the *analytisches Gebilde* of an *algebraic function*. In Sect. 3, we shall give another construction of this. Hence we recommend that readers who are interested in this basic example should skip the *analytische Gebilde* and proceed directly to Sect. 3.

One of the simplest examples of a "multivalued function" is the square root. By choosing, for example, the principal branch $\sqrt{z}$, we can obtain uniqueness, but we obtain only an analytic function on the slit plane $\mathbb{C}_-$. Other branches such as $-\sqrt{z}$ on $\mathbb{C}_-$ have equal validity. We would like to glue all possible branches together into a unique function. This is possible, but the domain of definition cannot then be a domain in the complex plane, but instead is a surface which lies over it. The following construction of the *analytisches Gebilde* leads to this surface.

2.1 Definition. *A **function element** is a pair (a, P) consisting of a complex number a and power series with center a,*

$$P(z) = \sum_{n=0}^{\infty} a_n (z - a)^n,$$

which has a positive radius of convergence.

In particular, this function element defines an analytic function on a small disk around a.

2.2 Definition (Weierstrass). *Let*

$$\alpha_0 : I \longrightarrow \mathbb{C}$$

*be a curve in the complex plane. A **regular allocation** of α_0 is a map which assigns to any $t \in I$ a function element*

$$(\alpha_0(t), P_t)$$

with center $\alpha_0(t)$. The following conditions have to be satisfied. Every $t_0 \in I$ admits an $\varepsilon = \varepsilon(t_0) > 0$ with the following property: if

$$t \in I, \quad |t - t_0| < \varepsilon,$$

then $\alpha_0(t)$ is in the interior of the disk of convergence of P_{t_0}, and P_t can be obtained by rearranging P_{t_0} according to powers of $z - \alpha_0(t)$.

This means that the function elements arise by successive analytic continuation.

2.3 Definition. *Let (a, P) and (b, Q) be two function elements. We say that (b, Q) can be obtained by analytic continuation from (a, P) when both are members of a regular allocation.*

We then call the two function elements equivalent and write $(a, P) \sim (b, Q)$. Obviously, this is an equivalence relation.

2.4 Definition. *An **analytisches Gebilde** in the sense of Weierstrass is a full equivalence class of function elements.*

Hence the *analytisches Gebilde* collects together all function elements which can be obtained by analytic continuation from a single function element.

The essential point of this notion is that equivalent function elements (a, P), (b, Q) can be different even if their centers are equal, i.a. $a = b$. Take, for example, $a = b = 1$. For P, we take the power series of the principal branch of the square root around 1, and for Q we take $-P$. The function element $(1, Q)$ can be obtained from $(1, P)$ by means of analytic continuation along a circle around 0.

In general, the result of an analytic continuation will depend on the choice of the path. This simple but fundamental fact can be considered as part of the idea of the Riemann surface.

In the following, $\mathcal{R}$ denotes a fixed *analytisches Gebilde*. There is a natural projection

$$p : \mathcal{R} \longrightarrow \mathbb{C}, \quad p((a, P)) = a$$

into the complex plane. As we have pointed out, this need not be injective.

We now introduce a topology on $\mathcal{R}$. Let $(a, P) \in \mathcal{R}$. We want to define, for some $\varepsilon > 0$, the notion of an "ε-neighborhood"

$$U_\varepsilon(a, P) \subset \mathcal{R}$$

of (a, P) in $\mathcal{R}$. We assume that ε is not greater than the radius of convergence of P. In this case P defines an analytic function on the usual ε-neighborhood $U_\varepsilon(a)$ of a. For every $b \in U_\varepsilon(a)$, we can expand this function into a power series P_b around b, for example by rearranging it according to powers of $z - b$.

Notation. Let (a, P) be a function element and let $\varepsilon > 0$ be a number which is not greater than the radius of convergence of P. The ε-neighborhood $U_\varepsilon(a, P)$ of (a, P) in $\mathcal{R}$ consists of all function elements (b, P_b), $b \in U_\varepsilon(a)$, where P_b is obtained from P by expanding around b.

Remark. The natural projection

$$U_\varepsilon(a, P) \longrightarrow U_\varepsilon(a)$$

is bijective.

When (a, P) is contained in $\mathcal{R}$, then of course $U_\varepsilon(a, P)$ is contained in $\mathcal{R}$.

2.5 Definition.   *A subset $U \subset \mathcal{R}$ is called* **open** *if any function element $(a, P) \in U$ admits an ε-neighborhood (where ε is not greater than the radius of convergence of P) such that*

$$U_\varepsilon(a, P) \subset U.$$

Let $(b, Q) \in U_\varepsilon(a, P)$. For small enough δ, we have

$$U_\delta(b, Q) \subset U_\varepsilon(a, P).$$

This implies the following statement.

2.6 Proposition.   *We obtain a topology on $\mathcal{R}$, by means of Definition 2.5. The ε-neighborhoods $U_\varepsilon(a, P)$ are open. The natural projection $p : \mathcal{R} \to \mathbb{C}$ gives a topological map*

$$U_\varepsilon(a, P) \longrightarrow U_\varepsilon(a).$$

Corollary. *The natural projection*

$$p : \mathcal{R} \longrightarrow \mathbb{C}$$

is locally topological.

(A continuous map $f : X \to Y$ is called locally topological if any point $a \in X$ admits an open neighborhood $U(a)$ such that the restriction of f defines a topological map from $U(a)$ onto an open neighborhood $V(b)$ of $b = f(a)$.)

2.7 Remark (Additional Remark on Proposition 2.6). *The space $\mathcal{R}$ is Hausdorff.*

Proof. Let (a, P), (b, Q) be function elements.

First case. $a \neq b$. We choose $\varepsilon > 0$ smaller than $|a - b|$ and smaller than the radii of convergence of P and Q. Then, trivially, $U_\varepsilon(a, P) \cap U_\varepsilon(b, Q) = \emptyset$.

Second case. $a = b$ but $P \neq Q$. We choose ε smaller than the radii of convergence of P and Q and again obtain disjoint neighborhoods. (Otherwise, we would find a point c in an open disk where P and Q converge such that the expansions of P and Q around c agree. But then P and Q coincide.) $\qquad\square$

Now we are going to construct an analytic atlas on $\mathcal{R}$.

The Analytisches Gebilde as a Riemann Surface

The construction rests on the following general fact.

2.8 Lemma. *Let*

$$f : X \longrightarrow Y$$

be a locally topological map of Hausdorff spaces. Assume that Y carries a structure in the form of a Riemann surface. Then X carries a unique structure in the form of a Riemann surface such that f becomes locally biholomorphic.

Proof. An open subset $U \subset X$ is called "small" if f maps U topologically onto an open set $f(U)$ and if there exists an analytic chart on Y,

$$f(U) \overset{\sim}{\longrightarrow} V \qquad (\subset \mathbb{C} \text{ open}).$$

The composition

$$U \longrightarrow f(U) \longrightarrow V$$

is a chart on X. Obviously, these charts are analytically compatible. Hence they define a structure in the form Riemann surface on X.

To prove uniqueness, we describe the elements $\varphi : U \to V$ of the maximal atlas. We can restrict ourselves to φ such that U is small in the sense that it is mapped biholomorphically onto the open set $f(U)$. Obviously, φ is biholomorphic (i.e. contained in the maximal atlas) if $\varphi \circ f^{-1} : f(U) \to V$ is biholomorphic. $\qquad\square$

By applying this lemma to $p : \mathcal{R} \to \mathbb{C}$, we obtain the announced structure of a Riemann surface on $\mathcal{R}$.

Curves on the Analytisches Gebilde

A *curve* on a topological space X is a continuous map

$$\alpha : I \longrightarrow X$$

of an interval $I \subset \mathbb{R}$ into X. We are interested mainly in the case where $I = [a, b]$ $(a < b)$ is a compact interval. Then $\alpha(a)$ is called the starting point and $\alpha(b)$ the endpoint of α. When both agree, α is said to be closed.

2.9 Lemma. *Let*

$$\alpha : I \longrightarrow \mathcal{R} \qquad (I \subset \mathbb{R} \ an\ interval)$$

be a curve in $\mathcal{R}$. The composition with the natural projection $p : \mathcal{R} \to \mathbb{C}$ defines a curve

$$\alpha_0 : I \longrightarrow \mathbb{C}.$$

The family

$$\alpha(t) =: (\alpha_0(t), P_t)$$

is a regular allocation of α_0. Conversely, if $(\alpha_0(t), P_t)$ is a regular allocation of a curve $\alpha_0 : I \to \mathbb{C}$, then

$$\alpha(t) = (\alpha_0(t), P_t)$$

is a curve in $\mathcal{R}$.

In other words:

Regular allocations in the sense of Weierstrass and curves in $\mathcal{R}$ are the same.

Obviously, the definition of the topology on $\mathcal{R}$ has been devised in such a way that Lemma 2.9 is true.

2.10 Lemma. *Let*

$$f : X \longrightarrow Y$$

be a locally topological map of topological spaces, let

$$\alpha_0 : [a, b] \longrightarrow Y \qquad (a < b)$$

be a curve in Y, and let $x_0 \in X$ be a point over $\alpha_0(a)$ (i.e. $f(x_0) = \alpha_0(a)$). There is at most one curve

$$\alpha : [a, b] \longrightarrow X$$

with the properties

a) $$f \circ \alpha = \alpha_0,$$
b) $$\alpha(a) = x_0.$$

We call α a lifting of α_0.

Proof of Lemma 2.10. Let β be a second lift of α_0 ($f \circ \beta = \alpha_0$, $\beta(a) = x_0$). We consider

$$t_0 := \sup\{t \in [a, b]; \quad \alpha(t) = \beta(t)\}$$

and make use of the fact that f maps a neighborhood of the point $\alpha(t_0)$ topologically onto a neighborhood of $\alpha_0(t_0)$. $\qquad\qquad\square$

The next statement follows from this topological fact.

2.11 Lemma. *Let*
$$\alpha_0 : [a, b] \longrightarrow \mathbb{C} \qquad (a < b)$$
be a curve in the complex plane and let $(\alpha_0(a), P)$ *be a function element whose center is the starting point* $\alpha_0(a)$ *of* α_0. *When there exists a regular allocation*

$$(\alpha_0(t), P_t)$$

of α_0 *which starts with*
$$P_a = P,$$

then this allocation is uniquely determined.

In particular, the end P_b of the allocation is uniquely determined (by α_0 and $P_a = P$). We can also say that the function element $(\alpha_0(b), P_b)$ arises by analytic continuation of $(\alpha_0(a), P_a)$ along the path α_0. Hence this analytic continuation is determined by the curve and the starting element if such a continuation exists at all.

The Analytisches Gebilde of an Analytic Function

Let
$$f : D \longrightarrow \mathbb{C}, \quad D \subset \mathbb{C} \text{ a domain}$$
be an analytic function on a (connected) domain. Let
$$\alpha_0 : [a, b] \longrightarrow D \qquad (a < b)$$
be a curve in D.

We denote the power expansion of f around a point $a \in D$ by f_a. This gives us a regular allocation

$$\alpha(t) := (\alpha_0(t), f_{\alpha_0(t)}).$$

In particular, all elements (a, f_a) are equivalent and hence are contained in one *analytisches Gebilde* $\mathcal{R}(f)$.

We call $\mathcal{R}(f)$ *the concrete Riemann surface which belongs to* f.

In simple language, this means that $\mathcal{R}(f)$ consists of all function elements (a, P) which can be obtained somehow (i.e. along a suitable path) from f by analytic continuation.

The map
$$D \longrightarrow \mathcal{R}(f), \quad a \longmapsto (a, f_a),$$
is then an *open embedding*, i.e. a biholomorphic map of D into some open part of $\mathcal{R}$.

Recall that there is a natural projection
$$p : \mathcal{R}(f) \longrightarrow \mathbb{C}, \quad (a, P) \longmapsto a.$$

The image of p is a domain in $\mathbb{C}$ which contains D. It is called the domain of definition of $\mathcal{R}(f)$. Besides the projection p, we can consider the obviously holomorphic map

$$F : \mathcal{R}(f) \longrightarrow \mathbb{C}, \quad F(a, P) = f(a).$$

The diagram

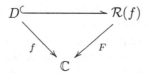

commutes. We can say that F "includes" the function f, but not only f. It includes all possible analytic continuations of f.

This can be expressed roughly as follows. All analytic continuations of f : $D \to \mathbb{C}$ are multivalued (because of the path dependence). But one can make them single-valued if one extends their domain of definition to a surface $\mathcal{R}(f)$ lying over the plane.

A Simple Example

We take for D a plane which is slit along the negative real half-line, and for $f(z)$ we take the principal value of the square root of z. We denote the corresponding concrete Riemann surface by $\mathcal{R}(\sqrt{\ })$. This consists of all function elements (a, P) such that $a \in \mathbb{C}^{\bullet}$ and $P(z)^2 = z$. Every point admits exactly two such elements $(a, P(z))$. Therefore $\mathcal{R}(\sqrt{\ })$ is a connected Riemann surface together with a holomorphic map

$$p : \mathcal{R}(\sqrt{\ }) \longrightarrow \mathbb{C}^{\bullet} \quad (a, P) \longmapsto a,$$

and is such that every point in $\mathbb{C}^{\bullet}$ has two inverse points. The same property is shared by the map $\mathbb{C}^{\bullet} \longrightarrow \mathbb{C}^{\bullet}$, $z \longmapsto z^2$. It is not difficult to show the following.

There exists a biholomorphic map

$$\mathcal{R}(\sqrt{\ }) \xrightarrow{\sim} \mathbb{C}^{\bullet}$$

such that the diagram

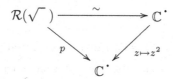

commutes.

By the way, this shows that $\mathcal{R}(\sqrt{})$ is biholomorphically equivalent to the Riemann surface which is obtained when one removes two points from the Riemann sphere (0 and ∞). In the next section, we shall see that this is a very general phenomenon. The Riemann surface of any algebraic function can be obtained from a *compact Riemann surface* by removing a finite set of points.

We can picture this construction as follows. Consider two copies of the punctured plane. One of them is equipped with the principal branch $\sqrt{z}$, and the other with its negative. To get continuous functions, we slit both planes along their negative real half-axes. But we leave an exemplar of the negative real line on both sides of the negative real line. We call these two half-lines the upper and the lower side. Now we glue the lower side of each of the two planes to the upper side of the other plane. It should be clear how the resulting shape can be identified with $\mathcal{R}(\sqrt{})$.

Exercises for Sect. I.2

1. Show that the set of all function elements (a, P), $a \in \mathbb{C}^{\bullet}$, $\exp(P(z)) = z$ builds an *analytisches Gebilde*. We denote it by $\mathcal{R}(\log)$ and call it the Riemann surface of the logarithm. It can be shown that this surface is biholomorphically equivalent to the plane, and that the map

$$\mathcal{R}(\log) \longrightarrow \mathbb{C}, \quad (a, P) \longmapsto P(a),$$

   is biholomorphic.

   This can be expressed as follows. Two analytic maps of Riemann surfaces

$$X \longrightarrow Y, \quad X' \longrightarrow Y'$$

   are called *isomorphic* if there exist biholomorphic maps $X \to X'$ and $Y \to Y'$ such that the diagram

$$
\begin{array}{ccc}
X & \longrightarrow & Y \\
\downarrow & & \downarrow \\
X' & \longrightarrow & Y'
\end{array}
$$

   commutes. In this sense, the maps

$$\mathcal{R}(\log) \longrightarrow \mathbb{C}^{\bullet}, \quad \mathbb{C} \longrightarrow \mathbb{C}^{\bullet},$$
$$(a, P) \longmapsto a, \quad z \longmapsto \exp(z),$$

   are isomorphic.

2. The observation in Exercise 1 is part of a more general phenomenon.

   Let $f : D \to D'$ be an analytic map from one domain $D \subset \mathbb{C}$ onto another D'. Assume that the derivative of f has no zeros in D. Consider the set of all function elements (b, P) with the properties

a) $b \in D'$;

b) $P(b) \in D$, $P(f(z)) = z$ in a neighborhood of $P(b)$.

Show that the set of all these function elements – we denote it by $\mathcal{R}(f^{-1})$ – is an *analytisches Gebilde*. The maps

$$f : D \longrightarrow D' \text{ and } \mathcal{R}(f^{-1}) \longrightarrow D' \; ((a, P) \longmapsto P(a))$$

are isomorphic.

3. A special case of a *connectedness theorem*, which we shall obtain in the next section, states that the set of all function elements (a, P) with the property

$$P(z)^4 + z^4 = 1$$

is an *analytisches Gebilde* $\mathcal{R}$. Show that this is true.

3. Show that every *analytisches Gebilde* has countable basis of its topology.

> *Hint.* Use the fact that $\mathbb{C}$ has a countable dense subset.

3. The Riemann Surface of an Algebraic Function

The simplest example of a "multivalued" function is the square root. It is a special example of an *algebraic function*. An analytic function

$$f : D \longrightarrow \mathbb{C}, \quad D \subset \mathbb{C} \text{ a domain,}$$

is called algebraic if there exists a polynomial of two variables

$$P(z, w) = \sum_{0 \le \mu, \nu \le N} a_{\mu\nu} z^\mu w^\nu$$

which is not identically zero and is such that

$$P(z, f(z)) = 0 \text{ (for all } z \in D).$$

Example. Let

$D := \mathbb{C} - \{x \in \mathbb{R}, \; x \le 0\}$ and $f(z) := \sqrt{z}$ (for example, the principal branch)

$(P(z, w) = w^2 - z)$.

In this section, we shall construct a *compact Riemann surface* from f. This will be done in two steps.

First step. We shall construct a finite subset $S \subset \mathbb{C}$ and a Riemann surface X_0 together with a holomorphic map

$$p : X_0 \longrightarrow \mathbb{C} - S,$$

such that the following properties are satisfied:

a) p is locally biholomorphic;

b) p is proper in the topological sense, i.e. the inverse image of an arbitrary compact set from $\mathbb{C} - S$ is compact.

For example, we can take the *analytisches Gebilde* $\mathcal{R}(f)$ for X_0 and take the natural projection for p. But it is also possible to obtain X_0 in a different way, namely as an *algebraic curve*. This is the approach which we prefer here. Hence the *analytisches Gebilde* described in section 2 can be avoided. Nevertheless, we give a sketch of how it can be used here.

Second step. Using the properties a) and b) singly, we construct a *compact* Riemann surface X and a holomorphic map $\bar{p} : X \to \bar{\mathbb{C}}$ such that the following properties are satisfied:

1) X_0 is an open subset of X, and the complement $X - X_0$ is a finite set.
2) The diagram

$$
\begin{array}{ccc}
X_0 & \subset & X \\
\downarrow{\scriptstyle p} & & \downarrow{\scriptstyle \bar{p}} \\
\mathbb{C} - S & \subset & \bar{\mathbb{C}}
\end{array}
$$

commutes.

This second step rests on a special case of *covering theory*, which we shall treat independently of general covering theory in this context. The statement is:

If

$$
f : X \longrightarrow \mathbb{E}^{\bullet} = \{q \in \mathbb{C}; \quad 0 < |q| < 1\}
$$

is a locally biholomorphic proper map of a connected Riemann surface into the punctured unit disk, then there exists a biholomorphic map

$$
\sigma : X \xrightarrow{\sim} \mathbb{E}^{\bullet}
$$

and a natural number e such that the diagram

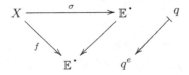

commutes.

　　As mentioned, this is a special case of general covering theory, which will be developed in full generality in Chap. III. Because of the great importance of algebraic functions, we shall treat the special case already here with a proof as simple as possible. This will be done in Appendix A of this section. In Appendix B, we shall formulate and prove a theorem on implicit functions for analytic functions of two variables.

　　A polynomial $P \in \mathbb{C}[z, w]$, $P \neq 0$,

$$
P(z, w) = \sum_{0 \le \mu, \nu \le N} a_{\mu\nu} z^{\mu} w^{\nu},
$$

of two complex variables is called *irreducible* if it cannot be written as a product of two nonconstant polynomials. Every polynomial can be written as a product of finitely many irreducible ones:

$$P = P_1 \cdots \cdot P_m, \quad P_j \text{ irreducible.}$$

If

$$f : D \longrightarrow \mathbb{C} \ (D \subset \mathbb{C} \text{ a domain})$$

is an analytic function with the property $P(z, f(z)) \equiv 0$, then

$$P_j(z, f(z)) \equiv 0 \text{ for some } j \qquad (1 \le j \le m).$$

For this reason, we may assume in what follows that P itself is irreducible.

We associate with P a *plane affine algebraic curve*

$$\mathcal{N} = \mathcal{N}(P) = \{(z, w) \in \mathbb{C} \times \mathbb{C}; \quad P(z, w) = 0\}.$$

3.1 Remark. *Let P be an irreducible polynomial which truly depends on w. For every point $a \in \mathbb{C}$, there exist only finitely many $b \in \mathbb{C}$ with the property $(a, b) \in \mathcal{N}$. In other words, the fibers of the natural projection*

$$p : \mathcal{N} \longrightarrow \mathbb{C}, \quad p(z, w) = z,$$

are finite.

Proof. We argue indirectly and assume that there exists a point $a \in \mathbb{C}$ such that the polynomial $w \mapsto P(a, w)$ has infinitely many zeros b. Then this polynomial must vanish. Reordering according to powers of $z - a$, we see that

$$P(z, w) = (z - a) \, Q(z, w).$$

Since P is irreducible, Q has to be constant. But then P would be independent of w. □

Our next goal is to show that the map p is locally topological outside a finite set of points (called branch points). For this, we need the following proposition.

3.2 Proposition. *Let $P \in \mathbb{C}[z, w]$ be an irreducible polynomial which truly depends on w. There are only finitely many solutions (a, b) of the equations*

$$P(a, b) = 0 = \frac{\partial P}{\partial w}(a, b).$$

For the proof, we make use of the *discriminant* $d_P(z)$ of the polynomial $w \mapsto P(z, w)$ for arbitrary $z \in \mathbb{C}$. For the definition of the discriminant and its basic properties, we refer to the algebraic appendix at the end of this book (Sect. VIII.3.1).

The discriminant is a polynomial in z which, because of the irreducibility of P, does not vanish identically. If (a, b) is a solution of the given equations, then the polynomial $w \mapsto P(a, w)$ has a multiple zero. This means that its discriminant vanishes, i.e. $d_P(a) = 0$. Since d_P has only finitely many zeros, there exist only finitely many a. Because of Remark 3.1, there exist only finitely many (a, b). $\qquad\qquad\qquad\qquad\qquad\qquad\qquad\qquad\qquad\qquad\qquad\qquad\square$

It is often useful for our purposes to reorder a polynomial $P(z, w)$ according to powers of w:
$$P(z, w) = a_n(z)w^n + \cdots + a_0(z).$$
The polynomials a_i are called the coefficients of P (with respect to the variable w). When P is different from 0, one can achieve $a_n \neq 0$. Then we call $a_n(z)$ the *highest coefficient* of P. The next statement follows immediately from Proposition 3.2.

3.3 Theorem. *Let $P \in \mathbb{C}[z, w]$ be an irreducible polynomial depending truly on w. There exists a finite set $S \subset \mathbb{C}$ with the following properties:*

a) *The zeros of the highest coefficient of P are contained in S.*
b) *Let $(a, b) \in \mathbb{C} \times \mathbb{C}$ be such that*
$$P(a, b) = 0 \text{ and } \frac{\partial P}{\partial w}(a, b) = 0;$$
then $a \in S$.

In the following, $P \in \mathbb{C}[z, w]$ always means an irreducible polynomial which depends truly on w, and $S \subset \mathbb{C}$ is a finite subset with the properties described in Theorem 3.3.

We define
$$X := \{(a, b) \in (\mathbb{C} - S) \times \mathbb{C}; \quad P(a, b) = 0\}.$$
This point set is obtained from the original affine algebraic curve by removing finitely many points. We equip X with the topology induced from $\mathbb{C} \times \mathbb{C}$.

3.4 Proposition. *The projection*
$$p : X \longrightarrow \mathbb{C} - S, \quad p(a, b) = a,$$
*is **locally topological and proper**.*

Corollary. *The space X admits a unique structure as its Riemann surface such that p is locally biholomorphic (Lemma 2.8).*

Additional remark. *The second projection*
$$q : X \longrightarrow \mathbb{C}, \quad q(a, b) = b,$$
is analytic as well.

Proof, first part. p is locally topological.

The proof follows immediately from a complex variant of a theorem of implicit functions which is well known in real analysis; see Appendix B. A proof of the additional remark can be obtained from that theorem too.

Second Part. p is proper.

The inverse image $A = p^{-1}(B)$ of a compact subset $B \subset \mathbb{C} - S$ is closed in $\mathbb{C} \times \mathbb{C}$, since B is closed in $\mathbb{C}$. Therefore it is sufficient to show that A is bounded.

For $(a, b) \in A$, we have $P(a, b) = 0$. By the definition of the exceptional set S, the highest coefficient of P has no zeros in $\mathbb{C} - S$. Hence it is bounded from below on the compact set B by a positive number. All coefficients are bounded from above on the compact set B. Now the claim follows from the following simple lemma.

3.5 Lemma. *Let n be a natural number and let $C > 0$ be a positive real number. There exists a positive real number $C' = C'(C, n)$ with the following property:*

Let

$$P(z) = a_n z^n + \cdots + a_0$$

be a polynomial (in $\mathbb{C}[z]$) whose coefficients satisfy

$$|a_i| \leq C \quad (0 \leq i \leq n), \quad |a_n| \geq C^{-1}.$$

Then any zero a of P satisfies the inequality

$$|a| \leq C'.$$

Proof. From $a_n a^n + \cdots + a_0 = 0$, it follows that

$$a_n a = -\left(a_{n-1} + \cdots + \frac{a_0}{a^{n-1}} \right).$$

In the case $|a| \geq 1$, we obtain

$$|a_n a| \leq nC \quad \text{or} \quad |a| \leq nC^2. \qquad \square$$

Alternative Construction of the Riemann Surface by Means of the Analytisches Gebilde (Sketch)

Let $\mathcal{R}_0$ be the set (a, Q) of function elements with the property

$$a \notin S, \quad P(z, Q(z)) \equiv 0.$$

Remark. $\mathcal{R}_0$ is an open part of the (disjoint) union of several *analytische Gebilde* and hence is a Riemann surface.

The proof follows easily from the fact that, by the principle of analytic continuation, the equation

$$P(z, Q(z)) \equiv 0$$

carries over to all function elements which are obtained from Q by analytic continuation.

On $\mathcal{R}_0$, we have two analytic functions

$$p_0 : \mathcal{R}_0 \longrightarrow \mathbb{C} - S, \quad p_0(a, Q) = a,$$
$$q_0 : \mathcal{R}_0 \longrightarrow \mathbb{C}, \quad\quad q_0(a, Q) = Q(a).$$

For $(a, Q) \in \mathcal{R}_0$, the point $(a, Q(a))$ is contained in the algebraic curve. In this way, we get a map

$$h : \mathcal{R}_0 \longrightarrow X, \quad (a, Q) \longmapsto (a, Q(a)).$$

The next result follows from the theorem of implicit functions.

3.6 Proposition. *The canonical map*

$$h : \mathcal{R}_0 \overset{\sim}{\longrightarrow} X$$

is biholomorphic.

Additional remark. *The diagrams*

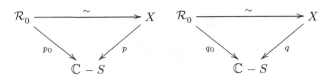

commute.   □

It is our goal to extend the Riemann surface X to a compact Riemann surface by adding finitely many points.

3.7 Proposition. *Let $X, \bar{Y}$ be Riemann surfaces, let $S \subset \bar{Y}$ be a finite set of points, let*

$$Y := \bar{Y} - S,$$

and let

$$f : X \longrightarrow Y$$

be a locally biholomorphic proper map. Then there exists a Riemann surface $\bar{X}$ which contains X as an open Riemann subsurface, and a holomorphic map

$$\bar{f} : \bar{X} \longrightarrow \bar{Y}$$

with the following properties:

1) *The complement $T = \bar{X} - X$ is finite.*
2) *The map $\bar{f}$ is proper.*
3) *The diagram*

$$\begin{array}{ccc} \bar{X} & \xrightarrow{\bar{f}} & \bar{Y} \\ \cup & & \cup \\ X & \xrightarrow{f} & Y \end{array}$$

is commutative.

We call $(\bar{X}, \bar{f})$ a **completion** of $(X, f, Y \subset \bar{Y})$. If $\bar{Y}$ is compact, then $\bar{X}$ is compact as well.

Proof. For each exceptional point $s \in S$, we choose an open neighborhood $U(s)$ which is biholomorphically equivalent to the unit disk $\mathbb{E}$ and such that

$$U(s) \cap U(t) = \emptyset \text{ for } s \neq t \quad \text{(both contained in } S\text{)}.$$

Obviously, the restriction $f^{-1}(U(s) - \{s\}) \to U(s) - \{s\}$ is proper, as is f. For any $s \in S$, the set $f^{-1}(U(s) - \{s\})$ decays into its connected components. Let $Z \subset f^{-1}(U(s) - \{s\})$ be such a connected component. It is open and closed in $f^{-1}(U(s) - \{s\})$. Since it is closed, the restriction $Z \to U(s) - \{s\}$ is proper as well. Hence it is surjective. From this and the fact that f is proper, we now obtain the result that there are only finitely many connected components.

From Proposition 3.17, we obtain a commutative diagram with biholomorphic rows:

$$\begin{array}{ccc} Z & \xrightarrow{\varphi_Z} & \mathbb{E}^{\cdot} \qquad q \\ f\downarrow & & \downarrow \qquad \uparrow \\ U(s) - \{s\} & \longrightarrow & \mathbb{E}^{\cdot} \qquad q^n \qquad (n = n(Z)). \end{array}$$

For any Z, we choose a symbol $a(Z)$ which is not contained in X and is such that

$$Z \neq Z' \implies a(Z) \neq a(Z').$$

Now we consider the set

$$\bar{Z} := Z \cup \{a(Z)\}.$$

We extend the map φ_Z to a bijective map

$$\varphi_{\bar{Z}} : \bar{Z} \longrightarrow \mathbb{E}$$

by means of

$$\varphi_{\bar{Z}}(a(Z)) := 0.$$

The topology of $\mathbb{E}$ can be carried over to $\bar{Z}$ (such that $\varphi_{\bar{Z}}$ becomes topological). Now we set

$$\bar{X} := X \cup \{a(Z), Z \text{ conncted component of } f^{-1}(U(s) - \{s\}), s \in S\}$$

and define a topology on $\bar{X}$. A subset $U \subset \bar{X}$ is said to be open if

a) $U \cap X$ is open;

b) $U \cap \bar{Z}$ is open for every Z.

Obviously, $\bar{X}$ becomes a Hausdorff space and X becomes an open subspace (such that its topology agrees with the induced topology of $\bar{X}$).

Next we extend the analytic atlas $\mathcal{A}$ of X to an analytic atlas $\bar{\mathcal{A}}$ of $\bar{X}$:

$$\bar{\mathcal{A}} := \mathcal{A} \cup \{\bar{\varphi}_Z\}$$

(Obviously, the charts $\bar{\varphi}_Z$ are analytically compatible with the analytic charts of X.)

Hence $\bar{X} = (\bar{X}, \bar{\mathcal{A}})$ is a Riemann surface. By use of

$$\bar{f}(\mathcal{Z}) = s \qquad (\mathcal{Z} \subset f^{-1}(s)),$$

we obtain an extension

$$\bar{f} : \bar{X} \longrightarrow \bar{Y},$$

which is obviously analytic. It is easy to see that it is proper. (We use the fact that f itself and $\mathbb{E} \to \mathbb{E}$, $q \mapsto q^n$ are proper.) $\qquad\qquad\square$

We are particularly interested in the case where $\bar{Y}$ is compact. Then $\bar{X}$ is a compactification of X by finitely many points. Now we show that such compactifications are unique.

3.8 Lemma. *Let X be a surface, let $S \subset X$ a finite subset, let $X_0 = X - S$, and let*

$$\bar{X} \supset X_0$$

be a compact space which contains X_0 as an open subspace. Assume also that the complement $\bar{X} - X_0$ is finite. Then there exists a continuous continuation

$$f : X \longrightarrow \bar{X}$$

of the identity id_{X_0}.

Corollary. *Let X_0 be a Riemann surface and let $\bar{X}, \tilde{X}$ be two compact Riemann surfaces which contain X_0 as an open Riemann subsurface. The complements $\bar{X} - X_0$, $\tilde{X} - X_0$ are assumed to be finite. Then there exists a biholomorphic map*

$$\varphi : \bar{X} \longrightarrow \tilde{X}, \quad \varphi|X_0 = \mathrm{id}_{X_0}.$$

Proof. We denote by

$$b_1, \ldots, b_n \qquad (b_i \neq b_j \text{ for } i \neq j)$$

the points of the complement $\bar{X} - X_0$. Then, for each $i \in \{1, \ldots, n\}$, we choose an open neighborhood

$$U(b_i) \subset \bar{X} \qquad (1 \leq i \leq n)$$

such that

$$U(b_i) \cap U(b_j) = \emptyset \text{ for } i \neq j.$$

Now, let a be a point in $X - X_0$. We want to construct its image point b in $\bar{X}$.

Claim. There exists a neighborhood $U(a) \subset X$ of a which, besides a, does not contain another point of S, and is such that

$$U(a) - \{a\} \subset U(b_1) \cup \cdots \cup U(b_n).$$

Proof of the claim. We argue indirectly: if the claim is false, then there exists a sequence

$$a_n \in X_0, \qquad a_n \longrightarrow a,$$

such that a_n is not contained in any $U(b_i)$. Since $\bar{X}$ is compact, we can assume (taking a subsequence) that (a_n) converges in $\bar{X}$. The limit must necessarily be one of the b_i, since (a_n) does not converge in X_0. But then almost all a_n would lie in $U(b_i)$. This proves the claim.

We may assume that $U(a)$ is open and connected. Since X is a surface, $U(a) - \{a\}$ is connected as well and hence is contained in precisely one $U(b_i)$:

$$U(a) - \{a\} \subset U(b_j) \text{ for one } j.$$

Since the neighborhoods $U(b_j)$ can be taken arbitrarily small, the extension of the identity defined by

$$\varphi(a) := b_j$$

is continuous at a.

The proof of the corollary is a consequence of the Riemann extension theorem: a continuous map between Riemann surfaces which is analytic outside a finite set of points is analytic everywhere. □

3.9 Theorem. *Let $P \in \mathbb{C}[z, w]$ be an irreducible polynomial which truly depends on w. There exists a **compact** Riemann surface $\bar{X}$ which contains the Riemann surface*

$$X = \{(a, b) \in (\mathbb{C} - S) \times \mathbb{C}, \ P(a, b) = 0\}$$

as an open Riemann subsurface. The complement $\bar{X} - X$ is finite. Both of the projections

$$p : X \longrightarrow \mathbb{C}, \quad p(a,b) = a,$$
$$q : X \longrightarrow \mathbb{C}, \quad q(a,b) = b,$$

admit (uniquely determined, of course) holomorphic extensions

$$\bar{p} : \bar{X} \longrightarrow \bar{\mathbb{C}} \quad (Riemann \ sphere),$$
$$\bar{q} : \bar{X} \longrightarrow \bar{\mathbb{C}}.$$

The triple $(\bar{X}, \bar{p}, \bar{q})$ is essentially unique.

Proof. All that remains to be proved is that q can be extended. (For this, we can assume that P truly depends on z. Otherwise, we would have $P(z) = C(z - a)$ and q would be constant.) Now we can interchange the roles of p and q.

There exists a finite subset $T \subset \mathbb{C}$ such that the canonical projection

$$q : X_0 \longrightarrow \mathbb{C} - T, \quad X_0 := \{(a,b) \in \mathbb{C} \times (\mathbb{C} - T), \ P(a,b) = 0\},$$

is locally topological and proper.

One can choose T large enough such that X_0 is a subset of X. The complement $X - X_0$ is a finite set. Because of the uniqueness of the compactification (Lemma 3.8, Corollary), it is sufficient to extend q to *some* compactification of X_0. Such a compactification is given by Proposition 3.7 (with q instead of p). □

3.10 Proposition. *The compact Riemann surface which is associated (by Theorem 3.9) with an irreducible polynomial is connected.*

Proof. The degree n of the polynomial $w \mapsto P(z, w)$, for $z \in \mathbb{C} - S$, is independent of z. We denote the zeros of this polynomial, in an arbitrary ordering, by $t_1(z), \ldots, t_n(z)$. We have

$$P(z, w) = a_n(z) \prod_{\nu=1}^{n} (w - t_\nu(z)).$$

The highest coefficient $a_n(z)$ is a polynomial in z. The points $x := (z, t_\nu(z))$ lie on the curve X. They are precisely those points which are mapped to z under p. Because $q(x) = t_\nu(z)$, we can write

$$P(z, w) = a_n(z) = \prod_{x \in X, \, p(x) = z} (w - q(x)).$$

We now give an indirect argument, where we assume that X is the union of two open nonempty subsurfaces, i.e. $X = X_1 \cup X_2$. Correspondingly, we decompose P as a product:

$$P(z, w) = P_1(z, w) P_2(z, w), \quad P_\nu(z, w) = \prod_{x \in X_\nu, \, p(x) = z} (w - q(x)), \quad z \in \mathbb{C} - S.$$

For fixed w, the functions $P_\nu(z, w)$ are analytic on $\mathbb{C} - S$, since p admits holomorphic local inversions here.

We want to show that the singularities $s \in S \cup \{\infty\}$ are inessential for any fixed w. Then $P_\nu(z, w)$ is, for any fixed w, a meromorphic function on $\bar{\mathbb{C}}$ and hence a rational function. Now we obtain the result that the $P_\nu \in \mathbb{C}(z)[w]$ are polynomials in w over the field of rational functions in z. This gives a contradiction to the irreducibility of P, since "Gauss's Lemma" states that if $P \in \mathbb{C}[z, w]$ is an irreducible polynomial in two variables, then P is also irreducible as a polynomial in the variable w over the field of rational functions $\mathbb{C}(z)$. (A proof will be given in the algebraic appendix of the book, Sect. VIII.2.8.)

It remains to prove that the singularities are inessential. This follows from the next statement.

3.11 Remark.   Let $p : Y \to \mathbb{E}$ be a proper analytic map of a Riemann surface Y onto the unit disk which is locally biholomorphic outside $p^{-1}(0)$. Furthermore, let $q : Y \to \bar{\mathbb{C}}$ be a meromorphic function whose values outside $p^{-1}(0)$ are different from ∞. Then the function

$$z \longmapsto Q(z, w) = \prod_{x \in X, \, p(x) = z} (w - q(x)) \quad (z \neq 0),$$

for every $w \in \mathbb{C}$, has an inessential singularity at the origin.

Proof. If we decompose Y into its connected components, then Q decomposes into a product. Hence we can assume that X is connected. Hence we can assume that $X = \mathbb{E}$ and $q(z) = z^n$. Now the statement is trivial. □

Proposition 3.10 has a corollary, which is of elementary nature:

3.12 Corollary.   Let (a, Q) and $(\tilde{a}, \tilde{Q})$ be two function elements with the property

$$P(z, Q(z)) \equiv 0 \quad P(z, \tilde{Q}(z)) \equiv 0 \quad (a, \tilde{a} \notin S).$$

Then there exists a curve which connects a and $\tilde{a}$ such that $\tilde{Q}$ is obtained from Q by analytic continuation of Q along this curve.

Appendix A. A Special Case of Covering Theory

3.13 Lemma.   *Let $f : X \to Y$ a locally topological and proper map of Hausdorff spaces. Each point $b \in Y$ has only finitely many preimages*

$$f^{-1}(b) = \{a_1, \ldots, a_n\} \qquad (a_i \neq a_j \text{ for } i \neq j).$$

There are open neighborhoods

$$b \in V \subset Y \text{ and } a_i \in U_i \subset X \qquad (1 \leq i \leq n)$$

with the following property:
1) $f^{-1}(V) = U_1 \dot{\cup} \cdots \dot{\cup} U_n$ *(disjoint union, i.e. $U_i \cap U_j = \emptyset$ for $i \neq j$).*
2) *The restriction of f gives a topological map*

$$U_i \overset{f}{\underset{\sim}{\longrightarrow}} V \qquad (1 \leq i \leq n).$$

Proof. Since f is locally topological, $f^{-1}(b)$ is a discrete subset. Since f is proper, it is also compact. These two facts together show the finiteness.

Now we choose pairwise disjoint open neighborhoods $a_i \in U_i' \subset X$. After a possible diminishment, we have the result that the restriction of f to U_i' gives a topological map from U_i' onto an open subset V'. Since f is proper, there exists an open neighborhood $a \in V \subset V'$ with the property $f^{-1}(V) \subset U_1' \cup \cdots \cup U_n'$ (see Remark 0.2). We define $U_i = U_i' \cap f^{-1}(V)$. (Then $U_1 \cup \cdots \cup U_n$ is the *full* inverse image of V.) $\square$

The following definition arises from Lemma 3.13.

3.14 Definition.   *A continuous map*

$$f : X \longrightarrow Y$$

of topological spaces is called a covering*) if every point $b \in Y$ admits an open neighborhood V, $b \in V \subset Y$, and any preimage $a \in X$ ($f(a) = b$) admits an open neighborhood $U(a)$, such that the followings conditions are satisfied:

1) $f^{-1}(V) = \displaystyle\bigcup_{f(a)=b} U(a)$ *(disjoint union).*
2) *The restriction of f induces, for each $a \in f^{-1}(b)$, a topological map*

$$U(a) \overset{\sim}{\longrightarrow} V.$$

---

*) In the literature, the term "covering" is not always used in this strict sense. Our notion of a covering agrees with what sometimes is called an "unlimited and unramified covering".

Proper and locally topological maps are examples of coverings. An example of a nonproper covering is given by

$$\mathbb{C} \longrightarrow \mathbb{C}^{\bullet}, \quad z \longmapsto e^z.$$

The key to the study of coverings is the so-called *path lifting*, as described below.

3.15 Proposition. *Let $f : X \to Y$ be a covering. For each point $x_0 \in X$ and for each curve*

$$\alpha : [a, b] \longrightarrow Y, \quad \alpha(a) = f(x_0) \quad (a < b),$$

which starts at $f(x_0)$, there exists a unique curve

$$\beta : [a, b] \longrightarrow X$$

with the properties
a) $f \circ \beta = \alpha$;
b) $\beta(0) = x_0$.

The curve β is called the lifting of α with starting point x_0 (over $\alpha(a)$).

Proof. The uniqueness of the lifting was proved in Lemma 2.10.

Existence of a lifting. A simple compactness argument shows that there exists a finite partition

$$a = a_0 < a_1 < \cdots < a_m = b$$

and, for each i, $0 \le i \le m$, an open neighborhood

$$\alpha(a_i) \in V_i \subset Y$$

such that:
1) V_i has the property mentioned in Definition 3.14 ($p^{-1}(V_i)$ decomposes into pairwise disjoint sets, which are mapped topologically onto V_i under f).
2) $\alpha[a_i, a_{i+1}] \subset V_i$.
Now we can lift

$$\alpha_i := \alpha|[a_i, a_{i+1}]$$

inductively in such a way that the starting point of the lift β_{i+1} equals the endpoint of β_i. The composition of the curves β_i gives the desired lift β.

$\square$

The same proof shows a little more, namely the following.

3.16 Theorem. *Let $f : X \to Y$ be a covering, let*

$$Q = I \times J, \quad I, J \subset \mathbb{R} \text{ intervals,}$$

be a (not necessarily compact) rectangle, let

$$H : Q \longrightarrow Y$$

be a continuous map, and let $q_0 \in Q$, $x_0 \in X$ be points with the property $H(q_0) = f(x_0)$. Then there exists a continuous map

$$\tilde{H} : Q \longrightarrow X$$

with the properties

a) $\quad f \circ \tilde{H} = H,$
b) $\quad \tilde{H}(q_0) = x_0.$

For the proof we can assume that Q is compact, since every rectangle can be written as the union of an ascending chain of compact rectangles. After division into four pieces, we can assume that q_0 is a vertex of Q. Now the proof is similar to that in the case of curve lifting. Instead of a partition, we use a decomposition into small subrectangles.

3.17 Proposition. *Let X be a connected Riemann surface and let*

$$f : X \longrightarrow \mathbb{E}^{\bullet} = \{q \in \mathbb{C}; \quad 0 < |q| < 1\}$$

be a locally biholomorphic and proper map of X into the punctured disk. Then there exists a natural number n and a biholomorphic map

$$\varphi : X \xrightarrow{\sim} \mathbb{E}^{\bullet}$$

such that the diagram

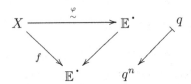

commutes.

Proof. We consider the upper half-plane $\mathbb{H}$ and the map

$$\mathrm{ex} : \mathbb{H} \longrightarrow \mathbb{E}^{\bullet}, \qquad z \longmapsto q := e^{2\pi i z}.$$

It is clear that this map is a (nonproper) covering. By Theorem 3.16, there exists a *continuous* lift

$$\mathrm{Ex} : \mathbb{H} \longrightarrow X \qquad (f \circ \mathrm{Ex} = \mathrm{ex}).$$

In fact, this lift is analytic, since both of the maps $\mathrm{ex} : \mathbb{H} \to \mathbb{E}^{\bullet}$ and $f : X \to \mathbb{E}^{\bullet}$ are locally biholomorphic. From the fact that both are coverings, we can easily deduce the following:

The map

$$\mathrm{Ex} : \mathbb{H} \longrightarrow \mathbb{E}^{\bullet}$$

is a covering.

The point now is that we have to know when two points $a, b \in \mathbb{H}$ have the same image in X (i.e. $\mathrm{Ex}(a) = \mathrm{Ex}(b)$). A necessary condition for this is

$$\mathrm{ex}(a) = \mathrm{ex}(b), \quad \text{i.e. } a = b + n, \quad n \in \mathbb{Z}.$$

3.18 Claim. *Let n be an integer. The set of all points $z \in \mathbb{H}$ for which*

$$\mathrm{Ex}(z) = \mathrm{Ex}(z + n)$$

is open in $\mathbb{H}$.

Corollary. *Since this set is also closed, for trivial reasons, the equation $\mathrm{Ex}(z) = \mathrm{Ex}(z + n)$ holds either for all $z \in \mathbb{H}$ or for none of them.*

Proof of the claim. Assume $\mathrm{Ex}(a) = \mathrm{Ex}(a + n)$. We consider open neighborhoods $U(a)$ of a and $U(a + n)$ of $a + n$, which are mapped topologically under Ex onto an open neighborhood V of $\mathrm{Ex}(a)$ and are such that V is mapped topologically under $f : X \to \mathbb{E}^{\bullet}$ onto an open neighborhood W of $\mathrm{ex}(a)$. Consider $z \in U(a)$ and $z + n \in U(a + n)$. Because of the periodicity of the exponential function, we have $\mathrm{ex}(z) = \mathrm{ex}(z + n) \in W$. The points $\mathrm{Ex}(z)$ and $\mathrm{Ex}(z + n)$ are contained in the neighborhood V, which is mapped injectively under f. Since their images under f agree, we obtain $\mathrm{Ex}(z) = \mathrm{Ex}(z + n)$. Hence there exists a full neighborhood of a in which we have $\mathrm{Ex}(z) = \mathrm{Ex}(z + n)$. $\qquad \square$

Now we consider the set $L \subset \mathbb{Z}$ of all integers with the following property: there exists $a \in \mathbb{H}$ such that

$$\mathrm{Ex}(a) = \mathrm{Ex}(a + n).$$

Because of the remark above, we then have

$$\mathrm{Ex}(z) = \mathrm{Ex}(z + n) \text{ for all } z \in \mathbb{H}.$$

Hence L is a subgroup of $\mathbb{Z}$. Every subgroup of $\mathbb{Z}$ is cyclic:

$$L = n\mathbb{Z}, \quad n \geq 0, \ n \in \mathbb{Z}.$$

Hence we obtain the following result.

There exists an integer $n \geq 0$ such that two points a, b from $\mathbb{H}$ have the same image iff

$$a \equiv b \bmod n \text{ (i.e. } a - b \in n\mathbb{Z}).$$

Clearly, $n \neq 0$. (Otherwise the map $\mathbb{H} \overset{\text{ex}}{\to} \mathbb{E}^{\bullet}$ would be proper, as $f : X \to \mathbb{E}$.)

Using the natural number n, we consider the surjective map

$$g : \mathbb{H} \longrightarrow \mathbb{E}^{\bullet}, \qquad z \longmapsto e^{2\pi i z / n}.$$

We know that

$$g(a) = g(b) \Longleftrightarrow \mathrm{Ex}(a) = \mathrm{Ex}(b).$$

Hence there exists a map

$$\varphi : \mathbb{E}^{\bullet} \longrightarrow X$$

such that the diagram

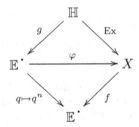

commutes.

Both of the maps f and $f \circ \varphi$ $(q \mapsto q^n)$ are proper and locally topological. From this, we obtain the result that φ is proper and locally topological as well. Obviously,

$$\varphi \text{ proper} \Longrightarrow \varphi(\mathbb{E}^{\bullet}) \text{ closed},$$
$$\varphi \text{ locally topological} \Longrightarrow \varphi \text{ open} \Longrightarrow \varphi(\mathbb{E}^{\bullet}) \text{ open}.$$

Since X by assumption is *connected,* we obtain $\varphi(\mathbb{E}^{\bullet}) = X$. Hence the map φ is bijective. Since it is continuous and open, it is topological. Since f and $f \circ \varphi$ are locally biholomorphic, it is biholomorphic. This proves Proposition 3.17.   □

Appendix B. A Theorem of Implicit Functions

Let $D \subset \mathbb{C} \times \mathbb{C}$ be an open subset. A function

$$f : D \longrightarrow \mathbb{C}$$

is called *analytic* if it satisfies the following two conditions:

1) f is continuously differentiable in the sense of real analysis. (Here, one has to identify $\mathbb{C}$ with $\mathbb{R}^2$ and $\mathbb{C}^2$ with $\mathbb{R}^4$.)
2) f is analytic in both variables, fixing the other variable.

We can then take the complex partial derivatives

$$z \longmapsto \frac{\partial f}{\partial z} \text{ and } w \longmapsto \frac{\partial f}{\partial w}$$

in an obvious way. (We denote the coordinates of $\mathbb{C} \times \mathbb{C}$ by (z, w).)

The theorem of implicit functions can be stated as follows:

Let $(a, b) \in D$ be a point with the properties

$$f(a, b) = 0, \quad \frac{\partial f}{\partial w}(a, b) \neq 0.$$

There exist open neighborhoods

$$a \in U \subset \mathbb{C}, \ b \in V \subset \mathbb{C}$$

with the following properties:

1) *$U \times V \subset D$.*
2) *For each point $z \in U$, there exists a unique point $\varphi(z) \in V$ such that*

$$f(z, \varphi(z)) = 0.$$

3) *The function*

$$\varphi : V \longrightarrow \mathbb{C}$$

is analytic.

As in the proof of the theorem of invertible functions ([FB], Theorem I.5.7), we reduce the proof to the analogous real case.

First of all, we have to verify that the assumptions of the real theorem of implicit functions are satisfied. This means that we have to show that the rank of the real Jacobi matrix (this is a real 4×2 matrix) is 2. But this is clear because this matrix contains the 2×2 matrix

$$\left(\begin{array}{cc} \dfrac{\partial \operatorname{Re} f}{\partial x} & \dfrac{\partial \operatorname{Re} f}{\partial y} \\[2mm] \dfrac{\partial \operatorname{Im} f}{\partial x} & \dfrac{\partial \operatorname{Im} f}{\partial y} \end{array}\right)\Bigg|_{(a,b)} \quad (z = x + iy)$$

as a submatrix and the determinant of this submatrix $|\partial f/\partial z|^2$ is different from zero by our assumption.

Now the claims 1) and 2) follow immediately from the real theorem of implicit functions. Instead of 3), so far we know only that φ is differentiable in the sense of real analysis. But from the formulae for the partial derivatives of φ (they follow from $f(z, \varphi(z)) = 0$ by means of the chain rule), we obtain the Cauchy-Riemann equations for φ. $\square$

Exercises for Sect. I.3

1. Let be $P(z) = \sqrt[5]{1 + z^4}$ the branch of the fifth root in a disk around $z = 0$ which is defined by the principal branch of the logarithm. Construct a closed curve starting and ending at 0 such that analytic continuation along this curve transforms the function element $(0, P)$ into $(0, \zeta P)$, with $\zeta = e^{2\pi i/5}$.

2. Show that the compact Riemann surface belonging to $P(z, w) = w^2 - z$ is biholomorphically equivalent to the Riemann sphere.

3. Show that the set of zeros of an irreducible polynomial $P \in \mathbb{C}[z, w]$ in $\mathbb{C} \times \mathbb{C}$ is connected.

4. Let Q be a polynomial of degree 3 or 4 without multiple zeros and let X be the Riemann surface which is associated with $P(z, w) = w^2 - Q(z)$, together with the projection $p : X \to \bar{\mathbb{C}}$. Then p has precisely four branch points, i.e. there are exactly four points with one preimage. The other points have two preimages. Compare this behavior with the mapping behavior of the Weierstrass $\wp$-function.

   (This gives a hint that X might be biholomorphically equivalent to a torus.)

5. Consider the Riemann surface associated with the polynomial

$$P(z, w) = w^4 - 2w^2 + 1 - z$$

together with the projection $p : X \to \bar{\mathbb{C}}$. Show that all points $z \in \bar{\mathbb{C}}$ besides $0, 1$, and ∞ have four preimages in X. Describe the mapping behavior for the exceptional points.

6. In [FB], in the appendix to Sect. V.3, we introduced the projective space $P^n(\mathbb{C})$ as the quotient space of $\mathbb{C}^{n+1} - \{0\}$. Show that this is a compact space (equipped with the quotient topology). We also introduced there the projective closure $\bar{\mathcal{N}}$ of an affine algebraic curve $\mathcal{N}$. Show that there exists a natural continuous map of the associated Riemann surface onto $\bar{\mathcal{N}}$.

7. Let $L \subset \mathbb{C}$ be a lattice. Show that the projection $\mathbb{C} \to \mathbb{C}/L$ is a covering.

8. Show that Proposition 3.17 remains true if $\mathbb{E}$ is replaced by $\mathbb{C}$.

9. Give an example of a locally topological map which is not a covering.

10. Is $\sin : \mathbb{C} \to \mathbb{C}$ a covering?

II. Harmonic Functions on Riemann Surfaces

In contrast to a domain $D \subset \mathbb{C}$, where rational functions already provide a big class of meromorphic functions, on a Riemann surface it is not possible without further effort to construct meromorphic functions, for example with a given finite set of poles. Constructive problems of this kind are central problems in the theory of Riemann surfaces. It turns out that it is easier to construct harmonic instead of analytic functions. So we are led to pick up a thread which we dropped very early on in the first volume [FB] (Chap. I, Sect. 5). Our treatment is very much based on that in the classic book by Nevanlinna [Ne].

We recall some basic facts about harmonic functions:

1) The real part of an analytic function is harmonic.

2) An analytic function on a domain $D \subset \mathbb{C}$ is determined up to an additive (pure imaginary) constant by its real part.

3) On an elementary domain (= simply connected domain), every harmonic function is the real part of some analytic function. As a consequence, harmonic functions *locally* are real parts of analytic functions.

4) The function
$$\mathbb{C}^{\cdot} := \mathbb{C} - \{0\} \longrightarrow \mathbb{C}, \qquad z \longmapsto \log|z|,$$
is harmonic. But on the whole $\mathbb{C}^{\cdot}$, it is not the real part of an analytic function.

In the first six sections, we shall deal with the *Dirichlet boundary value problem*. The question is whether, on a relatively compact domain $U \subset X$ of a Riemann surface, one can construct a harmonic function which takes given boundary values $f : \partial U \to \mathbb{R}$ when one approaches the boundary. In the case where X is the complex plane and U the unit disk, such a solution can be written down explicitly by means of the Poisson integral. The solution for other domains U in the plane or, even more generally, on a Riemann surface is more involved. It is not necessary for the theory of Riemann surfaces to solve the boundary value problem for arbitrary U. We need only a big enough class of such domains. What we need is the following:

Every Riemann surface X with a countable basis of the topology admits a sequence of open relatively compact open subsets $U_1 \subset U_2 \subset \cdots$, for which the boundary value problem is solvable and such that $X = \bigcup U_n$.

There are several methods for solving this problem. Probably the most powerful method is the Dirichlet principle. Here, the function in question is provided as the solution of an extremal problem. The Dirichlet principle is a powerful tool in other branches of analysis also. Hence it is treated in its own right textbooks about Riemann surfaces, [Fo1, Pf]. Techniques from functional analysis are needed for its use.

Another approach is that due to Perron. Perron's method uses families of "subharmonic functions" which are associated with a boundary value problem. The solution is obtained as the supremum of such a family.

Maybe the most elementary approach is given by the *Schwarz alternating method*, which we shall use in our approach. By this method, we obtain the result that the boundary value problem is solvable for the union of two domains if it is solvable for both of them and if the two domains have only finitely many boundary points in common. Every Riemann surface is the union of countably many "disks". Thus it is very easy to construct an exhaustion of the Riemann surface in the above sense.

The alternating method makes it necessary to admit boundary value distributions which are not continuous at a finite set of points. Thanks to a generalization of the maximum principle, which is treated in Nevanlinna's important book [Ne], it is sufficient to admit a finite set of exceptional points at which, besides a boundedness condition, nothing has to be demanded. When one takes the union of two domains, one usually gets kinks at the intersection points of the two boundaries of the original domains. These intersection points can be included in the set of exceptional points. So we only have to take care that they are finitely many. This means that no problems of topological nature will arise during the exhaustion of the surface . So the alternating method is very simple in this regard also.

This relatively simple solution of the boundary value problem is not the end of construction problems for harmonic functions, since we need harmonic functions on the whole domain and not only on open relatively compact subdomains. Even more importantly, on a compact connected Riemann surface every harmonic function turns out to be constant. Hence we have to modify our original task and admit singularities. So, we have to construct harmonic functions with prescribed singularities. The following problem arises.

Let $U \subset X$ be an open relatively compact subset of a Riemann surface, let $S \subset U$ be a finite subset, and let $u_0 : U - S \to \mathbb{R}$ be a harmonic function. Does there exist a harmonic function $u : X - S \to \mathbb{R}$ such that $u - u_0$ can be extended to a harmonic function on the whole X and such that u remains bounded in the complement of U?

The function u would then have the same singular behavior as u_0 but be harmonic on the whole of $X - S$. The function u_0 should be considered as a description of the desired singularity behavior of u. Hence we call u a solution of the singularity problem. Usually this function will not be unique; the boundedness condition only restricts the class of solutions if there is one. The singularity problem is not always solvable. For example, we shall derive from the residue theorem that for compact Riemann surfaces and for subsets U which are biholomorphically equivalent to a disk, a solution can exist only if u_0 is the real part of an analytic function. On the other hand, the singularity problem is trivially solvable for domains in the plane. For this reason, we divide the class of all Riemann surfaces into two subclasses:

A Riemann surface is called **positively bounded** *if the singularity problem is always solvable. Otherwise, it is called* **zero-bounded**.

Our central existence theorem states that a residue condition is sufficient even in the zero-bounded case to solve the singularity problem.

The solution of the singularity problem by means of the Schwarz alternating method is given in Sects. 8–11.

The various methods for the solution of singularity problems should not be valued differently. Each of them has its advantages and justifications. We shall come back to this in connection with the proof of the uniformization theorem in Chap. III.

For the solution of the singularity problem in the zero-bounded case, we need a quite general version of Stokes's theorem for rather general differentiable oriented surfaces. In the appendix to this chapter (Sect. 13), we shall introduce the necessary calculus of differential forms and include a proof of the Stokes formula.

A by-product of the alternating method is a proof of the nontrivial fact, due to T. Radó (1925), that connected Riemann surfaces always have a countable basis of their topology. Despite the fact that the practical use of the theorem is not very high – one simply could assume the countability –, we give a sketch of this result in an appendix to Sect. 6.

1. The Poisson Integral Formula

A function
$$u : D \to \mathbb{R}, \quad D \subset \mathbb{C} \text{ open},$$
is called *harmonic* if it is two times continuously differentiable (in the sense of real analysis) and if
$$\Delta u = \left(\frac{\partial^2}{\partial x^2} + \frac{\partial^2}{\partial y^2} \right) u = 0$$
holds.

We shall derive from *Cauchy's integral formula* for analytic functions the *Poisson integral formula* for harmonic functions and perform some simple manipulations of Cauchy's integral formula.

Let D be an open subset of the complex plane which contains the closed unit disk
$$\bar{\mathbb{E}} = \{ z \in \mathbb{C}; \ |z| \leq 1 \}.$$

Cauchy's integral formula states, for an analytic function $f : D \to \mathbb{C}$, that
$$f(z) = \frac{1}{2\pi i} \oint_{|\zeta|=1} \frac{f(\zeta)}{\zeta - z} \, d\zeta \qquad (z \in \mathbb{E}).$$

In particular, for $z = 0$ it follows that
$$f(0) = \frac{1}{2\pi} \int_0^{2\pi} f(\zeta(t)) \, dt, \quad \zeta(t) = e^{it}.$$

Taking the complex conjugate in the above formula, we get
$$\overline{f(0)} = \frac{1}{2\pi i} \oint_{|\zeta|=1} (\overline{f(\zeta)}/\zeta) \, d\zeta.$$

This gives

$$f(z) + \overline{f(0)} = \frac{1}{2\pi i} \oint_{|\zeta|=1} \left[\frac{f(\zeta)}{\zeta - z} + \frac{\overline{f(\zeta)}}{\zeta} \right] d\zeta$$

$$= \frac{1}{2\pi i} \oint_{|\zeta|=1} \frac{f(\zeta) + \overline{f(\zeta)}}{\zeta - z} \, d\zeta - \frac{z}{2\pi i} \oint_{|\zeta|=1} \frac{\overline{f(\zeta)}}{\zeta(\zeta - z)} \, d\zeta.$$

Claim.

$$\oint \frac{\overline{f(\zeta)}}{\zeta(\zeta - z)} \, d\zeta = 0.$$

Proof. The function $f(z)$ admits, in $\bar{\mathbb{E}}$, an expansion into a uniformly convergent power series. The claim hence has to be proved for

$$f(z) = z^n, \quad n \geq 0.$$

On the integration path, we have

$$\overline{f(\zeta)} = \zeta^{-n} \quad (\text{since } \zeta\bar{\zeta} = 1).$$

Hence we have to show, for $R = 1$, that

$$\oint_{|\zeta|=R} \frac{d\zeta}{\zeta^{n+1}(\zeta - z)} = 0 \quad (|z| < 1, \, n \geq 0).$$

By Cauchy's theorem, the integral does not change if R is enlarged. The claim follows from taking the limit $R \to \infty$. □

We have obtained the *modified Cauchy integral formula.*

1.1 Lemma. *Assume that the function f is analytic in an open neighborhood of the closed unit disk. Then*

$$f(z) + \overline{f(0)} = \frac{2}{2\pi i} \oint_{|\zeta|=1} \frac{\operatorname{Re} f(\zeta)}{\zeta - z} \, d\zeta.$$

Variant. *In the special case $z = 0$, we get*

$$\operatorname{Re} f(0) = \frac{1}{2\pi i} \oint_{|\zeta|=1} \frac{\operatorname{Re} f(\zeta)}{\zeta} \, d\zeta,$$

and hence

$$f(z) - i \operatorname{Im} f(0) = \frac{1}{2\pi i} \oint_{|\zeta|=1} \frac{\operatorname{Re} f(\zeta)}{\zeta} \frac{\zeta + z}{\zeta - z} \, d\zeta.$$

The modified Cauchy integral formula states that the values of $f(z)$, $z \in \mathbb{E}$, can be computed up to an imaginary constant from the values of the real part of f on the boundary of the unit disk. This formula is due to H. A. Schwarz (1870) and is called the *Schwarz integral formula*.

By assumption, D is an open set which contains the closed unit disk. Hence there exists $R > 1$ such that

$$U_R(0) \subset D.$$

Every harmonic function on $U_R(0)$ is the real part of an analytic function f. Taking the real part in the last formula of Lemma 1.1, we obtain the Poisson integral formula.

Poisson Integral Formula

1.2 Proposition (S. Poisson, 1810). *Let*

$$u : D \longrightarrow \mathbb{R}, \quad D \subset \mathbb{C} \ open, \quad \bar{\mathbb{E}} \subset D,$$

be a harmonic function on an open neighborhood of the closed unit disk. Then, for $z \in \mathbb{E}$, we have

$$u(z) = \frac{1}{2\pi} \int_0^{2\pi} u(\zeta(t)) K(\zeta(t), z) \, dt$$

where

$$\zeta(t) = e^{it}$$

and

$$K(w, z) = \mathrm{Re}\left(\frac{w+z}{w-z}\right).$$

In the special case $z = 0$, we get the so-called *midpoint property* of harmonic functions,

$$u(0) = \frac{1}{2\pi} \int_0^{2\pi} u(\zeta(t)) \, dt.$$

We call

$$K(w, z) = \mathrm{Re}\left(\frac{w+z}{w-z}\right) = \frac{|w|^2 - |z|^2}{|w-z|^2} > 0$$

the *Poisson kernel* of the unit disk.

The Poisson integral formula for harmonic functions is a similar tool to the Cauchy integral formula for analytic functions. An example of this is provided by the maximum principle.

The Maximum Principle

1.3 Lemma. *A harmonic function u on a domain D which attains a maximum, i.e. there exists a point*

$$a \in D \text{ such that } u(z) \le u(a) \text{ for all } z \in D,$$

is constant.

Proof. Let $a \in D$ be a point at which u attains its maximum $(u(z) \le u(a)$ for $z \in D$). The function

$$U(z) = u(a + rz), \quad r \text{ sufficiently small,}$$

is harmonic in an open neighborhood of the closed unit disk. It follows from the midpoint property that

$$u(a) = U(0) = \frac{1}{2\pi} \int\limits_0^{2\pi} u(a + r\zeta(t))\, dt \le \frac{1}{2\pi} \int\limits_0^{2\pi} u(a)\, dt.$$

Since equality must hold, we get $u(a + re^{it}) = u(a)$. Hence the function is constant in a full neighborhood of a. The set of all points $z \in D$ in which u has the maximal value $u(a)$, is open. Since it is closed for trivial reasons, we get $u(z) = u(a)$ for $z \in D$. $\square$

Exercises for Sect. II.1

1. Show that the product of two harmonic functions on an open subset of the plane is harmonic iff their gradients are orthogonal at any point of the domain of definition.

2. Let $\varphi : D \to D'$ be a conformal map between open subsets of the plane. Show that the harmonic functions on D' are in one-to-one correspondence with the harmonic functions on D (with respect to $u \mapsto u \circ \varphi$).

3. Show that every harmonic function on the whole plane $\mathbb{C}$ which is bounded from above (or from below) is constant.

    One can reduce this to Liouville's theorem or prove it directly by means of the Poisson integral formula.

4. Show that
$$\operatorname{Re} \frac{1 + re^{i\varphi}}{1 - re^{i\varphi}} = \frac{1 - r^2}{1 - 2r\cos\varphi + r^2}.$$

The Poisson integral formula can be written in the form

$$u(re^{i\varphi}) = \frac{1}{2\pi} \int\limits_0^{2\pi} u(e^{it}) \frac{1 - r^2}{1 - 2r\cos(t - \varphi) + r^2} \, dt.$$

5. Assume that $u : D \to \mathbb{R}$ is a nonconstant harmonic function on a domain $D \subset \mathbb{C}$. Show that the image $u(D)$ is open in $\mathbb{R}$. Use this for a proof of the maximum principle.

2. Stability of Harmonic Functions on Taking Limits

Harmonic functions have very good stability properties under limits. These properties rest on the Poisson integral formula (Sect. 1).

2.1 Remark.   *The Poisson kernel*

$$K(w, z) = \operatorname{Re} \frac{w + z}{w - z} \qquad (|w| = 1, \ |z| < 1)$$

is harmonic for every fixed w.

The proof is trivial since $(w + z)/(w - z)$ is analytic.   □

2.2 Corollary.   *Let*

$$f : [a, b] \longrightarrow \mathbb{R} \quad (a < b)$$

be a continuous function on a compact interval. The function

$$u(z) = \int\limits_a^b f(t) K(e^{it}, z) \, dt$$

is harmonic (on the unit disk $\mathbb{E}$).

The proof follows immediately from Remark 2.1 and the Leibniz criterion.
                                                                                      □

Since proper integrals are stable with respect to locally uniform convergence, we obtain the following theorem from Corollary 2.2 and the Poisson integral formula.

2.3 Theorem.   *Let (u_n) be a sequence of harmonic functions*

$$u_n : D \longrightarrow \mathbb{R} , \quad D \subset \mathbb{C} \text{ open,}$$

which converges uniformly. Then the limit function is harmonic as well.

The Harnack Inequality (A. Harnack, 1887)

2.4 Proposition. *Let u be a harmonic function on an open neighborhood of the compact disk $\overline{U_R(a)}$. Assume that*

$$u(z) \geq 0 \ for \ |z - a| \leq R.$$

For any number r such that $0 < r < R$ and for all z such that $|z - a| = r$, the Harnack equality

$$\frac{R - r}{R + r} u(a) \leq u(z) \leq \frac{R + r}{R - r} u(a).$$

holds.

The proof follows from the Poisson formula applied to

$$U(z) = u(a + Rz),$$

together with the trivial inequality

$$\frac{|w| - |z|}{|w| + |z|} \leq K(w, z) \leq \frac{|w| + |z|}{|w| - |z|} \qquad (|z| < |w|). \qquad \square$$

2.5 Corollary. *Let u be a harmonic function on an open neighborhood of the compact disk $\overline{U_R(a)}$ such that*
a) $u(a) = 0$;
b) $m \leq u(z) \leq M$ *for $z \in \overline{U_R(a)}$ (where m, M are real constants).*
Then

$$m\frac{2r}{R + r} \leq u(z) \leq M\frac{2r}{R + r} \ for \ |z - a| = r < R.$$

Proof. We apply Harnack's inequality (Proposition 2.4) to the functions $u(z) - m$ and $M - u(z)$. $\qquad \square$

By the way, because of the maximum principle, it is enough to know that the inequality b) holds on the boundary of the disk. The most important application of Harnack's inequality is the following statement.

Harnack's Principle

2.6 Proposition. *Let (u_n) be a monotonically increasing sequence of harmonic functions*

$$u_n : D \longrightarrow \mathbb{R}, \quad D \subset \mathbb{C} \ open,$$
$$u_1(z) \leq u_2(z) \leq \cdots \ for \ z \in D.$$

The set of all points $z \in D$ for which the sequence $(u_n(z))$ remains bounded is open and closed in D.

Corollary. *Let D be a (connected) domain. When the sequence $(u_n(z_0))$ converges for **some** $z_0 \in D$, then it converges for all $z \in D$ and the convergence is locally uniform. In particular, the limit function is harmonic.*

Proof. Since $u_n(z)$ can be replaced by $u_n(z) - u_1(z)$, we may assume that

$$u_n(z) \geq 0 \text{ for all } z \in D.$$

Now let $a \in D$ be a point such that $(u_n(a))$ is bounded, i.e.

$$u_n(a) \leq C.$$

It follows from Harnack's inequality that

$$u_n(z) \leq C \frac{R+r}{R-r} \qquad (r = |z - a|)$$

for all z in a full neighborhood of a. Hence the set of all $z \in D$ on which (u_n) remains bounded is open in D. By means of an estimation of $u(z)$ from below using Proposition 2.4, it can be shown analogously that the set of all points $z \in D$ on which (u_n) is unbounded is open as well.

It remains to prove the locally uniform convergence as stated in the corollary. Again this follows from Harnack's inequality, applied to the functions

$$u_m(z) - u_n(z) , \quad m \geq n.$$

Namely, let $a \in D$ be a given point; it then follows from Harnack's inequality that there exists a neighborhood U (say $U = U_{\frac{1}{2}R}(a)$) such that each $\varepsilon > 0$ admits an $N \in \mathbb{N}$ with

$$0 \leq u_m(z) - u_n(z) \leq \varepsilon \text{ for } m \geq n \geq N \text{ and } z \in U.$$

Hence the sequence (u_n) is a locally convergent Cauchy sequence. $\square$

Exercises for Sect. II.2

1. Show from Harnack's inequality that any harmonic function on the whole of $\mathbb{C}$ which is bounded from above or below is constant.

2. Let $\mathcal{H}$ be a nonempty set of harmonic functions on a domain $D \subset \mathbb{C}$. Assume that for two functions $u_1, u_2 \in \mathcal{H}$ there exists a function $u \in \mathcal{H}$ with the property

$$u \geq \max(u_1, u_2).$$

Assume also that there exists at least one point $a \in D$ such that the function values $u(a)$, $u \in \mathcal{H}$, remain bounded from above. Show that there exists a unique harmonic function $\tilde{u}$ with the property

$$\tilde{u}(z) = \sup\{u(z); \ u \in \mathcal{H}\}.$$

Hint. First, construct a sequence $u_n \in \mathcal{H}$ such that $u_n(a)$ converges increasingly to $\sup\{u(a);\ u \in \mathcal{H}\}$. The limit function of this sequence is harmonic, by Harnack's principle. One can deduce from the maximum principle that the limit function has the desired properties.

3. Let $M > 0$ and $\varepsilon > 0$ be positive numbers. Show that there exists a positive number $\delta > 0$ such that for any harmonic function

$$u : \mathbb{E} \longrightarrow \mathbb{R} \quad \text{with } |u(z)| \le M \text{ for all } z,$$

one has

$$|u(z_1) - u(z_2)| \le \varepsilon \text{ for } |z_1|, |z_2| \le 1/2 \text{ and } |z_1 - z_2| \le \delta.$$

4. Prove the following variant of Montel's theorem:

Any bounded sequence of harmonic functions has a locally uniform convergent subsequence.

3. The Boundary Value Problem for Disks

Let

$$f : (a, b) \longrightarrow \mathbb{R}, \quad a < b,$$

be a *bounded and continuous* function on an open bounded interval. Then the improper integral

$$\int_a^b f(t)\, dt$$

converges absolutely, since there exists a constant $C > 0$ with the property

$$\int_c^d |f(t)|\, dt \le C \text{ for } a < c < d < b.$$

Since the Poisson kernel $K(e^{it}, z)$ is bounded for $z \in \mathbb{E}$ as a function of t, the integral

$$u(z) = \int_a^b f(t) K(e^{it}, z)\, dt$$

exists.

We choose sequences

$$a < a_n < b_n < b, \quad \lim a_n = a,\ \lim b_n = b.$$

Obviously, the sequence

$$u_n(z) = \int_{a_n}^{b_n} f(t)K(e^{it}, z)\, dt$$

converges uniformly to $u(z)$. Because of Corollary 2.2 and Theorem 2.3, the function $u(z)$ is harmonic.

More generally, let

$$f : (a, b) \to \mathbb{R}$$

be a *bounded* function which is continuous with finitely many exceptions

$$a = a_1 < \cdots < a_n = b.$$

We then define

$$\int_a^b f(t)\, dt = \sum_{\nu=1}^{n-1} \int_{a_\nu}^{a_{\nu+1}} f(t)\, dt.$$

The function

$$u(z) = \int_a^b f(t)K(e^{it}, z)\, dt$$

is harmonic in the unit disk. Now we assume that the length of the integration interval is 2π. We are interested in the behavior of u when we approach a boundary point

$$z_0 = e^{it_0}, \quad t_0 \in (a, a + 2\pi).$$

3.1 Lemma.   *Let*

$$f : (a, a + 2\pi) \longrightarrow \mathbb{R}$$

be a bounded function which is continuous outside a finite set of points. Then the function

$$u(z) = \frac{1}{2\pi} \int_a^{a+2\pi} f(t)K(e^{it}, z)\, dt$$

is harmonic in the unit disk. Let $t_0 \in (a, a + 2\pi)$ be a point at which f is continuous, and let $z_0 = e^{it_0}$. Then

$$\lim_{z \to z_0, |z| < 1} u(z) = f(t_0).$$

Proof. The Poisson integral formula states, for $u \equiv 1$, that

$$\int\limits_{a}^{a+2\pi} K(e^{it}, z)\, dt = \int\limits_{0}^{2\pi} K(e^{it}, z)\, dt = 2\pi.$$

Hence, for constant f, the statement in Lemma 3.1 is true. For this reason, we may assume that $f(t_0) = 0$. We then have to show, for given $\varepsilon > 0$, that

$$|u(z)| < \varepsilon$$

in a full neighborhood of z_0. To prove this, we choose a small $\delta > 0$ such that

$$|f(t)| < \varepsilon/2 \text{ for } |t - t_0| < \delta, \quad t \in (a, a + 2\pi).$$

We can assume that

$$(t_0 - \delta, t_0 + \delta) \subset (a, a + 2\pi).$$

When t is not contained in $(t_0 - \delta, t_0 + \delta)$, we have

$$\lim_{z \to z_0} K(e^{it}, z) = 0,$$

where the convergence is uniform in t. It follows that

$$|u(z)| \leq \frac{\varepsilon}{2} + \frac{1}{2\pi} \int\limits_{t_0-\delta}^{t_0+\delta} |f(t)| K(e^{it}, z)\, dt$$

$$\leq \frac{\varepsilon}{2}\left(1 + \frac{1}{2\pi} \int\limits_{t_0-\delta}^{t_0+\delta} K(e^{it}, z)\, dt\right)$$

if z is sufficiently close to z_0. Now we have

$$\int\limits_{t_0-\delta}^{t_0+\delta} K(e^{it}, z)\, dt \leq \int\limits_{t_0-\pi}^{t_0+\pi} K(e^{it}, z)\, dt = \int\limits_{0}^{2\pi} K(e^{it}, z)\, dt = 2\pi.$$

So we obtain

$$|u(z)| \leq \varepsilon, \quad z \text{ close enough to } z_0,$$

as desired. $\quad\square$

The *solution of the boundary value problem for disks* follows from Lemma 3.1.

3.2 Proposition (H.A. Schwarz, 1872). *Let*

$$f : \partial\mathbb{E} \to \mathbb{R}$$

be a bounded function on the boundary of the unit disk which is continuous outside a finite set of points. Then the Poisson integral

$$u(z) = u_f(z) = \frac{1}{2\pi} \int_0^{2\pi} f(e^{it}) K(e^{it}, z) \, dt$$

defines a harmonic function with the following property: if $z_0 \in \partial\mathbb{E}$ is a boundary point at which f is continuous, then

$$\lim_{z \to z_0, z \in \mathbb{E}} u(z) = f(z_0).$$

Additional remark.

1) *In the case $f \equiv 1$, we have $u \equiv 1$.*
2) *It follows from $f \le g$ that $u_f \le u_g$.*

In particular, an estimate

$$m \le f \le M \qquad (m, M \in \mathbb{R})$$

implies a corresponding estimate for u,

$$m \le u \le M.$$

As a consequence, the harmonic functions constructed in Proposition 3.2 are bounded.

Let $D \subset \mathbb{C}$ be a bounded open subset and let $f : \partial D \to \mathbb{R}$ be a continuous function on its boundary. The *Dirichlet boundary value problem* is to find a continuous function on the closure $\bar{D}$ which is harmonic in D and agrees with f on the boundary. Proposition 3.2 provides a solution to this boundary problem in case of the unit disk. In the following sections, we shall formulate the boundary value problem more generally on Riemann surfaces and prove several general existence and uniqueness results.

Exercises for Sect. II.3

1. The Poisson integral formula has been proved for functions which are harmonic in an open neighborhood of the closed unit disk. Show that it is true more generally

for functions which are continuous on the closed unit disk and harmonic in the interior.

2. Let $D \subset \mathbb{C}$ be a domain. Assume that there exists a conformal map from D onto the unit disk which extends to a topological map $\bar{D} \to \bar{\mathbb{E}}$. Show that the Dirichlet boundary value problem for D is solvable.

3. Let u be a continuous function on the closure of the upper half-plane $\mathbb{H}$ in the Riemann sphere which is in harmonic in $\mathbb{H}$. Prove the following "Poisson integral formula" for the upper half-plane:

$$u(z) = \frac{y}{\pi} \int\limits_{-\infty}^{\infty} \frac{u(t)}{(x-t)^2 + y^2} dt \quad (z = x + iy \in \mathbb{H}).$$

4. Let $\varphi : \mathbb{R} \to \mathbb{R}$ be a bounded continuous function. Show that

$$u(z) = \frac{y}{\pi} \int\limits_{-\infty}^{\infty} \frac{\varphi(t)}{(x-t)^2 + y^2} dt$$

defines a harmonic function on $\mathbb{H}$ which converges to φ as one approaches the real axis.

5. Solve the boundary value problem for the unit disk and the boundary values

$$f(z) = \begin{cases} 1 & \text{if } |z| = 1 \text{ and } \operatorname{Im} z > 0, \\ 0 & \text{if } |z| = 1 \text{ and } \operatorname{Im} z < 0. \end{cases}$$

Result:

$$u(z) = 1 - \frac{1}{\pi} \operatorname{Arg}\left(\frac{1}{i}\frac{z-1}{z+1}\right).$$

6. Let D be an open bounded subset of the plane $\mathbb{C}$, and let u, v be two continuous functions on the closure $\bar{D}$ which are harmonic in D and which agree on the boundary ∂D. Show that they agree on D.

7. Let $D \subset \mathbb{C}$ be an open subset. A continuous function $h : D \to \mathbb{R}$ has the *midpoint property* if, for any closed disk $\bar{U}_r(a) \subset D$, the equation

$$h(a) = \frac{1}{2\pi} \int\limits_{0}^{2\pi} h(a + re^{it}) dt$$

is satisfied. So, harmonic functions have the midpoint property. Show the converse: continuous functions with the midpoint property are harmonic.

Hint. First show that functions with the midpoint property obey the maximum principle (Lemma 1.3). Then solve the boundary value problem for a disk $U_r(a)$ whose closure is contained in D and where the boundary values are given by h. Then show that this solution and h coincide in $U_r(a)$.

4. The Formulation of the Boundary Value Problem on Riemann Surfaces and the Uniqueness of the Solution

Let

$$u : U \longrightarrow \mathbb{R} \qquad (U \subset \mathbb{C} \text{ open})$$

be a harmonic function and let

$$\varphi : D \longrightarrow U \qquad (D \subset \mathbb{C} \text{ open})$$

be an analytic function. Then the function

$$\tilde{u}(z) = u(\varphi(z))$$

is harmonic too.

This is a statement of local nature. Hence we may assume, for the proof, that u is the real part of an analytic function f. Then

$$\tilde{u}(z) = \mathrm{Re}(f(\varphi(z))).$$

This simple observation allows us to extend the notion of a harmonic function to arbitrary Riemann surfaces.

4.1 Definition.   *A function $u : X \to \mathbb{R}$ on a Riemann surface is called* **harmonic** *at a point $a \in X$ if there exists an analytic chart*

$$\varphi : U \longrightarrow V \ , \quad a \in U,$$
$$\cap \qquad \cap$$
$$X \qquad \mathbb{C}$$

such that the function

$$u_\varphi = u \circ \varphi^{-1} : V \longrightarrow \mathbb{R}$$

is harmonic in an open neighborhood of $\varphi(a)$.

The initial remark in this section then shows that this is true for *all* analytic charts φ. Examples of harmonic functions are provided by the real parts of analytic functions.

4.2 Remark.   *Let u be a harmonic function on a connected Riemann surface which vanishes in a nonempty open subset of X. Then u vanishes everywhere.*

The simple proof of the identity theorem is left to the reader. We note only that the zero set of a harmonic function is usually not discrete, as the example $\log|z|$ shows.

Formulation of the Boundary Value Problem

A subset U of a topological space is called *relatively compact* if its closure is compact. A subset of $\mathbb{R}^n$ is relatively compact iff it is bounded.

4.3 Definition.   *Let $U \subset X$ be an open relatively compact subset of a Riemann surface. We assume that each connected component of U has infinitely many boundary points. Let*

$$f : \partial U \longrightarrow \mathbb{R}$$

be a bounded function on the boundary of U which is continuous outside a finite set of points.

   *A solution of the boundary value problem "(U, f)" is a **bounded** harmonic function*

$$u : U \longrightarrow \mathbb{R}$$

such that

$$\lim_{x \to a,\, x \in U} u(x) = f(a)$$

for all $a \in \partial U$ outside a finite set of points.

Remark. There then exists a finite set $\mathcal{M} \subset \partial U$ such that the extension

$$u : \bar{U} - \mathcal{M} \longrightarrow \mathbb{R},$$

$$u(x) = \begin{cases} u(x) & \text{for } x \in U, \\ f(x) & \text{for } x \in \partial U - \mathcal{M} \end{cases}$$

of u is continuous.

   We want to show that the solution, if it exists, is uniquely determined. For this, we need certain generalizations of the maximum principle (Lemma 1.3). An obvious generalization states the following.

4.4 Lemma.   *A harmonic function $u : X \to \mathbb{R}$ on a connected Riemann surface which attains its maximum is constant.*

This implies the following statement.

Variant of the Maximum Principle

4.5 Lemma.   *Let*

$$u : U \longrightarrow \mathbb{R}$$

be a harmonic function on an open relatively compact subset U of a Riemann surface X. We assume that no connected component of U is compact. We also assume that every boundary point $a \in \partial U$ admits a neighborhood $U(a) \subset X$ with the property

$$u(x) \geq 0 \text{ for all } x \in U(a) \cap U.$$

Then

$$u \geq 0.$$

Proof. We may assume that U is connected and nonempty, since the assumptions of Lemma 4.5 carry over to each connected component. Let

$$m := \inf_{x \in U} u(x) \geq -\infty \quad (-\infty \text{ is allowed}).$$

Since $\bar{U}$ is compact, there exists a sequence

$$x_n \in U, \quad u(x_n) \longrightarrow m \text{ for } n \longrightarrow \infty$$

that converges in $\bar{U}$. We can assume that u is not constant. Here we make use of the assumption that ∂U is not empty. The maximum principle shows that the limit $a = \lim x_n$ is contained in the boundary of U. But then

$$u(x_n) \geq 0 \text{ for almost all } n,$$

and hence $m \geq 0$. $\square$

The following variant of the maximum principle allows finitely many exceptional points on the boundary. As a consequence, we do not have to take care of finitely many "bad points" such as vertices and isolated boundary points. Hence we shall never be confronted with difficulties of topological nature.

4.6 Proposition. *Let*
$$u : U \longrightarrow \mathbb{R}$$

be a harmonic function on an open relatively compact subset of a Riemann surface which is bounded from below. Assume that each boundary component of U has infinitely many boundary points. Furthermore, assume that each point $a \in \partial U$ up to finitely many exceptions admits a neighborhood $U(a) \subset X$ with the property
$$u(x) \geq 0 \text{ for all } x \in U(a) \cap U.$$

Then
$$u \geq 0.$$

Corollary 1. *The solution of the boundary value problem, if it exists, is uniquely determined.*

Corollary 2. *Let "(U, f)" and "(U, g)" be two boundary value problems on the same U. Then*
$$f \leq g \Longrightarrow u_f \leq u_g.$$

In particular, an estimate

$$m \leq f \leq M \text{ (on } \partial U) \quad (m, M \in \mathbb{R})$$

carries over to an estimate for u_f:

$$m \le u_f \le M \ (\text{on } U).$$

Proof of Proposition 4.6. We can assume that U is connected and that u is not constant, since ∂U contains infinitely many boundary points. Since u is bounded from below, we have

$$m := \inf_{x \in U} u(x) > -\infty.$$

We want to give an indirect argument, and hence assume

$$m < 0.$$

Since $\bar{U}$ is compact, there exists a convergent sequence (x_n), $x_n \in U$, with the property

$$\lim_{n \to \infty} u(x_n) = m.$$

The maximum principle shows that

$$a = \lim_{n \to \infty} x_n \in \partial U$$

(and a is one of the finitely many exceptional points which are allowed by Proposition 4.6).

Now we choose an open neighborhood $U(a)$ such that there are no exceptional points in $U(a) \cap \partial U$ besides a. We can achieve the result that there exists an analytic chart

$$\varphi : U(a) \longrightarrow \mathbb{E} , \quad \varphi(a) = 0.$$

To improve the boundary, we diminish $U(a)$:

$$U'(a) := \{x \in U(a), \ |\varphi(x)| < 1/2\}.$$

We now modify u on $U'(a) \cap U$ in such a way that the assumptions of Proposition 4.6 are fulfilled for all boundary points (including a). We consider

$$u_\varepsilon(x) = u(x) - \varepsilon \log |\varphi(x)|, \qquad x \in U'(a) \cap U.$$

This function is harmonic. We are interested in its boundary behavior.

1) Obviously, for arbitrary $\varepsilon > 0$,

$$\lim_{x \to a} u_\varepsilon(x) = +\infty.$$

2) Let $b \in \partial(U \cap U'(a))$, $b \neq a$. A simple topological consideration shows that either

a)
$$b \in \overline{U'(a)} \cap \partial U, \quad b \neq a,$$

or

b)
$$b \in K := \bar{U} \cap \partial(U'(a)).$$

In the case a) we have, for positive ε,

$$u_\varepsilon(x) \geq u(x) \geq 0$$

in a sufficiently small neighborhood of b. Now we consider

$$\mu = \inf_{x \in U \cap \partial U'(a)} u(x).$$

Claim. $\mu > m$.

Proof of the claim. In the case where μ is not negative, we are already done because $m < 0$. Hence we assume $\mu < 0$. Then u has a minimum in $U \cap \partial U'(a)$. The claim now follows from the maximum principle, since by assumption u is not constant.

We can choose $\varepsilon > 0$ such that

$$u_\varepsilon(x) \geq M \text{ for } x \in U \cap \partial U'(a), \quad 0 > M > m \text{ suitable.}$$

Now we can apply Lemma 4.5 to the function

$$u_\varepsilon(x) - M \text{ on } U'(a) \cap U$$

and obtain
$$u_\varepsilon(x) \geq M \text{ for } x \in U'(a) \cap U.$$

This inequality can be applied to the sequence (x_n). Taking the limit, we obtain
$$m \geq M,$$

which contradicts the choice of M.   □

Proof of the corollaries. Let u and v be two solutions of the boundary value problem. Then, for each $\varepsilon > 0$, the functions $u - v + \varepsilon$ and $v - u + \varepsilon$ satisfy the assumptions of Proposition 4.6. Hence both are nonnegative on U. Since this is true for arbitrary $\varepsilon > 0$, we obtain $u = v$. The second corollary is proved similarly.

The assumption that any connected component of U has infinitely many boundary points is harmless, because of the following fact.

4.7 Remark. *Let X be a connected Riemann surface and let U be an open nonempty relatively compact subset which has only finitely many boundary points. Then X is compact and $X - U$ is a finite set.*

Proof. Let S be the set of boundary points of U. The set U is open and closed in $X - S$. By assumption, S is finite. Hence $X - S$ remains connected. We obtain $U = X - S$ and $\bar{U} = X$.   □

Exercises for Sect. II.4

1. Prove the identity theorem (Theorem 4.2) for harmonic functions.

2. Prove the following variant of Riemann's extension theorem:

   Let $D \subset \mathbb{C}$ be an open subset, let $a \in D$ be a given point, and let $u : D - \{a\} \to \mathbb{R}$ be a bounded harmonic function. Then the singularity a is removable, i.e. u is the restriction of a harmonic function on the whole of D.

   Hint. It can be assumed that D is the unit disk and that u extends as a continuous function to the closed disk. Consider u as the solution of a boundary problem, where a can be considered as an exceptional boundary point.

   A further proof of the Riemann extension theorem for harmonic functions comes from the following exercise.

3. Show that any harmonic function u in the punctured disk $\mathbb{E}^{\bullet}$ has the form

   $$u(z) = A \log |z| + \operatorname{Re} f(z)$$

   with a constant A and an analytic function in $\mathbb{E}^{\bullet}$.

   Hint. Take the residue of the analytic function $(\partial_1 - i\partial_2)(u)$ for A.

4. Prove the following variant of the Schwarz reflection principle.

   Let D be the intersection of the unit disk with the upper half-plane and let u be a harmonic function on D which tends to 0 when the real axis is approached. Show that the function

   $$U(z) = \begin{cases} u(z) & \text{for } \operatorname{Im} z > 0, \\ 0 & \text{for } \operatorname{Im} z = 0, \\ -u(\bar{z}) & \text{for } \operatorname{Im} z < 0 \end{cases}$$

   is harmonic in the full unit disk.

   Hint. It can be assumed that u has a continuous extension to the closure of D. We know what U has to look like at the boundary of the unit disk. Define U as the solution of the corresponding boundary problem. Show that U vanishes on the real axis, and make use of the uniqueness of the solution of the boundary problem for D to show that U and u agree in D.

5. Solution of the Boundary Value Problem by Means of the Schwarz Alternating Method

Since it is tedious to repeatedly point out the finitely many exceptional points of a solution of a boundary value problem and the boundedness property included in it, we shall adopt the following convention from this point onwards.

5.1 Notation.   *Let $U \subset X$ be an open subset of a Riemann surface, let*

$$u : U \longrightarrow \mathbb{R}$$

be a harmonic function, let $\Delta \subset \partial U$ be a subset of the boundary (usually the whole boundary), and let f be a function which is defined on Δ.

   The notation

$$u \geq f \text{ on } \Delta$$

means the following. Let ε be a positive number. Every point $a \in \Delta$ admits an open neighborhood $U(a)$ with the following two properties:

a) *u is bounded from below on $U \cap U(a)$.*
b) *With the exception of finitely many a, we have*

$$u(x) \geq f(a) - \varepsilon \text{ for } x \in U \cap U(a).$$

In the cases $u \geq f$ and $u \leq f$ (i.e. $-u \geq -f$) on Δ, we write

$$u = f \text{ on } \Delta.$$

If we are familiar with the notion of a limit superior, we can formulate a) and b) as follows:
a') $\limsup_{x \to a} u(x) > -\infty$ for all $a \in \Delta$;
b') $\limsup_{x \to a} u(x) \geq f(a)$ for almost all $a \in \Delta$.

   The condition "$u = f$ on Δ" means the following. If we approach an arbitrary boundary point, then u remains bounded. For almost all a on the boundary, the following holds:

$$\lim_{x \to a} u(x) = f(a).$$

When $U \subset X$ is an open relatively compact subset and

$$f : \partial U \longrightarrow \mathbb{R}$$

is a bounded function on the boundary which is continuous outside a finite set, then the *solution of the boundary value problem* "(U, f)" means a harmonic function $u : U \to \mathbb{R}$ such that

$$u = f \text{ on } \partial U.$$

We say that the boundary value problem on U is solvable if a solution exists for every f.

5.2 Theorem. *Let U, V be two open relatively compact subsets of a Riemann surface X. Assume that every connected component of $U \cup V$ has infinitely many boundary points.*

Assumption.

a) *$\partial U \cap \partial V$ is a finite set.*

b) *The boundary value problems for U and V are solvable.*

Claim. *Then the boundary value problem for $U \cup V$ is solvable.*

Proof. It follows from Remark 4.7 that each of the sets U, V, $U \cap V$ has infinitely many boundary points. We may assume that $U \cap V$ is not empty.

We decompose the boundaries

$$\partial U = \partial' U \cup \partial'' U,$$
$$\partial'' U = \partial U \cap V, \quad \partial' U = \partial U - \partial'' U$$

and, analogously,

$$\partial V = \partial' V \cup \partial'' V,$$
$$\partial'' V = \partial V \cap U, \quad \partial' V = \partial V - \partial'' V.$$

We have

$$\partial(U \cup V) = \partial' U \cup \partial' V.$$

This decomposition is nearly disjoint, i.e. $\partial' U$ and $\partial' V$ have only finitely many points (from the set $\partial U \cap \partial V$) in common. We also have the fact that the sets $\partial(U \cap V)$ and $\partial'' U \cup \partial'' V$ are equal up to finitely many points (also from the set $\partial U \cap \partial V$):

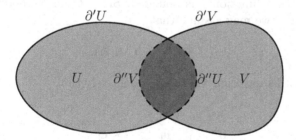

Now we construct, for a given boundary value distribution

$$f : \partial(U \cup V) \longrightarrow \mathbb{R},$$

two sequences of harmonic functions

$$u_n : U \longrightarrow \mathbb{R},$$
$$v_n : V \longrightarrow \mathbb{R} \quad (n = 0, 1, 2, \ldots).$$

The construction is done inductively in the order

$$u_0, v_0, u_1, v_1, \ldots$$

(always *alternating* in U and V). We start with

$$u_0 \equiv 0$$

and then define $v_0, u_1, \ldots$ as solutions of the following boundary conditions:

$$v_n = \begin{cases} f & \text{on } \partial' V, \\ u_n & \text{on } \partial'' V, \quad n \geq 0, \end{cases}$$

$$u_n = \begin{cases} f & \text{on } \partial' U, \\ v_{n-1} & \text{on } \partial'' U, \quad n > 0. \end{cases}$$

Now we show, step by step, the following:

1) The limits

$$u = \lim u_n, \quad v = \lim v_n$$

exist and are harmonic.
2) $u|U \cap V = v|U \cap V$.
3) The harmonic function

$$(u, v) : U \longrightarrow \mathbb{R},$$

which arises from gluing u and v, gives the solution of the boundary value problem "$(U \cup V, f)$".

1) By assumption, the function f is bounded. Since f can be modified by an additive constant, we may assume that

$$0 \leq f \leq C \qquad (C \in \mathbb{R}).$$

The corollary of Proposition 4.6 shows inductively that

$$0 \leq u_0 \leq u_1 \leq \cdots \leq C,$$
$$0 \leq v_0 \leq v_1 \leq \cdots \leq C.$$

By Harnack's principle, the limits

$$u = \lim u_n, \quad v = \lim v_n,$$

exist and are harmonic.
This proves 1).

2) Again using Proposition 4.6, we show that on $U \cap V$ the inequality

$$u_n \geq v_{n-1} \qquad (n \geq 1)$$

holds. Here we have to use the fact that the boundary of $U \cap V$ coincides with $\partial''U \cup \partial''V$ up to finitely many points. It follows from the above inequality that

$$u \geq v \quad \text{on } U \cap V$$

by taking limits, and then $u = v$ by symmetry.

3) We show (using Notation 5.1) that

$$u = f \quad \text{on } \partial'U$$

and analogously,

$$v = f \quad \text{on } \partial'V$$

and hence

$$(u, v) = f \quad \text{on } \partial(U \cup V).$$

For the proof, we consider the harmonic function

$$\omega : U \longrightarrow \mathbb{C}$$

which solves the boundary value problem

$$\omega = \begin{cases} 0 & \text{on } \partial'U, \\ C & \text{on } \partial''U. \end{cases}$$

From Proposition 4.6 we get

$$u_1 \leq u \leq u_1 + \omega \quad \text{on } U$$

and hence

$$u = u_1 = f \text{ on } \partial'U. \qquad \square$$

Some simple examples for which the boundary value problem can be solved are described below.

5.3 Remark. *Let $U \subset X$ be an open relatively compact subset of a Riemann surface. Assume that there exists a biholomorphic map onto the unit disk*

$$f : U \longrightarrow \mathbb{E}$$

which, outside a finite set of points, extends to a topological map of the boundary.

(This means that there exist finite subsets $\mathcal{M} \subset \partial U$, $\mathcal{N} \subset \partial\mathbb{E}$ and a topological extension

$$\bar{U} - \mathcal{M} \longrightarrow \bar{\mathbb{E}} - \mathcal{N}$$

of f). Then the boundary value problem is solvable on U.

The proof is trivial. $\qquad \square$

5.4 Corollary. *The boundary value problem is solvable on a disk two-gon,*

(A disk two-gon is the intersection of the interior of a disk with the interior or exterior of another disk such that neither of the disks is inside the other one.)

For the proof, we construct a conformal map from the disk two-gon onto the unit disk such that this map extends topologically to the closures. This is done in three steps; the details are left to the reader.

First step. Let a be a vertex of the two-gon. By means of the conformal map

$$z \longmapsto \frac{1}{z - a},$$

the two-gon is mapped onto an angle area:

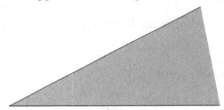

Second step. We can assume that the negative real axis is not contained in the angle area. We can then use

$$z^\alpha := e^{\alpha \log z} \quad (\log = \text{principal branch of the logarithm})$$

for a suitable α to get a conformal map onto a half-plane.

Third step . We can assume that the half-plane is the upper half-plane and then use the standard map $z \to (z - i)/(z + i)$ from the upper half-plane onto the unit disk. It is easy to see how these maps behave on the boundaries.   □

Now we prove a proposition by means of the alternating method (Theorem 5.2).

5.5 Proposition. *The boundary value problem is solvable for annuli*

$$r < |z| < R \qquad (0 < r < R < \infty).$$

Proof. Obviously, the annulus can be covered by annuli in such a way that Theorem 5.2 can be applied iteratively.

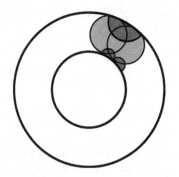

Covering of a part of an
annulus by three disk two-
gons and two disks. The
boundaries of any two of
them have only finitely
many points in common.

□

The Exceptional Points of the Boundary Value Problem

Let $U \subset X$ be an open relatively compact subset of a Riemann surface for
which the boundary value problem is solvable.

A boundary point $a \in \partial U$ is called an *exceptional point* (with respect to the
boundary value problem) if there is a boundary value problem "(U, f)" such
that f is continuous at a, but the equality

$$\lim_{x \to a} u(x) = f(a) \quad (u \text{ is the solution of the boundary value problem})$$

is *not* true.

The proof of Theorem 5.2 immediately shows the following.

5.6 Remark.   *We use the same notation as in Theorem 5.2. Let a be an
exceptional point of $U \cup V$. Then*

> *either a is an exceptional point of U,*
> *or   a is an exceptional point of V,*
> *or $a \in \partial U \cap \partial V$.*

It is in the nature of the alternating method that points from $\partial U \cap \partial V$ can
produce new exceptional points.

The boundary value problem for U is called *strictly solvable* if U has no
exceptional points. Then the solution extends continuously to the whole closure
$\bar{U}$. The next statement follows from Proposition 3.2.

The boundary value problem is strictly solvable for the unit disk.

We have shown that the boundary value problem is solvable for disk two-
gons. It is clear that the only possible exceptional points are the vertices.
(Looking a little closely, we shall see that they are not exceptional points, but
we shall not need this.)

The solution of the boundary value problem for an annulus was obtained
by covering the annulus with disks and disk two-gons and application of the
alternating method (Proposition 5.5). Since the disks and two-gons can be
changed a little, we can obtain an improvement using Remark 5.6.

5.7 Remark. Improvement of Remark 5.5. *The boundary value problem is strictly solvable (without exceptional points) for annuli.*

Exercises for Sect. II.5

1. Let D be an open square and let K be a compact square contained in D with edges parallel to the axes. Show that the boundary value problem for $(D - K, \mathbb{C})$ is solvable.

2. Show that the boundary value problem is solvable for $(\bar{\mathbb{C}} - [-1, 1], \bar{\mathbb{C}})$.

3. Show that any conformal map of a disk two-gon onto the unit disk extends continuously to the closures.

4. Show that there is a nonconstant continuous function on the Riemann sphere which is harmonic in the complement of the unit circle.

6. The Normalized Solution of the External Space Problem

In the following constructions, we make use of an *exhaustion* of the Riemann surface. For this we need the following assumption.

6.1 Assumption. *The Riemann surface X considered below has a countable basis of the topology.*

Actually, we shall see in the appendix to this section that every connected Riemann surface has a countable basis of the topology.

6.2 Definition. *A **disk** on a Riemann surface X is an analytic chart whose domain of values is the unit disk:*

$$\varphi : U \xrightarrow{\sim} \mathbb{E}.$$

We shall use the notation

$$U(r) = \{x \in U; \quad |\varphi(x)| < r\} \qquad (0 < r < 1),$$
$$\partial(r) = \partial(U(r)) = \{x \in U; \quad |\varphi(x)| = r\}.$$

Let

$$f : \partial(1/2) \longrightarrow \mathbb{R}$$

*be a continuous function. A **solution of the external space problem** is a function*

$$u : X - U(1/2) \longrightarrow \mathbb{R}$$

with the properties

1) *u is continuous;*
2) $u|\partial(1/2) = f$;
3) *u is harmonic in the interior* $(X - \overline{U(1/2)})$.

Usually, the solution u of the external space problem is not unique. Clearly, $\partial(1/2)$ is the boundary of $X - U(1/2)$, but this set needs not to be compact. So, at the moment, we know neither existence nor uniqueness.

In the following, we shall construct a distinguished, so-called *normalized* solution of the external space problem.

6.3 Proposition. *Let X be a Riemann surface with a countable basis of the topology and let $\varphi : U \to \mathbb{E}$ be a disk on X. There exists a unique map which assigns to each continuous function $f : \partial(1/2) \to \mathbb{R}$ a **normalized solution of the external space problem** which has the following minimality property:*

Assume $f \geq 0$ and let u be an arbitrary solution of the external space problem defined by f such that $u \geq 0$. Then $u(f) \leq u$.

Moreover:

1) *The assignment $f \mapsto u(f)$ is $\mathbb{R}$-linear.*
2) $f \leq g \Longrightarrow u(f) \leq u(g)$.

Special case. *In the case $f = 1$, the normalized solution has a special meaning. This solution is denoted by ω. We have*

$$\omega = 1 \ \text{on} \ \partial(1/2) \ \text{and} \ 0 \leq \omega \leq 1.$$

The uniqueness of the map is obvious. For the proof of its existence, we construct an exhaustion. Since X has a a countable basis of the topology and since it is locally compact, it is countable at infinity. This means that X can be written as a countable union of compact sets,

$$X = K_1 \cup K_2 \cup \ldots, \qquad K_n \ \text{compact}.$$

We use this for the construction of the exhaustion of X. (Actually, the solution is independent of the choice of the exhaustion because of its uniqueness.)

6.4 Proposition. *Let X be a Riemann surface with a countable basis of the topology, and let*

$$\varphi : U \longrightarrow \mathbb{E}$$

be a disk on X. There exists a sequence

$$U(3/4) \subset A_1 \subset A_2 \subset A_3 \subset \ldots$$

of relatively compact open subsets with the following properties:

1) $X = \bigcup_{n=1}^{\infty} A_n$.

2) *The boundary value problem for*

$$A_n - \overline{U(1/2)}, \qquad n = 1, 2, \ldots ,$$

   is solvable.

6.5 Remark. *We have*

$$\partial(A_n - \overline{U(1/2)}) = \partial(A_n) \cup \partial(1/2).$$

We call $\partial(A_n)$ the exterior boundary and $\partial(1/2)$ the interior boundary of $A_n - \overline{U(1/2)}$. The two are disjoint because of the assumption $U(3/4) \subset A_n$.

Proof. Let

$$X' := X - \overline{U(1/2)}.$$

Like X, X' also has a countable basis of the topology. For each point $a' \in X'$, we choose a disk

$$\varphi' : U' \longrightarrow \mathbb{E}; \quad U' \subset X', \ \varphi'(a') = 0.$$

Each compact subset can be covered by finitely many of the $U'(1/2)$.

Hence there exists a sequence

$$\varphi_n : U_n \longrightarrow \mathbb{E}$$

of disks in X' with the property

$$X' = \bigcup_{n=1}^{\infty} U_n(1/2).$$

With a suitable choice of a sequence of numbers

$$1/2 < r_n < 1$$

that will be fixed later, we define

$$A_n := U(3/4) \cup U_1(r_1) \cup \ldots \cup U_n(r_n).$$

These sets are relatively compact and they exhaust X. We have to investigate whether the boundary value problem is solvable for $A_n - \overline{U(1/2)}$. We have

$$A_n - \overline{U(1/2)} = \mathcal{R} \cup U_1(r_1) \cup \ldots \cup U_n(r_n),$$

where $\mathcal{R}$ denotes the annulus

$$\mathcal{R} = \{x \in U; \quad 1/2 < |\varphi(x)| < 3/4\}.$$

Since the boundary value problem has been solved for disks and annuli, it is sufficient – by the alternating method – to choose the numbers $r_n \in (1/2, 1)$ in such a way that

$$U_{n+1}(r_{n+1}) \quad \text{and} \quad \mathcal{R} \cup U_1(r_1) \cup \ldots \cup U_n(r_n)$$

share only finitely many boundary points. The following lemma gives an inductive construction for a sequence r_n with these properties.

6.6 Lemma. *Let*

$$\varphi : U \longrightarrow \mathbb{E}, \quad \psi : V \longrightarrow \mathbb{E}$$

be two disks on the Riemann surface X. If the two domains $U(r)$ and $V(r)$ have infinitely many boundary points in common, then their boundaries agree $(\partial U(r) = \partial V(r))$.

Proof. We assume that there are infinitely many boundary points, and show that

$$\partial V(r) \subset \partial U(r).$$

(Equality then follows from symmetry.) Since $\partial V(r)$ is compact, the intersection $\partial V(r) \cap \partial U(r)$ contains an accumulation point in $\partial V(r)$. We denote by $\mathcal{M} \subset \partial V(r)$ the set of all such accumulation points. This set is closed, for trivial reasons. If we can show that $\mathcal{M}$ is also open in $\partial V(r)$ we are done, because then, for reasons of connectedness, we have

$$\partial V(r) = \mathcal{M} \quad (\subset \partial U(r) \cap \partial V(r)).$$

It follows directly from the following simple remark that $\mathcal{M}$ is open.

6.7 Remark. *Let f be an analytic function on an open neighborhood of* $1 \in \mathbb{C}$. *Assume that there exists a sequence* (a_n) *of complex numbers with the following properties:*

a) $a_n \neq 1$, $\lim a_n = 1$;
b) $|a_n| = 1$;
c) $|f(a_n)| = 1$.

Then there exists a neighborhood U of 1 such that

$$a \in U, \quad |a| = 1 \Longrightarrow |f(a)| = 1.$$

For the proof, we consider the analytic function

$$g(z) = \overline{f(\bar{z})},$$

and

$$h(\varphi) = f(e^{i\varphi})g(e^{-i\varphi}) - 1.$$

This function is analytic in a small (complex!) disk around $\varphi = 0$. For *real* φ, we have

$$h(\varphi) = |f(e^{i\varphi})|^2 - 1.$$

By assumption, the zeros of h have 0 as an accumulation point. From the identity theorem for holomorphic functions, we get

$$h \equiv 0.$$

This proves the remark and hence Lemma 6.6.   □

Construction of the Solution of the Normalized External Space Problem

We now come to the proof of 6.3.

Proof of Proposition 6.3. We construct, for a given continuous boundary allocation

$$f : \partial(U(1/2)) \longrightarrow \mathbb{R}$$
$$\|$$
$$\text{``}|z| = 1/2\text{''},$$

a solution of the external space problem, i.e. a harmonic function

$$u = u(f) : X - \overline{U(1/2)} \longrightarrow \mathbb{R}$$

with boundary values f.

First of all, we consider the (uniquely determined) harmonic function

$$u_n : A_n - \overline{U(1/2)} \longrightarrow \mathbb{R}$$

with boundary values

$$u_n = f \text{ on the interior boundary } (\partial U(1/2)),$$
$$u_n = 0 \text{ on the exteriour boundary } (\partial A_n).$$

We notice that by Remarks 5.6 and 5.7, there will be no exceptional points on the interior boundary $\partial(1/2)$. Since f, by assumption, is continuous everywhere, we have

$$\lim_{x \to a} u_n(x) = f(a) \text{ for all } a \in \partial(1/2).$$

6.8 Lemma. *The sequence of solutions*

$$u_n = u_n(f) : A_n - \overline{U(1/2)} \longrightarrow \mathbb{R}$$

of the boundary value problem

$$u_n = f \text{ on the interior boundary } (\partial U(1/2)),$$
$$u_n = 0 \text{ on the exteriour boundary } (\partial A_n)$$

*converges *) locally uniformly to a bounded harmonic function*

$$u : X - \overline{U(1/2)} \longrightarrow \mathbb{R}.$$

We have

$$u = f \quad on \quad \partial U(1/2).$$

Proof. First, assume $f \geq 0$. By the maximum principle, all u_n are nonnegative, since this is true on the boundary. In particular, u_{n+1} is greater or equal to u_n on the open set $A_n - \overline{U(1/2)}$ since this true on the boundary. Hence the sequence of functions u_n is increasing. Now the statement about convergence in Lemma 6.8 follows from Harnack's principle. In the general case, we use the decomposition

$$f = \frac{1}{2}(f + |f|) - \frac{1}{2}(|f| - f).$$

It remains to study the behavior of $u = u(f)$ at the boundary. (The obvious argument

$$\lim_{x \to a} u(x) = \lim_{x \to a} \lim_{n \to \infty} u_n(x) = \lim_{n \to \infty} \lim_{x \to a} u_n(x) = f(a)$$

---

*) Actually, we have never introduced the notion of a convergent sequence of functions $f : U_n \to \mathbb{C}$ for a varying sequence of domains of definition U_n. In our case, this sequence is increasing. Hence, for an arbitrary point a of the union of the U_n, we have $a \in U_n$ for almost all n. Thus it is clear what the limit means.

is not correct, since it is usually not allowed to interchange limits.) In the first step, we consider the special case

$$f \equiv 1.$$

In this case we write

$$\omega_n := u_n(1), \qquad \omega := \lim \omega_n.$$

We have

$$0 \leq \omega_n(x) \leq \omega(x) \leq 1.$$

The claim about the behavior at the boundary of ω becomes clear if we take for fixed n (for example $n = 1$) the limit $x \to a$, where a is a boundary point. Since $\omega_1(x)$ then converges to 1, we get the same result for $\omega(x)$.

General case. It is sufficient to prove that

$$u(f) \geq f \quad \text{on} \quad \partial U(1/2).$$

Equality is then obtained by replacing f by $-f$.

For the proof of the inequality, we choose a constant $C > 0$ with the property

$$f + C \geq 0.$$

In this case (u_n) is an increasing sequence, and from

$$u_n(f + C) = f + C \quad \text{on} \quad \partial(U(1/2))$$

we obtain at least

$$u(f + C) \geq f + C \quad \text{on} \quad \partial(U(1/2)).$$

On the other hand, we have

$$u(f + C) = u(f) + C\omega.$$

Both of these together give

$$u(f) \geq f \quad \text{on} \quad \partial(U(1/2)),$$

as stated. The properties 1) and 2) in Proposition 6.3 are clearly satisfied.

□

Examples.

1) Let $X = \mathbb{C}$, $\quad \varphi = \mathrm{id}_{\mathbb{E}} : \mathbb{E} \to \mathbb{E}$, $\quad$ boundary value distribution $f \equiv 1$.

We choose the exhaustion

$$A_n = \{z; \quad |z| < n\}$$

and obtain

$$u_n(z) = \frac{\log|z/n|}{\log(1/2n)} = \frac{\log|z| - \log n}{-\log(2n)}.$$

Taking the limit $n \to \infty$, we obtain

$$u \equiv 1.$$

2) $X = \{z \in \mathbb{C}; \quad |z| < 2\}, \quad \varphi = \mathrm{id}_{\mathbb{E}} : \mathbb{E} \to \mathbb{E}$, boundary value distribution $f \equiv 1$.

We take the exhaustion

$$A_n = \{z; \quad |z| < 2 - 1/n\}$$

and obtain in this case

$$u_n(z) = \frac{\log|z/(2-1/n)|}{\log(1/(2(2-1/n)))}.$$

Taking the limit $n \to \infty$, we get

$$u(z) = \frac{\log|z/2|}{\log(1/4)}.$$

The normalized solution of the boundary value problem with respect to a constant boundary value distribution can, but need not, be constant.

Exercises for Sect. II.6

1. Let D be a bounded domain of the plane. Show that the normalized solution of the external space problem with respect to a boundary value distribution equal to constant 1 is not constant.

2. Let X be a compact Riemann surface and let S be a finite subset. Show that the normalized solution of the external space problem for a constant boundary value distribution is constant.

3. Choose a concrete disk in the upper half-plane and determine the explicit solution of the external space problem for the boundary value distribution $f \equiv 1$.

4. Show that the normalized solution of the external space problem for the boundary value distribution $f \equiv 1$ on the slit plane $\mathbb{C} - \{x \in \mathbb{R}; \ x \leq 0\}$ is not constant.

Appendix to 6. Countability of Riemann Surfaces

6.9 Theorem. *Every connected Riemann surface has a countable basis of the topology. As a consequence, it is countable at infinity, i.e. it can be written as the union of an ascending chain of a sequence of compact sets.*

In the usual constructions of Riemann surfaces, the countability is immediately clear. Hence, in practise, this relatively deep result is not necessary. We shall be brief with the proof. First of all, we recall the definition of countability:

A topological space X has a countable basis of the topology if there exists a sequence of open subsets U_1, U_2, U_3, ... such that each open subset can be written as a union of members of this sequence.

Example. A metric space (X, d) which admits a countable dense subset $S \subset X$ has a countable basis of the topology. For the proof, we consider the (countable) system of balls around points from S with rational radii.

First of all, we assume that on the connected Riemann surface X there exists a nonconstant analytic function f. Then we prove that X is metrizable.

Metrization of X

Let

$$\alpha : [0, 1] \longrightarrow X$$

be a piecewise smooth curve. We denote the Euclidean length of its image under f by

$$L(\alpha) := l(f \circ \alpha) \qquad (= \text{Euclidean length}).$$

Now, let a, b be from X. We define

$$d(a, b) := \inf_{\alpha} L(\alpha),$$

where α runs over all curves with starting point a and endpoint b. It is not difficult to show that this defines a metric on X which induces the given topology on X.

Construction of a Countable Dense Subset

For this purpose, we can remove a discrete subset from X, for example the set of all points where f is not locally topological. Hence we assume that

$$f : X \longrightarrow \mathbb{C}$$

is locally topological. We choose some point $a \in X$ and can assume that $f(a) = 0$.

6.10 Definition. *A point $b \in X$ is called* **rational** *if there exists a curve*

$$\alpha : [0,1] \longrightarrow X, \qquad \alpha(0) = a, \ \alpha(1) = b,$$

whose image curve $f \circ \alpha$ is a piecewise linear function with rational vertices.

The set of all of these piecewise linear curves is countable. Because of the uniqueness of the curve lifting, we obtain the result that the set of all curves α and hence the set of all rational points $b \in X$ is countable. We can also see easily that the set of rational points is dense in X. This proves the countability of (the topology of) X (if the function f exists).

We want to weaken this condition. So, we now assume only that on the connected Riemann surface there exists a nonconstant *harmonic* function

$$u : X \longrightarrow \mathbb{R}.$$

Then we construct a holomorphic map

$$p : \tilde{X} \longrightarrow X$$

of a connected Riemann surface $\tilde{X}$ onto X such that

$$\tilde{u} := u \circ p$$

is the real part of an analytic function. (Then $\tilde{X}$ and, as a consequence, X are countable.) It is possible to take the universal covering for $\tilde{X}$ (see the appendix to Chap. III). If we want to avoid using its existence, we can use the following construction.

Fix a point $a \in X$. In a small open neighborhood $U(a)$ the function u is the real part of an analytic function. It is easy to show the following.

If $\alpha : [0,1] \to X$ is a curve with starting point $a = \alpha(0)$, then f admits an analytic continuation along α. (It should be clear what this means.) In analogy to the *analytisches Gebilde,* we construct from all these continuations a "covering" $\tilde{X} \to X$, on which f becomes single-valued.

The essential point of the proof of the countability turns out to be the existence of a harmonic function.

Existence of a Harmonic Function

We choose a disk

$$\varphi : U \xrightarrow{\sim} \mathbb{E}$$

on our connected Riemann surface, and claim the following:

There exists a nonconstant harmonic function

$$u : X - \overline{U(1/2)} \longrightarrow \mathbb{R}.$$

Then $X - \overline{U(1/2)}$ and, as a consequence, also X have a countable basis of the topology. For the construction of u, we choose some continuous boundary value distribution

$$f : \partial U(1/2) \longrightarrow \mathbb{R}, \quad 0 \leq f \leq 1,$$

which is different from 0. We want to construct u such that

$$u = f \text{ on } \partial U(1/2).$$

The problem is that we do not know that X has a countable basis of the topology. Therefore we consider open connected subsurfaces $Y \subset X$ such that $U \subset Y \subset X$ with a countable basis of the topology. We call them "countable subsurfaces" for short. Clearly, X is the union of all the countable Y:

$$X = \bigcup_{\substack{U \subset Y \subset X \\ Y \text{ "countable"}}} Y.$$

For each Y (and the given disk φ), we denote by

$$u_Y : Y - U(1/2) \longrightarrow \mathbb{R}$$

the normalized solution of the external space problem

$$u_Y | \partial(1/2) = f.$$

It is easy to see that

$$Y \subset Y' \quad \Longrightarrow \quad u_Y \leq u_{Y'} \quad (\text{on } Y).$$

All u_Y are bounded by 1. Hence the function

$$u := \sup_Y u_Y$$

is well defined on the whole of $X - U(1/2)$.

Claim. u is continuous on $X - U(1/2)$.

Proof. First we show that, for any $a \in X - U(1/2)$, there exists a countable Y with $u(a) = u_Y(a)$. Since the supremum of a set of real numbers can be obtained as the limit of a sequence of real numbers in this set, there exists for each $a \in S$ a sequence of countable surfaces $a \in Y_1 \subset Y_2 \subset \ldots$ such that

$$u(a) = \lim u_n(a), \quad u_n := u_{Y_n}.$$

The union $Y = Y_1 \cup Y_2 \cup \cdots$ is countable too. Because $u_n \leq u_Y \leq u$, we have $u(a) = u_Y(a)$ for this Y.

Now let S be a countable subset of X. For each $a \in S$, there exists a countable $Y(a)$ such that $u_{Y(a)}(a) = u(a)$. We now denote the union of these $Y(a)$ by Y. Then we have a countable Y such that $u(a) = u_Y(a)$ for all $a \in S$.

This shows that the restriction of $u|S$ to any countable subset is continuous on S. As a consequence, u is sequence continuous. Continuity is a local property and, locally, X looks like an open subset of the plane. Hence sequence continuity implies continuity.

Claim. u is harmonic on $X - \overline{U(1/2)}$.

Proof of the claim. We consider open subsets $U \subset X - \overline{(U(1/2))}$ which admit a countable dense subset S. We know that there exists a countable $Y \supset S$ such that $u = u_Y$ on S. Since u is continuous, we get $u = u_Y$ on U. This shows that u is harmonic on U. Since the sets U cover $X - \overline{(U(1/2))}$, we obtain the result that u is harmonic there. This completes the proof of the countability.   □

7. Construction of Harmonic Functions with Prescribed Singularities: The Bordered Case

Let X be a Riemann surface and let $A \subset X$ be an open relatively compact subset. We assume that each boundary component of A has infinitely many boundary points. (For example, this excludes the case where X is compact and $X = A$.)

We assume that a disk φ is given in A. Recall that this is an analytic chart which maps an open subset $U \subset A$ onto the unit disk:

$$\varphi : U \to \mathbb{E}.$$

For a number $r \in (0,1)$, we introduce the notation

$$U(r) = \{x \in U; \quad |\varphi(x)| < r\},$$
$$\overline{U(r)} = \{x \in U; \quad |\varphi(x)| \le r\}.$$

We then consider the "annulus"

$$\mathcal{R} = U(3/4) - \overline{U(1/2)} \qquad (1/2 < |z| < 3/4),$$
$$\bar{\mathcal{R}} = \overline{U(3/4)} - U(1/2) \qquad (1/2 \le |z| \le 3/4)$$

and, finally, make the following assumption.

7.1 Assumption. *The boundary value problem is solvable for the domain*

$$A - \overline{U(1/2)}.$$

Under this assumption (and the assumption that each boundary component of A has infinitely many boundary points), we prove the following proposition.

7.2 Proposition. *Let*

$$u_0 : \bar{\mathcal{R}} \to \mathbb{R}$$

be a continuous function on the closed annulus $\bar{\mathcal{R}}$ which is harmonic in the interior $\mathcal{R}$. There exists a unique bounded harmonic function

$$u : A - \overline{U(1/2)} \to \mathbb{R}$$

with the following properties:

a) *The harmonic function $u - u_0$ (a priori defined on $\mathcal{R}$) extends to $U(3/4)$ as a harmonic function.*

b) *$u = 0$ on ∂A.*

Condition a) can be expressed roughly as follows: the harmonic function u has the same singular behavior in the "hole" $\overline{U(1/2)}$ as u_0.

Proof. The uniqueness of the solution is clear, since the difference of two solutions is harmonic on A and vanishes along ∂A.

Existence. (Compare with the proof of Proposition 3.2.) We construct a sequence of harmonic functions which are defined alternatingly on the two domains

$$A - \overline{U(1/2)} \text{ and } U(3/4).$$

We define

$$\alpha := \partial A$$

and

$$\partial(r) := \{x \in U; \quad |\varphi(x)| = r\}.$$

We then have

$$\partial(A - \overline{U(1/2)}) = \alpha \cup \partial(1/2),$$
$$\partial(U(3/4)) = \partial(3/4).$$

We now define sequences of harmonic functions

$$u_n : A - \overline{U(1/2)} \longrightarrow \mathbb{R}, \quad n \geq 1,$$
$$v_n : U(3/4) \longrightarrow \mathbb{R}, \quad n \geq 0,$$

in the ordering

$$v_0, u_1, v_1, u_2, v_2, \ldots,$$

by the conditions

$$v_0 = 0,$$
$$u_n = \begin{cases} v_{n-1} + u_0 & \text{on } \partial(1/2), \\ 0 & \text{on } \alpha, \end{cases}$$
$$v_n = u_n - u_0 \quad \text{on } \partial(3/4) \qquad (n \geq 1).$$

We have to show that both sequences converge. For this, we estimate the difference $u_{n+1} - u_n$ on the line $\partial(3/4)$. We claim the following:

There exist constants

$$C \geq 0 \text{ and } 0 \leq q < 1$$

with the property

$$|u_{n+1} - u_n| \leq Cq^n \quad \text{on } \partial(3/4) \qquad (n \geq 0).$$

Proof. The function $u_1 - u_0$ is continuous on $\partial(3/4)$. Hence we find $C \geq 0$ such that

$$|u_1 - u_0| \leq C \quad \text{on } \partial(3/4).$$

To construct q, we consider the harmonic function w on $A - \overline{U(1/2)}$ which solves the boundary value problem

$$w = \begin{cases} 0 & \text{on } \alpha = \partial A, \\ 1 & \text{on } \partial(1/2). \end{cases}$$

7.3 Claim. *There exists a constant $q < 1$ such that*

$$w(x) \leq q \quad \text{for } x \in \partial(3/4).$$

Because of the maximum principle (and the compactness of $\partial(3/4)$), we have to show that w is not constant 1 on the connected component of $A - \overline{U(1/2)}$ which contains $\partial(3/4)$. This follows from the boundary behavior ($w = 0$ on ∂A) and the assumption that each connected component has infinitely many boundary points.

The numbers C and q having been defined, we prove the inequality

$$|u_{n+1} - u_n| \leq Cq^n \quad \text{on } \partial(3/4)$$

by induction on n. By construction of the sequences (u_n), (v_n), we have

$$v_n - v_{n-1} = u_n - u_{n-1} \quad \text{on } \partial(3/4).$$

This shows that

$$|v_n - v_{n-1}| \leq Cq^{n-1} \quad \text{on } \partial(3/4).$$

Since the functions v_n are harmonic on the whole of $U(3/4)$, this inequality must be true on the whole of $U(3/4)$, and, in particular, on $\partial(1/2)$. From the construction of the two sequences, it again follows that

$$u_{n+1} - u_n = v_n - v_{n-1} \quad \text{on } \quad \partial(1/2)$$

and hence
$$|u_{n+1} - u_n| \leq Cq^{n-1} \quad \text{on} \quad \partial(1/2).$$

Hence the inequality
$$|u_{n+1} - u_n| \leq Cq^{n-1}\omega$$
is true on the boundary of $A - \overline{U(1/2)}$. By the maximum principle, it is true in the interior too. We obtain

$$|u_{n+1} - u_n| \leq Cq^{n-1}q = Cq^n \quad \text{on} \quad \partial(3/4),$$

as has been stated.

It now follows from the maximum principle that

$$|v_{n+1} - v_n| \quad (= |u_{n+1} - u_n| \quad \text{on} \quad \partial(3/4)) \quad \leq Cq^n \text{ on the whole of } U(3/4).$$

Since v_n extends continuously to the boundary of $U(3/4)$, the inequality holds on $\overline{U(3/4)}$. We apply this inequality on $\partial(1/2)$ and obtain

$$|u_{n+1} - u_n| \leq Cq^{n-1} \quad \text{on} \quad \partial(1/2).$$

Again we apply the maximum principle, and obtain the result that the last inequality is true on $A - \overline{U(1/2)}$. Since u_n has a continuous extension to the boundary $\partial(1/2)$, it is true on the set $A - U(1/2)$. It follows from the Weierstrass majorization theorem that the series

$$u := \sum(u_{n+1} - u_n) \quad (\text{on } A - U(1/2)),$$
$$v := \sum(v_{n+1} - v_n) \quad (\text{on } \overline{U(3/4)})$$

converge uniformly and the limit functions

$$u - u_0 = \lim u_n, \quad v = \lim v_n$$

are continuous and harmonic in the interiors.

To prove Proposition 7.2, it remains to show that $u - u_0$ and v coincide in the intersection $\mathcal{R}$ of the two domains, i.e.

$$\lim_{n \to \infty} u_n = \lim_{n \to \infty} v_n.$$

As we have seen, the two limits are uniform on the closure $\bar{\mathcal{R}}$. Hence they represent continuous functions there. It is sufficient to show that u and v coincide on the boundary of $\mathcal{R}$, and hence on $\partial(1/2)$ and $\partial(3/4)$. This immediately follows from the boundary behavior of the u_n and v_n.

It remains to show that u vanishes on the exterior boundary α. For this, we consider the harmonic function ω on $A - \overline{U(1/2)}$ which solves the boundary value problem $\omega = 1$ on $\partial(1/2)$ and $\omega = 0$ on α. If we take an upper bound of u_0 for C, then it follows from the maximum principle that

$$-C\omega \leq u_n, v_n \leq C\omega.$$

This inequality carries over to the limits and shows that they are 0 on α.

$\square$

Exercises for Sect. II.7

1. Consider the disk
    $$A = \{z; \; |z| < 2\} \subset \mathbb{C}.$$
    Determine a harmonic function u on $A - \{0\}$ which extends continuously to zero at the boundary of A and is such that $u(z) + \log|z|$ extends harmonically to zero. Interpret this as a solution in the sense Proposition 7.2.

2. Consider an arbitrary finite subset S of the upper half-plane and construct a harmonic function u on $\mathbb{H} - S$ which cannot be extended continuously to any point of S and is such that u can be extended as a continuous function to the closure of $\mathbb{H}$ in the Riemann sphere.

3. In the proof of Proposition 7.2, we showed that
    $$\lim_{n \to \infty} u_n = \lim_{n \to \infty} v_n.$$
    Another proof runs as follows. It is enough to treat the case $u_0 \geq 0$. But then
    $$v_0 \leq u_1 \leq v_1 \leq u_2, \leq v_2 \leq \ldots$$
    This gives $0 \leq v_n - u_n \leq u_{n+1} - u_n$ and hence the claim. Provide the details of this proof.

8. Construction of Harmonic Functions with a Logarithmic singularity: The Green's Function

A harmonic function on a connected compact Riemann surface is constant. Hence we are led to allow singularities for harmonic functions. In Sect. 7 we performed such a construction for relatively open subsets $U \subset X$ with a *true* boundary. We are led to exhaust X by a sequence of such domains and to take a limit. In this section, we describe such an exhaustion and apply it to the simplest case where the limit works. By definition, this case is the *hyperbolic case*. The function which we shall obtain, is called the Green's function.

It is our goal to construct, on a punctured Riemann surface (X, a), a harmonic function $u : X - \{a\} \to \mathbb{R}$ with a singularity at a that is as simple as possible. We describe what may be simplest the type of singularity.

8.1 Definition. *A harmonic function*

$$u : U^{\bullet} = U - \{0\} \longrightarrow \mathbb{R} , \quad 0 \in U \subset \mathbb{C} \ \text{open},$$

*is called **logarithmically** singular (at 0) if*

$$u(z) + \log |z|$$

extends as a harmonic function to U.

We want to extend this notion to Riemann surfaces. For this, we have to check its conformal invariance.

8.2 Remark. *Let*

$$\varphi : U \longrightarrow V , \quad U, V \subset \mathbb{C} \ \text{open},$$
$$0 \longmapsto 0,$$

be a biholomorphic map between open neighborhoods of the origin in the complex plane. The harmonic function

$$\log |z| - \log |\varphi(z)| \qquad (z \in U - \{0\})$$

has a removable singularity at 0 (i.e. it extends to a harmonic function on U).

Proof. We use the formula

$$\log |\varphi(z)| - \log |z| = \mathrm{Re} \log \frac{\varphi(z)}{z}.$$

(The function $\varphi(z)/z$ has a removable singularity at $z = 0$ and is different from 0 there. Hence it admits a holomorphic logarithm in a small neighborhood of 0.) □

8.3 Definition. *Let X be a Riemann surface and let $a \in X$ be a point. A harmonic function*

$$u : X - \{a\} \longrightarrow \mathbb{R}$$

*is called **logarithmically singular** at a if there exists an analytic chart*

$$\varphi : U \xrightarrow{\sim} V$$
$$\cup \qquad \cup$$
$$a \longmapsto 0$$

such that the transported function u_φ on V is logarithmically singular at 0 in the sense of Definition 8.1.

Because of Remark 8.2, this condition is independent of the choice of φ.

Green's Function

For the rest of this section, we make the following assumption.

8.4 Assumption. *The Riemann surface X is connected but not compact.*

From Remark 4.7, we know that each open relatively compact subset of X has infinitely many boundary points.

For a given point $a \in X$, we consider an exhaustion

$$a \in A_1 \subset A_2 \subset \cdots$$

as described in Proposition 6.4. By Proposition 7.2, there exist harmonic functions

$$u_n : A_n - \{a\} \longrightarrow \mathbb{R}$$

with the following properties:

a) u_n is logarithmically singular at a;
b) $u_n = 0$ on ∂A_n.

We want to try to take the limit $n \to \infty$, and for this purpose we point out the following:

1) The function u_n is positive in a small neighborhood of a. Since it is zero on the "exterior" boundary ∂A_n, we get

$$u_n \geq 0 \text{ on the whole of } A_n - \{a\}.$$

2) For $m > n$, the function $u_m - u_n$ is harmonic in the whole of A_n (and also in a) and not negative on the boundary ∂A_n (because of 1)). It follows that

$$u_m \geq u_n \text{ for } m \geq n.$$

3) Now we assume that the sequence $u_n(x)$ remains bounded at at least one point $x \in X$ (it is defined at the given point for almost all n). By Harnack's principle, the limit

$$u(x) = \lim_{\substack{n \to \infty \\ (n \geq n(x))}} u_n(x)$$

then exists for all $x \in X - \{a\}$ and defines a harmonic function

$$u : X - \{a\} \longrightarrow \mathbb{R}$$

with the following properties:

a) $u \geq 0$;
b) u is logarithmically singular at a.

The property b) results if we apply Harnack's principle to $u_n - u_1$ in the domain A_1.

8.5 Notation.   *The set*
$$\mathcal{M}_a = \mathcal{M}_a(X)$$
consists of all harmonic functions $v : X - \{a\} \longrightarrow \mathbb{R}$ with the following properties:

a) $v \geq 0$;
b) v *is logarithmically singular at* a.

If the above limit works, the set $\mathcal{M}_a$ is not empty. The converse statement is also true! For if $v \in \mathcal{M}_a$, then

$$v - u_n \geq 0 \quad \text{on } \partial A_n$$

and, by the maximum principle, on the whole of A_n. Hence the sequence $(u_n(x))_{n \geq n(x)}$ remains bounded for each $x \in X$.

We also obtain
$$u := \lim u_n \leq v.$$

This is true for arbitrary $v \in \mathcal{M}_a$. Hence the function u is minimal in $\mathcal{M}_a$.

So, we obtain the following result.

8.6 Proposition.   *If the set $\mathcal{M}_a$ (as defined in Notation 8.5) is not empty, then it contains a unique minimal element u, i.e.*

$$u \leq v \text{ for all } v \in \mathcal{M}_a.$$

8.7 Definition.   *The minimal element of $\mathcal{M}_a$ – in the case where $\mathcal{M}_a$ is not empty – is called the **Green's function** of X with respect to a. We denote the Green's function (when it exists) by*

$$G_a : X - \{a\} \longrightarrow \mathbb{R}.$$

Example. The Green's function on the unit disk $\mathbb{E}$ with respect to $a = 0$ exists, since the function $- \log |z|$ is contained in $\mathcal{M}_a$.

The minimality property of the Green's function admits a refinement.

8.8 Remark.   *Let S be a discrete subset of X which contains a. Let u be a nowhere negative harmonic function on $X - S$ which is logarithmically singular at a. Then*
$$G_a \leq u.$$

The same proof as in Proposition 8.6 works if one takes into consideration the fact that the maximum principle admits finitely many exceptional points.

8.9 Definition. *The Riemann surface X is called* **hyperbolic,** *if the Green's function exists for every point $a \in X$.*

For example, the unit disk is hyperbolic, since the Green's function exists for one point and the group of biholomorphic transformations $\mathbb{E} \to \mathbb{E}$ acts transitively.

The plane $\mathbb{C}$ is not hyperbolic, as the example 2) at the end of Sect. 6 shows.

Exercises for Sect. II.8

1. Show that the function $-\log|z|$ is the Green's function of the unit disk with respect to the origin.

2. Let $S \subset \mathbb{C}$ be a finite set. Show that the Riemann surface $\mathbb{C} - S$ is not hyperbolic.

3. Show that every bounded domain of the plane is hyperbolic.

4. Show that the slit plane $\mathbb{C}_-$ is hyperbolic.

9. Construction of Harmonic Functions with a Prescribed Singularity: The Case of a Positive Boundary

Again we assume that X is a connected Riemann surface with a countable basis of the topology and that

$$\varphi : U \longrightarrow \mathbb{E} \qquad (U \subset X \text{ open}),$$
$$a \longmapsto 0,$$

is a disk.

In this section, we shall use the normalized solution

$$u = u(f) : X - \overline{U(1/2)} \longrightarrow \mathbb{R}$$

of an external space problem (see Proposition 6.3)

$$f : \partial(1/2) \longrightarrow \mathbb{R}.$$

The case

$$f \equiv 1$$

is of special importance. In this case, the solution is denoted by

$$\omega \qquad (0 \leq \omega \leq 1).$$

9.1 Definition. *The Riemann surface X has a **zero boundary** with respect to the disk $\varphi : U \to \mathbb{E}$ if the normalized solution ω of the boundary value problem $f \equiv 1$ on $\partial(1/2)$ is constant:*

$$\omega \equiv 1.$$

*Otherwise, we say that X has a **positive boundary** (with respect to φ).*

It is clear that compact surfaces have a zero boundary. Later (Proposition 11.5) we shall see that the notions of a zero boundary and a positive boundary are independent of the choice of the disk φ.

Motivation for the Notion of a "Zero Boundary"

The function ω was constructed as the limit of a sequence of functions ω_n which have value 1 on an interior boundary $\partial(1/2)$ and vanish on an exterior boundary $\partial(A_n)$. When the sequence ω_n converges to 1, we may imagine that the exterior boundary $\partial(A_n)$ loses its power with increasing n. So ω becomes constant 1. We may imagine that the sequence $\partial(A_n)$ tends to an "ideal boundary" which has no ability to influence ω. (The ideal boundary is often spoken about in the old literature. The function $1 - \omega$ is sometimes called the "harmonic measure". There is no need for us to make these notions precise.)

As in Sect. 7, we consider a continuous function

$$u_0 : \bar{\mathcal{R}} = \{x \in U; \quad 1/2 \leq |\varphi(x)| \leq 3/4\}$$

on a closed annulus $\bar{\mathcal{R}}$ which is harmonic in the interior. We look for a harmonic function

$$u : X - \overline{U(1/2)} \longrightarrow \mathbb{R}$$

such that $u - u_0$ extends to a harmonic function on $U(3/4)$. There are two obvious ways to do this:

1) We exhaust X by domains A_n and try to obtain u as the limit u_n of the solution in the bounded case (see Sect. 7, $u_n = 0$ on the exterior boundary $\partial(A_n)$.)

This way leads us to the construction of the Green's function.

2) We try to apply directly the alternating method which we used for the proof of Proposition 7.2. We just replace the functions u_n by the normalized solutions of the exterior boundary problem.

This is the way in which we shall now go. Let

$$u_n : X - \overline{U(1/2)} \longrightarrow \mathbb{R}, \quad n \geq 1,$$
$$v_n : U(3/4) \longrightarrow \mathbb{R}, \quad n \geq 0,$$

be the following inductively defined sequences of functions:

a) $v_0 = 0$;

b) u_n, v_n are the solutions of the boundary value problems

$u_n = v_{n-1} + u_0 \quad$ on $\partial(1/2)$ (external space problem; see Proposition 6.3),
$v_n = u_n - u_0 \quad$ on $\partial(3/4)$.

Now we assume that X has a positive boundary with respect to the given disk. Then

$$w(x) < 1 \quad \text{on } X - \overline{U(1/2)}.$$

Hence there exists a number $0 < q < 1$ with the property

$$w(x) \leq q \quad \text{for } x \in \partial(3/4).$$

This inequality is precisely what we need to transfer the proof of Proposition 7.2 to the present case. It replaces Lemma 7.3, which we had to use there.

This gives the following statement.

9.2 Proposition. *Assume that the Riemann surface X has a positive boundary. Then, for every continous function that is harmonic in the interior,*

$$u_0 : \bar{\mathcal{R}} \longrightarrow \mathbb{R},$$

there exists a harmonic function

$$u : X - \overline{U(1/2)} \longrightarrow \mathbb{R}$$

with the following properties:

a) *$u - u_0$ extends to a harmonic function on $U(3/4)$;*
b) *u is bounded on $X - \overline{U(3/4)}$.*

The construction also shows that in the case $u_0 \geq 0$, we can obtain $u \geq 0$.

Corollary. *When the Riemann surface has a positive boundary, the Green's function G_a exists.*

In the case of a zero boundary ($\omega = 1$), this proof of convergence fails. However, under suitable assumptions, it may be possible that the sequence converges nevertheless. We shall see later, by means of a more subtle proof of convergence, that the sequence converges if u_0 is the real part of an analytic function in the annulus (Proposition 11.9). Under this additional assumption, the analogue of Proposition 9.2 is true in the case of a zero boundary also. Moreover, we shall see that in the case of a zero boundary, this condition is necessary.

Exercises for Sect. II.9

1. Show that every compact Riemann surface has a zero boundary (with respect to any disk).

2. Let be $S \subset X$ be a finite subset of a Riemann surface. Show that $X - S$ has a zero boundary (with respect to any disk).

3. Let S be a countable and closed subset of a compact Riemann surface. Show that $X - S$ has a zero boundary.

4. Give an example of an infinite S in Exercise 3.

10. A Lemma of Nevanlinna

Let
$$u : D \longrightarrow \mathbb{R}, \quad D \subset \mathbb{C} \text{ open,}$$

be a harmonic function. It follows from the Cauchy-Riemann differential equations that the function
$$f(z) = \frac{\partial u}{\partial x} - i \frac{\partial u}{\partial y}$$

is analytic. The following two conditions are equivalent:
1) u is the real part of an analytic function F.
2) f admits a primitive.

The simple proof is left to the reader (see Exercises 1–3). (The relation between f and F is given by $F' = f$.)

Now, let
$$D = \{z \in \mathbb{C}; \quad r < |z| < R\} \qquad (r < R)$$

be an annulus. An analytic function f on D admits a primitive if the (-1)th coefficient of the Laurent expansion is zero. (The primitive is then obtained by differentiating the Laurent series term by term.) This condition is equivalent to
$$\oint_{|\zeta|=\varrho} f(\zeta) \, d\zeta = 0,$$

where ϱ is a number between r and R. (It is sufficient that this condition is true for one ϱ; it is then true for all ϱ.) We obtain the following result.

10.1 Remark. *Let*

$$u : D \longrightarrow \mathbb{C}, \qquad D = \{z \in \mathbb{C}; \quad r < |z| < R\} \qquad (r < R),$$

be a harmonic function on an annulus. The following conditions are equivalent:
a) *u is the real part of an analytic function F.*
b) *The analytic function*
$$f = \frac{\partial u}{\partial x} - i \frac{\partial u}{\partial y}$$

   admits a primitive.
c) *For one (all) $\varrho \in (r, R)$, we have*

$$\oint_{|\zeta|=\varrho} f(\zeta) \, d\zeta = 0.$$

Supplement. *If the conditions a)–c) are satisfied, then the integral*

$$\int\limits_0^{2\pi} u(\varrho e^{i\varphi})\, d\varphi \qquad (r < \varrho < R)$$

is independent of ϱ.

It remains to prove the supplement. The integral in the supplement equals the real part of the line integral

$$\frac{1}{i} \oint\limits_{|\zeta|=\varrho} \frac{F(\zeta)\, d\zeta}{\zeta},$$

and this independent of ϱ by Cauchy's theorem. □

Recall that a *disk* on a Riemann surface X is a biholomorphic map

$$\varphi : U \longrightarrow \mathbb{E} \text{ (unit disk)}$$

of an open subset $U \subset X$ onto the unit disk. We define

$$\mathbb{E}(r) = \{q \in \mathbb{E};\ \ |q| < r\},$$
$$U(r) = \varphi^{-1}(\mathbb{E}(r)) \qquad (0 < r < 1).$$

10.2 Proposition. *Let X be a **compact** Riemann surface and let*

$$\varphi : U \longrightarrow \mathbb{E}$$

be a disk in X. Furthermore, let

$$u : X - \overline{U(1/2)} \longrightarrow \mathbb{R}$$

be a harmonic function.

Claim. *The restriction of u to the "annulus" $U - \overline{U(1/2)}$ ("$1/2 < |z| < 1$") is the real part of an analytic function.*

Corollary. *Consider*

$$u_\varphi(z) = u(\varphi^{-1}(z)) \qquad (1/2 < |z| < 1).$$

The integral

$$\int\limits_0^{2\pi} u_\varphi(\varrho e^{i\varphi})\, d\varphi \qquad (1/2 < \varrho < 1)$$

is independent of ϱ.

As a special case, we obtain the following result.

There is no harmonic function

$$u : X - \{a\} \longrightarrow \mathbb{R} \quad \text{(X compact!)}$$

with a logarithmic singularity at a.

Proof of Proposition 10.2. We need Stokes's theorem, as formulated and proved in the appendix to this chapter. There, we also introduce the differential calculus that we use below. The important fact is that a differential form (a 1-form) can be associated with a harmonic function $u : X \to \mathbb{R}$ on a Riemann surface in the following way. Since, locally, every harmonic function can be written as the real part of an analytic function, there exists an open covering

$$X_0 = \bigcup U_i$$

and holomorphic functions

$$F_i : U_i \longrightarrow \mathbb{C} \text{ such that } u|U_i = \operatorname{Re}(F_i).$$

The difference $F_i - F_j$ is locally constant in the intersection $U_i \cap U_j$. Therefore the differentials

$$\omega_i := dF_i \quad \text{(on } U_i)$$

agree in the intersections. For this reason, there exists a differential ω on the whole X which coincides with ω_i in U_i. Because $d \circ d = 0$, we obtain

$$d\omega = 0.$$

It is easy to describe ω in local coordinates. Let

$$\varphi : U \longrightarrow V$$

be an analytic chart; then

$$\omega_\varphi = \left(\frac{\partial u_\varphi}{\partial x} - i \frac{\partial u_\varphi}{\partial y} \right) dz \quad (dz = dx + i dy).$$

Because of the importance of this construction, we fix it as follows.

10.3 Remark. *Let $u : X \to \mathbb{R}$ be a harmonic function on a Riemann surface. There exists a differential ω which, on charts, φ is given by*

$$\omega_\varphi = \left(\frac{\partial u_\varphi}{\partial x} - i \frac{\partial u_\varphi}{\partial y} \right) dz.$$

We have $d\omega = 0$.

After this preparation, we come to the proof of Proposition 10.2.

Proof of Proposition 10.2. We consider the differential ω which belongs to u. By the definition of a line integral (Remark 13.4), we have

$$\oint \omega = \oint f(\zeta)d\zeta \qquad \text{(notation as in Remark 10.1).}$$

Hence Proposition 10.2 states that

$$\oint \omega = 0,$$

where the integral is taken along the inverse image (with respect to ψ) of the circle

$$\varrho e^{i\varphi} \qquad (0 \leq \varphi \leq 2\pi)$$

for some $\varrho \in (1/2, 1)$. This circle is the boundary of the domain

$$M = X - \overline{U(\varrho)},$$

and the integral in question equals the negative of the boundary integral (Remark 13.20),

$$\int_{\partial M} \omega \quad \left(= - \oint_{\text{``}|\zeta|=\varrho\text{''}} \omega\right).$$

By assumption, X is compact. Hence M is open and relatively compact. Stokes's theorem (Theorem 13.19) applies and we obtain

$$\int_{\partial M} \omega = \int_M d\omega = 0,$$

since $d\omega = 0$. □

For the proof of the uniformization theorem, we need a variant of Proposition 10.2 for certain noncompact Riemann surfaces.

Readers who are interested mainly in the theory of compact Riemann surfaces can skip the rest of this section, since we shall not use uniformization theory for the theory of compact Riemann surfaces.

10.4 Lemma (Nevanlinna's Lemma)*).   *Let X be a (not necessarily compact) Riemann surface and let $Y \subset X$ be an open relatively compact subset of X. Furthermore, a disk Y such that*

$$\varphi : U \xrightarrow{\sim} \mathbb{E} , \quad U \subset Y \text{ open,}$$

---

*) This lemma is implicitly contained in [Ne], Chap. VI, in the proof of Lemma 6.22. The application of Stokes's theorem to the function u_m there is problematic, since in the solution of the boundary value problem a finite number of exceptional points are allowed. It is difficult to control the derivatives of u_m there. Hence the proof of Nevanlinna's lemma becomes somewhat longer in our presentation.

and a harmonic function

$$u : Y - \overline{U(1/2)} \longrightarrow \mathbb{R}$$

with the following properties are given:

a) $u \geq 0$;
b) $u = 0$ on ∂Y.

Claim. *If we set*

$$f = \frac{\partial u_\varphi}{\partial x} - i\frac{\partial u_\varphi}{\partial y},$$

then

$$i \oint_{|\zeta|=\varrho} f(\zeta)d\zeta \geq 0 \qquad (1/2 < \varrho < 1).$$

(In particular, this expression is real.)

Note. Proposition 10.2 is a special case of Nevanlinna's lemma: We can apply Lemma 10.4 to $Y = X$ (compact), and the assumption b) is then trivially true. We can also assume that u is bounded, since the inner annulus can be enlarged a little. Since the integral does not change if u is replaced by $u + C$ $(C \in \mathbb{R})$, we can also assume a). Now we get

$$i \int_{|\zeta|=\varrho} f(\zeta)\, d\zeta \geq 0.$$

Since we can replace u by $-u$, equality must hold, and we can apply Remark 10.1.

□

The rest of this section will be taken up by the proof of Lemma 10.4.

Proof of Lemma 10.4. We may assume that u does not vanish identically in the annulus "$1/2 < |z| < 1$".

Again we consider the associated differential ω (see Remark 10.3). The claim states that

$$i \int_\alpha \omega \geq 0, \qquad \alpha(t) = \varphi^{-1}(\varrho e^{it}), \quad 0 \leq t \leq 2\pi.$$

First of all, we show that the integral is real. For this we use

$$\omega' := -i(\omega - du).$$

In local coordinates, this is computed as

$$\omega'_\varphi = -\frac{\partial u_\varphi}{\partial y}dx + \frac{\partial u_\varphi}{\partial x}dy.$$

Obviously, the integral of ω' is real. The integral of du along a closed curve vanishes (Theorem 13.18). Hence we have

$$\mathrm{i} \oint \omega = - \oint \omega' \qquad \in \mathbb{R}.$$

We want to apply Stokes's Theorem to the Riemann surface $Y_0 := Y - \overline{U(1/2)}$. For this, we need suitable open and relatively compact subsets $B \subset Y_0$. Since Y_0 is relatively compact, we have only to take care that the boundary of B (taken in X) is disjoint with the boundary of Y_0 The boundary of Y_0 is the union of the interior boundary $\partial(1/2)$ and the exterior boundary ∂Y.

To avoid the interior boundary, we choose a fixed ϱ and consider

$$Y' = Y - \overline{U(\varrho)}.$$

By assumption, u vanishes on the exterior boundary ∂Y. We choose $\varepsilon > 0$ and consider
$$B(\varepsilon) := \big\{\, x \in Y'; \quad u(x) > \varepsilon \,\big\}.$$

We would like to have that the boundary $B(\varepsilon)$ is relatively compact in Y_0. This would be the case if u were continuous (and hence identically zero) on the whole exterior boundary ∂Y. But our convention (Notation 5.1) allows finitely many exceptional points on the boundary. This fact requires a modification similar to that we used in the proof of the maximum principle (Lemma 4.5).

By Propositions 6.3 and 9.2, there exists for a given point $a \in \bar{Y}$ an open neighborhood
$$\bar{Y} \subset A \subset X$$
$$\uparrow$$
$$\text{open}$$

and a harmonic function $h_0 : A - \{a\} \to \mathbb{R}$, $h_0 \leq 0$, which is logarithmically singular at a. Summing up, we obtain for each finite set $\mathcal{M} \subset \bar{Y}$ a harmonic function
$$h : A - \mathcal{M} \longrightarrow \mathbb{R} \qquad (\bar{Y} \subset A \subset X \text{ open})$$

with the properties
a) $h(x) \leq 0$ for $x \in A - \mathcal{M}$;
b) $\lim_{x \to a} h(x) = -\infty$ for $a \in \mathcal{M}$.

We apply this to the finite set of exceptional points on the boundary of Y (into which u cannot be extended continuously).

Now we consider instead of $B(\varepsilon)$ the modified domain

$$B(\varepsilon, h) := \big\{\, x \in Y'; \quad u(x) + h(x) > \varepsilon \,\big\}.$$

The domain $B(\varepsilon, h)$ is relatively compact in Y_0. The boundary points satisfy $|\varphi(x)| = \varrho$ or $u(x) + h(x) = \varepsilon$. We want to keep these two sets disjoint. By the maximum principle, u has no zeros on the circular line $|\varphi(x)| = \varrho$. We can choose ε so small that $u(x) > \varepsilon$ on the circular line. The function h can be multiplied by an arbitrary positive number. Hence we can arrange that $u(x) + h(x) > \varepsilon$ on the circular line. Now it is clear that the boundary of $B(\varepsilon, h)$ is the union of two distinct parts, the "interior" and the "exterior":

a) $\qquad |\varphi(x)| = \varrho$ (interior boundary);

b) $\quad u(x) + h(x) = \varepsilon$ (exterior boundary).

We want to apply Stokes's theorem on $B(\varepsilon, h)$. For this, we require that the exterior boundary of $B(\varepsilon, h)$ is smooth.

The theorem of implicit functions of real analysis shows the following.

10.5 Remark. *Let*

$$f : U \longrightarrow \mathbb{R}, \quad 0 \in U \subset \mathbb{C} \ open,$$

be a $\mathbb{C}^\infty$ *function with the properties*

a) $\quad f(0) = 0;$

b) $\quad \left(\dfrac{\partial f}{\partial x}(0), \dfrac{\partial f}{\partial y}(0) \right) \neq (0,0).$

Then the origin is a smooth boundary point of

$$U^+ := \{x \in U; \quad f(x) > 0\}.$$

We also have the following fact.

10.6 Remark. *Let*

$$f : U \longrightarrow \mathbb{R}, \quad U \subset \mathbb{C} \ a \ domain,$$

be a nonconstant **harmonic** *function. The set of all points in which both derivatives of* f *vanish is discrete in* U.

This is clearly true, since

$$\frac{\partial f}{\partial x} - i\frac{\partial f}{\partial y}$$

is an analytic function. The two remarks above show that the boundary of $B(\varepsilon, h)$ is smooth outside a countable set of ε. We can avoid such ε and such find domains to which Stokes's theorem applies. We want to apply it to the differential

$$\omega_h' = -\frac{\partial(u+h)}{\partial y}\, dx + \frac{\partial(u+h)}{\partial x}\, dy.$$

Recall that our claim in Lemma 10.4 means that

$$\oint \omega' \geq 0 \qquad \left(\omega' = -\frac{\partial u}{\partial y}\, dx + \frac{\partial u}{\partial x}\, dy\right).$$

Obviously, the differentials ω'_h are closed, $d\omega'_h = 0$. It follows from Stokes's theorem that

$$0 = \int_{Y(\varepsilon)} d\omega'_h = \int_{\partial Y(\varepsilon)} \omega'_h = \int_{\text{exterior boundary}} \omega'_h + \int_{\text{interior boundary}} \omega'_h.$$

10.7 Claim. *We have*

$$\int_{\text{exterior boundary}} \omega'_h \geq 0.$$

It follows from this that

$$- \int_{\text{interior boundary}} \omega'_h \geq 0.$$

The interior boundary is parametrized by α. The boundary integral is the negative of the curve integral (see Remark 13.20). Replacing h by h/n and taking the limit $n \to \infty$, we obtain

$$\int_\alpha \omega' \geq 0.$$

That is precisely what we have to prove.

Proof of Claim 10.7 (and hence of Nevanlinna's lemma, Lemma 10.4).
We use the following notation.

10.8 Notation. *Let $U \subset X$ be an open subset, let a be a smooth boundary point of U, and let ω be a differential which is defined in an open neighborhood of a. The differential ω is called **nonnegative along the boundary** at a if there exists an oriented differentiable chart*

$$\begin{array}{ccc} \varphi : U(a) & \longrightarrow & V \\ \cup & & \cup \\ a & \longmapsto & 0 \end{array}$$

with the property

$$\varphi(U(a) \cap U) = V \cap \mathbb{H},$$
$$\varphi(\partial U(a) \cap U) = V \cap \mathbb{R},$$

such that the following condition is satisfied:

$$f(0) \geq 0 \quad for \quad \omega_\varphi = f\, dx + g\, dy.$$

This condition is independent of the choice of φ: if φ is replaced by another chart, then $f(0)$ is multiplied by a nonnegative number. Now we assume that the conditions for integrability are fulfilled, i.e. that $\mathrm{Support}(\omega)\cap\partial U$ is compact and is contained in the smooth part of the boundary.

We obtain the following result immediately from the definition of the boundary integral.

10.9 Remark.   *When ω is nonnegative along the boundary (i.e. at each smooth point of the boundary), then*

$$\int_{\partial U}\omega \geq 0.$$

Hence our claim is proved, if ω'_h is not negative along the exterior boundary. This statement follows from the following simple local criterion.

10.10 Criterion.   *Let $U \subset \mathbb{C}$ be an open subset, let a be a smooth boundary point of U, and let u be a C^∞ function, defined in an open neighborhood of $U(a)$.*

Assumption.

a) $u \equiv C$ (= constant) on $(\partial U)\cap U(a)$;
b) $u \geq C$ on $U\cap U(a)$.

Then the differential

$$\omega = -\frac{\partial u}{\partial y}\,dx + \frac{\partial u}{\partial x}\,dy$$

is nonnegative along the boundary of U.

Proof. After shrinking of $U(a)$, we can choose an orientation-preserving diffeomorphism

$$\varphi : U(a) \longrightarrow V \subset \mathbb{C} \text{ open}$$
$$\cup \qquad \cup$$
$$a \longmapsto 0$$

with the property

$$\varphi(U(a)\cap U) = V\cap\mathbb{H},$$
$$\varphi(U(a)\cap \partial U) = V\cap\mathbb{R}.$$

We would like to have that the Jacobi map

$$J(\varphi,a) : \mathbb{R}^2 \longrightarrow \mathbb{R}^2 \quad (=\mathbb{C})$$

is a similarity transformation, i.e. multiplication by a complex number. To enforce this, we may replace φ by its composition with an $\mathbb{R}$-linear map of the form

$$(x,y) \longmapsto (\alpha x + \beta y, \gamma y) = B(x,y), \quad \alpha > 0, \quad \gamma > 0.$$

Such maps transform the upper half-plane into itself. Now we use the following simple fact:

If

$$A : \mathbb{R}^2 \longrightarrow \mathbb{R}^2$$

is any linear map with a positive determinant, then there exist real numbers $\alpha > 0, \beta, \gamma$ such that $B \cdot A$ is a similarity transformation.

Now a simple computation shows that taking the differential

$$u \longmapsto \omega = \omega(u)$$

is compatible with similarity transformations

$$m_b : \mathbb{C} \longrightarrow \mathbb{C} \qquad (b \in \mathbb{C}^{\bullet}),$$
$$z \longmapsto bz,$$

i.e.

$$\omega(u \circ m_b) = m_b^*(\omega).$$

Therefore we can assume that the Jacobi matrix $J(\varphi, a)$ is the unit matrix. Now the differential

$$\omega = -\frac{\partial u}{\partial y} \, dx + \frac{\partial u}{\partial x} \, dy$$

can be transformed into V very easily. Let

$$\tilde{u} = u_\varphi \quad (= u \circ \varphi^{-1})$$

and

$$\tilde{\omega} = \omega_\varphi = f \, dx + g \, dy.$$

From the definition of the transformation of a differential, we get simply

$$f(0) = -\frac{\partial \tilde{u}}{\partial y}(0).$$

Now the function $\tilde{u}$ is constant C on the real axis (close to 0) but not greater than C in the upper half-plane. Taking the differential quotient, we see that

$$\frac{\partial \tilde{u}}{\partial y}(0) \leq 0 \quad \text{and hence} \quad f(0) \geq 0,$$

which was to be proved.   □

Exercises for Sect. II.10

1. Let u be a harmonic function on an open subset of the plane. Show that $f(z) = \partial u/\partial x - i\partial u/\partial y$ is an analytic function.

2. Let u be the real part of an analytic function F on an open subset of the plane. Show that
$$F' = \frac{\partial u}{\partial x} - i\frac{\partial u}{\partial y}.$$

3. Let u be a harmonic function on an open subset of the plane such that $f = \partial u/\partial x - i\partial u/\partial y$ has a primitive F. Show that, up to an additive constant, u is the real part of F.

4. Let u be a harmonic function on a Riemann surface. Check directly the compatibility of the family
$$\omega_\varphi = \left(\frac{\partial u_\varphi}{\partial x} - i\frac{\partial u_\varphi}{\partial y}\right)dz$$
and prove in this way that it defines a differential.

11. Construction of Harmonic Functions with a Prescribed Singularity: The Case of a Zero Boundary

Again we assume that X is a connected Riemann surface and that
$$\varphi : U \longrightarrow \mathbb{E} \qquad (U \subset X \text{ open}),$$
$$a \longmapsto 0,$$

is a disk. In this section, we assume that (X, φ) is has a zero boundary, which means that the normalized solution
$$\omega : X - \overline{U(1/2)} \longrightarrow \mathbb{R}$$

of the boundary problem
$$\omega = 1 \quad \text{on } \partial(1/2)$$

with respect to this disk is constant 1. For example, this is the case if X is compact.

The Extended Maximum Principle for Surfaces with a Zero Boundary

11.1 Lemma.   *Assume that X has a zero boundary (with respect to φ). Furthermore, let*
$$u : X - \overline{U(1/2)} \longrightarrow \mathbb{R}$$
be a harmonic function bounded from below which extends continuously to the boundary $\partial U(1/2)$ and is such that
$$u \geq 0 \quad \text{on } \partial U(1/2).$$

Then
$$u \geq 0 \text{ everywhere.}$$

Proof. We consider an exhaustion
$$U(3/4) \subset A_1 \subset A_2 \subset \ldots$$
as in Sect. 6. By definition, ω is the limit of a sequence ω_n. By assumption,
$$u \geq -C$$
for a suitable constant $C \geq 0$. The inequality
$$u \geq -C(1 - \omega_n)$$
holds on the boundary of $A_n - \overline{U(1/2)}$ and hence also in the interior. The claim follows from the assumption that $\omega_n \to 1$ by taking the limit $n \to \infty$. □

11.2 Corollary.   *If u is bounded, then u is the normalized solution of a boundary value on $\partial U(1/2)$.*

We now generalize Proposition 10.2 from the compact to the zero-bounded case.

11.3 Lemma.   *Assume that X has a zero boundary and that $u = u(f)$ is the normalized solution of a (continuous) boundary value problem f on $\partial U(1/2)$. Then u is the real part of an analytic function in the annulus $\mathcal{R}$ ("$1/2 < |z| < 3/4$").*

Proof. Consider the differential ω which is associated with u
$$(\text{locally, } \omega = \frac{\partial u}{\partial x} \, dx - i\frac{\partial u}{\partial y} \, dy).$$
The statement in Lemma 11.3 is equivalent to
$$\int_{\partial U(\varrho)} \omega = 0 \quad (1/2 < \varrho < 3/4).$$
Nevanlinna's lemma (Sect. 10) and the construction of the normalized solution u show that
$$i \int_{\partial U(\varrho)} \omega \geq 0.$$
Equality must hold, since we can replace u by $-u$. □

11.4 Proposition.   *The Riemann surface X has a positive boundary (with respect to φ) if and only if the Green's function G_a ($\varphi(a) = 0$) exists.*

Proof. We already know that the Green's function exists in the case of a positive boundary. So assume that X has a zero boundary. We argue indirectly and assume that the Green's function exists. We have

$$G_a(x) \geq \delta > 0 \text{ for } x \in \overline{U(1/2)} - \{a\}.$$

This equality follows everywhere from the extended maximum principle. Now consider the sequence of functions (u_n) which approximate G_a via the exhaustion (A_n). We know that $G_a(x) - \delta - u_n \geq 0$, first on the boundary of A_n and then on A_n. Taking the limit $n \to \infty$, we get $-\delta \geq 0$, in contradiction to the construction of δ. $\qquad\square$

For the construction of the Green's function, it is not necessary to take the center of the given disk for a. Any other point can be taken as the center. It follows from Proposition 11.4 that the set of all points in X that admit a Green's function is open. The same argument shows that the set of all points which do not admit a Green's function is open as well. This shows the following fact.

11.5 Proposition.   *If a Riemann surface has a positive boundary with respect to one disk φ, then this is the case for all disks. In particular, a Green's function exists for all points if it exists for one.*

So we see that the properties "zero-bounded" and "positive bounded" are intrinsic properties of the Riemann surface and we can omit the "with respect to". Even more, from Proposition 11.4 we obtain the following result.

11.6 Proposition.   *For a connected Riemann surface, the following two conditions are equivalent:*

1) *The surface has a positive boundary.*
2) *The surface is hyperbolic.*

Riemann surfaces a with zero boundary admit the following variant of Liouville's theorem.

11.7 Proposition.   *Any bounded harmonic function on a Riemann surface with a zero boundary is constant.*

Proof. Let u be bounded and harmonic on X. We can assume that $u(a) = 0$. In a small disk around a, we have

$$-\varepsilon \leq u \leq \varepsilon \qquad (\varepsilon > 0 \text{ given}).$$

The extended maximum principle shows that

$$-\varepsilon \leq u \leq \varepsilon. \qquad\square$$

11.8 Corollary.   *Any bounded analytic function on a Riemann surface with a zero boundary is constant.*

We have now collected together the means to extend the decisive existence theorem, under more restrictive conditions, to the case of surfaces with a zero boundary. In general, we have the following statement.

11.9 Proposition. *Let X be a connected Riemann surface with a distinguished disk. Assume that in a closed annulus (defined by $1/2 \leq |z| \leq 3/4$), a continuous function u_0, which is harmonic in the interior, is given.*

Assumption. *u_0 is, in $\mathcal{R}$, the real part of an analytic function.*

Claim. *There exists a harmonic function*

$$u : X - \overline{U(1/2)} \longrightarrow \mathbb{R}$$

with the following properties:
a) *$u - u_0$ extends to a harmonic function on $U(3/4)$.*
b) *u is bounded in $X - \overline{U(1/2)}$.*

Proof. If X has a positive boundary, this already has been proved (Proposition 9.2). So we assume that X has a zero boundary. As in the proof of Proposition 9.2, we consider the functions

$$u_n : X - \overline{U(1/2)} \longrightarrow \mathbb{R} \quad n \geq 1,$$
$$v_n : U(3/4) \longrightarrow \mathbb{R}, \quad n \geq 0,$$

that are defined by $v_0 \equiv 0$ and the boundary value conditions

$$u_n = v_{n-1} + u_0 \quad \text{on } \partial(1/2) \quad \text{(normalized solution)},$$
$$v_n = u_n - u_0 \quad \text{on } \partial(3/4),$$

inductively (alternating). Since we do not have the number q $(0 < q < 1)$ which arises from ω in the case of a positive boundary, we now need a completely different proof for the convergence. This is based on the following fact:

$$v_{n+1}(0) = v_n(0).$$

Proof. For simplicity, we denote the function

$$v \circ \varphi^{-1} : \overline{\mathbb{E}(3/4)} \longrightarrow \mathbb{R}$$

by v again. The difference will be made clear by the notation for the variables: In X we use $a, x, \ldots$, and in $\mathbb{E}$ we use $\zeta, z, \ldots$.

The *midpoint property* of harmonic functions states that

$$v_{n+1}(0) = \frac{1}{2\pi} \int_0^{2\pi} v_{n+1}\left(\frac{3}{4}e^{it}\right) dt$$

(we first shrink R and then take a limit). It follows from the boundary behavior

$$v_n = u_n - u_0 \quad \text{on } \partial(3/4)$$

that

$$v_{n+1}(0) = \int_0^{2\pi} u_{n+1}\left(\frac{3}{4}e^{it}\right) dt - \int_0^{2\pi} u_0\left(\frac{3}{4}e^{it}\right) dt.$$

We have assumed that u_0 is the real part of an analytic function on $\mathcal{R}$. From Nevanlinna's lemma (Lemma 10.4), we obtain the same property for u_{n+1}. Hence we obtain

$$v_{n+1}(0) = \int_0^{2\pi} u_{n+1}\left(\frac{1}{2}e^{it}\right) dt - \int_0^{2\pi} u_0\left(\frac{1}{2}e^{it}\right) dt.$$

Now, from the boundary behavior

$$u_{n+1} = v_n + u_0 \quad \text{on } \partial(1/2),$$

we get

$$v_{n+1}(0) = \int_0^{2\pi} v_n\left(\frac{1}{2}e^{it}\right) dt = v_n(0) \text{ (midpoint property)}.$$

This finishes the proof of the fact. $\square$
   We also have to use the following simple fact.

11.10 Fact.   *There is a constant $0 < q < 1$ such that every harmonic function*

$$v : \mathbb{E}(3/4) \longrightarrow \mathbb{R}$$

with the properties
a) $v(0) = 0$,
b) $|v(z)| \leq C$,
satisfies the inequality

$$|v(z)| \leq qC \text{ for } |z| = 1/2.$$

The proof follows from Harnack's inequality (Proposition 2.4).
   Now we choose a constant C with the property

$$|u_2 - u_1| \leq C \quad \text{on } \partial(1/2).$$

By the maximum principle, this inequality remains true everywhere, in particular on $\partial(3/4)$. We get

$$|v_2 - v_1| \leq C \quad \text{on } \partial(3/4)$$

and hence in $U(3/4)$. Fact 11.10 shows that

$$|u_3 - u_2| = |v_2 - v_1| \leq Cq \quad \text{on } \partial(1/2).$$

By induction on n, we get

$$|u_{n+1} - u_n| \leq q^n C \quad \text{on } X - \overline{U(1/2))},$$
$$|v_n - v_{n-1}| \leq q^n C \quad \text{on } U(3/4).$$

Now we have proved the decisive inequality, which allows us to imitate the
proof of Proposition 7.2 that u_n, v_n converge locally uniformly to harmonic
functions u and v, such that $u - u_0$ and v agree on $\mathcal{R}$. $\square$

Exercises for Sect. II.11

1. Let U be a nonempty open relatively compact subset of a Riemann surface X.
   Show that $X - \bar{U}$ is hyperbolic.

2. Show that any open and connected subset of a hyperbolic Riemann surface is
   hyperbolic.

3. Let $f : X \to Y$ be a nonconstant analytic map between Riemann surfaces. Show
   that if Y is hyperbolic, then so is X.

12. The Most Important Cases of the Existence Theorems

In this section, we collect together the existence theorems for harmonic functions which will be used in the following treatment.

The function $(z - 1)/(z + 1)$ takes values on the negative real axis only if z
is contained in $[-1, 1]$. Hence the principal branch of the logarithm

$$\mathbb{C} - [-1, 1] \longrightarrow \mathbb{C}, \qquad z \longmapsto \text{Log } \frac{z - 1}{z + 1},$$

is an analytic function. Both its real and its imaginary part are interesting
harmonic functions. Taking the real part, we obtain the following result.

12.1 Remark. *The function*

$$\text{Log}\,|z-1| - \text{Log}\,|z+1|$$

is the real part of an analytic function in $\mathbb{C} - [-1,1]$.

Using this function, we are lead to the following fundamental existence theorem for harmonic functions with logarithmic singularities.

12.2 Theorem. *Let* a, b *be two distinct points on a connected Riemann surface. There exists a harmonic function*

$$u := u_{a,b} : X - \{a, b\} \longrightarrow \mathbb{C}$$

with the following properties:

a) u *is logarithmically singular at* a;
b) $-u$ *is logarithmically singular at* b;
c) u *is bounded "at infinity", i.e. bounded in* $X - [U(a) \cup U(b)]$, *where* $U(a)$ *and* $U(b)$ *are neighborhoods of* a, b.

Proof. When the two points a, b are sufficiently close (such that they correspond to the points ± 1 with respect to a suitable disk "$|z| < 2$"), then the claim follows from the fundamental existence theorem (Proposition 11.9). But the assumption that "a and b are close" is not necessary. By joining a and b with a curve, we obtain by means of a simple compactness argument a chain of points

$$a = a_0, a_1, \ldots, a_n = b,$$

where any two consecutive points are close. We then define

$$u_{a,b}(x) := \sum_{i=1}^{n} u_{a_{i-1},a_i}$$

and obtain a function with the properties a)–c). $\qquad\square$

A different existence theorem is obtained if we take instead of the real part of $\text{Log}\,((z-1)/(z+1))$ its imaginary part (the real part of the analytic function $-i\,\text{Log}\,((z-1)/(z+1))$). The imaginary part of the principal branch of the logarithm is the principal branch of the argument. This makes a jump of 2π when the negative real axis is crossed. Analogously, $\text{Arg}((z+1)/(z-1))$ makes a jump by 2π when $(-1,1)$ is crossed (see Exercise 1). Hence we have the following result.

12.3 Remark. *The function*

$$\text{Arg}\left(\frac{z+1}{z-1}\right) \qquad \textit{(principal value of the argument)}$$

is the real part of an analytic function in $\mathbb{C} - [-1,1]$. *It cannot be extended continuously to any point of* $[-1,1]$.

From the fundamental existence theorem (Proposition 11.9) we obtain the following result.

12.4 Theorem. *Let X be a connected Riemann surface and let $\varphi : U \to$ $\{z;\ |z| < 2\}$ be a "disk of radius 2". We denote by C the inverse image of $[-1, 1]$ in X. There exists a bounded harmonic function $u : X - C \to \mathbb{R}$ such that*

$$u^\varphi(z) - \mathrm{Arg}\Big(\frac{z+1}{z-1}\Big)$$

extends to the whole disk as a harmonic function. The function u cannot be extended as a continuous function to any point of C.

Exercises for Sect. II.12

1. Give a mathematically rigorous formulation of the statement that $\mathrm{Arg}((z+1)/(z-1))$ makes a jump of 2π when $(-1, 1)$ is passed through, and prove this statement.

2. Let $S \subset X$ be a finite subset of a Riemann surface which contains at least two elements. Show that there exists a harmonic function u on $X - S$ such that for suitable constants C_s, the functions $C_s u$ have logarithmic singularities at the points of S.

3. Is there a harmonic function on $\mathbb{C}$ which is logarithmically singular at ∞?

13. Appendix to Chapter II. Stokes's Theorem

Stokes's theorem was an important tool in the proof of the central existence theorems in the case of a Riemann surface with a zero boundary. We shall also use this theorem in the following treatment. The Cauchy integral theorem can be considered as special case of Stokes's theorem. For this reason, it seems to be appropriate to include a complete proof, even though it needs some technical effort even in the two-dimensional case.

I Local Theory of Differential Forms

We start with the notion of a differential in the complex plane.

13.1 Definition. *Let D be an open subset of the complex plane.*
1) *A 0-form is a continuous function $f : D \to \mathbb{C}$.*
2) *A 1-form is a pair of continuous functions $f, g : D \to \mathbb{C}$.*
3) *A 2-form is a continuous function $f : D \to \mathbb{C}$.*

(Hence 0- and 2-forms are the same in the local theory. The situation is different on surfaces.)

"ν-forms" are also called "differential forms of degree ν". Sometimes 1-forms are simply called *differentials*.

We can add ν-forms in an obvious way (componentwise) and multiply them by complex-valued functions.

Let 1_D and 0_D be the functions "constant 1" and "constant 0", respectively on D. We define

a) the 1-forms

$$dx := (1_D, 0_D), \quad dy := (0_D, 1_D)$$

b) and the 2-form

$$dx \wedge dy := 1_D,$$

and obtain in this way the usual notation for differential forms

$$(f, g) = f\, dx + g\, dy \quad \text{for 1-forms,}$$
$$f = f\, dx \wedge dy \quad \text{for 2-forms.}$$

The symbol $dx \wedge dy$ can be generalized to the *alternating product* of two differentials as follows.

13.2 Definition. *The alternating product of two differentials is defined by*

$$(f_1 dx + g_1 dy) \wedge (f_2 dx + g_2 dy) := (f_1 g_2 - f_2 g_1) dx \wedge dy.$$

We can also define

$$dz := dx + i\, dy$$

and obtain

$$\boxed{(f, if) = f(z)\, dz.}$$

A ν-form is called *differentiable* if it is continuously partially differentiable infinitely often in the sense of real analysis.

Notation.

$$A^\nu(D) = \text{set of all differentiable } \nu\text{-forms on } D.$$

The Exterior Derivative, Local Case

We define

$$d : A^0(D) \longrightarrow A^1(D),$$

$$d(f) = \left(\frac{\partial f}{\partial x}, \frac{\partial f}{\partial y}\right) = \frac{\partial f}{\partial x}\, dx + \frac{\partial f}{\partial y}\, dy$$

and

$$d : A^1(D) \longrightarrow A^2(D),$$

$$d(f\, dx + g\, dy) = \left(\frac{\partial g}{\partial x} - \frac{\partial f}{\partial y}\right) dx \wedge dy.$$

Then, obviously,

$$d(df) = 0 \text{ for } f \in A^0(D).$$

For an *analytic* function f, we obtain from the Cauchy–Riemann equations

a) $d(f) = f' \cdot dz$;
b) $d(f \cdot dz) = 0$.

Transformation of Differential Forms: The Local Case

Let

$$\varphi : D \longrightarrow D', \quad D, D' \subset \mathbb{C} \text{ open,}$$

be an (infinitely often) differentiable map. We want to define the pulled-back differential form $\varphi^*\omega$ on D' for a differential form ω on D. This should be a map

$$\varphi^* : A^\nu(D') \longrightarrow A^\nu(D).$$

We denote the coordinates of D by $z = x + iy$ and those of D' by $w = u + iv$.
1) *0-forms.* Let f be a 0-form ($=$ function) on D'. We define

$$\varphi^*(f) = f \circ \varphi \quad (\text{and hence } \varphi^*(f)(z) = f(\varphi(z))).$$

2) *1-forms.* Let $\omega = f\, du + g\, dv$ be a 1-form on D'. We define

$$\varphi^*\omega = \varphi^*(f)\varphi^*(du) + \varphi^*(g)\varphi^*(dv),$$

where

$$\varphi^*(du) = d\varphi_1, \quad \varphi^*(dv) = d\varphi_2 \quad (\varphi = \varphi_1 + i\varphi_2).$$

Using the notation

$$\varphi^*(\omega) = \tilde{f}\, dx + \tilde{g}\, dy,$$

this means

$$\begin{pmatrix} \tilde{f}(z) \\ \tilde{g}(z) \end{pmatrix} = J(\varphi, z)^t \begin{pmatrix} f(\varphi(z)) \\ g(\varphi(z)) \end{pmatrix}.$$

Here J is the real Jacobi matrix and J^t its transpose, i.e.

$$J(\varphi, z)^t = \begin{pmatrix} \dfrac{\partial \varphi_1}{\partial x} & \dfrac{\partial \varphi_2}{\partial x} \\[2mm] \dfrac{\partial \varphi_1}{\partial y} & \dfrac{\partial \varphi_2}{\partial y} \end{pmatrix}.$$

Special case. Assume that the map φ is *analytic*. It follows from the Cauchy–Riemann differential equations that

$$\varphi^*(dw) = \varphi' dz$$

or, in general,

$$\varphi^*(f \, dw) = \varphi^*(f) \, \varphi' dz.$$

3) *2-forms.* Let $\omega = f \, du \wedge dv$ be a 2-form on D'. We define

$$\varphi^*(\omega) = \varphi^*(f) \, \varphi^*(du \wedge dv)$$

and

$$\varphi^*(du \wedge dv) := \varphi^*(du) \wedge \varphi^*(dv).$$

A simple calculation shows that

$$\varphi^*(du \wedge dv) = \det J(\varphi, \cdot) dx \wedge dy.$$

(Here $\det J(\varphi, \cdot)$ denotes the function $z \mapsto \det J(\varphi, z)$). By means of the *chain rule*, it is easy to prove the following statement.

13.3 Remark (The pullback is natural).
1) *We have*
$$\varphi^*(\omega \wedge \omega') = \varphi^*(\omega) \wedge \varphi^*(\omega').$$

2) *Let*

$$D \xrightarrow{\varphi} D' \xrightarrow{\psi} D'', \quad D, D', D'' \subset \mathbb{C} \text{ open,}$$

be differentiable maps and let ω be a differential form on D''. Then

$$(\psi \circ \varphi)^* \omega = \varphi^*(\psi^* \omega).$$

Evaluation of Differential Forms: Local Case

1) A 0-form f is evaluated at a point $a \in D$. The result is the *function value* $f(a)$.

2) A 1-form $f \, dx + g \, dy$ has to be evaluated on a piecewise smooth curve

$$\alpha : [a, b] \longrightarrow D \quad (a < b)$$

The result is the *line integral*

$$\int_\alpha (f, g) := \int_a^b [f(\alpha(t))\dot\alpha_1(t) + g(\alpha(t))\dot\alpha_2(t)]\, dt.$$

Here
$$\alpha_1 = \operatorname{Re}\alpha, \qquad \alpha_2 = \operatorname{Im}\alpha$$

are the components of α.

(To be precise, the integrand is not defined at points where α is not smooth. We get around this difficulty by cutting the curve into pieces or by using the more general notion of an integral from Sect. 3, where finitely many points are allowed at which the integrand is not continuous.)

13.4 Remark.   *For differentials of the special form $\omega = f\,dz = f(dx + i\,dy)$, we have*

$$\int_\alpha \omega = \int_\alpha f(\zeta)\,d\zeta = \int_a^b f(\alpha(t))\dot\alpha(t)dt.$$

This is the usual line integral in complex analysis, as described for example in [FB], Chap. I.

By means of the chain rule, we can easily prove the following statement.

13.5 Remark.   *Let*

$$\varphi : D \longrightarrow D' \qquad (D, D' \subset \mathbb{C} \; open)$$

be a differentiable map, let

$$\alpha : [a, b] \longrightarrow D \qquad (a < b)$$

be a piecewise smooth curve, and let ω be a 1-form on D'. Using the notation

$$\omega = f\,du + g\,dv,$$
$$\varphi^*\omega = \tilde{f}\,dx + \tilde{g}\,dy,$$
$$\tilde\alpha = \varphi \circ \alpha \; (= image\ of\ \alpha\ in\ D'),$$

we obtain the relation

$$\tilde{f}(\alpha(t))\dot\alpha_1(t) + \tilde{g}(\alpha(t))\dot\alpha_2(t) = f(\tilde\alpha(t))\dot{\tilde\alpha}_1(t) + g(\tilde\alpha(t))\dot{\tilde\alpha}_2(t).$$

Corollary.

$$\int_\alpha \varphi^*\omega = \int_{\varphi \circ \alpha} \omega.$$

For the proof, we one simply express the derivative of $\varphi(\alpha(t))$ by means of the chain rule in terms of the derivative of α and then partial derivatives of φ. $\qquad\qquad\square$

We can formulate this simple remark in a somewhat sloppy way as follows:
The pullback of 1-forms is compatible with taking the line integral.
3) A 2-form f is evaluated on an open subset $U \subset D$ as a *surface integral*. For the existence of the surface integral, we assume that

$$K = \bar{U} \cap \text{support}(f)$$

is compact. Here

$$\text{support}(f) = \overline{\{a \in D, \ f(a) \neq 0\}}.$$

(The topological closure has to be taken in D.) Under this assumption, the integral

$$\int_U f(z) \, dx \wedge dy \ \left(:= \int_U (\text{Re } f)(z) \, dx \, dy + \text{i} \int_U (\text{Im } f)(z) \, dx \, dy \right)$$

exists as Lebesgue integral. (If we prefer to work with the Riemann integral, which is sufficient for our purposes, we need sharper assumptions, for example that K is Jordan measurable.) From the transformation formula for surface integrals, we obtain the following result.

13.6 Remark. *Let*

$$\varphi : D \xrightarrow{\sim} D' \qquad (D, D' \subset \mathbb{C})$$

be an orientation-preserving diffeomorphism (i.e. a bijective map which is differentiable in both directions and whose functional determinant is positive everywhere). Then

$$\int_U \varphi^*(\omega) = \int_{\varphi(U)} \omega$$

for any 2-form ω on D' and any open subset $U \subset D$ such that the condition for integrability is fulfilled.

Hence the pullback of a 2-form is compatible with the surface integral in the case of an orientation-preserving diffeomorphism.

13.7 Example. *A biholomorphic map φ is orientation-preserving, since it follows from the Cauchy–Riemann differential equation that*

$$|\varphi'(z)|^2 = \det J(\varphi, z).$$

Here J denotes the real functional determinant.

II Differential Forms on Differentiable Surfaces

Let X be a topological surface and let $\mathcal{D}$ be an atlas on X. We call $\mathcal{D}$ *differentiable* if the chart transformation

$$\psi \circ \varphi^{-1} : \varphi(U_\varphi \cap U_\psi) \longrightarrow \psi(U_\varphi \cap U_\psi)$$
$$\cap \qquad\qquad \cap$$
$$\mathbb{C} \qquad\qquad \mathbb{C}$$

is a diffeomorphism (in the sense of real analysis) for any two charts φ, ψ from $\mathcal{D}$. Two differentiable atlases are said to be differentiably compatible if their union is a differentiable atlas too. A differentiable surface is a pair consisting of a topological surface X and a full class of equivalent differentiable atlases. Hence each differentiable atlas $\mathcal{D}$ on X defines a differentiable surface, which we denote by $(X, \mathcal{D})$ for simplicity. Any Riemann surface can also be considered to be a differentiable surface.

Some basic notions about differentiable surfaces are analogous to the case of Riemann surfaces. For example:

1) If $U \subset X$ is an open subset of a differentiable surface $X = (X, \mathcal{D})$, then U – equipped with the restricted structure $\mathcal{D}|U$ – is a differentiable surface too.
2) A map $f : (X, \mathcal{D}_X) \to (Y, \mathcal{D}_Y)$ between differentiable surfaces is called differentiable if, for each of two charts $\varphi \in \mathcal{D}_X$ and $\psi \in \mathcal{D}_Y$, the function

$$\psi \circ f \circ \varphi^{-1} : \varphi\big(U_\varphi \cap f^{-1}(U_\psi)\big) \longrightarrow \mathbb{C}$$

   is continuously differentiable infinitely often in the sense of real analysis. (Of course, this depends only on the equivalence classes of the atlases.)
3) If f is bijective and f and f^{-1} are both differentiable, then f is called a *diffeomorphism*.
4) A *differentiable chart* φ on a differentiable manifold $(X, \mathcal{D})$ is a diffeomorphism

$$\varphi : U_\varphi \longrightarrow V_\varphi$$
$$\cap \qquad \cap$$
$$X \qquad \mathbb{C}$$

   from an open subset of X onto an open subset of $\mathbb{C}$. The set of all differentiable charts is a differentiable atlas which contains $\mathcal{D}$. It is the largest atlas which is equivalent to $\mathcal{D}$.

The Notion of a Differential Form on a Differentiable Surface

13.8 Definition. *A differential form of degree ($\nu \in \{0,1,2\}$) on a differentiable atlas $\mathcal{D}$ is a family $\omega = (\omega_\varphi)_{\varphi \in \mathcal{D}}$ of ν-forms*

$$\omega_\varphi \text{ on } V_\varphi \quad (\varphi : U_\varphi \longrightarrow V_\varphi$$
$$\cap \qquad \cap$$
$$X \qquad \mathbb{C})$$

*such that for any two charts $\varphi, \psi \in \mathcal{D}$, the **compatibility condition***

$$(\psi \circ \varphi^{-1})^* \omega_\psi = \omega_\varphi$$

is satisfied.

It is possible to extend differential forms to the maximal atlas in a unique way.

13.9 Remark. *Any differential form on a differentiable atlas extends uniquely to the maximal atlas (= atlas of all differentiable charts).*

Proof. The proof follows from the transitivity formula (Remark 13.3). We shall merely sketch the proof. Let (ω_φ) be a differential form on $\mathcal{D}$ and let $\psi : U \to V$ be an arbitrary differentiable chart. We have to define a function f_ψ on V. Let a be an arbitrary point from U and let $b = \varphi(a)$. We choose a chart $\varphi : U_\varphi \to V_\varphi$ in $\mathcal{D}$ whose domain of definition contains a. The compatibility formula dictates how f_ψ has to be defined on the part $\psi(U \cap V_\varphi)$. This defines $f_\psi(b)$. It follows from Remark 13.3 that this value is independent of the choice of φ. So f_ψ is defined. The rest is simple. $\square$

13.10 Definition. *A differential form of degree ν on a differentiable surface is a differential form on the atlas of all differentiable charts.*

Because of Remark 13.9, it is enough to define the differential form on a sub-atlas.

We also have the notion of a differential form on an open subset of a differentiable surface, since it carries the natural structure of a differentiable surface.

A ν-form (ω_φ) is called *differentiable* if all components ω_φ are differentiable ($\varphi \in \mathcal{D}$ is enough).

Notation.

$$A^\nu(X) = \text{set of all differentiable } \nu\text{-forms on } X.$$

Pullback of Differential Forms: The Global Case

Let

$$f : (X, \mathcal{D}_X) \longrightarrow (Y, \mathcal{D}_Y)$$

be a differentiable map of differentiable surfaces. For a ν-form ω on Y, we want to define the pulled-back ν-form $f^*\omega$ on X. In this way, we get maps

$$f^* : A^\nu(X) \longrightarrow A^\nu(Y) \qquad (\nu = 0, 1, 2).$$

To do this, we consider

$$\varphi : U_\varphi \longrightarrow V_\varphi \text{ a chart from } \mathcal{D}_X.$$

We assume that this chart is "small" in the sense that there exists a chart

$$\psi : U_\psi \longrightarrow V_\psi \text{ from } \mathcal{D}_Y$$

such that $V_\varphi \subset U_\psi$. This is sufficient for our purpose, since the small charts define an atlas. We can then consider the differentiable map

$$f \circ \varphi^{-1} : V_\varphi \longrightarrow V_\psi$$

and obtain a differential form on V_φ by pulling back. The compatibility condition can easily be verified.   □

In the case of a diffeomorphism f, the map $\omega \mapsto f^*\omega$ defines an isomorphism

$$f^* : A^\nu(Y) \overset{\sim}{\longrightarrow} A^\nu(X).$$

The naturalness of the alternating product allows us to generalize it to surfaces, as follows.

13.11 Definition and Remark. *By*

$$(\omega \wedge \omega')_\varphi := \omega_\varphi \wedge \omega'_\varphi,$$

a map

$$A^1(X) \times A^1(X) \longrightarrow A^2(X), \quad (\omega, \omega') \longmapsto \omega \wedge \omega',$$

is defined. It is compatible with a pullback,

$$f^*(\omega \wedge \omega') = f^*(\omega) \wedge f^*(\omega'),$$

for differentiable maps $f : X \to Y$ and differentials $\omega, \omega' \in A^1(Y)$.

Evaluation of Differential Forms: The Global Case

1) Let $(f_\varphi)_{\varphi \in \mathcal{D}}$ be a 0-form. We assign a function

$$f : X \longrightarrow \mathbb{C}.$$

The value of this function at $a \in X$ is defined by

$$f(a) := f_\varphi(\varphi(a)),$$

where φ is a chart in $\mathcal{D}$ whose domain of definition contains a. The compatibility relation states that this definition is independent of the choice of the chart φ. The assignment

$$(f_\varphi)_{\varphi \in \mathcal{D}} \longmapsto f$$

gives a bijection between the set of 0-forms and the set of continuous functions on X.

In the following, we shall identify a 0-form with the corresponding function.

In particular, we have

$$A^0(X) \text{ "="} \mathbb{C}^\infty(X)$$
$$:= \text{set of differentiable maps } f : X \longrightarrow \mathbb{C}.$$

2) *Line integrals.* Let

$$\omega = (\omega_\varphi), \quad \omega_\varphi = f_\varphi \, dx + g_\varphi \, dy,$$

be a 1-form on X and let

$$\alpha : [a, b] \longrightarrow X \qquad (a < b)$$

be a piecewise smooth curve. (It is clear how this notion has to be defined "via charts.") We define the function

$$h : [a, b] \longrightarrow \mathbb{C}.$$

Let $t_0 \in [a, b]$. We choose a differentiable chart $\varphi : U_\varphi \to V_\varphi$ from $\mathcal{D}$ such that its domain of definition contains $\alpha(t_0)$. We transport α into V_φ, i.e.

$$\beta(t) = \varphi(\alpha(t)) \qquad (t \text{ varies in a small neighborhood of } t_0),$$

and then we define

$$h(t_0) := f_\varphi(\beta(t))\dot{\beta}_1(t) + g_\varphi(\beta(t))\dot{\beta}_2(t).$$

It follows from Remark 13.5 that this definition is independent of the choice of the chart φ. Now the line integral

$$\int_\alpha \omega := \int_a^b h(t)\, dt$$

is well defined.

3) *Surface integrals.* Surface integrals can be defined only on *oriented* differentiable surfaces.

A differentiable atlas $\mathcal{D}$ is called *oriented* if the chart transformation

$$\psi \circ \varphi^{-1} : \varphi(U_\varphi \cap U_\psi) \longrightarrow \psi(U_\varphi \cap U_\psi)$$

is orientation-preserving for any two charts $\psi, \varphi \in \mathcal{D}$, which means that it has a positive functional determinant. Two oriented atlases are called *orientation-equivalent* if their union is oriented. An *oriented differentiable surface* is a topological surface together with a full class of orientation-equivalent atlases.

For an analytic function, the real functional determinant is the square of the modulus of its complex derivative. Hence analytic atlases are oriented. Hence Riemann surfaces can be considered as oriented differentiable surfaces.

Riemann surfaces are **oriented** *differentiable surfaces!*

Any open subset of an oriented differentiable surface $(X, \mathcal{D})$ is oriented as well (by means of the restricted atlas).

A diffeomorphism

$$f : (X, \mathcal{D}_X) \longrightarrow (Y, \mathcal{D}_Y)$$

between oriented differentiable surfaces is called *orientation-preserving* if, for each of two charts $\varphi \in \mathcal{D}_X$ and $\psi \in \mathcal{D}_Y$,

$$\psi \circ f \circ \varphi^{-1}$$

is orientation-preserving (i.e. has a positive functional determinant).

By an orientation-preserving differentiable chart φ on an oriented differentiable surface $(X, \mathcal{D})$, we understand an orientation-preserving diffeomorphism

$$\varphi : U \longrightarrow V$$

from an open subset $U \subset X$ onto an open subset V of the plane $\mathbb{C}$. The set of all orientation-preserving differentiable charts is the largest atlas which is orientation-equivalent to $\mathcal{D}$.

III Surface Integrals on Differentiable Surfaces

Let ω be a differential form on a differentiable surface. We say that ω vanishes at a point $a \in X$ if, for some differentiable chart

$$\begin{array}{ccc} \varphi : U & \longrightarrow & V, \quad a \in U, \\ \cap & & \cap \\ X & & \mathbb{C}, \end{array}$$

ω_φ vanishes at $\varphi(a)$. Clearly, this condition is independent of the choice of the chart. Hence the support of ω,

$$\operatorname{support}(\omega) = \overline{\{a \in X , \, \omega \text{ does not vanish at } a\}},$$

is well defined.

Now we assume that X is oriented and that ω is a 2-form (on the whole of X). We want to define the surface integral

$$\int_U \omega$$

of ω along an open subset $U \subset X$. For this, we need a *condition for integrability:* the intersection

$$K := \bar{U} \cap \operatorname{support}(\omega)$$

is compact.

First we treat a special case, in which there exists an orientation-preserving chart

$$\begin{array}{ccc} \varphi : U_\varphi & \longrightarrow & V_\varphi \\ \cap & & \cap \\ X & & \mathbb{C} \end{array}$$

such that the compact set K is contained in U_φ. In this case it seems natural to define

$$\int_U \omega := \int_{V_\varphi} \omega_\varphi.$$

We have to show that this definition is independent of the choice of φ. This follows immediately from the transformation formula (Remark 13.6) for surface integrals and the compatibility condition (Definition 13.8) for the family (ω_φ).

In general, we have to break U into pieces which are contained in charts and then to piece together the integral of ω over U. The easiest way to manage this is to use the technique of *partition of unity*.

13.12 Definition.   *Let K be a compact subset of the differentiable surface X and let*

$$K \subset U_1 \cap \cdots \cap U_n \subset X$$

be a covering of K by finitely many open subsets of X. A partition of unity (of K with respect to the given covering) is an n-tuple of differentiable functions

$$\varphi_\nu : X \longrightarrow \mathbb{C} \qquad (1 \le \nu \le n)$$

with the following properties:
a) $0 \le \varphi_\nu \le 1.$
b)   *The support of φ_ν is compact and contained in U_ν.*
c) $\sum_{\nu=1}^{n} \varphi_\nu(a) = 1$ *for all $a \in K$.*

We shall now prove the existence of partitions of unity for suitable coverings.

By a *(differentiable) disk* on an oriented surface X, we understand an orientation-preserving diffeomorphism

$$\varphi : U_\varphi \longrightarrow \mathbb{E} \text{ (unit disk)}$$

of an open relatively compact subset $U_\varphi \subset X$ onto the unit disk $\mathbb{E}$. We use the notation

$$U'_\varphi := \{a \in U\,,\ |\varphi(a)| < 1/2\}.$$

A simple compactness argument shows the following.

13.13 Remark.   *For any compact subset $K \subset X$ of a differentiable surface, there exist finitely many disks*

$$\varphi_\nu : U_\nu \longrightarrow \mathbb{E}$$

with the property

$$K \subset U'_1 \cup \cdots \cup U'_n \qquad (\subset U_1 \cup \cdots \cup U_n).$$

If X is oriented, one can achieve the result that all φ_ν are orientation-preserving.

Now we shall prove the following proposition.

13.14 Proposition.   *The covering (by disks)*

$$K \subset U_1 \cup \cdots \cup U_n,$$

as described in Remark 13.13, admits a partition of unity.

Proof. For the construction, we still need a covering of the boundary of $U_1' \cup \cdots \cup U_n'$. For this, we choose disks

$$\varphi_\nu : U_\nu \longrightarrow \mathbb{E}, \quad \nu = n+1, \dots, N,$$

with the following properties:

a) $U_\nu \cap K = \emptyset \quad (n < \nu \leq N)$;
b) $\bar{U}_1' \cup \cdots \cup \bar{U}_n' \subset U_1' \cup \cdots \cup U_N'$.

Now we use, without proof, the existence of a differentiable function

$$h : \mathbb{R} \longrightarrow \mathbb{R}$$

with the properties

a) $h(t) = 0$ for $t \geq 3/2$,
b) $h(t) = 1$ for $t \leq 1/2$,
c) $0 \leq h \leq 1$.

(For our purpose, a twice continuously differentiable function is sufficient.)

We then define the differentiable function

$$H_\nu : X \longrightarrow \mathbb{C} \quad (1 \leq \nu \leq N), \qquad H_\nu(a) = \begin{cases} h(|\varphi_\nu(a)|) & \text{for } a \in U_\nu, \\ 0 & \text{else} \end{cases}$$

and, for $\nu \in \{1, \dots, n\}$,

$$h_\nu(a) = \begin{cases} \dfrac{H_\nu(a)}{H_1(a) + \cdots + H_N(a)} & \text{if } H_\nu(a) \neq 0, \\ 0 & \text{else.} \end{cases}$$

By the construction of $U_{n+1}, \dots, U_N$, the support of H_ν is contained in the interior of the support of $H_1 + \cdots + H_N$. Hence the functions $h_1, \dots, h_n$ are differentiable. We also have, for $a \in K$,

$$\sum_{\nu=1}^{n} h_\nu(a) = \frac{\sum_{\nu=1}^{n} H_\nu(a)}{\sum_{\nu=1}^{N} H_\nu(a)} = 1,$$

since $H_{n+1}, \dots, H_N$ vanish on K by the construction of $U_{n+1}, \dots, U_N$.   □

Now we are close to the announced definition of the integral of a 2-form ω on an oriented surface X over an open subset $U \subset X$. We assume that the integrability condition is satisfied, i.e. the set

$$K = \text{support}(\omega) \cap \bar{U}$$

is assumed to be compact.

We choose a partition of unity $h_1, \ldots, h_n$ on K, where the support of h_ν is contained in the domain of definition U_ν of a chart. Then the integral $h_\nu \omega$ over U is well defined. So, we can try the definition

$$\int\limits_U \omega := \sum_{\nu=1}^n \int\limits_U h_\nu \omega.$$

All that we have to show is that the right-hand side is independent of the choice of the partition of 1. So let $\tilde{h}_1, \ldots, \tilde{h}_{\tilde{n}}$ be another partition. It is enough to show that

$$\sum_\nu \int h_\nu \omega = \sum_{\mu,\nu} \int \tilde{h}_\mu h_\nu \omega,$$

since the right-hand side is symmetric in $\tilde{h}_\mu$ and h_ν. Actually, we have, for each individual ν,

$$\sum_\mu \int\limits_U \tilde{h}_\mu h_\nu \omega = \int\limits_U \sum_\mu \tilde{h}_\mu h_\nu \omega = \int\limits_U h_\nu \omega. \qquad \square$$

IV Boundary Integrals (A Variant of Line Integrals)

Let U be an open relatively compact subset of an oriented surface X. If U is a reasonable domain, then its boundary is the union of the images of finitely many smooth curves without double points

$$\alpha_\nu : [0,1] \longrightarrow X, \quad 1 \le \nu \le n,$$

which can be oriented in such a way that U is "to the left":

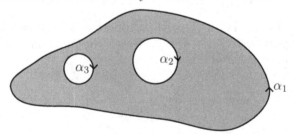

If ω is a 1-form on X, we would like to define

$$\int\limits_{\partial U} \omega := \sum_{\nu=1}^n \int\limits_{\alpha_\nu} \omega.$$

Actually, this concept of a boundary integral is technically complicated. But there is another approach to the boundary integral which is easier.

So, let X be an oriented differentiable surface and let $U \subset X$ be an open subset. A boundary point $a \in \partial U$ is called *smooth* if there exists an orientation-preserving chart

$$\varphi : U_\varphi \longrightarrow V_\varphi$$
$$\cup \qquad \cup$$
$$a \longmapsto 0$$

with the following properties:

a) $\varphi(U_\varphi \cap U) = V_\varphi \cap \mathbb{H}$,
b) $\varphi((\partial U_\varphi) \cap U) = V_\varphi \cap \mathbb{R}$:

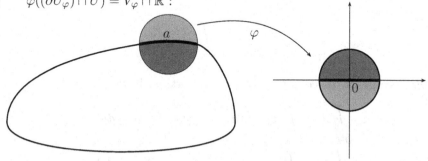

By the way, b) follows from a).

The set of all smooth boundary points of U is denoted by

$$\partial^0 U \subset \partial U.$$

It is an open subset of the whole boundary.

Now let ω be a 1-form on X. We assume that

$$K := \partial U \cap \text{support}(\omega)$$

is compact and contained in the smooth part of the boundary, i.e.

$$K \subset \partial^0 U.$$

Under these assumptions, we want to define

$$\int_{\partial U} \omega.$$

1) *Special case.* $\omega = f\,dx + g\,dy$ is a 1-form with compact support on $X = \mathbb{C}$ and U is the upper half-plane $\mathbb{H}$. In this case, we define

$$\int_{\partial \mathbb{H}} \omega = \int_{\mathbb{R}} f(x, 0)\,dx.$$

(Actually, the integral on the right-hand size is a proper integral. It can be understood as the line integral of ω along the segment from $-C$ to C, $C > 0$ sufficiently large).

This "prototype" of a boundary integral has an important *invariance property.*

13.15 Lemma. *Let $\omega, \tilde{\omega}$ be two 1-forms on $\mathbb{C}$ with compact support and let $U, \tilde{U} \subset \mathbb{C}$ be open subsets with the property*

$$\text{support}(\omega) \subset U, \quad \text{support}(\tilde{\omega}) \subset \tilde{U}.$$

Finally, let $\varphi : U \longrightarrow \tilde{U}$ be an orientation-preserving diffeomorphism with the property

$$\varphi(U \cap \mathbb{H}) = \tilde{U} \cap \mathbb{H}, \qquad \varphi(U \cap \mathbb{R}) = \tilde{U} \cap \mathbb{R},$$

and with

$$\varphi^*(\tilde{\omega}|\tilde{U}) = \omega|U.$$

Claim.

$$\int_{\partial \mathbb{H}} \omega = \int_{\partial \mathbb{H}} \tilde{\omega}.$$

Proof. Because of the condition for the support, we have

$$\int_{\partial \mathbb{H}} \omega = \int_{U \cap \mathbb{R}} f(x,0)\, dx \qquad (\omega = f\, dx + g\, dy),$$

$$\int_{\partial \mathbb{H}} \tilde{\omega} = \int_{\tilde{U} \cap \mathbb{R}} \tilde{f}(x,0)\, dx \qquad (\tilde{\omega} = \tilde{f}\, dx + \tilde{g}\, dy).$$

By restricting φ, we obtain a diffeomorphism

$$\varphi_0 : U \cap \mathbb{R} \longrightarrow \tilde{U} \cap \mathbb{R}.$$

The formula $\varphi^* \tilde{\omega} = \omega$ shows that

$$f(x,0) = \tilde{f}(\varphi_0(x), 0)\varphi_0'(x).$$

Hence the claimed identity follows from the transformation formula for one-dimensional integrals if we know that φ_0 is strictly increasing ($\varphi_0'(x) > 0$).

All that we need is the following simple result.

13.16 Remark. *Let*

$$\varphi : U \longrightarrow \tilde{U} \qquad (U, \tilde{U} \subset \mathbb{C} \text{ open})$$

be an orientation-preserving diffeomorphism and let

$$\varphi(U \cap \mathbb{H}) = \tilde{U} \cap \mathbb{H},$$
$$\varphi(U \cap \mathbb{R}) = \tilde{U} \cap \mathbb{R}.$$

Then

$$\frac{\partial \varphi}{\partial x}(a) \quad (= \varphi_0'(a)) > 0 \text{ for all } a \in U \cap \mathbb{R}. \qquad \square$$

2) *Another special case.* Again we assume that ω is a 1-form on an arbitrary oriented differentiable surface and that $U \subset X$ is an open subset such that support$(\omega) \cap \partial U$ is compact and contained in the smooth part of the boundary of U.

In addition, we assume that there is an oriented chart

$$\varphi : U_\varphi \longrightarrow V_\varphi$$

with the following properties:

$$\text{support}(\omega) \subset U_\varphi, \quad \varphi(U_\varphi \cap U) = V_\varphi \cap \mathbb{H}, \quad \varphi(U_\varphi \cap \partial U) = V_\varphi \cap \mathbb{R}.$$

Then ω_φ extends to a 1-form with compact support on the whole of $\mathbb{C}$. Hence we can define

$$\int_{\partial U} \omega := \int_{\partial \mathbb{H}} \omega_\varphi.$$

The above invariance property shows that this definition is independent of the choice of φ.

3) *General definition.* By means of a partition of unity $h_1, \ldots, h_n$ on $\partial U \cap$ support(ω), we now define the integral of the 1-form ω along the boundary of U. The partition has to be chosen in such a way that the support of h_ν is contained in the domain of definition of a chart with the properties a) and b) above. We then define

$$\int_{\partial U} \omega = \sum_{\nu=1}^{n} \int_{\partial U} h_\nu \omega.$$

As in the case of the surface integral, we show that this definition os independent of the choice of the partition of unity.

V Stokes's Theorem

In the following, all differential forms are assumed to be differentiable. In the local theory, we introduced the operators

$$d : A^0(D) \longrightarrow A^1(D),$$
$$A^1(D) \longrightarrow A^2(D) \qquad (D \subset \mathbb{C} \text{ open}).$$

The following is a simple consequence of the chain rule.

13.17 Remark. *Let*

$$\varphi : D \longrightarrow D' \qquad (D, D' \subset \mathbb{C} \text{ open})$$

be a differentiable map and let ω be a 0- or 1-form on D'. Then

$$\varphi^*(d\omega) = d(\varphi^* \omega).$$

Corollary. *If $(X, \mathcal{D})$ is a differentiable surface, then*

$$d((\omega_\varphi)) = (d\omega_\varphi)$$

defines maps

$$d : A^0(X) \longrightarrow A^1(X),$$
$$d : A^1(X) \longrightarrow A^2(X).$$

Now we have prepared everything that we need to formulate and prove Stokes's theorem for surfaces. We can regard this theorem as a generalization of the main theorem of differential and integral calculus. It can be formulated as follows.

13.18 Stokes's theorem for curves. *Let $\alpha : [a, b] \to X$ be a piecewise smooth curve on a differentiable surface X and let f be a differentiable function on X. Then*

$$\int_\alpha df = f(\alpha(b)) - f(\alpha(a)).$$

Proof. We can assume that X is the complex plane. We now use the formula

$$\frac{d}{dt} f(\alpha(t)) = \frac{\partial f}{\partial x}(\alpha(t))\dot\alpha_1(t) + \frac{\partial f}{\partial y}(\alpha(t))\dot\alpha_2(t)$$

and the main theorem of differential and integral calculus. $\square$

13.19 Stokes's theorem for surfaces. *Let X be an oriented differentiable surface and let ω be a differential on X. Let $U \subset X$ be an open subset with the following properties:*
1)   *support$(\omega) \cap \bar{U}$ is compact;*
2)   *support$(\omega) \cap \partial U$ is contained in the smooth part of the boundary.*
Then

$$\int_{\partial U} \omega = \int_U d\omega.$$

Proof. We choose a partition of unity $h_1, \ldots, h_n$ on support$(\omega) \cap \bar{U}$ with the following property. Let $\nu \in \{1, \ldots, n\}$. Then either support$(h_\nu) \cap \partial U = \emptyset$ or there exists an orientation-preserving chart $\varphi : U_\varphi \to V_\varphi$ with

$$\varphi(U_\varphi \cap U) = V_\varphi \cap \mathbb{H}, \quad \varphi(U_\varphi \cap \partial U) = V_\varphi \cap \mathbb{R}, \quad \text{support}(h_\nu) \subset U_\varphi.$$

It is sufficient to prove Stokes's theorem for $h_\nu\omega$ instead of ω. By the definition of the surface and boundary integrals, we only have to consider two standard situations:

First case. $X = \mathbb{C}, \quad U = \mathbb{H}$.

Second case. $X = \mathbb{C}, \quad U = \mathbb{C}$ (empty boundary).

Since the considered 1-form $\omega = f\,dx + g\,dy$ on $\mathbb{C}$ has compact support, we can consider the second case as a special case of the first one. It remains to treat the first case. The formula states that

$$\int\limits_{0}^{\infty}\int\limits_{-\infty}^{\infty}\left(\frac{\partial g}{\partial x}-\frac{\partial f}{\partial y}\right)dx \wedge dy = \int\limits_{-\infty}^{\infty} f(x,0)\,dx$$

(f and g are differentiable functions with compact support on $\mathbb{C}$).

It follows from the main theorem of differential and integral calculus that

$$\int\limits_{-\infty}^{\infty}\frac{\partial g}{\partial x}\,dx = g(x,y)\Big|_{x=-C}^{x=C} = 0$$

(y fixed, C sufficiently large). For the same reason,

$$-\int\limits_{0}^{\infty}\frac{\partial f}{\partial y}\,dy = f(x,0).$$

Since the order of integration can be changed, the claimed formula is again a trivial consequence of the main theorem of differential and integral calculus.

$\square$

VI Some Variants

Integration Along a Boundary Component

The boundary ∂U (assumptions as in the case of the theorem of Theorem 13.19) is not necessarily connected. There may exist a disjoint decomposition

$$\partial U = \partial_1 U \cup \partial_2 U$$

into nonempty closed subsets. Then one can define the integral $\int_{\partial_1 U}$ of ω along $\partial_1 U$ alone. Formally, this can be reduced to the previous case by replacing X by $X - \partial_2 U$. We have

$$\int\limits_{\partial U_1}\omega + \int\limits_{\partial U_2}\omega = \int\limits_{\partial U}\omega.$$

The Connection Between Boundary Integrals and the Usual Type of Line Integrals

We have mentioned that the boundary integral introduced above can also be seen as a line integral of the usual type. We shall occasionally use this in the very simple case of a disk.

13.20 Remark. *Let U be an open subset of the complex plane which contains the unit circle, and let ω be a differential on U. We set*

$$U^+ = \{z \in U,\ |z| < 1\},$$
$$U^- = \{z \in U,\ |z| > 1\}.$$

We then have

$$\int_{\partial U^+} \omega = -\int_{\partial U^-} \omega = \oint_{|\zeta|=1} \omega.$$

Here $\oint_{|\zeta|=1} \omega$ denotes a line integral of the usual type along the circle e^{it}, $0 \leq 1 \leq 2\pi$. The proof is simple and will be skipped.

A Cauchy Integral Theorem

The following special case of Stokes's theorem is a variant of the Cauchy integral theorem.

Let $U \subset \mathbb{C}$ be an open relatively compact subset of the complex plane with a smooth boundary and let f be a holomorphic function on an open neighborhood of $\bar{U}$. We have

$$\int_{\partial U} f(z)\,dz = 0.$$

In the case of an annulus $r < |z| < R$, this variant states that

$$\oint_{|\zeta|=R} f(\zeta)\,d\zeta = \oint_{|\zeta|=r} f(\zeta)\,d\zeta.$$

Note. Stokes's theorem admits generalizations which we shall not use. For example, it is not necessary that ω is defined and differentiable on the whole of X. It is sufficient to know that ω is differentiable in the interior U, if ω and $d\omega$ can be extended continuously to the boundary. Even this can be weakened in terms of integrability conditions. Finally, it can be allowed that the support of ω contains finitely many nonsmooth points. Such generalizations can be reduced by means of smoothing functions to the "smooth case" (Theorem 13.19). We just formulate an extreme special case which can easily be treated in this way.

13.21 Remark (Stokes's integral theorem for triangles). *Let Δ be a triangle in the plane and let $\partial\Delta$ be the triangular path which runs through the boundary of Δ anticlockwise. Let ω be a differential in the interior of Δ such that the components of ω and $d\omega$ extend continuously to Δ. Then*

$$\int_{\Delta} d\omega = \int_{\partial\Delta} \omega.$$

III. Uniformization

By a *uniformization* of a Riemann surface X, we understand an analytic map $\varphi : D \to X$ of a domain $D \subset \bar{\mathbb{C}}$ of the Riemann sphere onto X with additional properties*). If such a map is given, any meromorphic function $f : X \to \bar{\mathbb{C}}$ can be pulled back to a meromorphic function $F = f \circ \varphi : D \to \bar{\mathbb{C}}$ on D, which has the advantage that the function f is related to the usual type of of function of a complex variable. The meromorphic functions F on D which one obtains in this way are precisely those which have the invariance property

$$\varphi(z_1) = \varphi(z_2) \implies F(z_1) = F(z_2).$$

After a Riemann surface has been uniformized, i.e. $\varphi : D \to X$, the meromorphic functions on X are in one-to-one correspondence with meromorphic functions on the domain D in the Riemann sphere, with certain invariance properties.

In uniformization theory, it is shown that for any (connected) Riemann surface there exists a domain D in the Riemann sphere and a surjective holomorphic map $\varphi : D \to X$. Of course, the pair (D, φ) is not unique. In uniformization theory, it is not only shown that such pairs exist. We can construct a pair with the following most beautiful properties:

a) The result that φ is a *covering* (Definition I.3.14) can be achieved. In particular, φ is then locally biholomorphic.

b) The result that D is the full Riemann sphere $\bar{\mathbb{C}}$, the full plane $\mathbb{C}$, or the unit disk $\mathbb{E}$ can be achieved.

These strong properties determine (D, φ) in an essentially unique way.

In the case of a torus $X = \mathbb{C}/L$, the natural projection $\varphi : \mathbb{C} \to X$ is such a uniformization map. The meromorphic functions on X are in one-to-one correspondence with the invariant meromorphic functions on X, where invariance in this case just means periodicity of F with respect to L.

By the way, a torus never admits a biholomorphic map onto a domain in the Riemann sphere, since such a domain would be the full sphere. But a torus and the Riemann sphere are not conformally equivalent.

The elegance and simplicity of the theory of elliptic functions ([FB], Chap. V) give rise to the question of whether arbitrary Riemann surfaces admit a uniformization. First of all, we should mention that every Riemann surface can be uniformized locally: each point admits an open neighborhood which can be mapped biholomorphically onto an open domain of the plane. Such maps have been called "analytic charts". Sometimes they are called "local uniformizers". But uniformization theory deals with the question of *global uniformization*.

---

*) There is another aspect of uniformization theory, closely related its historical development, which we shall treat at the end of Sect. 2.

Uniformization theory consists of two parts. In a (comparably easy) purely topological part, it is shown that for every connected Riemann surface there exists a *simply
connected Riemann surface* $\tilde{X}$ together with a locally biholomorphic map $\varphi : \tilde{X} \to X$
which is a covering in the topological sense (Definition I.3.14). Such a simply connected covering is essentially unique and has the following important property.

For any two points $x, y \in \tilde{X}$ with $\varphi(x) = \varphi(y)$, there exists a unique biholomorphic
map $\gamma : \tilde{X} \to \tilde{X}$ with the properties $\varphi \circ \gamma = \gamma$ and $\gamma(x) = y$. The set of all these
biholomorphic maps γ is a group Γ, called the *deck transformation group*. Two points
$x, y \in \tilde{X}$ have the same image in X iff there is a deck transformation with the property
$y = \gamma(x)$. The surface X can be reconstructed from $\tilde{X}$ and the deck transformation
group, i.e.
$$X = \tilde{X}/\Gamma.$$
The pair $(\tilde{X}, \Gamma)$ is essentially uniquely determined by X. The group Γ turns out to
be isomorphic to the fundamental group of X.

For the example of the torus $X = \mathbb{C}/L$, the natural projection $\varphi : \mathbb{C} \to X$ is the
universal covering. The deck transformations are the translations $z \mapsto z + \omega$, $\omega \in L$.
Hence the deck transformation group is isomorphic to L and hence to $\mathbb{Z} \times \mathbb{Z}$.

As we mentioned, the existence and uniqueness of the universal covering are purely
topological matters. In an appendix to this chapter, we shall describe the construction
of the universal covering.

The (deeper) function-theoretic part of uniformization theory lies in the proof of
the following theorem:

Uniformization theorem

*Every simply connected Riemann surface is conformally equivalent to
exactly one of the following Riemann surfaces:*

a) *the unit disk* $\mathbb{E}$;
b) *the complex plane* $\mathbb{C}$;
c) *the Riemann sphere* $\overline{\mathbb{C}}$.

If the two parts of uniformization theory are combined, we obtain the result that
any connected Riemann surface admits a locally biholomorphic covering $\varphi : D \to X$,
where D is one of the three standard domains. We can reconstruct X from D by
identifying two points which can be transformed into each other by an element of the
deck transformation group Γ. The meromorphic functions on X correspond to the
Γ-invariant meromorphic functions on D.

This leads to a rough classification of Riemann surfaces, depending on whether
the unit disk, the complex plane, or the Riemann sphere is the universal covering. For
a complete classification, we need to describe all possible deck transformation groups.
This is easy in the last two cases. The Riemann sphere admits only one deck transformation group; this consists of the identity. So, up to biholomorphy, there exists
only one Riemann surface whose universal covering is the Riemann sphere, namely
the Riemann sphere itself. In the case of the plane, the deck transformation groups
are groups of translations $z \mapsto z + \omega$, where ω runs over the elements of a discrete

subgroup $L \subset \mathbb{C}$. Examples are the tori. The majority of Riemann surfaces have the unit disk as their universal covering. Instead of the unit disk, one can take the upper half-plane, since the two are conformally equivalent. Biholomorphic automorphisms of the upper half-plane are given by Möbius transformations $z \mapsto Mz$, $M \in \mathrm{SL}(2, \mathbb{R})$. The group of biholomorphic automorphisms is isomorphic to the group $\mathrm{SL}(2, \mathbb{R})/\pm E$. The deck transformation groups correspond to certain subgroups; these correspond uniquely to subgroups of $\mathrm{SL}(2, \mathbb{R})$ which contain the negative unit matrix. It turns out that any such a subgroup corresponds to a deck transformation group iff it is discrete and if it does not contain elliptic elements. Two such subgroups lead to biholomorphically equivalent Riemann surfaces if they are conjugated in $\mathrm{SL}(2, \mathbb{R})$. Now the classification problem for Riemann surfaces has been reduced to the classification of discrete subgroups without elliptic elements of the group $\mathrm{SL}(2, \mathbb{R})$, or, more precisely, to their conjugacy classes. The latter problem is very difficult. Hence uniformization theory does not mark the end of the theory of Riemann surfaces.

In subsequent chapters, which are concerned particularly with the theory of *compact Riemann surfaces,* we shall not make use of uniformization theory. Only the applications to the theory of modular forms are related to uniformization. One might have the impression that the applications of uniformization theory are not very impressive in comparison with the effort involved. Nevertheless, the central theorems of uniformization theory are important results of classical complex analysis. And, in any case, we shall derive Picard's little and big theorems from the uniformization theorem.

1. The Uniformization Theorem

To separate the function-theoretic part of uniformization from the topological part in a clean way, we shall first give a substitute for the notion of simple connectedness, where we introduce the notion of an *elementary Riemann surface.* In the topological appendix at the end of this chapter, we shall give an introduction to covering theory. It will then turn out that elementary surfaces and simply connected surfaces are the same.

1.1 Definition.   *A connected Riemann surface is called **elementary** if the following condition is satisfied.*
Let

$$X = \bigcup_{j \in J} U_j, \quad U_j \subset X \ open,$$

be an open covering and let

$$f_j : U_j \longrightarrow \bar{\mathbb{C}}$$

be a family of invertible (i.e. not identically vanishing on some connected component of U_j) meromorphic functions. Assume that

$$|f_j/f_k| = 1 \ on \ U_j \cap U_k.$$

Then there exists a meromorphic function

$$f : X \longrightarrow \mathbb{C}$$

such that

$$|f/f_j| = 1 \quad on \ U_j \ for \ all \ j \in J.$$

In Appendix C of this chapter, we shall prove, in connection with covering theory, the *monodromy theorem* for simply connected Riemann surfaces. This monodromy theorem immediately implies that simply connected Riemann surfaces are elementary in the sense of Definition 1.1.

Conversely, elementary surfaces are simply connected, as our proof of the uniformization theorem will show. The situation is comparable to the case of the Riemann mapping theorem, which we treated in [FB]. We refer to the characterization of elementary domains given there ([FB], Chap. IV, Appendix C). One of these characterizations states that a domain is simply connected if and only if any analytic function without zeros admits an analytic square root. It is easy to show that elementary surfaces in the sense of Definition 1.1 have this property (see Exercise 1 for this section). So we have already obtained the equivalence of "elementary" and "simply connected" for domains in the plane. In particular, the Riemann mapping theorem is true for domains in the plane which are elementary in the sense of Definition 1.1. In this section, we shall actually prove the following theorem.

1.2 Theorem. *Any elementary Riemann surface is biholomorphically equivalent to the unit disk* $\mathbb{E}$, *to the plane* $\mathbb{C}$, *or to the Riemann sphere* $\bar{\mathbb{C}}$.

Since each of the three standard domains is simply connected, we see that any elementary Riemann surface is simply connected. In connection with the monodromy theorem mentioned above, we obtain the following theorem.

1.3 Theorem (the uniformization theorem, P. Koebe, and H. Poincaré, 1907). *Any simply connected Riemann surface is biholomorphically equivalent to the unit disk* $\mathbb{E}$, *to the plane* $\mathbb{C}$, *or to the Riemann sphere* $\bar{\mathbb{C}}$.

Proof of Theorem 1.2. In the following, X denotes an elementary Riemann surface. We shall prove that there exists an injective holomorphic map

$$f : X \longrightarrow \bar{\mathbb{C}}.$$

Then X defines a biholomorphic map of X onto the domain $f(X)$, and we can apply the Riemann mapping theorem if $f(X)$ is different from $\bar{C}$. Actually, with some extra work, one can avoid using the Riemann mapping theorem here (see Exercise 2).

The hyperbolic and the zero-bounded case are treated separately.

The Positive-Bounded (= Hyperbolic) Case

We shall see later that a biholomorphic map onto the unit disk

$$f : X \to \mathbb{E}$$

exists. We assume that this has already been proved. Then

$$G_a(x) := -\log|f(x)|$$

is the Green's function of X with respect to $a := f^{-1}(0)$ (because of its invariant characterization; see Definition II.8.7). We obtain

$$|f(x)| = e^{-G_a(x)}.$$

This gives a hint of how the function f has to be constructed.

For every point, there exists the Green's function

$$G_a : X - \{a\} \longrightarrow \mathbb{R}.$$

Claim. *There exists a holomorphic function*

$$F_a : X \longrightarrow \mathbb{C}$$

with the property

$$|F_a(x)| = e^{-G_a(x)} \text{ for } x \neq a.$$

In particular, we have

$$F_a(a) = 0, \quad |F_a(x)| < 1 \text{ for all } x.$$

With regard to Definition 1.1, it is sufficient to construct, for each point $b \in X$, an open neighborhood $U(b)$ and a holomorphic function

$$F : U(b) \longrightarrow \mathbb{C}$$

with

$$|F(x)| = e^{-G_a(x)} \quad \text{for } x \in U(b) \ (x \neq a).$$

First case. $b \neq a.$

With suitable choice of $U(b)$, the harmonic function G_a is the real part of an analytic function f in $U(b)$, and we can define

$$F := e^{-f}.$$

Second case. $b = a$.

Since G_a is logarithmically singular at a, it is sufficient (taking into account the first case) to consider the special case

$$X = \mathbb{E}, \quad a = 0, \quad G_a(z) = -\log|z|.$$

In this case we take

$$F(z) = z.$$

This proves the claim. $\square$

Now we fix a point $a \in X$. It is our aim to show that the function F_a is injective. For the proof, we consider for an arbitrary point $b \neq a$ the function

$$F_{a,b}(x) := \frac{F_a(x) - F_a(b)}{1 - \overline{F_a(b)}F_a(x)}.$$

Its most important properties are:

a) $F_{a,b}$ is analytic in X.
b) $|F_{a,b}| < 1$.
c) $F_{a,b}$ has a zero at $x = b$, let us say of order $k \in \mathbb{N}$.
d) $F_{a,b}(a) = -F_a(b)$.

The function $F_{a,b}$ (obtained from the Green's function for a) is related to the Green's function for b; namely, we make the following claim.

Claim.

$$|F_{a,b}(x)| = |F_b(x)| \text{ for all } x \in X.$$

(This claim can be directly verified for $\mathbb{E}$.)

For the *proof*, we observe that the function

$$u(x) := -\frac{1}{k}\mathrm{Log}\,|F_{a,b}(x)|$$

is nonnegative and harmonic outside some discrete subset. It has a logarithmic singularity at $x = b$. The extremal property of the Green's function (Remark II.8.8) shows

$$G_b(x) \leq u(x).$$

Exponentiating this inequality, we obtain (because $k \geq 1$)

(∗) $$\frac{|F_{a,b}(x)|}{|F_b(x)|} \leq 1 \text{ for all } x.$$

If we specialize this inequality to $x = a$, we obtain from d)

$$|F_a(b)| \leq |F_b(a)|.$$

Since we can exchange the roles of a and b, equality must hold. (By the way, we have obtained the remarkable symmetry relation $G_a(b) = G_b(a)$.) In the inequality $(*)$, the equality holds for at least one x, namely $x = a$. Then, by the maximum principle, equality must hold everywhere. This finishes the proof.

□

The claim shows that

$$F_{a,b}(x) \neq 0 \ \text{ for } x \neq b,$$

and hence

$$F_a(x) \neq F_a(b) \text{ for } x \neq b.$$

Since b was arbitrary, we obtain the claimed injectivity of F_a.   □

The domain $F_a(X)$ is bounded and simply connected. The Riemann mapping theorem shows that it is biholomorphically equivalent to the unit disk. It is possible to avoid the Riemann mapping theorem here and to prove directly that

$$F_a(X) = \mathbb{E}.$$

At any rate, we have seen the following:

Up to conformal equivalence, the unit disk is the unique simply connected hyperbolic Riemann surface.

The Zero-Bounded Case.

In this case, we have "Liouville's theorem" (Corollary II.11.8):

Any bounded analytic function on a zero-bounded Riemann surface is constant.

The function $f : X \to \bar{\mathbb{C}}$ that we are going to construct now would be, in the case $X = \mathbb{C}$ or $X = \bar{\mathbb{C}}$, the map

$$f(z) := \frac{z - 1}{z + 1},$$

which implies

$$\log|f(z)| = \log|z - 1| - \log|z + 1|.$$

An analogue for arbitrary X exists by the existence theorem (Theorem II.12.2):

There exists a harmonic function

$$u := u_{a,b} : X - \{a, b\} \longrightarrow \mathbb{C}$$

with the following properties:

a) *u is logarithmically singular at a;*
b) *$-u$ is logarithmically singular at b;*

c) *u is bounded "at infinity", i.e. in $X - [U(a) \cup U(b)]$, where $U(a)$ and $U(b)$ are arbitrary neighborhoods of a, b.*

We now make use of the assumption that X is elementary. Analogously to the hyperbolic case, we construct by means of Definition 1.1 an analytic function

$$f_{a,b} : X - \{a,b\} \longrightarrow \mathbb{C}$$

with the property

$$|f_{a,b}| = e^{u_{a,b}}.$$

From an estimate

$$|u_{a,b}| \leq C \qquad \text{(away from } a, b\text{)},$$

we obtain the estimate

$$e^{-C} \leq |f_{a,b}| \leq e^{C}.$$

So we have proved the following statement:

Let X be an elementary zero-bounded Riemann surface. For any two distinct points $a, b \in X$, there exists an analytic function

$$f_{a,b} : X - \{a,b\} \longrightarrow \mathbb{C}$$

with the following properties:

1) *$f_{a,b}$ has a zero of order one at a and a pole of order one at b.*
2) *For any two neighborhoods $U(a), U(b)$ of a, b, there exists a constant $C > 0$ with the property*

$$C^{-1} \leq |f_{a,b}(x)| \leq C \text{ for } x \notin U(a) \cup U(b).$$

In particular, $f_{a,b}$ outside $\{a,b\}$ has neither poles nor zeros.

Up to a constant factor, the function $f_{a,b}$ is uniquely determined, since the quotient of two such functions is holomorphic and bounded on X, and hence constant by Liouville's theorem.

Now we fix two points and consider $f = f_{a,b}$ as a meromorphic function on X:

$$f : X \longrightarrow \bar{\mathbb{C}}.$$

Claim. *The function $f : X \to \bar{\mathbb{C}}$ is injective.*

Let c be a third point, different from a and b. We have to show that the function

$$f(z) - f(c)$$

has only one zero, namely $z = c$.

For the *proof*, we consider

$$g(z) := \frac{f(z) - f(c)}{f_{c,b}(z)}.$$

Obviously, this function is holomorphic and bounded on X (since it is bounded "away from a, b, c"). Hence it is constant:

$$f(z) - f(c) = \lambda f_{c,b}(z) \qquad (\lambda \neq 0).$$

The only zero of the function on the right-hand side is at $z = c$. This is what we wanted to prove.   □

The domain $f(X)$ is an elementary domain on the Riemann sphere. If X is compact, we are done, since $f(X) = \bar{\mathbb{C}}$. Otherwise, there exists a point $p \in \bar{\mathbb{C}}$ such that $f(X)$ is contained in $\bar{\mathbb{C}} - \{p\} \cong \mathbb{C}$. By the Riemann mapping theorem, an elementary domain which is contained properly in $\mathbb{C}$ is conformally equivalent to the unit disk and thus hyperbolic. We obtain $f(X) \cong \mathbb{C}$. We shall see in Exercise 2 how one can avoid the use of the Riemann mapping theorem here.

Historical Comments on the Uniformization Theorem

The first complete proof of the uniformization theorem was given by Koebe and Poincaré at about the same time in 1907 [Koe, Po]. It was the highlight of a development which had lasted for more than 50 years. This started with the Riemann mapping theorem, which had already appeared in Riemann's doctoral thesis of 1851, but only under the assumption that the boundary of the given simply connected, bounded domain D had smoothness properties. Under this restriction, Riemann proved a finer result, namely that the mapping function extends to a topological map of the closures. Riemann's proof used a solution of the Dirichlet boundary value problem for D. He made use of the Dirichlet principle, which states that a not necessarily harmonic solution of the boundary value problem is in fact harmonic if it minimizes the functional

$$\int_D \left(\left(\frac{\partial f}{\partial x} \right)^2 + \frac{\partial f}{\partial y}^2 \right) dx\, dy.$$

His working hypothesis that such a minimum exists was criticized by Weierstrass, who gave examples of functionals for which such a minimum does not exist. Around the turn of the century, nearly 50 years after Riemann's thesis, Hilbert invalidated this criticism by giving a proof of the Dirichlet principle.

The efforts to prove the Riemann mapping theorem were closely related to efforts to prove more general mapping theorems for Riemann surfaces. Since the Dirichlet principle seemed not to be suitable for solving such problems, other potential-theoretic methods were developed. Many notable mathematicians

contributed to these developments; it is not possible to give an appreciation
of them all here. The "alternating method" of Neumann and Schwarz, which
we treated in detail in Chap. II, was of decisive importance for uniformization
theory. These developments allowed the solution of many existence problems
for harmonic functions. For example, Osgood, using ideas of Poincaré and
Harnack, succeeded in 1900 in proving the existence of the Green's function
on bounded domains with smoothness assumptions for the boundary. This
was the essential step in the proof of the Riemann mapping theorem, as we
have seen in the proof of the uniformization theorem in the hyperbolic case.
That the Riemann mapping theorem is a consequence of the existence of the
Green's function was proved by Koebe. Now the door was open for a proof
of the general uniformization theorem. Koebe proceeded in his paper of 1907
as follows. He exhausted an arbitrary simply connected Riemann surface by
an ascending chain U_n of relatively compact, simply connected subsets with a
good boundary. In analogy to the theorem of Osgood mentioned above, there
exists the Green's function G_n on U_n with respect to some point $a \in U_n$. This
function can be used to construct a biholomorphic map from U_n onto a disk.
The final mapping function is constructed by a limiting procedure. Whether
the unit disk or the full plane is the final image depends on the value of

$$c_n := \lim_{z \to a} (G_n(z) + \log |z|).$$

By the maximum principle, the sequence is monotonically increasing. We obtain the disk if this sequence is bounded; otherwise we obtain the plane.

What is nowadays called the uniformization theorem is only the culmination
of several other theorems. Several investigations have been concerned with the
question of the extension of the uniformization map to the boundary, if there is
indeed a boundary. Other investigations have concerned the uniformization of
surfaces which are not simply connected. Several new proofs of the uniformization theorem have been developed. After the great success of the alternating
method, the Dirichlet principle was established again, as we have already mentioned. For a proof of the uniformization theorem which rests on this method,
we refer to the classic publication by [We] and to the book by Forster [Fo2].
Another method to attack the boundary value problem, developed by Perron
in 1928, can be used to prove the uniformization theorem, but there is no space
here to explain this.

But it is not only potential-theoretic methods that have been used to prove
the uniformization theorem. In 1917, Bieberbach, using work of Plemelj and
Koebe, succeeded in giving a pure function-theoretic proof in the spirit of
Weierstrass complex analysis, just as we gave a pure function-theoretic proof
of the Riemann mapping theorem in the first volume. We conclude with the
remark that new investigations of the Ricci flow on Riemann manifolds have
led to a new proof of the uniformization theorem.

Exercises for Sect. III.1

1. Show that any analytic function f without zeros on an elementary Riemann surface in the sense of Definition 1.1 has an analytic square root g.

    In the following exercises, we shall see how the Riemann mapping theorem can be avoided in the proof of the uniformization theorem. It can then considered as a special case.

2. Prove, without making use of the Riemann mapping theorem, that in the proof of Theorem 1.3 in the zero-bounded case one has $f(X) = \bar{\mathbb{C}}$ or $f(X) = \bar{\mathbb{C}} - \{b\}$.

    Hint. Any simply connected domain in the plane can be mapped onto a bounded domain in the plane and hence is hyperbolic. For the construction of such a conformal map, see the first two simple steps in the proof of the Riemann mapping theorem ([FB], Theorem IV.4.5).

3. Let D be a simply connected domain which contains the origin and is contained in the unit disk but is different from the unit disk. Show that the Green's functions of D and $\mathbb{E}$ with respect to the origin are different.

    Hint. If D and $\mathbb{E}$ are different, there exists a boundary point of D which is contained in $\mathbb{E}$. Consider, for this boundary point, an analytic function

    $$\psi : D \longrightarrow \mathbb{E}, \quad \psi(0) = 0, \quad \lim_{z \to a} |\psi(z)| = |\sqrt{a}|$$

    (see [FB], Lemma IV.4.6). From the extremal property of the Green's function, we obtain

    $$G(z) \le - \log |\psi(z)|.$$

    But this inequality is false for the Green's function of the unit disk $(- \log |z|)$.

4. Prove, without making use of the Riemann mapping theorem, that in the proof of Theorem 1.2 in the hyperbolic case one has $f(X) = \mathbb{E}$.

    Hint. It follows easily from the construction of f that $- \log |z|$ is the Green's function of the image domain $f(X)$. The rest follows from the previous exercise.

2. A Rough Classification of Riemann Surfaces

Any Riemann surface is biholomorphic to a quotient $\tilde{X}/\Gamma$, with respect to a freely acting group of biholomorphic transformations, of a simply connected Riemann surface $\tilde{X}$. This will be shown in the appendix to this chapter on covering theory on covering theory (Proposition 5.24). By the uniformization theorem, we can achieve the result that the universal covering $\tilde{X}$ is one of the standard domains (the sphere, plane or unit disk). So we can divide Riemann surfaces into three groups, depending on whether the universal covering is the sphere, the plane, or the unit disk.

The Riemann Sphere as the Universal Covering

Any biholomorphic automorphism of the Riemann sphere is a Möbius transformation ([FB], Chap. III. appendices to Sects. 4 and 5, and Exercise 7)

$$z \longmapsto \frac{az+b}{cz+d}, \quad \begin{pmatrix} a & b \\ c & d \end{pmatrix} \in \mathrm{GL}(2,\mathbb{C}).$$

Each Möbius transformation has at least one fixed point. The only group of biholomorphic transformations of the Riemann sphere which acts freely is the group consisting of the identity alone. So, we see the following.

2.1 Remark. *Up to biholomorphy, there exists only one Riemann surface with a compact universal covering, namely the Riemann sphere.*

The Plane as the Universal Covering

Any biholomorphic automorphism of the plane is affine ([FB], Chap. III. appendices to Sects. 4 and 5, and Exercise 5):

$$z \longmapsto az + b,$$

where a is different from zero. If a is different from 1, there is a fixed point. Hence a freely operating group of biholomorphic transformations consists only of translations $z \mapsto z + b$. The set of all occurring b is a subgroup L of $\mathbb{C}$. It is easy to show that such a group acts freely iff L is discrete. Recall that there are three types of discrete subgroups: either L consists only of 0, L is cyclic, or L is a lattice (see Lemma VI.1.1).

2.2 Proposition. *A Riemann surface which admits $\mathbb{C}$ as a universal covering is conformally equivalent to one of the following three types:*
1) *The plane $\mathbb{C}$ itself ($L = \{0\}$).*
2) *The punctured disk $\mathbb{C}^{\cdot}$. ($L = \mathbb{Z}b$, $b \neq 0$. In this case the map $z \mapsto \exp(2\pi i z/b)$ gives a biholomorphic map from $\mathbb{C}/L$ onto the punctured plane.)*
3) *The tori $\mathbb{C}/L$, L a lattice.*

Of course, the plane is not conformally equivalent to any other type, since it is the only simply connected surface in this list. Also, tori cannot be conformally equivalent to any other surface in the list, since they are the only compact surface in the list.

But two tori can be conformally equivalent. We recall the following fact.

Two tori are conform equivalent iff their lattices can be transformed into each other by multiplication by a complex number. This is the case iff the j-invariants agree. Each complex number occurs as a j-invariant.

So, we have found a complete description of the biholomorphy classes of all Riemann surfaces whose universal covering is not the unit disk.

We now have to study the Riemann surfaces whose universal covering is the unit disk. Instead of the unit disk, we can take the upper half-plane, since the two are conformally equivalent. Recall that the group $\mathrm{Bihol}(\mathbb{H})$ of all biholomorphic automorphisms of $\mathbb{H}$ is $\mathrm{SL}(2,\mathbb{R})/\{\pm E\}$ ([FB], Sect. V.7, Exercise 6). A matrix $M \in \mathrm{SL}(2,\mathbb{R})$ acts on $\mathbb{H}$ by means of

$$z \longmapsto Mz = \frac{az+b}{cz+d}, \quad M = \begin{pmatrix} a & b \\ c & d \end{pmatrix}.$$

The matrix is determined by the map up to the sign.

The subgroups of $\mathrm{Bihol}(\mathbb{H})$ are in one-to-one correspondence with the subgroups of $\mathrm{SL}(2,\mathbb{R})$ which contain the negative unit matrix. We have to investigate two things:

1) Let Γ, Γ' be two subgroups of $\mathrm{SL}(2,\mathbb{R})$ which act freely on $\mathbb{H}$. When are the Riemann surfaces $\mathbb{H}/\Gamma$, $\mathbb{H}/\Gamma'$ biholomorphically equivalent?

2) Which subgroups of $\mathrm{SL}(2,\mathbb{R})$ (more precisely, their images in $\mathrm{Bihol}(\mathbb{H})$) act freely?

The first question is simple to answer. The answer follows immediately from the universal property of the universal covering (Remark 5.25):

Let Γ, Γ' be two subgroups of $\mathrm{SL}(2,\mathbb{R})$ which act freely on $\mathbb{H}$. Assume that both contain the negative unit matrix. The Riemann surfaces

$$\mathbb{H}/\Gamma, \quad \mathbb{H}/\Gamma'$$

are biholomorphically equivalent iff both groups are conjugated:

$$\Gamma' = L^{-1}\Gamma L, \quad L \in \mathrm{SL}(2,\mathbb{R}).$$

The second question is more involved. We need some more notations:

a) A subset $\Gamma \subset \mathrm{SL}(2,\mathbb{R})$ is *discrete* if the intersection of Γ with each compact subset of $\mathrm{SL}(2,\mathbb{R})$ is finite.

b) A subgroup $\Gamma \subset \mathrm{SL}(2,\mathbb{R})$ *acts discontinuously* if, for two compact sets K_1, K_2, the set
$$\{M \in \Gamma; \quad M(K_1) \cap K_2 \neq 0\}$$
is finite. Here we can assume that $K_1 = K_2$, since both can be replaced by $K_1 \cup K_2$.

2.3 Lemma. *A subgroup $\Gamma \subset \mathrm{SL}(2,\mathbb{R})$ is discrete iff it acts discontinuously.*

Proof. 1) Assume that the group acts discontinuously. We choose a compact subset $K \subset \mathrm{SL}(2,\mathbb{R})$. Its image under the map
$$p : \mathrm{SL}(2,\mathbb{R}) \longrightarrow \mathbb{H}, \quad M \longmapsto M(\mathrm{i}),$$
is also compact. Obviously,
$$M \in K \Longrightarrow M(\mathrm{i}) \in p(K),$$
and this set is finite.

2) We need an important property of the map p. As we shall show at the end of this proof, this map is surjective and proper. Now let $K \subset \mathbb{H}$ be a compact subset and let $\mathcal{K} \subset \mathrm{SL}(2,\mathbb{R})$ be its inverse image under p. We have
$$M(K) \cap K \neq 0 \Longrightarrow M \in \mathcal{K}\mathcal{K}^{-1}.$$
The latter set is the image of the compact set $\mathcal{K} \times \mathcal{K}$ under the continuous map $(x,y) \mapsto xy^{-1}$ and hence is compact too. □

We still have to prove the announced statement that the map p is surjective and proper.

2.4 Lemma. *The map*
$$p : \mathrm{SL}(2,\mathbb{R}) \longrightarrow \mathbb{H}, \quad M \longmapsto M(\mathrm{i}),$$
is surjective and proper.

Proof. The surjectivity follows from the formula
$$z = \begin{pmatrix} 1 & x \\ 0 & 1 \end{pmatrix} \begin{pmatrix} \sqrt{y} & 0 \\ 0 & \sqrt{y}^{-1} \end{pmatrix} (\mathrm{i}).$$

The proof of the properness is based on the fact that the stabilizer of the point i is the special orthogonal group:
$$\mathrm{SO}(2,\mathbb{R}) := \{M \in \mathrm{SL}(2,\mathbb{R}); \quad M'M = E\} = \{M \in \mathrm{SL}(2,\mathbb{R}); \quad M(\mathrm{i}) = \mathrm{i}\}.$$

This can be shown by an easy computation. The special orthogonal group is bounded, since its rows have Euclidean length one, and is also closed, and hence compact. Now we can show that p is proper, i.e. that the inverse image of a compact set is compact. Since our spaces have a countable basis of the topology, it is sufficient to prove "sequence compactness". Hence we have to show the following:

Let $M_n \in \mathrm{SL}(2, \mathbb{Z})$ be a sequence such that $z_n = M_n(i)$ has an accumulation value in the upper half-plane. Then M_n has an accumulation value.

By assumption, the sequence

$$P_n = \begin{pmatrix} 1 & x_n \\ 0 & 1 \end{pmatrix} \begin{pmatrix} \sqrt{y_n} & 0 \\ 0 & \sqrt{y_n}^{-1} \end{pmatrix}$$

has an accumulation value. We can assume that it converges. We have $M_n = P_n N_n$, with orthogonal matrices N_n. Since the orthogonal group is compact, the sequence N_n and hence M_n have an accumulation value. $\square$

It is clear that freely acting groups also act discontinuously. The converse is false.

2.5 Definition.   *A subgroup $\Gamma \subset \mathrm{SL}(2, \mathbb{R})$ acts in a fixed-point-free way if there is no element $M \in \Gamma$, $M \neq \pm E$, which has a fixed point in the upper half-plane.*

A simple consideration, which can be left to the reader, gives the following result.

2.6 Lemma.   *A subgroup $\Gamma \subset \mathrm{SL}(2, \mathbb{R})$ acts freely on the upper half-plane iff it acts discontinuously and in a fixed-point-free way.*

A matrix $M \in \mathrm{SL}(2, \mathbb{R})$ which is different from $\pm E$ has a fixed point in the upper half-plane if the modulus of its trace is less than or equal to two ([FB], Proposition VI.1.7). Such matrices are called *elliptic*.

Putting all of this together, we can say the following.

2.7 Proposition.   *The biholomorphy classes of Riemann surfaces with the universal covering $\mathbb{E}$ are in one-to-one correspondence with the conjugacy classes of discrete subgroups of $\mathrm{SL}(2, \mathbb{R})$ which contain the negative unit matrix but no elliptic element.*

The theory of these subgroups is very difficult, and its value for the theory of Riemann surfaces is limited.

We conclude this section by describing some examples of groups. The elliptic modular group $\mathrm{SL}(2, \mathbb{Z})$ is certainly discrete, but it is not fixed-point-free, as we know ([FB], Chap. VI). Hence we consider the principal congruence subgroup

$$\Gamma[q] = \mathrm{Kernel}(\mathrm{SL}(2, \mathbb{Z}) \longrightarrow \mathrm{SL}(2, \mathbb{Z}/q\mathbb{Z})).$$

Since this does not always contain the negative unit matrix, we prefer to take

$$\tilde{\Gamma}[q] = \Gamma[q] \cup -\Gamma[q].$$

Looking at the classification of elliptic fixed points ([FB], Proposition VI.1.8), we see that $\tilde{\Gamma}[q]$ contains no elliptic fixed points in the case $q > 1$. In this way, we obtain a series of deck transformation groups of Riemann surfaces.

Finally, we treat a historical aspect of uniformization theory. Let D be one the three standard domains and let Γ be a group of biholomorphic transformations which acts freely on D. By an *automorphic function*, we understand a Γ-invariant meromorphic function on D here. The automorphic functions are in one-to-one correspondence with the meromorphic functions on D/Γ. Assume now that an analytic function F of two complex variables, defined on a domain $U \subset \mathbb{C} \times \mathbb{C}$, is given. Even the case of a polynomial is of great interest. By a *uniformization* of F, we understand a pair of automorphic functions f, g, with respect to a suitable (D, Γ), such that the following two conditions are satisfied:

1) We have identically
$$F(f(t), g(t)) = 0,$$
   where $t \in D$ runs through all points of D such that f and g have no poles.

2) Every point (z, w) of the zero set
$$\mathcal{N} := \{(z, w) \in D \times D; \quad F(z, w) = 0\},$$
   up to a discrete set, can be written in the form $(z, w) = (f(t), g(t))$.

So uniformization in this sense means *parametrization by means of automorphic functions*.

For example,
$$z^2 + w^2 = 0 \qquad (F(z, w) = z^2 + w^2 - 1)$$
can be uniformized in this sense by the functions $f(t) = \sin t$ and $g(t) = \cos t$. The corresponding domain is $D = \mathbb{C}$; the group Γ consists of the translations $z \mapsto z + 2\pi i k$, $k \in \mathbb{Z}$.

Another example is given by the equation
$$w^2 = 4z^3 - g_2 z - g_3, \qquad g_2^3 - 27 g_3^2 \neq 0.$$

We know from the theory of elliptic functions that there exists a lattice $L \subset \mathbb{C}$ such that this equation is uniformized by $f(t) = \wp(t)$ and $g(t) = \wp'(t)$. Here $\wp$ denotes the Weierstrass $\wp$-function of the lattice L.

In general, one can obtain a uniformization by automorphic functions as follows. First of all, one can show that there exists a discrete subset $S \subset \mathcal{N}$ of the zero set such that the complement $X_0 = \mathcal{N} - S$ admits a natural structure in the form of a Riemann surface. "Natural" should imply that the projections $f(z, w) = z$ and $f(z, w) = w$ are analytic functions. We have proved this here for polynomials; the general case can be treated in a similar way. We shall skip this, since the case of polynomials is interesting enough. Sometimes the surface X_0 can be extended by finitely many points, as we have seen in the theory of algebraic functions. Therefore we assume, in general, that there is a Riemann surface X and a finite subset $T \subset X$, such that $X - T$ and X_0 are biholomorphically equivalent. For simplicity, we assume

$X - T = X_0$. We also assume that f and g are meromorphic on X. In the choice of X and T, there may be freedoms such that there may exist different uniformizations for the same P.

Now we make use of the uniformization theorem which we have developed. The surface X can be represented as the quotient D/Γ of one of the three standard domains by a freely acting group Γ of biholomorphic transformations. The two functions f and g now correspond to automorphic functions on D, which, for simplicity, we again denote by f and g. In this way, we obtain a uniformization of F

$$F(f(t), g(t)) = 0$$

by automorphic functions.

Exercises for Sect. III.2

1. Let S be a nonempty finite subset of a torus $X = \mathbb{C}/L$. Show that the universal covering of the surface $X - S$ is the unit disk.

2. Let U, V be two simply connected open subsets of the plane. Show that each connected component of $U \cap V$ is simply connected.

    Hint. Use the fact that the condition of being "simply connected" can be characterized by means of the winding number (see [FB], Chap. IV, Appendix C).

    Show that the analogous statement for other surfaces, for example $\bar{\mathbb{C}}$, instead of $\mathbb{C}$ is false.

3. Let X be a Riemann surface, and let G be a finite group of biholomorphic automorphisms which have a common fixed point a. Show that there exists an analytic chart

$$\varphi : U \to \mathbb{E}, \quad a \in U \subset X,$$

    such that U is invariant under G. Show that the transformed group $G_\varphi = \varphi G \varphi^{-1}$ consists of all rotations

$$z \longmapsto e^{2\pi i \nu z/n}; \quad 0 \leq \nu < n,$$

    where n denotes the order of G.

    Hint. Use the result of the previous exercise and the Riemann mapping theorem.

4. The previous exercise admits a very simple solution in the following important special case. Let $X = \mathbb{H}$ be the upper half-plane and assume that G consists of Möbius transformations $z \mapsto (az + b)(cz + d)^{-1}$, $\left(\begin{smallmatrix} a & b \\ c & d \end{smallmatrix}\right) \in \mathrm{SL}(2, \mathbb{R})$. Obtain the solution.

    Hint. Replace $\mathbb{H}$ by the unit disk and a by the origin. Now use the fact that every biholomorphic transformation of the unit disk with fixed point 0 is a rotation $z \mapsto \zeta z$, $|\zeta| = 1$. Every finite group of roots of unity has the form that has been described in the previous exercise.

5. Let G be the group of rotations $z \to \zeta z$, $\zeta^n = 1$ of the unit disk of order n and denote the quotient space by $\mathbb{E}/G$. Show that there exists a map $\varphi : \mathbb{E}/G \to \mathbb{E}$ such that the diagram

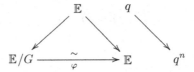

commutes. The map φ is topological and defines a structure on $\mathbb{E}/G$ in the form of a Riemann surface.

6. Let X be a Riemann surface and let Γ be a group of biholomorphic transformations of X with the following properties:

   1) X/Γ is a Hausdorff space.

   2) The stabilizer Γ_a is finite for any $a \in X$.

   3) Every $a \in X$ admits a neighborhood $U(a)$ with the property

   $$\gamma(U(a)) \cap U(a) \neq \emptyset \Longrightarrow \gamma \in \Gamma_a.$$

   Show that $U(a)$ can be chosen such that it is invariant under Γ_a and such that the map

   $$U(a)/\Gamma_a \longrightarrow X/\Gamma$$

   is an open embedding (= homeomorphism onto an open subset).

   Construct a structure in the form of a Riemann surface on the quotient space X/Γ, such that the natural projection $X \to X/\Gamma$ is analytic. Show that a structure with this property is unique.

8. Show that a subgroup of $\mathrm{SL}(2, \mathbb{R})$, or more precisely, its image in Bihol $\mathbb{H}$ has the properties 1)–3) of the previous exercise iff it is discrete. For example, all subgroups of $\mathrm{SL}(2, \mathbb{Z})$ are discrete.

7. Show that the j-function defines a biholomorphic map

$$j : \mathbb{H}/\,\mathrm{SL}(2, \mathbb{Z}) \longrightarrow \mathbb{C}.$$

3. Picard's Theorems

A beautiful application of the theory of Riemann surfaces is provided by Picard's little and big theorems.

By the Casorati-Weierstrass theorem, the image of a nonconstant function in the complex plane is dense in $\mathbb{C}$. Picard's little theorem states further that all values, with at most one exception, are taken.

3.1 Theorem (Picard's little theorem) (Picard 1879). *Every analytic function*

$$f : \mathbb{C} \longrightarrow \mathbb{C} - \{0, 1\}$$

is constant.

The proof rests on the fact that the universal covering of the twice-punctured plane $\mathbb{C} - \{0, 1\}$ is biholomorphically equivalent to the unit disk. This follows from the classification of the Riemann surfaces with universal coverings $\mathbb{C}$ and $\bar{\mathbb{C}}$ in the previous section.

There is only one surface in this list which needs a little consideration, namely the punctured plane $\mathbb{C}^{\bullet}$. Actually, the two surfaces are topologically inequivalent, as can be seen in an elementary way. The fundamental group of $\mathbb{C}^{\bullet}$ is cyclic but that of the twice-punctured disk is not. Even simpler may be the proof that the two are not biholomorphically equivalent. This can be seen from the groups of biholomorphic automorphisms. The group of biholomorphic automorphisms of $\mathbb{C}^{\bullet}$ contains infinitely many rotations. But the group of biholomorphic automorphisms of $\mathbb{C} - \{0, 1\} = \bar{\mathbb{C}} - \{0, 1, \infty\}$ is finite. First of all, it is clear that such an automorphism does not have an essential singularity at the three points $0, 1, \infty$. Hence it is a restriction of a Möbius transformation. Such a Möbius transformation has to permute the three points. Since we know that a Möbius transformation is determined by its values on three points, we get an embedding into (actually an isomorphism onto) the group of permutations of three elements. It particular, the group of automorphisms is finite.

The function f can be lifted to an analytic map of the universal coverings,

$$F : \mathbb{C} \longrightarrow \mathbb{E}.$$

By Liouville's theorem, F must be constant.   □

Picard's big theorem states that any analytic map

$$f : \mathbb{E} - \{0\} \longrightarrow \mathbb{C} - \{0, 1\}$$

with an essential singularity at 0 is constant. The proof needs a generalization of Montel's theorem ([FB], Theorem IV.9). Before we can formulate this, we have to make a comment on the notion of local uniform convergence.

3.2 Remark.   *Let $D, D' \subset \mathbb{C}$ be open subsets of the plane and let*

$$f_n : D \longrightarrow D'$$

be a sequence of continuous functions which converges locally uniformly to a function

$$f : D \longrightarrow \mathbb{C}.$$

Then, for each point $a \in D$ and each neighborhood $V(b) \subset D'$ of the image point $b = f(a)$, there exists a neighborhood $a \in U(a) \subset D$ such that $U(a)$ is mapped by all f_n ($n \in \mathbb{N}$) and by f into $V(b)$.

This small observation, whose proof can be skipped, allows us to generalize the notion of locally uniform convergence to Riemann surfaces.

3.3 Definition.   *Let X, Y be surfaces. A sequence of continuous functions converges locally uniformly to a continuous function*

$$f : X \longrightarrow Y$$

if each point $a \in X$ admits charts

$$\varphi : U_\varphi \longrightarrow V_\varphi, \quad a \in U_\varphi \subset X,$$
$$\psi : U_\psi \longrightarrow V_\psi, \quad b \in U_\psi \subset Y,$$

such that U_φ is mapped by all f_n and by f into U_ψ and such that the sequence

$$\psi \circ f_n \circ \varphi^{-1} : V_\varphi \longrightarrow V_\psi$$

converges locally uniformly to $\psi \circ f \circ \varphi^{-1}$.

It is clear that for open subsets X, Y of the plane, we obtain the usual notion of locally uniform convergence.

3.4 Theorem (generalized Montel's theorem).   *Let X, Y be two Riemann surfaces. Assume that the universal covering of Y is biholomorphically equivalent to the unit disk. Let*

$$f_n : X \longrightarrow Y$$

be a sequence of analytic functions. We assume that $(f_n(a))$ has a convergent subsequence in Y for at least one point $a \in X$. Then f_n admits a locally uniform convergent subsequence.

If X is an open domain in the plane and Y is a disk, then we obtain the usual Montel's theorem. (The assumption that $(f_n(a))$ has a convergent subsequence is inessential, since it is satisfied for any $a \in X$ if one replaces the disk by a slightly larger disk.)

Proof of Theorem 3.4. We may assume that the sequence $f_n(a)$ converges to $b \in Y$. We consider the universal coverings $\tilde{X} \to X$ and $\lambda : \mathbb{E} \to Y$. Let $\tilde{a} \in \tilde{X}$ be a preimage of a and $\tilde{b} \in \mathbb{E}$ be a preimage of b. We lift f_n to a sequence

$$F_n : \tilde{X} \longrightarrow \mathbb{E}$$

of analytic functions. This can be done in such a way that $F_n(\tilde{a})$ is mapped to a given preimage of $f_n(a)$. Hence we can arrange that $F_n(\tilde{a})$ converges to $\tilde{b}$. By the usual Montel's theorem, F_n has a locally convergent subsequence. We can assume that this is the full sequence. The limit function F maps $\tilde{a}$ into $\mathbb{E}$. By the maximum principle, $\tilde{X}$ is mapped into $\mathbb{E}$ (and not only into the closure). Now we can compose F with the covering map $\lambda : \mathbb{E} \to Y$. Like $\lambda \circ F_n$, the map $\lambda \circ F$ is invariant under the deck transformation group of $\tilde{X} \to X$. Hence it is the lift of a function $f : X \to Y$. Clearly, f_n converges to f locally uniformly. $\square$

3.5 Theorem (Picard's big theorem) (Picard 1879).   *Let*

$$f : \mathbb{E}^{\bullet} = \mathbb{E} - \{0\} \longrightarrow \mathbb{C}$$

be an analytic map with an essential singularity at 0. Then f takes all values of the complex plane with at most one exceptional point.

Before the proof, we make a remark:

Let $q_n \in \mathbb{E}^{\bullet}$ be a null sequence and let $f : \mathbb{E}^{\bullet} \to \mathbb{C}$ be an analytic function. Assume that $f_n(q) = f(q_n q)$ converges locally uniformly in the unit disk. Then f has a removable singularity at 0.

Proof of the remark. Since the sequence converges locally uniformly, it is bounded on the circle $|q| = 1/2$:

$$|f(q_n q)| \leq C, \quad |q| = 1/2.$$

This means that f is bounded on a sequence of circles whose radii converge to 0. By the maximum principle, f is bounded between each of those circles (by C). This shows that f is bounded in $0 < |z| \leq 1/2$. By Riemann's removability theorem, 0 is a removable singularity. $\square$

After this preparation, we can give the proof of Picard's big theorem.

Proof of Theorem 3.5. We have to show that an analytic function

$$f : \mathbb{E}^{\bullet} \longrightarrow \mathbb{C} - \{0, 1\}$$

cannot have an essential singularity. We argue indirectly and assume that 0 is an essential singularity. By the Casorati-Weierstrass theorem, there exists a null sequence q_n, $0 < |q_n| < 1/2$, such that $f(q_n)$ converges to an arbitrarily given point b of the plane. We choose for b an arbitrary point which is different from 0 and 1. The sequence of functions

$$f_n : \mathbb{E}^{\bullet} \longrightarrow \mathbb{C} - \{0, 1\},$$
$$f_n(q) := f(2q_n q),$$

converges for at least one point (namely $1/2$). By the generalized Montel theorem, it has a locally uniformly convergent subsequence. But then, by our remark, 0 is a removable singularity. This gives a contradiction.   □

In the proof of Picard's big theorem, we used uniformization theory: we used the uniformization theorem to prove that there exists a holomorphic covering map

$$\lambda : \mathbb{E} \longrightarrow \bar{\mathbb{C}} - \{0, 1, \infty\}.$$

If we have a direct construction for such a function, we can avoid the uniformization theorem. It is a remarkable fact that the theory of modular functions gives such a function. The principal congruence subgroup of level two, $\Gamma[2]$, of the full elliptic modular group acts on the upper half-plane. There is an explicit construction of a modular function λ which maps $\mathbb{H}/\Gamma[2]$ conformally onto the three-times-punctured Riemann sphere. This shows that $\mathbb{H}$ is the universal covering of the three-times-punctured sphere and that $\Gamma[2]/\{\pm E\}$ is the deck transformation group. For the construction of λ, we refer to the exercises.

Since $\Gamma[2]$ is a normal subgroup of the elliptic modular group Γ, the factor group $\Gamma/\Gamma[2]$ acts as a group of biholomorphic transformations on $\mathbb{H}/\Gamma[2]$. This factor group is isomorphic to the permutation group S_3. This is in accordance with the determination of the automorphism group of the three-times-punctured sphere mentioned above.

Exercises for Sect. III.3

1. Let $S \subset \mathbb{C}$ be a finite subset of the plane and let Γ be the fundamental group of $\mathbb{C} - S$ (with respect to some base point). Construct a surjective homomorphism

$$\Gamma \longrightarrow \mathbb{Z}^n, \quad n = \#S.$$

Deduce that the twice-punctured plane and the once-punctured plane are not homeomorphic.

2. The following exercise contains a proof that the fundamental group of the twice-punctured plane is not commutative. Show, step by step, the following:

a) The map
$$f : \mathbb{C} - \{0\} \longrightarrow \mathbb{C}, \quad z \longmapsto z^4 + z^{-4}$$
is proper.

b) Let T be the set of complex numbers which consist of the six roots of unity of order 6 and the origin, and let $S = \{-1, 2\}$. The restriction of f defines a proper and locally biholomorphic map
$$f_0 : \mathbb{C} - T \longrightarrow \mathbb{C} - S.$$

c) The transformations
$$z \longmapsto iz \text{ and } z \longmapsto \frac{1}{z}$$
are deck transformations which do not commute.

3. Let $S \subset \bar{\mathbb{C}}$ be a finite subset of the sphere. Show that the fundamental group $\bar{\mathbb{C}} - S$ is commutative iff $\#S \leq 2$.

4. In [FB], Sect. VI.5, we introduced the Jacobi theta series $\vartheta, \widetilde{\vartheta}, \widetilde{\widetilde{\vartheta}}$. Use the results proved there to show that the function
$$\lambda(z) = \frac{\widetilde{\vartheta}(z)^4}{\widetilde{\widetilde{\vartheta}}(z)^4}$$
is invariant under the principal congruence group $\Gamma[2]$. It generates the field of modular functions. It induces a biholomorphic map from $\mathbb{H}/\Gamma[2]$ onto the three-times-punctured Riemann sphere. Which three points are missing?

(The function λ is Klein's lambda function.)

4. Appendix A. The Fundamental Group

Let X be an arcwise connected topological space. We want to construct, for each point $a \in X$, a group, called the *fundamental group* $\pi(X, a)$. Its elements are homotopy classes of closed curves with base point a. The isomorphy type of this group does not depend on the choice of the base point. Hence one speaks of *the* fundamental group of X. The space X is simply connected if and only if the fundamental group is trivial.

In this section, all curves use the unit interval,
$$\alpha : [0, 1] \longrightarrow X$$
for the parametrization. So, the *composition* of two curves
$$\begin{array}{l} \alpha : [0, 1] \longrightarrow X, \\ \beta : [0, 1] \longrightarrow X, \end{array} \quad \text{with } \alpha(1) = \beta(0),$$

is defined by

$$\gamma : [0,1] \longrightarrow X, \qquad \gamma(t) = \begin{cases} \alpha(2t) & \text{for } 0 \le t \le 1/2, \\ \beta(2t-1) & \text{for } 1/2 \le t \le 1. \end{cases}$$

Notation. $\gamma = \alpha \cdot \beta$.

The *reciprocal*, or inverse curve,

$$\alpha^- : [0,1] \longrightarrow X$$

of α is defined by

$$\alpha^-(t) := \alpha(1-t) \quad \text{for } 0 \le t \le 1.$$

For $\alpha(0) = a$ and $\alpha(1) = b$, the curve α^- runs from b to a.

There are two important equivalence relations for curves α, β which have the same starting point and the same endpoint.

1) The *parameter equivalence*. This means that there exists a topological map

$$\tau : [0,1] \longrightarrow [0,1], \quad \tau(0) = 0, \ \tau(1) = 1,$$

with

$$\beta\left(\tau(t)\right) = \alpha(t).$$

Notation. $\alpha \sim \beta$ (parameter equivalence).

2) The *homotopy* (see also [FB], Chap. IV, Definition A3).

4.1 Definition.   *Let*

$$\alpha, \beta : [0,1] \longrightarrow X$$

be two curves with the same starting point and the same endpoint

$$a = \alpha(0) = \beta(0), \quad \alpha(1) = \beta(1) = b.$$

A **homotopy** *between α and β is a continuous map*

$$H : [0,1] \times [0,1] \longrightarrow X$$

with the following properties:

a) *The curves*

$$\alpha_s : [0,1] \longrightarrow X, \quad \alpha_s(t) = H(t,s) \qquad (s \in [0,1]),$$

all have the same starting point a and the same endpoint b.

b) $\alpha_0 = \alpha, \quad \alpha_1 = \beta.$

We call α and β homotopic if there exists a homotopy between α and β.

Notation. $\alpha \hat{\sim} \beta$ *(homotopy).*

Intuitively, a homotopy H is a continuous deformation of α into β such that the starting points and endpoints are kept fixed:

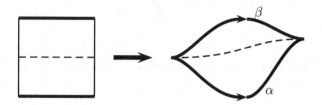

4.2 Definition. *A **closed** curve $\alpha : [0,1] \to X$ is called **nullhomotopic** if it is homotopic to the constant curve*

$$\beta(t) = \alpha(0) \quad (= \alpha(1)), \ t \in [0,1].$$

We now formulate some simple facts. The proofs are left to the reader.

1) Parameter equivalence and homotopy equivalence are equivalence relations.

2) Parameter equivalence implies homotopy:

$$\alpha \sim \beta \implies \alpha \,\hat{\sim}\, \beta.$$

(Consider $H(t,s) = \alpha\left(t(1-s) + s\tau(t)\right)$.)

3) Let

$$\alpha, \beta, \gamma : [0,1] \longrightarrow X$$

be three curves with the property

$$\alpha(1) = \beta(0), \quad \beta(1) = \gamma(0).$$

Then

$$(\alpha \cdot \beta) \cdot \gamma \sim \alpha \cdot (\beta \cdot \gamma), \quad \text{and hence} \quad (\alpha \cdot \beta) \cdot \gamma \,\hat{\sim}\, \alpha \cdot (\beta \cdot \gamma).$$

4) Let α, α' and β, β' be two pairs of homotopic curves. Assume $\alpha(1) = \beta(0)$. Then the curves $\alpha \cdot \beta$ and $\alpha' \cdot \beta'$ are homotopic.

5) The curve $\alpha \cdot \alpha^-$ is nullhomotopic, as the homotopy

$$H(t,s) = \begin{cases} \alpha(2t(1-s)), & 0 \le t \le 1/2, \\ \alpha(2(1-t)(1-s)), & 1/2 \le t \le 1, \end{cases}$$

shows.

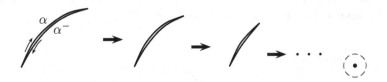

6) Two curves α, β with the same starting points and endpoints are homotopic iff $\alpha \cdot \beta^-$ is nullhomotopic:

$$(\alpha \sim \beta \Longleftrightarrow \alpha \cdot \beta^- \sim \beta \cdot \beta^-)$$

(see the figure above). We choose a point $a \in X$ and denote by

$$\mathcal{S}(X, a)$$

the set of all closed curves with base point a. Homotopy defines an equivalence relation on $\mathcal{S}(X, a)$. The set of equivalence classes is denoted by

$$\pi(X, a) := \mathcal{S}(X, a)/\overset{\sim}{}.$$

We shall sometimes denote the homotopy class of a curve α by $[\alpha]$. Our observations so far show the following fact.

4.3 Remark. *The definition*

$$[\alpha] \cdot [\beta] = [\alpha \cdot \beta] \; \text{for } \alpha, \beta \in \mathcal{S}(X, a)$$

is independent of the choice of the representatives and defines a composition on $\pi(X, a)$. In this way, $\pi(X, a)$ is equipped with the structure of a group. Its unit element is the homotopy class of the constant curve (with starting point and endpoint a). The inverse of $[\alpha]$ is $[\alpha]^{-1} = [\alpha^-]$.

We call $\pi(X, a)$ the *fundamental group* of X with respect to the base point a. Other names are the Poincaré-group and the first homotopy group. Since higher homotopy groups can also be defined, we also use the notation $\pi_1(X, a)$ for the fundamental group. We now investigate how $\pi(X, a)$ depends on the base point a.

Let $a, b \in X$ be two points. We connect them by some curve

$$\gamma : [0, 1] \longrightarrow X, \quad \gamma(0) = a, \; \gamma(1) = b.$$

We obviously have the following fact.

4.4 Remark. *The map*

$$d_\gamma : \pi(X, a) \longrightarrow \pi(X, b),$$
$$[\alpha] \longmapsto [\gamma^- \cdot \alpha \cdot \gamma]$$

defines an isomorphism of groups.

Hence, for (arcwise connected) topological spaces X, the fundamental group $\pi(X, a)$ is independent of the choice of a up to isomorphism. So, we sometimes write $\pi(X)$ instead of $\pi(X, a)$. But this has to be done with caution, since there is no distinguished (canonical) isomorphism. (It depends on the choice of γ.)

4.5 Definition. *A topological space X is called **simply connected** if it is arcwise connected and if any two curves with the same starting points and endpoints are homotopic.*

The notion of "nullhomotopic" is connected with the following notion of "fillable".

4.6 Definition. *A continuous map of the circular line*

$$\alpha : \partial\mathbb{E} \longrightarrow X$$

*is called **fillable** if it admits a continuous extension to the closed unit disk*

$$A : \bar{\mathbb{E}} \longrightarrow X.$$

The same definition can be used with the square $Q = [0, 1] \times [0, 1]$ instead of the disk $\bar{\mathbb{E}}$. There is no difference, because there exists a topological map of Q onto $\bar{\mathbb{E}}$ which also maps the boundaries topologically.

4.7 Remark. *Let X be an arcwise connected space. The following conditions are equivalent:*

1) *X is simply connected (in the sense of Definition 4.5).*
2) *$\pi(X, a) = \{e\}$ for some $a \in X$.*
3) *$\pi(X, a) = \{e\}$ for all $a \in X$.*
4) *Every continuous map*

$$\alpha : \partial\mathbb{E} \longrightarrow X$$

is fillable.

(e denotes the unit element of the fundamental group.)

Proof. The equivalence of 1), 2), and 3) is already clear. Hence it is sufficient to prove the equivalence of 2) and 4).

2) $\Rightarrow$ 4). Let $\alpha : \partial\mathbb{E} \to X$ be a continuous map. The curve

$$\beta(t) = \alpha\left(e^{2\pi i t}\right)$$

is nullhomotopic. Let $H(t,s) = \beta_s(t)$ be a deformation of β into the point $\beta(0)$. Then

$$A\left(re^{2\pi i t}\right) = \beta_{1-r}\left(e^{2\pi i t}\right), \quad 0 \le r, t \le 1,$$

defines a filling of α.

4) $\Rightarrow$ 2). Let α be a closed curve. By assumption, there exists a filling $A :$ $\bar{\mathbb{E}} \to X$ with $A(e^{2\pi i t}) = \alpha(t)$. The map

$$H(t,s) = A\left(s + (1-s)e^{2\pi i t}\right)$$

is a homotopy which contracts α to a point.

Examples of Fundamental Groups

1) Let $D \subset \mathbb{R}^n$ be star-shaped with center $a \in D$. Then $\pi(D, a) = \{e\}$, since

$$H(t,s) = (1-s)\alpha(t) + sa$$

defines a homotopy between α and the constant curve $\beta(t) = a$ for $t \in [0,1]$. Hence every star domain and, in particular, $\mathbb{R}^n$ itself is simply connected for each $n \in \mathbb{N}$.

2) In [FB], Chapt. IV, Proposition A10, we showed that the fundamental group of the punctured disk is isomorphic to $\mathbb{Z}$. The isomorphism is given by

$$\pi(\mathbb{C}^\bullet, 1) \longrightarrow \mathbb{Z},$$

$$[\alpha] \longmapsto \text{winding number of } \alpha \text{ around } 0.$$

But the fundamental group of a topological space is commutative only in very exceptional cases. For example, the fundamental group of the twice-punctured plane is not commutative (see Exercise 2 in Sect. III.3). (One can show more, namely that it is a free group with two generators.) There are several possibilities to extract a commutative "part" from a group, for example as follows.

Let G be any group; then the set of all homomorphisms $\mathrm{Hom}(G, \mathbb{R})$ of G into the additive group of real numbers is a linear subspace of the vector space of all maps from G into $\mathbb{R}$. The vector space

$$H^1(X, a) = \mathrm{Hom}(\pi(X, a), \mathbb{R})$$

is called the (first) *cohomology group* of X (with coefficients in $\mathbb{R}$). In principle, this group depends on the choice of the base point a. If b is a further base point, we can consider some curve γ which combines a with b By means of γ, we obtain an isomorphism (see Remark 4.4)

$$\pi(X, a) \longrightarrow \pi(X, b).$$

This isomorphism usually depends on the choice of γ. But it is easy to show that the induced isomorphism

$$H^1(X, b) \longrightarrow H^1(X, a)$$

is independent of the choice of γ. We call this the *canonical* isomorphism. Since we can identify the cohomology groups for different base points by means of a canonical isomorphism, we simply write

$$H^1(X) = H^1(X, a).$$

Frequently, we write $H^1(X, \mathbb{R})$ instead of $H^1(X)$ to indicate that we have used $\mathbb{R}$ as the coefficient domain. In principle, one can take any abelian group instead of $\mathbb{R}$ as the domain of coefficients. But then the cohomology group is only a group and not a vector space.

The dimension of the vector space $H^1(X, \mathbb{R})$ (it can be infinite) is an important invariant of the topological space X. For topologically equivalent spaces, the invariants coincide. If X and Y are spaces with different invariants, then they cannot be homeomorphic. To illustrate this, we now give an example but without proof.

Let $S \subset \mathbb{C}$ be a subset consisting of n complex numbers. It can be shown that

$$H^1(\mathbb{C} - S, \mathbb{R}) \cong \mathbb{R}^n.$$

If T is another finite subset, then $\mathbb{C} - S$ and $\mathbb{C} - T$ can be homeomorphic only if S and T contain the same number of points. (It can be shown that they are then homeomorphic.) We shall treat a special case in an exercise for Sect. 6.

5. Appendix B. The Universal Covering

The following construction of the universal covering is closely related historically to the theory of Riemann surfaces. In the theory of Riemann surfaces, a surface is first slit in such a way that a surface which is topologically equivalent to a plane domain is obtained. Usually, infinitely many copies of this domain are pasted together along the slit boundaries to obtain the "universal covering".

This can easily be visualized in the case of a torus. A torus can be obtained by gluing opposite edges of a rectangle together. The edges define two closed curves on the torus. If the torus is slit along these curves, the rectangle is obtained. The universal covering of the torus is a plane, which can be imagined as a space obtained from infinitely many rectangles, glued together in an obvious way. In a similar way, in Chap. IV we shall slit a Riemann surface in such a way that a $4p$-gon is obtained.

The purely topological background came into being only step by step. The universal covering, in the sense in which is understood today, was constructed for the first time in the well-known book [We] by Weyl.

We have already introduced the idea of a covering in connection with the construction of the Riemann surface of an algebraic function, and we have treated an extreme special case of covering theory. For the sake of completeness, we repeat the basic facts here.

5.1 Definition. *A locally topological map*

$$f : Y \longrightarrow X$$

*is called a **covering** if any point $b \in Y$ admits an open neighborhood $V(b)$ with the following property: any preimage $a \in Y$ of b admits an open neighborhood $U(a)$ such that the full inverse image $f^{-1}(V(b))$ is the disjoint union of all $U(a)$, i.e.*

$$f^{-1}(V(b)) = \bigcup_{f(a)=b} U(a) \qquad (a \neq a' \Longrightarrow U(a) \cap U(a') = \emptyset),$$

and such that each $U(a)$ is mapped topologically by f onto $V(b)$.

The covering property has an immediate consequence. If we assign to each point $a \in X$ the total number of points in the fiber over a (this number can be infinite), i.e.

$$a \longmapsto \#f^{-1}(a) \leq \infty,$$

then this map is obviously locally constant (constant on $V(b)$). This shows the following.

5.2 Remark.   *Let $Y \to X$ be a covering of a connected space X; then each point $a \in X$ has the same number of preimages. In particular, the covering is surjective if Y is not empty.*

The common number described above is called the *degree* or the *sheet number* of the covering. The latter term should not be misunderstood in the sense that Y is a disjoint union of different sheets which can be distinguished.

The typical example of a covering of degree n is $\mathbb{C}^{\bullet} \to \mathbb{C}^{\bullet}$, $q \mapsto q^n$. An example of a covering of infinite degree is $\exp : \mathbb{C} \to \mathbb{C}^{\bullet}$. In both cases we can take for $V(b)$ a plane which is slit along a half-line not passing through b.

An important tool for the study of coverings is the lifting of curves. We recall the notion of curve lifting.

5.3 Definition.   *A continuous map $f : Y \to X$ of topological spaces has the* **curve-lifting property** *if, for any curve*

$$\alpha : [0, 1] \longrightarrow X$$

and for any point $b \in Y$ over $\alpha(0)$ (i.e. $f(b) = \alpha(0)$), there exists a lift curve β with starting point b, i.e.

$$\alpha = f \circ \beta, \quad \beta(0) = b.$$

Sometimes one has to lift not only single curves but also homotopies, as below.

5.4 Definition.   *A continuous map*

$$f : Y \longrightarrow X$$

has the **homotopy-lifting property** *if, for any continuous map*

$$H : [0, 1] \times [0, 1] \longrightarrow X$$

and for any point $b \in Y$ with $f(b) = H(0, 0)$, there exists a continuous map

$$\tilde{H} : [0, 1] \times [0, 1] \longrightarrow Y$$

with the properties

$$f \circ \tilde{H} = H, \quad \tilde{H}(0, 0) = b.$$

We know that for locally topological maps, the lift $\tilde{H}$ is uniquely determined by H and b. The case where H is a homotopy is of special interest. This means that both of the maps maps

$$s \longmapsto H(0, s), \quad s \longmapsto H(1, s)$$

are constant.

5.5 Remark. *Let $f : Y \to X$ be a locally topological map which possesses the homotopy-lifting property. Using the notation of Definition 5.4, we have the result that if H is a homotopy, then $\tilde{H}$ is a homotopy too.*

Proof. By

$$s \longmapsto \tilde{H}(0, s), \quad s \longmapsto \tilde{H}(1, s),$$

we define continuous maps from the unit interval into the inverse images of the two points $H(0,0) = H(0,s)$, $H(1,1) = H(1,s)$. If f is locally topological, both sets are discrete. Hence both maps are constant. $\qquad\qquad\square$

We have already shown the following fact (Proposition I.3.15 and Theorem I.3.16).

5.6 Proposition. *Coverings have the curve-lifting and the homotopy-lifting property.*

For the more advanced properties of coverings, we need conditions on the underlying topological spaces which are satisfied for surfaces for trivial reasons. One of these properties is the following:

*A space X is called **locally arcwise connected** if any neighborhood U of an arbitrary point $a \in X$ contains an arcwise connected neighborhood $a \in V \subset U$.*

In the following, we tacitly assume that all spaces occurring are arcwise connected and locally arcwise connected.

The curve-lifting and the homotopy-lifting properties are special cases of the following lifting property.

5.7 Proposition. *Let*

$$f : Y \longrightarrow X$$

be a covering and let

$$g : Z \longrightarrow X$$

*be a continuous map of a **simply connected** space Z into X, and let $c \in Z$ and $b \in Y$ be two points with the same image point in X, i.e. $f(c) = g(b)$. Then there exists a unique continuous "lifting"*

$$h : Z \longrightarrow Y$$

with the properties

$$f \circ h = g, \quad h(c) = b.$$

Proof. We connect an arbitrary point $z \in Z$ with c by a curve:

$$\alpha : [0,1] \longrightarrow Z, \qquad \alpha(0) = c, \; \alpha(1) = z.$$

We then consider the image curve $g \circ \alpha$ in X and denote its unique lift for the starting point b by

$$\beta : [0,1] \longrightarrow \tilde{X}.$$

We define
$$h(z) := \beta(1).$$
The homotopy-lifting property together with the simple connectedness shows that $h(z)$ is independent of the choice of the connecting curve β.

It remains to show that h is continuous. Here we have to make use of the locally arcwise connectedness. Let us show continuity at a given point z_0.

Let $x_0 = g(z_0)$ and $y_0 = h(z_0)$. We choose an open neighborhood $V(y_0)$ which is mapped topologically under f onto a neighborhood $U(x_0)$ of x_0. After that, we choose an arcwise connected neighborhood $W(z_0)$ such that $g(W(z_0)) \subset U(x_0)$. We claim that $h(W(z_0))$ is contained in $V(y_0)$. (This shows continuity of h at z_0, since $V(y_0)$ can be chosen arbitrarily small.) Let $z \in W(z_0)$ be an arbitrary point. The image point $y = h(z)$ can be constructed as follows: we connect z_0 with z in $W(z_0)$ and consider the image curve in $U(x_0)$. This curve can be lifted to a curve in Y with starting point y_0. Since, for trivial reasons, it can be lifted to $V(y_0)$, this lift is contained in $V(y_0)$. In particular, the endpoint, i.e. $h(z)$, must be contained in $V(y_0)$.

5.8 Proposition. *Let $f : Y \to X$ be a covering of a simply connected space X. Then f is trivial, i.e. a topological map.*

This follows easily from Proposition 5.7. □

In the proof of Proposition 5.7, the only part of the covering property that we used was that f is a locally topological map that possesses the curve- and homotopy-lifting properties. In particular, Proposition 5.8 holds under this weaker property. Since this will play a role in the proof of the monodromy theorem, we state this fact specifically here.

5.9 Supplement (to Proposition 5.8). *Proposition 5.8 is true for all locally topological f with the curve- and homotopy-lifting properties.*

In the following, we shall construct, under certain assumptions, a very distinguished covering, called the *universal covering X.*

5.10 Definition. *A covering of an arcwise connected space X,*
$$f : \tilde{X} \longrightarrow X,$$
*is called **universal** if $\tilde{X}$ is simply connected.*

The following proposition justifies this notion.

5.11 Proposition. *Let*
$$f : \tilde{X} \longrightarrow X$$
be a universal covering and let
$$g : Y \longrightarrow X$$

be an arbitrary covering. Then there exists a covering

$$h : \tilde{X} \longrightarrow Y$$

such that the diagram

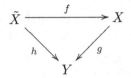

commutes $(g \circ f = h)$.

5.12 Supplement.   *Let $a \in \tilde{X}$ and $b \in Y$ be points with the same image in X, then the map can be constructed in such a way that*

$$h(a) = b.$$

The map is uniquely determined by this property.

Proof. The existence of a continuous map h with the claimed properties follows from Proposition 5.7. It is easy to show that h is a covering if both f and g are coverings.   □

5.13 Definition.   *Let*

$$f : Y \longrightarrow X$$

*be a covering. A **deck transformation** is a topological map*

$$\gamma : Y \longrightarrow Y$$

such that

$$\gamma \circ f = f.$$

The truth of the following remark is immediately clear.

5.14 Remark.   *The set of all deck transformations of a covering is a group (with respect to composition of maps).*

The group of all deck transformations $f : Y \to X$ is the *deck transformation group*.

5.15 Definition.   *A covering $f : Y \to X$ is called a **Galois covering** if for any two points $a, b \in Y$ with the same trace point $f(a) = f(b)$ there exists a deck transformation $\gamma : Y \to Y$ with $\gamma(a) = b$.*

In the case where Y is simply connected, there exists by Proposition 5.11 a unique covering map $\gamma : Y \to Y$ with $\gamma(a) = b$ and $\gamma \circ f = f$. Since the roles of a and b can be exchanged, we see that γ is a topological map and hence a deck transformation. We obtain the following fact.

5.16 Remark.   *Universal coverings are Galois coverings.*

We now collect together the basic properties of deck transformation groups.

5.17 Definition.   *Let Γ be a group of topological self-maps of a topological space X. The group acts **freely** on X if the following conditions are satisfied:*

1) *For any two points $a, b \in X$, there exist neighborhoods $U(a), U(b)$ with the property*
$$\gamma(U(a)) \cap U(b) \neq \emptyset \Longrightarrow \gamma(a) = b.$$

2) *If an element $\gamma \in \Gamma$ has a fixed point $a \in X$, $\gamma(a) = a$, then γ is the identity.*

5.18 Remark.   *The deck transformation group of a covering $f : Y \to X$ acts freely.*

Proof. It is sufficient to prove 1), since 2) is contained in the statement about uniqueness in Proposition 5.7. Let us first assume that the points a and b have different trace points. Since f is locally topological, we can choose disjoint neighborhoods $U(a)$ and $U(b)$ in Y such that their images in X remain disjoint. Then, obviously, $\gamma(U(a)) \cap U(b) = \emptyset$. In the case where the trace points are equal, $c = f(a) = f(b)$, we choose a small connected neighborhood $V(c)$ in X whose inverse image is the union of pairwise disjoint neighborhoods $U(y)$ of the preimages $y \in Y$, $f(y) = c$, which are mapped topologically under f onto $V(c)$. If γ is a deck transformation, then $\gamma(V(a))$ is contained in the union of all $V(y)$, $f(y) = c$. If $\gamma(U(a)) \cap U(b)$ is not empty, then for reasons of connectedness we must have $\gamma(U(a)) = U(b)$. It follows from $\gamma \circ f = f$ that $\gamma(a) = b$.   □

Now let Γ be a group of topological self-maps of a topological space X. Two points $a, b \in X$ are said to be equivalent with respect to Γ if there exists an element
$$\gamma \in \Gamma, \qquad \gamma(a) = b.$$

The quotient space by this equivalence relation is denoted by X/Γ.

5.19 Remark.   *Let Γ be a group of topological self-maps of X which acts freely. Then the natural projection*

$$p : X \longrightarrow X/\Gamma$$

is a Galois covering with deck transformation group Γ.

Proof. For an arbitrary given point $a \in X$, we choose an open neighborhood $U(a)$ such that $\gamma(U(a)) \cap U(a) = \emptyset$ for $\gamma \neq e$. Then the image $V(b) = p(U(a))$ is an open neighborhood of $b = p(a)$ with the property

$$p^{-1}(V(b)) = \bigcup_{\gamma \in \Gamma} \gamma(U(a)).$$

The decomposition on the right-hand side is disjoint and each individual $\gamma(U(a))$ is mapped topologically onto $V(b)$. The rest is clear.   □

There is a close relation between the *universal covering* and the *fundamental group*. This connection leads to a proof of the existence of the universal covering.

Let $f : \tilde{X} \to X$ be a universal covering. We take a point $\tilde{a} \in \tilde{X}$ and denote its image point by $a = f(\tilde{a})$. We want to assign to an element γ of the deck transformation group Γ an element of the fundamental group $\pi(X, a)$. For this, we connect the points $\tilde{a}$ and $\gamma(\tilde{a})$ by a curve α,

$$\alpha : [0,1] \longrightarrow \tilde{X}, \qquad \alpha(0) = \tilde{a},\ \alpha(1) = \gamma(\tilde{a}).$$

Because of the simple connectedness, the homotopy class of this curve is uniquely determined. Hence its image in X defines a well-determined element of the fundamental group $\pi(X, a)$. The constructed map $\Gamma \to \pi(X, a)$ is surjective because of the curve-lifting property and because of Proposition 5.7. Because of Proposition 5.6, it is also injective. So we obtain the following result.

5.20 Remark.   *Let $\tilde{X} \to X$ be a universal covering, let $\tilde{a}$ be a distinguished point in $\tilde{X}$, and let a be its image point in X. The constructed map*

$$\Gamma \longrightarrow \pi(X, b)$$

is an isomorphism.

> *The fundamental group of a space and the deck transformation group of its universal covering are isomorphic.*

This observation suggests a construction of the universal covering. Its realization demands certain conditions on the space X, which are satisfied in the case of our interest, connected surfaces. What we need is the following:

*A space X is called **sufficiently connected** if it is arcwise connected and if each neighborhood U of an arbitrary point $a \in X$ contains a simply connected open neighborhood $V \subset U$.*

We shall assume this property for the rest of this section.

5.21 Proposition.   *Every sufficiently connected space admits a universal covering.*

Proof. We fix a point $a \in X$. For an arbitrary point x of X, we denote by $\langle a, x \rangle$ the set of all homotopy classes of curves with starting point a and endpoint x. Then we consider the set $\tilde{X}$ of all pairs

$$(x, A), \quad x \in X, \ A \in \langle a, x \rangle.$$

We call the homotopy class A a *marking* of the point x. So, the elements of $\tilde{X}$ are points from X which are equipped with a marking. There is a natural projection (if we ignore the marking)

$$f : \tilde{X} \longrightarrow X,$$
$$(x, A) \longmapsto x.$$

It is our aim to equip $\tilde{X}$ with a topology in such a way that $f : \tilde{X} \to X$ is a universal covering.

Let $U \subset X$ be a simply connected subset and let A be a marking of some point $x_0 \in U$. Then every other point $x \in U$ can be marked in a unique way as follows. We choose a curve β inside U with starting point x_0 and endpoint x. The homotopy class B of $\alpha \cdot \beta$ is unique, since U is simply connected. We equip x with the marking B. The set of all $(x, B) \in \tilde{X}$ that is obtained in this way is a subset $W = W(U, (x_0, A))$. The natural projection maps this subset bijectively onto U.

It seems natural to topologize $\tilde{X}$ in such a way that the sets W are open and the map f defines a topological map $W \to U$. This leads to the following definition:

A subset $\tilde{U} \subset \tilde{X}$ is called open iff the following condition is satisfied:

Let (x_0, A) be a point in $\tilde{U}$. There exists a simply connected neighborhood $x_0 \in U \subset X$ such that $W(U, (x_0, A))$ is contained in $\tilde{U}$.

The following three properties are obvious:

1) This condition defines a topology on $\tilde{X}$.
2) The space $\tilde{X}$ is connected and Hausdorff.
3) The projection f is locally topological (in particular, it is continuous).

It remains to be proved that:

4) The map $f : \tilde{X} \to X$ is a covering.
5) $\tilde{X}$ is simply connected.

Proof of 4). Let $x_0 \in X$ be an arbitrary point. We choose a simply connected open neighborhood U of x_0. We obtain the decomposition

$$f^{-1}(U) = \bigcup_A W(U, (x_0, A)) \qquad \text{(disjoint union)},$$

where A runs over all markings of x_0. In this way, we can verify the covering property.

Proof of 5). First of all, we must show that $\tilde{X}$ is connected. This is the case and can easily be proved. We shall skip the proof, since it is sufficient for the proof of the present proposition to replace $\tilde{X}$ by a connected component. It remains to show that any closed curve $\tilde{\alpha}$ in $\tilde{X}$ is nullhomotopic. For this, we can assume that the base point of $\tilde{\alpha}$ is a, marked with the homotopy class of the constant curve $\beta(t) = a$. The image of $\tilde{\alpha}$ is a closed curve $\alpha = f \circ \tilde{\alpha}$ in X. Then $\tilde{\alpha}(t) = (\alpha(t), A_t)$, where A_t is the homotopy class of a curve which joins a to $\alpha(t)$:

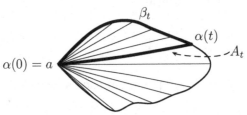

There is a distinguished curve from a to $\alpha(t)$, namely the "restriction" β_t,

$$\beta_t(s) = \alpha(st),$$

which runs from a to $\alpha(t)$ inside α. The proof rests on the following claim.

The curve β_t is contained in the homotopy class of A_t.

Proof of the claim. The claim is true for $t = 0$. Making use of the continuity of $\tilde{\alpha}$ and the definition of the topology on $\tilde{X}$, we can see that β_t is in the homotopy class of A_t for sufficiently small t. Now we can use a standard argument which works in such situations: we consider the supremum of all t with the claimed property. The above argument shows that this supremum is equal to 1.

Since we have assumed that the curve $\tilde{\alpha}$ is closed, the starting and end markings must agree. This means that α is nullhomotopic. Therefore there exists a family of closed curves α_s $(0 \leq s \leq 1)$, all with starting point and endpoint a, which deform α into a constant curve $\alpha_0 = \alpha$, $\alpha_1(t) = a$. Let s be fixed. We mark each point $\alpha_s(t)$ as above, considering the restriction $\alpha_s|[0,t]$ and reparametrizing it to $[0,1])$. This gives a lifting of α_s to a curve $\tilde{\alpha}_s$ on $\tilde{X}$. Obviously, this is a homotopy which contracts $\tilde{\alpha}$ to a point. $\square$

An important consequence of the existence of a universal covering is the following.

5.22 Proposition. *A sufficiently connected topological space is simply connected if and only if each connected covering is trivial (i.e. a topological mapping).*

We get a new characterization of the notion of a "covering":

A map $f : Y \to X$ of a topological space Y into a sufficiently connected space X is a covering if and only if, for each simply connected open neighborhood

$U \subset X$, each connected component of the inverse image $f^{-1}(U)$ is mapped topologically by f onto U.

We conclude this section with some special aspects of coverings of Riemann surfaces. There are no additional difficulties.

We recall the following trivial result (Lemma I.2.8):

Let $f : Y \to X$ be a locally topological map of a surface Y into a Riemann surface X. Then the surface Y carries a unique structure in the form of a Riemann surface such that f is locally biholomorphic.

In particular, the universal covering of a Riemann surface is a Riemann surface as well. For trivial reasons, the deck transformations are biholomorphic.

Let $Y \to X$ be a locally topological map of Riemann surfaces. An analytic structure of Y will not usually induce an analytic structure in X. But in the following special situation this is the case.

5.23 Remark. Let Γ be a group of biholomorphic self-maps of a Riemann surface X which acts freely. Then the quotient space X/Γ carries a unique structure in the form of a Riemann surface such that the natural projection $p : X \to X/\Gamma$ is locally biholomorphic.

We have proved this for the complex plane $X = \mathbb{C}$ and a group of translations $z \mapsto z + \omega$, where ω runs over a lattice. The generalization brings no new difficulties with it, and so we shall state it only briefly:

An open subset $U \subset X$ is said to be small if the projection p maps it topologically onto an open set $U' \subset X/\Gamma$, and if it is the domain of definition of an analytic chart $U \to V$. By inverting p, we obtain a topological chart $U' \to V$. The set of all these charts defines an analytic atlas. $\square$

From the topological covering theory, we obtain the following proposition.

5.24 Proposition. For every Riemann surface X, there exists a simply connected Riemann surface $\tilde{X}$ and a freely acting group Γ of biholomorphic automorphisms of $\tilde{X}$ such that X and $\tilde{X}/\Gamma$ are biholomorphic equivalent. The pair $(\tilde{X}, \Gamma)$ is essentially unique.

The latter statement means that if $(\tilde{X}', \Gamma')$ is another pair with this property, then there exists a biholomorphic map $\varphi : \tilde{X}' \to \tilde{X}$ with the property $\Gamma' = \varphi^{-1}\gamma\varphi$.

It is worthwhile to formulate a special case of the uniqueness property, as below.

5.25 Remark. Let X be a simply connected Riemann surfaces and let Γ, Γ' be two freely acting groups of biholomorphic transformations of X. The Riemann surfaces X/Γ, X/Γ' are biholomorphic iff the groups Γ, Γ' are conjugated in the group of all biholomorphic transformations of X.

6. Appendix C. The Monodromy Theorem

The monodromy theorem is a classical result of complex analysis which has a purely topological background. For this reason, we treat it in this *topological* appendix. The topological background is the following proposition.

6.1 Proposition. *A locally topological map $f : Y \to X$ between sufficiently connected spaces is a covering if and only if has the curve-lifting property.*

Proof. The decisive step is to show that f has the homotopy-lifting property. So, let $H : [0,1] \times [0,1] \to X$ be a continuous map and let $b \in Y$ be a point over $a = H(0,0)$. We have to lift H to a continuous map $\tilde{H} \to Y$ with the property $\tilde{H}(0,0) = b$. Using the curve-lifting property, we first define $\tilde{H}(0,y)$. Next, we define $\tilde{H}(x,y)$ for fixed y by lifting the curve $x \mapsto H(x,y)$ with respect to the starting point $\tilde{H}(0,y)$. In this way, a map $\tilde{H} : [0,1] \times [0,1] \to Y$ is defined. It remains to prove its continuity. By construction, it is continuous on the left edge and on horizontal lines. In the case where the image of $\tilde{H}$ is contained in an open subset of Y which is mapped topologically onto its image, the map $\tilde{H}$ coincides with the lift which results from inverting f. This consideration shows that $\tilde{H}$ is continuous in some neighborhood of the left edge and then, by a compactness argument, on a rectangle $[0,1] \times [0,\varepsilon)$, $0 < \varepsilon \le 1$. Moreover, it then follows that $\tilde{H}$ is continuous on the closure $[0,1] \times [0,\varepsilon]$. We simply consider an open neighborhood of (x,ε) which is mapped topologically under f. After that, we consider the supremum of all ε and show that $\varepsilon = 1$ as usual.

Now the rest of the proof of Proposition 6.1 runs as follows. Consider a lifting $\tilde{f} : \tilde{Y} \to \tilde{Y}$ onto the universal covering (using Proposition 5.7). By Supplement 5.9, it is topological. Now it easily follows that f is a covering. $\qquad\square$

The monodromy theorem is a uniqueness statement about the analytic continuation of function elements. We introduced the notion of a function element in connection with the *analytisches Gebilde*. Here we want to use this notion in a slightly more general form:

1) We want to consider *meromorphic* function elements, i.e. we admit poles.
2) The base space for the function elements is not the complex plane but, more generally, a Riemann surface.

This generalization is harmless and brings no new problems with it.

By a *function element* on a Riemann surface X we understand a pair (a,f) consisting of a point $a \in X$ and a meromorphic function

$$f : U(a) \longrightarrow \bar{\mathbb{C}}$$

on some open neighborhood of a. Two function elements (a,f) and (b,g) are considered to be equal if $a = b$ and if f and g agree in a small neighborhood of $a = b$. More precisely, this means that we have to consider an equivalence relation. When we want to emphasize this, we write $[a, f]$ instead of (a, f) to denote the equivalence class.

Let

$$\alpha : I \longrightarrow X, \quad I \subset \mathbb{R} \text{ interval},$$

be a curve in X. Assume that we have fixed a function element $(\alpha(t), f_t)$ for each $t \in I$. The family of these function elements is called a *regular allocation* of I if, for each $t_0 \in I$, there exists an open neighborhood $U = U(\alpha(t_0))$ of $\alpha(t_0)$ and a meromorphic function

$$f : U \longrightarrow \bar{\mathbb{C}}$$

such that

$$[\alpha(t), f] = [\alpha(t), f_t]$$

for all t from a sufficiently small neighborhood of $t_0 \in I$. Two function elements (a, f) and (b, g) are said to be equivalent if there exist a curve α connecting a and b and a regular allocation $(\alpha(t), f_t)$ with the property $f_0 = f$ and $f_1 = g$.

Let $\mathcal{R}$ be a full equivalence class of function elements on X. As in the case $X = \mathbb{C}$ (where we considered only holomorphic function elements, which makes no difference to the argument), we can equip $\mathcal{R}$ with a structure in the form of a Riemann surface with the following properties:

1) The map

$$p : \mathcal{R} \longrightarrow X, \quad (a, f) \longmapsto a,$$

   is locally biholomorphic.
2) The function

$$\mathcal{R} \longrightarrow \bar{\mathbb{C}}, \quad (a, f) \longmapsto f(a),$$

   is meromorphic.
3) The curves in $\mathcal{R}$ are in one-to-one correspondence with the regular allocations in X.

We obtain:

4) A regular allocation $(\alpha(t), f_t)$ of a curve

$$\alpha : [0, 1] \longrightarrow X$$

   is uniquely determined by the starting element $(\alpha(0), f_0)$.

(One says that $(\alpha(1), f_1)$ arises by analytic continuation from $(\alpha(0), f_0)$ along α.)

5) *Assumption.* Let (a, f_a) be a function element that can be analytically continued along each curve which starts from a. Then the map

$$p : \mathcal{R} \longrightarrow X$$

has the *curve-lifting property.* By Proposition 6.1, it is covering. By Proposition 5.22, this covering is trivial if X is simply connected.

These topological considerations contain the following "old-fashioned" monodromy theorem which is contained in the lectures of Weierstrass.

6.2 Theorem (the monodromy theorem).   *Let (a, f_a) be a function element on a simply connected Riemann surface X which can be analytically continued along any curve starting from a. Then there exists a meromorphic function*

$$f : X \longrightarrow \bar{\mathbb{C}}$$

with

$$[a, f_a] = [a, f].$$

As an application of the monodromy theorem, we shall now show that simply connected Riemann surfaces are elementary in the sense of Definition 1.1:

So, let

$$X = \bigcup U_i$$

be an open covering of X and let

$$f_i : U_i \longrightarrow \bar{\mathbb{C}}$$

be a family of invertible meromorphic functions with the property

$$|f_i/f_j| = 1 \text{ on } U_i \cap U_j.$$

Let a be a point which is contained in the intersection of two sets of the covering, i.e. $a \in U_i \cap U_j$. If $U \subset U_i \cap U_j$ is a connected open neighborhood of a, then f_i and f_j agree in U up to a *constant* factor of absolute value 1.

It follows easily from this property that one of the function elements (a, f_i), for some fixed chosen i and $a \in U_i$, can be analytically continued along every curve starting from a. By the monodromy theorem, there exists a meromorphic function f on X with $[a, f] = [a, f_i]$. The principle of analytic continuation implies that $|f/f_j| = 1$ for all j.

6.3 Proposition.   *A simply connected Riemann surface is elementary in the sense of Definition 1.1.*

Conversely, the uniformization theorem (Theorem 1.2) shows that elementary surfaces are simply connected.

Exercises for the Appendices to Chap. III

1. Let $S \subset \mathbb{C}$ be a finite subset. Show that every biholomorphic self-map of $\mathbb{C} - S$ is the restriction of a Möbius transformation.

2. Let $P(z)$ be a nonconstant polynomial and let M a Möbius transformation with the property $P(M(z)) = P(z)$ for all z. Show that M is affine, i.e. $M(z) = az + b$.

3. Construct a polynomial P of degree > 1 such that every Möbius transformation M with the property $P(M(z)) = P(z)$ is the identity. (This means that the matrix M is a scalar multiple of the unit matrix.) Show that such a polynomial is of degree at least three.

4. Let P be a nonconstant polynomial. Suppose that we choose a finite set $S \subset \mathbb{C}$ such that its inverse image $T = P^{-1}(S)$ under P contains all points at which the derivative of P vanishes. Show that $P : \mathbb{C} - T \longrightarrow \mathbb{C} - S$ is a covering.

5. Construct, by means of the results of Exercises 1–4, a non-Galois covering.

6. Let $\tilde{X} \to X$ be a universal covering with deck transformation group Γ. Let $Y \to X$ be an arbitrary covering. Suppose that we choose a covering $\tilde{X} \to Y$ as in Proposition 5.11.

    Show that the covering $\tilde{X} \to Y$ is Galois, and that its deck transformation group Γ_0 is a subgroup of Γ. Show that the degree of $Y \to X$ is finite iff the index of Γ_0 in Γ is finite, and then that the index and the degree agree. Show that the covering $Y \to X$ is Galois iff Γ_0 is a normal subgroup of Γ. Show that in this case, the factor group Γ / Γ_0 is canonically isomorphic to the deck transformation group of $Y \to X$.

7. Show that any subgroup of index 2 is a normal subgroup.

8. Show, by means of the results of Exercises 6 and 7, that every covering of degree 2 is Galois.

9. An analytic function f on a simply connected Riemann surface has an analytic logarithm.

    Give two proofs:
    a) Use the monodromy theorem.
    b) Integrate the differential df/f.

IV. Compact Riemann Surfaces

This big chapter is devoted to the theory of *compact* Riemann surfaces. Tori are examples of compact Riemann surfaces. This means that we generalize the theory of elliptic functions here. A compact Riemann surface can be associated with any algebraic function, and in this way we obtain all compact Riemann surfaces. The compact Riemann surfaces achieve the same result for the integration of algebraic functions as does the theory of elliptic functions for the elliptic integrals. The triumph of the theory of Riemann surfaces was that it made the "integrals of the first kind" understandable and solved the so-called Jacobi inversion problem. We have to go a long way to achieve this aim. At the end, we shall arrive at the best-known theorems of the theory of Riemann surfaces, such as the the *Riemann–Roch theorem, Abel's theorem,* and the *Jacobi inversion theorem.* On the way, we must also understand the topology of compact Riemann surfaces. We shall treat the topological classification completely here.

1. Meromorphic Differentials

Here, we will reformulate the central existence theorem for Riemann surfaces and, in this connection use the language of *meromorphic differentials* instead of *harmonic functions* with singularities. (After this reformulation, potential-theoretic methods can be dispensed with.)

1.1 Definition. *A holomorphic differential ω on an open subset $U \subset \mathbb{C}$ is a differential of the special form $\omega = f(z)\,dz$ with a **holomorphic** function $f : U \to \mathbb{C}$.*

So, holomorphic differentials are special differentials as introduced in the appendix of Chap. II (Sect. II.13). They are of the form $f\,dz = f\,dx + if\,dy$, with a holomorphic function f. The rules which we developed there for differentials of the form $f\,dx + g\,dy$ become much simpler for holomorphic differentials and holomorphic transformations. Once more, we collect together the basic rules for calculations with holomorphic differentials:

1) The holomorphic differentials on an open domain $D \subset \mathbb{C}$ are in one-to-one correspondence with the holomorphic functions on D.

The transformation formula for holomorphic differentials under holomorphic transformations is very simple:

2) If $\varphi : U \longrightarrow V$, $U, V \subset \mathbb{C}$ open, is a holomorphic map, and $\omega = g(w)\,dw$ is

184

a holomorphic differential on V, then

$$\varphi^*\omega := g\left(\varphi(z)\right) \cdot \varphi'(z)\,dz.$$

This "pullback" is transitive.

3) The total derivative of a holomorphic function is

$$df := f'(z)\,dz.$$

4) Holomorphic differentials are closed, i.e. their total derivative is 0.

This follows, for example, from the fact that holomorphic functions locally admit primitives.

5) Let $(X, \mathcal{A})$ be a Riemann surface. Since X can then be considered as a differentiable surface, the notion of a differential form and, in particular, of differentials (Definition II.13.10) can be used. Because of Remark II.13.9, the components ω_φ have to be defined only for a subatlas, for example the atlas of all analytic charts or a subatlas $\mathcal{A}$ of it. A differential (ω_φ) is called *holomorphic* if ω_φ is holomorphic for all analytic charts. It is sufficient to demand this for all φ in the subatlas $\mathcal{A}$. This leads us to the following very simple direct description of holomorphic differentials on Riemann surfaces.

1.2 Remark. *A holomorphic differential* $\omega = (\omega_\varphi)$ *on a Riemann surface* $(X, \mathcal{A})$ *is a map which assigns to each analytic chart* $\varphi : U_\varphi \to V_\varphi$ *a holomorphic differential*

$$\omega_\varphi = f_\varphi\,dz$$

such that for any two analytic charts φ, ψ *the formula*

$$(\psi \circ \varphi^{-1})^*\omega_\varphi = \omega_\psi$$

holds.

6) If

$$f : (X, \mathcal{A}) \longrightarrow (Y, \mathcal{B})$$

is an analytic map of Riemann surfaces, then for every holomorphic differential ω on Y, the pulled-back differential $f^*\omega$ on X is holomorphic too.

Notation. $\Omega(X)$ is the set of all holomorphic differentials on X. This is a $\mathbb{C}$-vector space, and also a module over the ring $\mathcal{O}(X)$ of holomorphic functions on X.

Meromorphic Differentials

Let X be a Riemann surface, let $S \subset \mathbb{C}$ a be discrete subset, and let ω be a holomorphic differential on $X - S$. For each $s \in S$, we can choose an analytic map $\varphi : U \to V$, $s \mapsto 0$ with $U \cap S = \{s\}$. Then $U - \{s\} \to V - \{0\}$ is an analytic chart on $X - S$, and, with respect to this chart, ω is of the form $f(z)\,dz$ with a holomorphic function on $V - \{0\}$. It may happen that f has an inessential singularity at 0, i.e. f defines a meromorphic function on V. The compatibility condition for the ω_φ implies that this condition is independent of the choice of φ. The same is true for

$$\mathrm{Ord}(\omega, s) := \mathrm{Ord}(f, 0).$$

If this number is negative, we call s a pole of ω of order $-\mathrm{Ord}(\omega, s)$.

1.3 Definition. *A **meromorphic differential** ω on a Riemann surface X is a holomorphic differential*

$$\omega \in \Omega(X - S),$$

where $S \subset X$ is a discrete subset of X. The points of S are assumed to be poles of ω.

For an analytic chart $\varphi : U_\varphi \to V_\varphi$, we define in an obvious way the local component

$$\omega_\varphi = f_\varphi(z)\,dz,$$

with a function f_φ that is meromorphic on V_φ. So, the meromorphic differential can be considered as a family of meromorphic functions $f_\varphi : V_\varphi \to \bar{\mathbb{C}}$ with the usual compatibility relations.

Notation.

$$\mathcal{K}(X) = \text{set of all meromorphic differentials on } X.$$

We define algebraic operations on $\mathcal{K}(X)$ similarly to the case of meromorphic functions.

1.4 Remark. *Meromorphic functions can be added to and multiplied by meromorphic functions, i.e.*
$\mathcal{K}(X)$ *is a module over the ring $\mathcal{M}(X)$ of meromorphic functions.*
There is a map (the "total differential")

$$d : \mathcal{M}(X) \longrightarrow \mathcal{K}(X), \qquad f \longmapsto d(f).$$

(In "local coordinates", $d(f) = f'(z)\,dz$.)

Now let ω_0 be a meromorphic differential on the Riemann surface X which does not vanish identically on any nonempty open subset. Let ω be another meromorphic differential, let

$$\varphi : U_\varphi \longrightarrow V_\varphi$$

be an analytic chart on X, and let

$$\omega_\varphi = g_\varphi \, dz, \qquad \omega_{0,\varphi} = h_\varphi \, dz.$$

Then

$$f_\varphi(x) := \frac{g_\varphi\left(\varphi(x)\right)}{h_\varphi\left(\varphi(x)\right)}, \qquad x \in U_\varphi,$$

is a meromorphic function on U_φ. The functions f_φ coincide on the intersection of two charts, since the transformation factors cancel. Hence they define a meromorphic function on the whole X. This shows the following.

1.5 Lemma. *Let ω_0 be a meromorphic differential on the Riemann surface X which does not vanish identically on any open subset. Then each meromorphic differential is of the form*

$$\omega = f\omega_0,$$

with a meromorphic function f.

In other words, the map

$$\mathcal{M}(X) \xrightarrow{\sim} \mathcal{K}(X), \qquad f \longmapsto f\omega_0,$$

is bijective.

Notation. $\qquad f := \dfrac{\omega}{\omega_0}.$

In analogy to meromorphic functions, we can make the following remark.

1.6 Remark. *A meromorphic differential on a connected Riemann surface vanishes identically if it vanishes on some open nonempty subset.*

The Residue

The residue of an analytic function satisfies certain transformation formulae (see [FB], Sect. III.6, Exercise 10).

Let $\varphi : U \to V$ be a biholomorphic map between open subsets of the plane and let f be a meromorphic function on V. For any $a \in U$, the transformation formula

$$\boxed{\operatorname{Res}(f(w); \varphi(a)) = \operatorname{Res}(\varphi'(z)f(\varphi(z)); a)}$$

holds.

This follows easlily from the representation

$$\operatorname{Res}_a \omega = \frac{1}{2\pi i} \oint f(z) \, dz,$$

where the integral is taken along a small circle around a, and the transformation formula for line integrals. Another proof can be given by means of the series expansions of f and φ'.

The factor $\varphi'(z)$ that occurs in this transformation formula prohibits a meaningful definition of the residue of a meromorphic function. But since this factor appears in the compatibility property of a meromorphic differential, it is possible to give a meaningful definition of the residue of a meromorphic differential.

1.7 Remark and Definition. *Let* $\omega = (\omega_\varphi)$ *be a meromorphic differential on a Riemann surface X and let a be a point in X. We choose some analytic chart φ whose domain of definition contains a. The expression*

$$\mathrm{Res}(\omega; a) := \mathrm{Res}(f_\varphi; \varphi(a)) \qquad (\omega_\varphi = f_\varphi \, dz)$$

*is independent of the choice of the chart. It is called the **residue of the differential** ω at a.*

Now we come to the construction of meromorphic differentials. We have to make use of the existence theorems for harmonic functions given in Chap. II. At the beginning of Sect. II.10, we mentioned that one can attach to a harmonic function u on an open subset of the plane a holomorphic function

$$f := \frac{\partial u}{\partial x} - \mathrm{i}\frac{\partial u}{\partial y}.$$

If u is the real part of an analytic function F (which is always locally the case), then $\omega = dF = f(z)\, dz$. From this we can easily deduce the following (compare Remark II.10.3).

1.8 Remark. *Let u be a harmonic function on a Riemann surface X. If we assign to an analytic chart $\varphi : U \to V$ the differential*

$$\omega_\varphi := \left(\frac{\partial u_\varphi}{\partial x} - \mathrm{i}\frac{\partial u_\varphi}{\partial y} \right) dz \qquad (u_\varphi = u \circ \varphi^{-1}),$$

we obtain a holomorphic differential.

We call ω the differential which is associated with u.

Now we shall discuss the question of how far the poles of a meromorphic differential on a compact Riemann surface can be prescribed. On a compact surface, of course, only finitely many poles are possible.

1.9 Proposition (residue theorem). *Let ω be a meromorphic differential on a **compact** Riemann surface X. Then the sum of all residues of ω is 0.*

Proof. For each pole $a \in X$, we choose a disk

$$\varphi_a : U_a \longrightarrow \mathbb{E}, \qquad a \longmapsto 0.$$

We can assume that these disks are pairwise disjoint. The open subset

$$U = X - \bigcup_{a \text{ pole}} \overline{U_a(1/2)}$$

is relatively compact (since X is compact) and has a smooth boundary. The general Stokes's theorem gives, because $d\omega = 0$,

$$0 = \int_{\partial U} \omega = - \sum_a \int_{\partial U_a(1/2)} \omega = -2\pi i \sum_a \operatorname{Res}_a \omega. \qquad \Box$$

Actually, the residue theorem is the only restriction on the existence of a meromorphic differential. A *central existence theorem* states the following.

1.10 Theorem. *Let $S \subset X$ be a finite subset of the Riemann surface X. Assume that for each point $a \in S$ an open neighborhood $U(a)$ and a meromorphic differential ω_a on $U(a)$ are given. We assume that the neighborhoods are pairwise disjoint and that ω_a is holomorphic on $U(a) - \{a\}$. We also assume that*

$$\sum_{a \in S} \operatorname{Res}_a \omega_a = 0.$$

Then there exists a meromorphic differential ω on X which, outside S, has no poles and is such that $\omega - \omega_a$ extends holomorphically to $U(a)$.

Proof.

First case. All residues are zero.

Let

$$f : \mathbb{E}^{\bullet} \longrightarrow \mathbb{C}$$

be an analytic function on the punctured unit disk whose residue at 0 vanishes. Then f admits a holomorphic primitive F (by termwise integration of the Laurent series). If u is the real part of F, then the differential associated with u is $f(z) \, dz$, since we have

$$\frac{\partial u}{\partial x} - i \frac{\partial u}{\partial y} = f.$$

The central existence theorem (Theorem II.12.2) shows that if the residue of ω_a vanishes at a, then there exists a harmonic function

$$h : X - \{a\} \longrightarrow \mathbb{C}$$

such that the associated differential $w(h) \in \Omega(X - \{a\})$ has the property that

$$w(h) - w_a \qquad (\text{on } U(a) - \{a\})$$

has a removable singularity at a.

This proves Theorem 1.10 if all residues of w_a vanish at a.

Second case. Let $S \subset X$ be a finite subset. Assume that for each $s \in S$ there is given a complex number a_s such that

$$\sum_{s \in S} a_s = 0.$$

Claim. There exists in X a meromorphic differential $w \in \Omega(X - S)$ such that

$$\mathrm{Res}_s\, w = a_s \quad \text{for } s \in S$$

and such that all poles are of order one.

Proof. It is sufficient to treat the case in which S consists of two points s, s' and where

$$a_s = -a_{s'} = 1.$$

The general existence theorem (Theorem II.12.2) gives the existence of a harmonic function

$$u : X - \{s, s'\} \longrightarrow \mathbb{C}$$

such that u is logarithmically singular at s and $-u$ is logarithmically singular at s'. The associated differential has the desired property because, in the case $u(z) = -\mathrm{Log}(z)$, we have

$$\frac{\partial u}{\partial x} - \mathrm{i}\frac{\partial u}{\partial y} = -\frac{1}{z}.$$

This completes the proof of Theorem 1.10.   □

By division of two meromorphic differentials with different poles, we can now construct nontrivial meromorphic functions. In this way, we see that our existence theorems about harmonic functions imply the following fundamental existence theorem for meromorphic functions.

1.11 Theorem.   *On any Riemann surface, there exists a nonconstant meromorphic function.*

For a nonconstant meromorphic function f on a connected Riemann surface, we can consider the meromorphic differential df. Any other meromorphic differential can be written in the form $g\,df$, with a further meromorphic function. Hence meromorphic functions and differentials are closely tied together.

Exercises for Sect. IV.1

1. Let $L \subset \mathbb{C}$ be a lattice and let $X = \mathbb{C}/L$ be the associated torus. The meromorphic differentials on X are in one-to-one correspondence with differentials of the form $f(z)\,dz$, where f is an elliptic function.

   From this, deduce:

   1) The residue theorem (Proposition 1.9) implies the third Liouville theorem.

   2) The vector space of holomorphic differentials is one-dimensional.

2. Let $f(z)$ be a holomorphic function in $|z| > r$ $(r > 0)$. The differential $f(z)\,dz$ has a removable singularity at $\infty \in \bar{\mathbb{C}}$ iff

   $$f\left(-\frac{1}{z}\right) z^{-2}$$

   has a removable singularity at the origin. Deduce from this that on the Riemann sphere, there exists no nonvanishing holomorphic differential.

3. Let X be a Riemann surface and let Γ be a group of biholomorphic automorphisms of X which acts freely. We denote the natural projection by $p : X \to X/\Gamma$. Show that the map

   $$\omega \longmapsto \tilde{\omega} = p^* \omega$$

   gives a bijection between the set of all holomorphic (or meromorphic) differentials on X/Γ and the set of all Γ-invariant holomorphic (or meromorphic) differentials on X. ("Γ-invariant" means that $\gamma^* \tilde{\omega} = \tilde{\omega}$ for all $\gamma \in \Gamma$.)

4. Let $D \subset \mathbb{C}$ be a domain and let $M \in \mathrm{SL}(2, \mathbb{C})$ be a Möbius transformation which leaves D invariant. Show that a meromorphic differential $f(z)\,dz$ on D is invariant under M iff

   $$f(Mz) = (cz + d)^2 f(z).$$

5. Let X be a compact Riemann surface.

   A *differential of the first kind* is a differential which is holomorphic on X. An *elementary differential of the second kind* is a meromorphic differential with precisely one pole. An *elementary differential of the third kind* is a meromorphic differential which has two simple poles and no other poles.

   Show that any meromorphic differential is the sum of finitely many elementary differentials and a differential of the first kind. How far is this decomposition unique?

2. Compact Riemann Surfaces and Algebraic Functions

In the following, X is a connected and compact Riemann surface. From the existence theorem of the last section, we deduce the following.

2.1 Proposition. *Let $S \subset X$ be a finite subset. Assume that for each $s \in S$ a complex number $b_s \in \mathbb{C}$ is given. There exists a meromorphic function*

$$f : X \longrightarrow \bar{\mathbb{C}} \text{ such that } f(s) = b_s \text{ for } s \in S.$$

Proof. Let

$$g(z) = \alpha \, z^{-2} + \text{ higher terms},$$

$$h(z) = \beta \, z^{-2} + \text{ higher terms}$$

be two Laurent series which converge in a punctured disk around the origin. Assume that $\beta \neq 0$. Then the function $g(z)/h(z)$ has a removable singularity at the origin, and its value there is α/β. This little observation together with Theorem 1.10 shows that Proposition 2.1 can be proved by dividing two suitable meromorphic differentials. $\square$

Ramification Points

Let $f : X \to Y$ be a nonconstant holomorphic map between connected Riemann surfaces. A point $a \in X$ is called a ramification point if there is no open neighborhood which is mapped biholomorphically onto its image. (Sometimes the image $b = f(a)$ in Y is also called a ramification point; a is a ramification point "upstairs" and b a ramification point "downstairs".) We know from Remark I.1.17 that any analytic map f is locally of the form $q \mapsto q^n$. We call n the ramification order. Hence a ramification point is present if and only if $n > 1$. If the map $f : X \to Y$ is proper, there is a better description of ramification points. The case $Y = \mathbb{E}$ is of special importance. We assume that a is the only possible ramification point and that it lies over $0 \in \mathbb{E}$. Then $X - \{a\} \to \mathbb{E}^{\bullet}$ is a locally biholomorphic and proper map. We can apply the result of covering theory stated in Proposition I.3.17. A slight variant states the following.

2.2 Proposition. *Let $f : X \to \mathbb{E}$ be a proper, holomorphic map between connected Riemann surfaces, with the only possible ramification point being a, which lies over the origin. Then there exists a biholomorphic map*

$$\varphi : X \xrightarrow{\sim} \mathbb{E}$$

such that the diagram

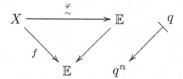

commutes.

Proof. It follows from covering theory (Proposition I.3.17) that there exists a biholomorphic map $X - \{a\} \to \mathbb{E}^{\boldsymbol{\cdot}}$ with the corresponding property. The extension $a \mapsto 0$ gives an obviously continuous map $X \to \mathbb{E}$. The Riemann removability theorem of standard complex analysis shows that it is analytic. But, as in standard complex analysis, we can show that bijective holomorphic maps between Riemann surfaces are biholomorphic. $\qquad\square$

Now we treat the important special case where $f : X \to Y$ is a nonconstant map of *compact* Riemann surfaces. Let b be some point of Y. It has only finitely many preimages $a_1, \dots a_n$. We denote by $k_1, \dots, k_n$ the ramification orders of each of these points.

Notation. *The point b has*

$$k_1 a_1 + \cdots k_n a_n$$

preimages, if we count them with multiplicities.

2.3 Proposition. *Let $f : X \to Y$ be a nonconstant proper, analytic map between connected compact Riemann surfaces. Each point has the same number of preimages if one counts with multiplicities.*

Proof. We assign to each b the number of preimages, counted with multiplicity. It is sufficient to show that this map is locally constant. For this, we can assume that Y is the unit disk. Now we apply Proposition 2.2 to each connected component of $f^{-1}(V)$. We have to note that the restrictions to the connected components remain proper (since they are closed). $\qquad\square$

2.4 Definition. *The degree of a nonconstant analytic, proper map $f : X \to Y$ between connected compact Riemann surfaces is the number of preimages (counted with multiplicity) of a point of Y.*

Another way to define the degree is as follows. We denote by T_0 the set of ramification points, by $S = f(T_0)$ its image under f, and by $T = f^{-1}(S)$ its inverse image. These are finite sets. The map $X - T \to Y - S$ is proper and locally biholomorphic and hence is a covering. We can consider the covering degree in the sense of Remark III.5.2. This coincides with the degree in the sense of Definition 2.4, since all multiplicities are one.

At any rate, our discussion of the degree rests ultimately on a purely topological result, namely the classification of coverings of $\mathbb{E}^{\cdot}$. Because of the great importance of this result, we shall mention for the special case $Y = \bar{\mathbb{C}}$ a completely different, purely function-theoretic proof. So, let $f : X \to \bar{\mathbb{C}}$ be a meromorphic function. It is sufficient to show that f has the same number of poles and zeros, since we can then apply this result to $f - C$. First of all, we notice that the order $\mathrm{Ord}(f, a)$ of f at a zero agrees with the ramification order. For a pole, the ramification order is $-\mathrm{Ord}(f, a)$. Hence the claim is

$$\sum_{a \in X} \mathrm{Ord}(f, a) = 0.$$

But we have the formula

$$\mathrm{Ord}(f, a) = \mathrm{Res}\left(\frac{df}{f}; a\right),$$

and so the claim follows from the residue theorem (Proposition 1.9).

2.5 Lemma. *Let*

$$f : X \longrightarrow \bar{\mathbb{C}} \qquad (X \ compact, \ connected)$$

be a nonconstant meromorphic function of degree d. Any other meromorphic function $g : X \to \bar{\mathbb{C}}$ satisfies a relation

$$\sum_{\nu=0}^{d} R_\nu(f) g^\nu = 0.$$

Here $R_\nu : \bar{\mathbb{C}} \longrightarrow \bar{\mathbb{C}}$ are rational functions.

Supplement. *Assume that there exists a point $b \in \bar{\mathbb{C}} - S$ which is not the image of a ramification point and is such that the restriction of g is injective on the fiber $f^{-1}(b)$. Then R_d is different from zero.*

Proof. We choose some finite subset $S \subset \bar{\mathbb{C}}$ which contains the images of the ramification points and is such that the set of poles of g is contained in $T := f^{-1}(S)$. Then we define the function

$$F(z, w) := \prod_{f(b)=z} (w - g(b)) = \sum_{\nu=o}^{d} R_\nu(z) w^\nu, \quad w \in \mathbb{C}, \ z \in \mathbb{C} - S.$$

From the fact that the restriction of f,

$$f : X - T \longrightarrow \bar{\mathbb{C}} - S$$

is a covering, we can show easily that the functions

$$R_\nu : \bar{\mathbb{C}} - S \longrightarrow \mathbb{C}$$

are holomorphic.

Claim. *The functions R_ν are meromorphic on the whole of $\bar{\mathbb{C}}$.*
(This proves Lemma 2.5, because for trivial reasons we have

$$\sum R_\nu \left(f(z) \right) g(z)^\nu = F \left(f(z), g(z) \right) = 0.)$$

The proof of the claim is a consequence of the following statement.

2.6 Remark. *Let $f : X \to Y$ be a proper surjective holomorphic map of Riemann surfaces and let*

$$S \subset Y, \quad T := f^{-1}(S) \subset X$$

be a discrete subset. We denote the covering degree of

$$X - T \longrightarrow Y - S$$

by d. We then have:

1) *A function $R : Y \to \bar{\mathbb{C}}$ is meromorphic if and only if its composition with f $R \circ f : X \to \bar{\mathbb{C}}$ is meromorphic.*

2) *Let g be a meromorphic function whose poles are contained in T. Additionally, let $S(z_1, \ldots, z_d)$ be a symmetric polynomial in d variables (i.e. invariant under permutations of the variables). Then the function*

$$G(x) := S \left(g(x_1), \ldots, g(x_d) \right), \quad x \in X - T, \; f^{-1}(f(x)) = \{x_1, \ldots, x_d\},$$

extends to a meromorphic function on X.

The second part of the remark is trivial if the map $f : X \to Y$ is "Galois". This means:

There exists a finite group Γ of holomorphic automorphisms of X such that

$$f^{-1} \left(f(x) \right) = \{\gamma(x); \quad \gamma \in \Gamma\} \text{ for } x \in X.$$

Namely, in this case, the function G is a polynomial in the meromorphic functions $g \circ \gamma$, $\gamma \in \Gamma$. An example of a Galois map is the "ramification element"

$$\mathbb{E} \longrightarrow \mathbb{E}, \quad z \longmapsto z^d.$$

Here the elements of Γ are multiplication by roots of unity of order d. The proof of the remark (and also of the first part) follows from the classification of the ramification points, because f is locally equivalent to the ramification element.

Proof of the supplement to Lemma 2.5. We apply part 2) of the above remark to the symmetric polynomial

$$S(z_1, \ldots, z_d) = \prod_{1 \leq i < j \leq d} (z_i - z_j)^2$$

and obtain the following result.

When the restriction of g on one fiber $f^{-1}(b)$ ($b \in \bar{\mathbb{C}} - S$) is injective, then this is true for all fibers with finitely many exceptions.

We can express the property described in the supplement briefly but meaningfully as follows.

Notation. (Assumptions as in the supplement.) *The function g is injective on the **generic fiber** of f.*

Now the proof of the supplement is clear too. We choose a point $x \in X$ that is general enough. Then the polynomial

$$P(z) := \sum_{\nu=0}^{d} R_\nu \left(f(z) \right) z^\nu$$

is different from zero and has d distinct roots (namely, the values of g on $f^{-1} \left(f(z) \right)$. This implies that $R_d \left(f(x) \right) \neq 0$. $\qquad \square$

The Field of Meromorphic Functions

The field of meromorphic functions $\mathcal{M}(X)$ contains the field of constant functions. For simplicity, we identify a complex number with the corresponding constant function. So, $\mathcal{M}(X)$ can be considered as an extension field of

$$\mathbb{C} \subset \mathcal{M}(X).$$

The simplest case is the Riemann sphere $X = \bar{\mathbb{C}}$. In this case $\mathcal{M}(X)$ is the field of rational functions

$$\mathcal{M}(\bar{\mathbb{C}}) = \mathbb{C}(z).$$

Let $\mathbb{C} \subset K$, where K is an arbitrary field extension of $\mathbb{C}$. Then any element $f \in K$, $f \notin \mathbb{C}$, is transcendental over $\mathbb{C}$. This means that for any polynomial $P \in \mathbb{C}[z]$ which is different from 0, $P(f)$ is different from zero too. This follows from the fundamental theorem of algebra, since P already has all its possible roots on $\mathbb{C}$. In particular, for an arbitrary rational function

$$R = \frac{P}{Q}; \quad P, Q \in \mathbb{C}[z], \quad Q \neq 0,$$

the expression

$$R(f) = \frac{P(f)}{Q(f)}$$

is well defined. The map $f \to R(f)$ defines an isomorphism of fields

$$\mathbb{C}(z) \xrightarrow{\sim} \mathbb{C}(f) \subset K$$

onto the subfield of K which is generated by f. A field extension is called *finite*
(or, sometimes, *finite algebraic*) if L is a finite-dimensional K-vector space.

2.7 Definition.   *A field $K \supset \mathbb{C}$ is called an **algebraic function field of
one variable** if there exists an element $f \in K$, $f \notin \mathbb{C}$, such that the extension*

$$K \supset \mathbb{C}(f)$$

is finite.

We use the following theorem from elementary algebra.

2.8 Theorem (theorem of the primitive element).   *Let K be a field of
characteristic zero, and let $L \supset K$ be a finite extension of degree d. Then there
exists an element $f \in L$ such that the powers*

$$1, f, \dots, f^{d-1} \qquad (d := \dim_K L)$$

form a basis of L over K.

This theorem implies the following statement.

*Let K be a field of characteristic zero and let $L \supset K$ be some field extension.
Assume that there exists a natural number $d > 0$ such that any element $f \in L$
satisfies an algebraic equation*

$$f^d + a_{d-1} f^{d-1} + \dots + a_1 f + a_0 = 0 \text{ with } a_j \in K \ (0 \le j \le d).$$

Then the extension is finite.

Otherwise, one could construct an increasing chain of finite extensions of K,

$$K \subsetneq K_1 \subsetneq K_2 \subsetneq \dots \subsetneq L.$$

The degrees

$$d_j := \dim_K K_j$$

become arbitrarily large. On the other hand, the assumption together with the
theorem of the primitive element shows that

$$d_j \le d. \qquad \qquad \square$$

2.9 Proposition.   *The field of meromorphic functions on a compact connected
Riemann surface is an algebraic function field of one variable. More precisely,
we can say that if $f \in \mathcal{M}(X)$ is an arbitrary nonconstant meromorphic function
of degree d and $g \in \mathcal{M}(X)$ is a meromorphic function which is injective on the
generic fiber of f, then*

$$\mathcal{M}(X) = \mathbb{C}(f) \oplus \mathbb{C}(f)g \oplus \dots \oplus \mathbb{C}(f)g^{d-1}.$$

The existence of f and g is ensured by the existence theorem (Proposition 2.1).

The *proof* is clear. One the one hand, we have

$$\dim_{\mathbb{C}(f)} \mathcal{M}(X) \leq d$$

(by Lemma 2.5 and the theorem of the primitive element, Theorem 2.8). On the other hand, the elements $1, g, \ldots, g^{d-1}$ are linearly independent because of the supplement to Lemma 2.5. Hence they must be a basis.   □

Examples.

1) Let $L \subset \mathbb{C}$ be a lattice and let $X := \mathbb{C}/L$ be the corresponding torus. The Weierstrass $\wp$-function has a degree ($=$ order) of two. Its derivative is injective on the generic fiber (otherwise $\wp'$ would be even). We obtain

$$\mathcal{M}(X) = \mathbb{C}(\wp) \oplus \wp'\mathbb{C}(\wp),$$

in accordance with the theory of elliptic functions ([FB], Theorem V.3.3).

2) Let X be the Riemann surface of an algebraic function. The field of meromorphic functions is generated by "the two projections" p and q; more precisely,

$$\mathcal{M}(X) = \bigoplus_{\nu=0}^{d-1} \mathbb{C}(p)q^{\nu}.$$

Now we have collected together all of the tools that we need to show that each compact Riemann surface is the Riemann surface of an algebraic function.

With the notation of Lemma 2.5, we have

$$g^d = \sum_{\nu=0}^{d-1} R_\nu(f)g^\nu$$

with suitable rational functions R_ν. Multiplication by a common denominator shows that there exists a polynomial $P(z, w) \in \mathbb{C}[z, w]$ with the following properties:

a) $P(f, g) = 0$.

b) For almost all z, the degree of P as a polynomial in z is d.

Additionally, we can achieve the following result.

c) The coefficients $a_\nu(z)$ of the polynomial

$$P(z, w) = \sum_{\nu=0}^{d} a_\nu(z)w^\nu$$

do not have a common divisor in $\mathbb{C}[z]$.

The the polynomial P is then irreducible. We denote by

$$X(P) := \{\, (z, w) \in \mathbb{C} \times \mathbb{C}; \quad P(z, w) = 0 \,\}$$

the algebraic curve which is associated with this curve. We choose a finite set $S \subset \bar{\mathbb{C}}$ large enough that the poles of f and g are contained in $T := F^{-1}(S)$, to obtain a map

$$X - T \longrightarrow X(P),$$

$$x \longmapsto (f(x), g(x)).$$

If S is taken large enough, then this map will be injective. We denote its image by $X_0(P)$. The complement of $X_0(P)$ in $X(P)$ consists of finitely many points. *The two maps f and g in the following commutative diagram,*

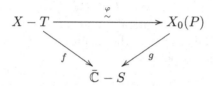

are locally biholomorphic. In particular, φ is holomorphic and hence biholomorphic.

Making use of the uniqueness of the compactification of a Riemann surface (Lemma I.3.8), we obtain the following result.

2.10 Proposition. *Every connected compact Riemann surface is biholomorphically equivalent to the Riemann surface which is associated with an irreducible polynomial $P(z, w) \in \mathbb{C}[z, w]$.*

As we have shown in Propostion 2.9, the field of meromorphic functions of a compact Riemann surface is an algebraic function field of one variable. Conversely, we can show that each algebraic function field of one variable is isomorphic to the field of meromorphic functions of a suitable compact Riemann surface. Here we say that two algebraic function fields of one variable are isomorphic if there exists a field isomorphism $\sigma : K \xrightarrow{\sim} L$ which is the identity on $\mathbb{C}$.

2.11 Proposition. *Any algebraic function field of one variable is isomorphic to the field of meromorphic functions of a suitable Riemann surface.*

Proof. Let $K \supset \mathbb{C}$ be an algebraic function field of one variable, let $f \in K$ be a nonconstant element such that K is algebraic over $\mathbb{C}(f)$, and let g be an associated primitive element, i.e.

$$K = \bigoplus_{\nu=0}^{d-1} \mathbb{C}(f) g^{\nu}.$$

As we have seen in the proof of Proposition 2.9, there then exists a polynomial $P(z, w)$ with the property

$$P(f, g) = 0.$$

The associated Riemann surface has the desired property. □

Homomorphisms of Function Fields

Let $\mathbb{C} \subset K$, $\mathbb{C} \subset L$ be two algebraic function fields of one variable. By a *homomorphism*

$$\varphi : K \longrightarrow L,$$

we understand a map with the properties

a) $\varphi(C) = C$ for $C \in \mathbb{C}$;

b) $\varphi(f + g) = \varphi(f) + \varphi(g)$, $\varphi(f \cdot g) = \varphi(f) \cdot \varphi(g)$.

Such a homomorphism is automatically injective. If φ is also surjective, then φ is an *isomorphism*. Let Y be another connected compact Riemann surface. Every nonconstant holomorphic map

$$f : X \longrightarrow Y$$

induces a homomorphism

$$\varphi = f^* : \mathcal{M}(Y) \longrightarrow \mathcal{M}(X), \qquad g \longmapsto g \circ f,$$

of the function fields. Every homomorphism is of this form.

2.12 Proposition.   *Let X, Y be two connected compact Riemann surfaces. Every homomorphism of function fields of one variable*

$$\varphi : \mathcal{M}(Y) \longrightarrow \mathcal{M}(X)$$

is induced by a unique holomorphic nonconstant map

$$h : X \longrightarrow Y,$$

which means that $\varphi = h^*$.

Corollary.   *Two connected compact Riemann surfaces are biholomorphically equivalent iff their function fields are isomorphic.*

Proof. We choose a nonconstant element $f \in \mathcal{M}(Y)$ and denote its image in $\mathcal{M}(X)$ by $\tilde{f} = \varphi(f)$. We choose primitive elements $g, \tilde{g}$;

$$\mathcal{M}(Y) = \sum_{\nu=0}^{d} \mathbb{C}(f)g^{\nu}, \quad \mathcal{M}(X) = \sum_{\nu=0}^{\tilde{d}} \mathbb{C}(\tilde{f})\tilde{g}^{\nu}.$$

We have

$$\varphi(g) = \sum_{\nu=0}^{d} R_\nu(\tilde{f})\tilde{g}^\nu,$$

with certain rational functions R_ν. We can achieve the result that they are polynomials (since $\tilde{g}$ can be multiplied by a polynomial in $\tilde{f}$). Finally, we consider irreducible polynomials P, $\tilde{P}$ in two variables such that

$$P(f,g) = 0, \ \tilde{P}(\tilde{f},\tilde{g}) = 0.$$

The assignment

$$(z,w) \longmapsto \left(z, \sum_{\nu=0}^{d} R_\nu(z)w^\nu\right)$$

defines a map $\tilde{\mathcal{N}} \to \mathcal{N}$ between the algebraic curves which are associated with $\tilde{P}$ and P. The diagram

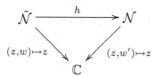

is commutative. We obtain the following result:

There exist finite point sets $T_1 \subset X$, $T_2 \subset Y$, $S \subset \bar{\mathbb{C}}$ and a holomorphic map

$$h : X - T_1 \longrightarrow Y - T_2$$

such that the diagram

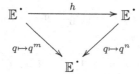

commutes, and such that $f, \tilde{f}$ are proper.

It remains to show that h extends holomorphically to the whole of X. By means of the classification of ramification points, this can be reduced to the following local situation.

Let $h : \mathbb{E}^{\cdot} \longrightarrow \mathbb{E}^{\cdot}$ be a holomorphic function such that the diagram commutes. Then h has a removable singularity at $q = 0$ and we have $h(0) = 0$. This can easily be shown by means of the Riemann removability theorem or by simple calculation with the Laurent series of h.   □

Propositions 2.11 and 2.12 mean roughly the following:

The theory of compact Riemann surfaces and the theory of function fields of one variable are equivalent.

Nonalgebraic Construction of Compact Riemann Surfaces

Let X be a Riemann surface and let $U \subset X$, $V \subset X$ be two open disjoint subsets together with a biholomorphic map

$$\varphi : U \xrightarrow{\sim} V.$$

We want to glue the two sets together inside X "via" φ. For this purpose, we introduce the following equivalence relation:

Two points $x, y \in X$ are called equivalent iff

$$\begin{aligned} either \quad & x = y \\ or \quad & x \in U \quad and \quad y = \varphi(x) \\ or \quad & y \in U \quad and \quad x = \varphi(y). \end{aligned}$$

We denote the quotient space by this equivalence relation by

$$Y := X / \sim .$$

Obviously, the natural map

$$X \longrightarrow Y$$

is locally topological.

We make the assumption that Y is Hausdorff. Then Y carries a structure in the form of a Riemann surface, such that the map $X \to Y$ is locally biholomorphic. The *proof* is trivial.

(By the way, this claim shows the importance of the Hausdorff property for the theory of Riemann surfaces. If we where to abandon this requirement, we would be able to construct very badly behaved objects. For example, we could take for X the disjoint union of two copies of the complex plane and glue them together along the punctured planes. The result would be a complex plane with a doubled origin.)

Example. We consider, on the Riemann surface X, two disjoint analytic charts

$$\psi : U \longrightarrow V, \quad \psi' : U' \longrightarrow V', \quad U, U' \subset X,$$

where V and V' are assumed to contain disks of radius 2 around 0. Then we punch two holes in the surface X:

$$X' = X - \{x \in U; \quad |\psi(x)| > 1/2\} - \{y \in U'; \quad |\psi'(y)| \le 1/2\}.$$

Now we consider the „annuli" in V and V'

$$\frac{1}{2} < |z| < 2.$$

We denote the complements of their inverse images under ψ, ψ' by

$$R \subset U,\ R' \subset U'.$$

Now we want to glue R and R'. In a first approach, we might think of taking $\psi^{-1} \circ \psi'$ as the gluing function which is induced by the identity on $\mathbb{E}$. But the result would not be Hausdorff. The situation changes if we interchange the roles of the inner and outer boundaries of R and R'. This happens if we take as the gluing function

$$\varphi : R \xrightarrow{\sim} R',$$
$$\varphi(x) = \psi'^{-1}\left(\psi(x)^{-1}\right).$$

This is a biholomorphic map, and the quotient space Y is now Hausdorff, as the reader may show.

Intuitively, we have punched two holes in X and connected them by a handle:

For example, we could take the Riemann sphere for X. The result is then a sphere with a handle, and therefore an object which is topologically equivalent to a torus. We might conjecture that Y is biholomorphically equivalent to a torus $\mathbb{C}/L$. As we shall see later, this is actually true. (One also could use the uniformization theorem to obtain a less obvious proof of this.)

One could ask the question of how the j-invariant could be computed. The answer to this obvious question is unknown.

Anyhow, these nonalgebraic constructions show that the theorem that any compact Riemann surface comes from an algebraic function is highly nontrivial. So, we should not be surprised that its proof uses fundamental potential-theoretic existence theorems.

Exercises for Sect. IV.2

1. Let f be a holomorphic function on the punctured unit disk $\mathbb{E} - \{0\}$. Assume that the function $z^{n-1}f(z^n)$ has a removable singularity at the origin for some natural number n. Show that f also has a removable singularity at the origin.

2. Consider, for a natural number n, the map
$$f : \mathbb{E} \longrightarrow \mathbb{E}, \quad z \longmapsto w := z^n.$$
Show that the pullback of a differential $g(w)\,dw$ under f equals
$$nz^{n-1}g(z^n)\,dz \quad (= f^*(g(w)\,dw)).$$

3. Let $f : X \to Y$ be a surjective and proper analytic map of Riemann surfaces and let ω be a meromorphic differential on Y. Show that ω is holomorphic iff its pullback $f^*\omega$ is holomorphic (on X).

4. Let $f : X \to \bar{\mathbb{C}}$ be a nonconstant meromorphic function on a connected Riemann surface. Determine the poles and zeros of the differential df.

    Answer. Poles of df occur only at the poles of f. If a is a pole of order n of f, then a is a pole of order $n + 1$ of df.

    Zeros are located at the ramification points of f. The zero order of df equals the ramification order minus one.

3. The Triangulation of a Compact Riemann Surface

A well-known result of topology states that any surface with a countable basis of its topology can be triangulated. The proof of this theorem is not simple. For Riemann surfaces, it is simpler but still difficult enough. In Sect. 2 we have shown that *compact Riemann surfaces* can be represented as ramified coverings of the sphere. Using this fact, we shall derive a very simple proof of the existence of a triangulation of a compact Riemann surface.

Polyhedra

We choose a standard triangle in the plane; to be concrete, we take the convex hull of the points 0, 1, i,

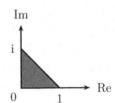

$$\Delta = \{z \in \mathbb{C}; \quad y \geq 0,\ x \geq 0;\ x + y \leq 1\}.$$

In this section, we denote the segment between two *complex* numbers a, b by

$$[a, b] := \{z; \quad z = a + t(b - a),\ 0 \leq t \leq 1\}.$$

The three points 0, 1, i are called the *vertices*, and the segments

$$[0,1], \quad [1,i] \text{ and } [i,0]$$

are called the *edges* of the standard triangle Δ.

3.1 Definition.   *A **triangle** φ in a topological space X is a topological map φ from Δ onto a (compact) subset $\Delta^\varphi \subset X$,*

$$\varphi : \Delta \xrightarrow{\sim} \Delta^\varphi \subset X.$$

1) We call Δ^φ the triangular area which underlies φ.
2) The images of the edges of Δ are called the edges of φ.
3) The images of the vertices of Δ are called the vertices of φ.

So triangular areas, edges, and vertices are point sets, but the triangle itself is a map.

3.2 Definition.   *A **(finite) polyhedron** is a pair $(X, \mathcal{M})$ consisting of a topological space X and a finite set $\mathcal{M}$ of triangles in X with the following properties:*

1) *We have*

$$X = \bigcup_{\varphi \in \mathcal{M}} \Delta^\varphi.$$

2) *Let $\varphi \neq \psi$ be two different triangles in $\mathcal{M}$. Then there are three possibilities for the intersection $\Delta^\varphi \cap \Delta^\psi$:*

   a) *It is empty.*
   b) *It consists of one joint vertex.*
   c) *It consists of one joint edge.*

3) *Three pairwise different triangles cannot share a joint edge.*

A simple example of a polyhedron is obtained from a closed disk by dividing it into $k \geq 3$ regular sectors.

3.3 Definition.   *Let $(X, \mathcal{M})$ be a polyhedron. We call $\mathcal{M}$ a **triangulation** of X if each vertex of $\mathcal{M}$ belongs to two triangles of $\mathcal{M}$.*

It is not very difficult to show that a topological space which admits a triangulation is a (compact) surface. We do not need this; our aim is to show the converse, namely that any compact Riemann surface admits a triangulation.

3.4 Proposition. *Any compact Riemann surface X can be triangulated.*

Supplement. *Let $S \subset X$ be a finite subset. It is possible to construct a triangulation with the following properties:*

1) *Each point of S is a vertex of the triangulation.*
2) *For each point $s \in S$, there exists a disk $U \xrightarrow{\sim} \mathbb{E}$ such that the triangular areas with vertex s are in one-to-one correspondence with the k sectors ($k \geq 3$ is suitable) of the disk of radius $1/2$ in $\mathbb{E}$ (compare the above example).*

First of all, it is rather clear that the proposition is true for the sphere. For example, we can start with a regular tetrahedron decomposition:

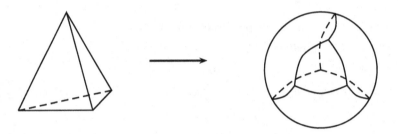

The edges of the triangles involved can be assumed to be segments of large circles. In the first step, we arrange a tetrahedron triangulation such that the points of S are contained in the interiors of the triangles (i.e. they are not contained in an edge). Then we take a small "disk-like" regular triangle around each point of S which is completely contained in the interior of the tetrahedron triangle containing s. Then we integrate this small triangle into the tetrahedron triangulation in some way, for example as indicated in the figure below. (So, we can get $k = 3$ in the case of the sphere.)

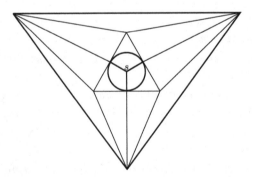

Now we come to the case of a general Riemann surface. Let $f : Y \to X$ be a nonconstant holomorphic map between compact Riemann surfaces. We assume that the proposition has been proved for X. We then prove it for Y. We choose a triangulation of X with the properties of the supplement to

Proposition 3.4, where S is the set of the downstairs ramification points. Now, on each edge, we choose an inner point (not necessarily the midpoint) and refine the triangulation as indicated in the following figure:

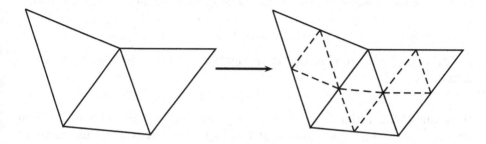

By applying this refinement construction several times, we obtain the result that the disk described in Proposition 3.4 is so small that its inverse image is a disjoint union of disks and f in each of these disks is of the form $q \mapsto q^d$. We can also assume that each triangle of the triangulation is contained in a disk which is small in this sense.

Now we can lift the triangulation of X to Y in an obvious way. A triangle

$$\varphi : \Delta \longrightarrow Y$$

belongs to the triangulation which has to be constructed iff the composition

$$f \circ \varphi : \Delta \longrightarrow X$$

belongs to the given triangulation of X.

It should be rather clear that this gives a triangulation of Y. We call this the "lifted triangulation". The proof needs only the following small consideration for the "ramification elements"

$$f : \mathbb{E} \longrightarrow \mathbb{E}, \quad q \longmapsto q^d.$$

Assume that the unit disk $\mathbb{E}$ downstairs has been decomposed into k sectors $\Delta_1, \ldots, \Delta_k$ by means of the roots of unity of order k. If, correspondingly, the unit disk upstairs is divided into kd sectors, then f maps each of these sectors topologically onto one of the sectors downstairs:

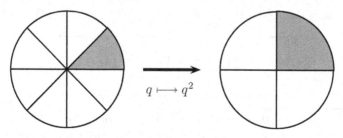

Appendix to 3. The Riemann–Hurwitz Ramification Formula

Let $(X, \mathcal{M})$ be a triangulated surface, let E be the number of vertices, let K be the number of edges, and let D be the number of triangles. The alternating sum

$$e := E - K + D$$

is called the Euler number of the triangulation. The topological genus p is defined by

$$e = 2p - 2.$$

We shall see later that p is a nonnegative integer for triangulated Riemann surfaces. A well-known theorem of topology states that the Euler number (and hence the genus) is a topological invariant. Different triangulations of a given surface lead to the same Euler number (and the same genus). Later we shall see by means of the Riemann–Roch theorem that homeomorphic compact Riemann surfaces always have the same Euler number, which is independent of the chosen triangulations. We will see also the converse:

Two compact Riemann surfaces are homeomorphic if and only if their Euler numbers agree.

For this reason, it is important to have simple methods for the computation of the Euler number. For this purpose we establish the *Riemann–Hurwitz ramification formula*.

Let $f : X \to Y$ be a holomorphic nonconstant map between (connected) compact Riemann surfaces. Of particular interest is the case $Y = \bar{\mathbb{C}}$. Let $S \subset Y$ be the set of downstairs ramification points of f and let $T := f^{-1}(S)$ be its inverse image in X. We recall that the map f is given in terms of suitable analytic charts around $t \in T$ and $f(t) \in S$ by the formula $z \mapsto z^n$, with a suitable number $n = n_t$. This number n_t is called the ramification order of f at t. In our normalization, it is 1 iff f is unramified at t.

3.5 Theorem. The Riemann–Hurwitz ramification formula. *Let $f : X \to Y$ be a holomorphic map of degree n between connected compact Riemann surfaces, let $S \subset Y$ be the set of downstairs ramification points of f, and let $T := f^{-1}(S)$ be the set of their preimages in X. We have (at least for suitable triangulations**) *the result that the genera $p(X)$, $p(Y)$ satisfy the relation*

$$p(X) = n\,(p(Y) - 1) + 1 + \frac{1}{2} \sum_{t \in T} (n_t - 1).$$

---

*) The condition in parantheses "at least for suitable triangulations", which occurs several times in what follows, can be deleted as soon as we have proved the invariance of the genus of the choice of a triangulation.

In the special case of the sphere $Y = \bar{\mathbb{C}}$, *we can start with a triangulation with* $p(Y) = 0$.

Corollary. *The number* $\sum_{t \in T}(n_t - 1)$ *is even if* $p(X)$ *and* $p(Y)$ *are integral (which is always true).*

Proof. We start with a triangulation as in the proof of Proposition 3.4 and lift it, as described there, to a triangulation of X. If the triangulation of Y has

$$E \text{ vertices, } K \text{ edges, and } D \text{ triangles,}$$

then the lifted triangulation obviously has nK edges and nD triangles, but the number of vertices is not nE; owing to the ramification, this number decreases to

$$nE - \sum_{t \in T}(n_t - 1).$$

We obtain

$$2 - 2p(Y) = E - K + D,$$

$$2 - 2p(X) = nE - nK + nD - \sum_{t \in T}(n_t - 1).$$

This shows the ramification formula.

In the case $Y = \bar{\mathbb{C}}$, the tetrahedron triangulation leads to $e = 2$, and hence $p(Y) = 0$. The refinements which we have used obviously do not change $p(Y)$. Hence it is clear that we can start with a triangulation with $p(Y) = 0$.   □

Again we point out that, by Euler's polyhedron theorem, $p(\bar{\mathbb{C}}) = 0$ holds for *every* triangulation.

Example. Let

$$P(z) = a_n z^n + \ldots + a_0$$

be a polynomial without multiple zeros. We consider the Riemann surface of the algebraic function

$$w^2 = P(z).$$

From the construction of the associated compact Riemann surface $X \to \bar{\mathbb{C}}$, we can deduce:

1) The zero set S of the polynomial P is the set of finite ramification points.
2) ∞ is a ramification point iff m is odd. (This would follow also from the corollary to Theorem 3.5, since the ramification order can only be one or two.)

Every ramification point has exactly one preimage, and the ramification order is two in each case. The total degree is two. Since the genus of the sphere is 0 (at least for suitable triangulations), we obtain for the genus p of X (at least for suitable triangulations)

$$p = 2(0 - 1) + 1 + \begin{cases} n/2, & \text{if } n \text{ even} \\ (n+1)/2 & \text{if } n \text{ odd.} \end{cases}$$

So we obtain the following result.

3.6 Theorem. *The genus p of the compact Riemann surface associated with the algebraic function*

$$w^2 = P(z),$$

where $P(z)$ is a polynomial of degree n without multiple zeros, is (at least for suitable triangulations)

$$p = \begin{cases} (n-2)/2 & \text{if } n \text{ is even,} \\ (n-1)/2 & \text{if } n \text{ is odd.} \end{cases}$$

For example, if the degree of P is 3 or 4, we obtain $p = 1$. As we shall see later, this means that the surface is topologically a torus. This fits the theory of elliptic functions in [FB], Chap. V, where we have shown that the Weierstrass $\wp$-function defines a bijective map from a torus onto a certain projective curve which is related to a polynomial of degree 3.

Exercises for Sect. IV.3

1. Show that for every integer $p \geq 0$ there exists a (triangulated) compact Riemann surface of genus p.

2. Construct, for a torus $X = \mathbb{C}/L$, triangulations with $p = 1$ in two different ways:
   a) Construct the triangulation directly geometrically, starting with a fundamental parallelogram.
   b) Consider the Weierstrass $\wp$-function $\wp : X \to \bar{\mathbb{C}}$ and study its ramification behavior.

3. A Riemann surface X is called *hyperelliptic* if there exists a meromorphic function $f : X \to \bar{\mathbb{C}}$ of degree 2. Show that the following statements are equivalent:
   a) X is hyperelliptic.
   b) The field K of meromorphic functions on X can be written as an extension of degree 2 of a rational function field, $K \supset \mathbb{C}(z)$.
   c) There exists a polynomial without multiple roots, such that X is biholomorphically equivalent to the Riemann surface of $w^2 = P(z)$.

4. Show that the genus of the compact Riemann surface associated with $w^n + z^n = 1$ is $(n-1)(n-2)/2$. (These surfaces are called "Fermat curves".)

4. Combinatorial Schemes

The topological nature of the space underlying a polyhedron is given by finitely many items of combinatorial data which describe how the triangles are located with respect to each other. We want to give a formal description of this.

Let $\mathcal{P}$ be a *finite set*. We introduce the following terminology:

1) A *vertex* in $\mathcal{P}$ is an element $a \in \mathcal{P}$.

2) An *edge* K in $\mathcal{P}$ is a subset of two elements from $\mathcal{P}$.

3) A *triangle* D in $\mathcal{P}$ is a subset consisting of three elements from $\mathcal{P}$.

Let D be a triangle in $\mathcal{P}$. Then every subset consisting of two elements is called a vertex of D. So each triangle has three edges. If a is an element of a triangle D or of a vertex K, then we call a a vertex of D or of K. So each triangle has three vertices and each edge has two vertices.

4.1 Definition. *A **combinatorial scheme** $\mathcal{S}$ is a pair*

$$\mathcal{S} = (\mathcal{P}, \mathcal{D})$$

consisting of a finite set $\mathcal{P}$ and a set of triangles $\mathcal{D}$ in $\mathcal{P}$, such that the following conditions are satisfied:

1) *Each point $P \in \mathcal{P}$ is a vertex of at least one triangle in $\mathcal{D}$:*

$$\mathcal{P} = \bigcup_{D \in \mathcal{D}} D.$$

2) *There are at most two triangles in $\mathcal{D}$ which share a given edge.*

Orientation

Let I be a finite set which contains at least two elements. An ordering of I is a bijective map

$$\alpha : \{1, \ldots, n\} \xrightarrow{\sim} I.$$

Two orderings

$$\alpha, \beta : \{1, \ldots, n\} \xrightarrow{\sim} I$$

are said to be *orientation-equal* if the permutation

$$\beta^{-1} \circ \alpha : \{1, \ldots, n\} \longrightarrow \{1, \ldots, n\}$$

is even.

An *orientation* of I is a full class of orientation equal orderings.

Since the group of even permutations (the alternating group) has index 2 in the full group of permutations ($n \geq 2$), I admits exactly two different

orientations. Here we need the concept of orientation only for sets of two or
three elements. In these cases we can avoid permutation groups and take the
following as the definition:

1) An orientation of a set of two elements is an ordering of this set as an
   ordered pair.
2) A set $\{a, b, c\}$ of three elements has the following two orientations:

$$[a, b, c] := \{ (a, b, c);\ (b, c, a);\ (c, a, b) \},$$
$$[b, a, c] := \{ (b, a, c);\ (a, c, b);\ (c, b, a) \}.$$

Let $J \subset I$ be a subset which is obtained by removing one element of I. So, we
have $\#J = n - 1$. Now we assume $n \geq 3$. It is possible to restrict an ordering
of I to J in an obvious way. Since we need this only in the case $n = 3$, we
define it in this case directly as follows. We consider the orientation $[a, b, c]$ on
the three-element set $\{a, b, c\}$. The orientations on the two-element subsets are
given by the ordered pairs (a, b), (b, c), (c, a).

In the following, we shall consider orientations on the set
of three vertices of a triangle. Intuitively, we think of this
as a direction running around the triangle.

4.2 Definition. *An **orientation** of a combinatorial scheme $(\mathcal{P}, \mathcal{D})$ is a
map that assigns to each triangle $D \in \mathcal{D}$ an orientation such that the following
condition is satisfied:*

*If K is an edge in $\mathcal{P}$ which belongs to two different triangles
D, D', then D and D' induce the two different orientations
on K.*

Example. Let $n \geq 3$ be a natural number. Consider $n + 1$ points

$$0, P_1, \ldots, P_n.$$

The n subsets $\{0, P_\nu, P_{\nu+1}\}$ $(1 \leq \nu \leq n)$ $(P_{n+1} := P_1)$ are assumed to be the
triangles. We obtain a combinatorial scheme, called the *n-gon*.

If we equip each of the n triangles of the n-gon with
the orientation

$$[0, P_\nu, P_{\nu+1}] \qquad (1 \leq \nu \leq n;\ P_{n+1} := P_1),$$

we get an orientation of the n-gon.

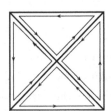

With each polyhedron $(X, \mathcal{M})$ there is associated a combinatorial scheme

$$\mathcal{S} = \mathcal{S}(X, \mathcal{M}) = (\mathcal{P}, \mathcal{D})$$

in a natural way. The points in $\mathcal{S}$ are the vertices of triangles from $\mathcal{M}$. Each triangle $\varphi \in \mathcal{M}$ induces a (combinatorial) triangle $D = D(\varphi)$ in $\mathcal{P}$, namely the set of the three vertices of φ. By definition, $\mathcal{D}$ is the set of triangles which are obtained in this way. So, we have a surjective map

$$\mathcal{M} \longrightarrow \mathcal{D}, \quad \varphi \longmapsto D(\varphi).$$

Since two different triangles in $\mathcal{M}$ cannot share three vertices, this map is injective, and hence bijective.

4.3 Definition. *An **orientation of a polyhedron** is an orientation of the associated combinatorial scheme.*

Intuitively, this means that each triangle obtains a direction that indicates how it has to be surrounded. In particular, each edge gets a direction. Triangles with a joint edge induce the two different directions on the edge.

4.4 Proposition. *Any compact Riemann surface admits an oriented triangulation.*

Proof. The tetrahedron triangulation of the sphere is orientable. The refinement and lifting constructions which we used to construct a triangulation all preserve orientations.   □

By the way, a better result is true: if a surface admits one orientable triangulation, then all triangulations are orientable (compare Exercise 4).

In what follows, we shall consider only oriented combinatorial scheme and oriented polyhedra. For combinatorial schemes, there is a natural notion of isomorphism.

4.5 Definition. *An **isomorphism between (oriented) combinatorial schemes***

$$f : (\mathcal{P}, \mathcal{D}) \overset{\sim}{\longrightarrow} (\mathcal{P}', \mathcal{D}')$$

is a bijective map $f : \mathcal{P} \longrightarrow \mathcal{P}'$ such that f and f^{-1} map triangles onto triangles, such that their orientation is preserved..

Similarly, there is a notion of isomorphism of polyhedra.

4.6 Definition. *An isomorphism*

$$F : (X, \mathcal{M}) \longrightarrow (X', \mathcal{M}')$$

of (oriented) polyhedra is a topological map

$$F : X \longrightarrow X'$$

such that F and F^{-1} map triangular areas onto triangular areas, edges onto edges, and vertices to vertices. The orientation of the three vertices of a triangle has to be preserved.

An isomorphism

$$F : (X, \mathcal{M}) \longrightarrow (X', \mathcal{M}')$$

of oriented polyhedra induces in a natural way an isomorphism of the associated combinatorial schemes,

$$f : \mathcal{S}(X, \mathcal{M}) \longrightarrow \mathcal{S}(X', \mathcal{M}').$$

We need a reverse result.

4.7 Proposition. *Let $(X, \mathcal{M})$, $(X', \mathcal{M}')$ be two oriented polyhedra and let $(\mathcal{P}, \mathcal{D})$, $(\mathcal{P}', \mathcal{D}')$ be the associated combinatorial schemes. Each isomorphism*

$$f : (\mathcal{P}, \mathcal{D}) \longrightarrow (\mathcal{P}', \mathcal{D}')$$

is induced by an isomorphism of the polyhedra

$$F : (X, \mathcal{M}) \longrightarrow (X', \mathcal{M}').$$

In particular, X and X' are homeomorphic when the "combinatorial data agree" (in the sense that the associated combinatorial schemes are isomorphic).

Proof. We take the triangular surfaces of the polyhedron $(X, \mathcal{M})$ in an arbitrary order $\Delta_1, \ldots, \Delta_n$, and denote the corresponding triangular surfaces $(X', \mathcal{M}')$ by $\Delta'_1, \ldots, \Delta'_n$.

Claim. There exist topological maps

$$F_j : \Delta_j \longrightarrow \Delta'_j, \quad 1 \le i \le n,$$

with the following two properties:

1) Vertices and edges of Δ_j are mapped under f onto the corresponding vertices and edges of Δ'_j. The orientation of the three vertices is preserved.

2) If Δ_j and a "precursor" Δ_i, $i < j$, have a nonempty intersection, then F_j and F_i agree on this intersection.

Proof of the claim. The maps $F_1, \ldots, F_n$ are constructed inductively. We assume that $F_1, \ldots, F_i$, $i < j$, have already been constructed, such that the properties 1) and 2) are correspondingly satisfied. We then have to construct a topological map $F_j : \Delta_j \to \Delta'_j$, such that it coincides on certain vertices and edges with given topological maps. (In the worst case, all triangular areas which have a nonempty intersection with Δ_j are precursors.) The existence of the map F_j follows easily from the following statement about the standard triangle.

Let h be a topological map of the boundary of the standard triangle onto itself which permutes the vertices. Then h extends to a topological map from the whole standard triangle onto itself.

Proof. First of all, it is easy to see that there exists a homeomorphism of Δ which maps the boundary onto itself and induces a prescribed permutation of the vertices. For example, a reflection along the diagonal permutes the vertices 1 and i. Hence we can assume, without loss of generality, that the three vertices are fixed. We can also assume that the map h is the identity on two of the three edges, since we can compose h with three maps of this type. Now we are in the following situation:

Let Δ be the standard triangle with vertices $0, 1, i$ and let $h : [0,1] \to [0,1]$ be a topological map which fixes 0 and 1. Then h can be extended to a topological map $H : \Delta \to \Delta$ which is the identity on the two remaining edges.

The proof is simple. We construct h in such a way that it defines an affine map from the segment $[i, t]$ for $t \in [0, 1]$ onto the segment $[i, h(t)]$.   □

Proof of Proposition 4.7. Now the maps F_i can be glued together to form a map $F : X \to X'$. It is easy to show that this map is topological (compare Exercise 5).   □

Our next goal is to show that each combinatorial scheme can be realized by a polyhedron. This polyhedron will be constructed by gluing several triangular areas together.

Gluing of Spaces

The gluing of spaces is based on the "quotient topology". Let X be a topological space and $\sim$ an equivalence relation; we can then define the quotient space (Sect. I.0.3).

We now consider a slightly more general case, where R is an arbitrary relation. This means only that R is a subset of $X \times X$. Then xRy just mean that $(x, y) \in R$. We can then consider the equivalence relation which is generated by R. This is the smallest equivalence relation which contains R, and can be defined as the intersection of all equivalence relations containing R. In the following, we denote by X/R the quotient space of X with respect to this equivalence relation. As a rule, X/R will not be Hausdorff even if X is. In our applications, the Hausdorff property of X/R will will always be clear, and we shall tacitly assume that it has been proved.

Now let X and Y be two Hausdorff spaces with two subsets $A \subset X$ and $B \subset Y$. We also assume that a topological map $f : A \to B$ is given: we assume that X and Y are disjoint, and define the topology on $X \cup Y$ in such a way that X and Y are open (and closed) subsets. Now we consider the relation

$$R := \{(a, f(a); \quad a \in A\}.$$

We can consider the quotient space $Z = (X \cup Y)/R$. We say that Z arises by gluing X and Y along A and B by means of f. We have natural maps $X \to Z$, $Y \to Z$, which are injective and continuous. If A and B are closed, then these maps are also closed (the images of closed sets are closed). We then obtain the result that the maps

$$X \longrightarrow Z, \quad Y \longrightarrow Z, \quad A \longrightarrow Z, \quad B \longrightarrow Z$$

are topological onto their images.

4.8 Proposition. *Any oriented combinatorial scheme is isomorphic to a combinatorial scheme which is associated with a polyhedron.*

Corollary. *The isomorphy classes of oriented combinatorial schemes and oriented polyhedra are in one-to-one correspondence.*

Proof. We equip the (finite) set $\mathcal{D}$ of triangles of the given combinatorial scheme with the discrete topology (every subset is open). The Cartesian product of $\mathcal{D}$ and the standard triangle Δ,

$$\mathcal{X} := \mathcal{D} \times \Delta,$$

is a compact space. It is the disjoint union of $\Delta_D := \{D\} \times \Delta$. One can think of Δ_D as a copy of the standard triangle Δ, so that $\mathcal{X}$ is the disjoint union of finitely many copies of the standard triangle. The three vertices of the standard triangle are oriented counterclockwise, i.e. $[0, 1, i]$. For each $D \in \mathcal{D}$, we choose a bijective orientation-preserving map from D onto the vertices of Δ. Using the natural bijective map $\Delta \xrightarrow{\sim} \Delta_D$, we can look at it also as a bijective map from D onto the three vertices of Δ_D (the points which correspond to $0, 1, i$). We define in $\mathcal{X}$ a certain relation R, whose purpose is to glue certain vertices and edges.

1) Let D, D' be two different triangles in $\mathcal{D}$ which share precisely one vertex. If $a \in \Delta_D$ and $b \in \Delta'_D$ are the corresponding vertices, we want to glue them. Hence we define aRb.

2) Let D, D' be two distinct triangles from $\mathcal{D}$ which share an edge. We choose a topological (for example an affine) map $f : K \to K'$ of the corresponding edges of Δ_D and Δ'_D such that the starting point and endpoint of K are mapped to the endpoint and starting point, respectively, of K'. We then also define $aRf(a)$ for $a \in K$.

It is easy to check that the quotient space

$$X := \mathcal{X}/R$$

is Hausdorff. Since it is the image of a compact space, it is compact. For each triangle $D \in \mathcal{D}$, we get a triangle

$$\varphi : \Delta \longrightarrow X,$$

namely the composition of the natural maps

$$\Delta \longrightarrow \Delta_D \longrightarrow \mathcal{X} \longrightarrow \mathcal{X}/R.$$

It should be clear that this construction gives a polyhedron with the desired properties. $\qquad\square$

Exercises for Sect. IV.4

1. let $(X, \mathcal{M})$ be a polyhedron. Show that if X is connected, then $(X, \mathcal{M})$ has either no or precisely two "opposite" orientations.

2. Let $\varphi : V \longrightarrow V'$ be a biholomorphic map between two simply connected domains of the plane, let α be a closed curve in V, and let $a \in V$ be a point which is surrounded by the curve with winding number one. Show that the image curve $\varphi \circ \alpha$ also surrounds $\varphi(a)$ with winding number $+1$.

3. Let $\varphi : \bar{\mathbb{E}} \to \mathbb{C}$ be an injective continuous map of the closed unit disk into the complex plane. The winding number of the curve
$$\alpha(t) = \varphi(\exp(2\pi i t)) \qquad (0 \le t \le 1)$$
around a point $b = \varphi(a)$, $a \in \mathbb{E}$ is ± 1 independent of a. Formulate an analogous statement for the standard triangle instead of $\bar{\mathbb{E}}$.

    Hint. The fundamental group of $(\bar{\mathbb{E}}, b)$ is generated by α. Therefore the winding number of α divides the winding number of any other curve around b.

4. Show that any triangulation of a compact Riemann surface is orientable.

    Hint. Since one can take suitable refinements, there is no loss of generality in assuming that the union of two triangles with a joint edge is contained in the domain of definition of an analytic chart. Define the orientation of a triangle in such a way that the winding numbers with respect to the analytic chart around inner points are $+1$. This possible because of the result of Exercise 3. It follows from the result of Exercise 2 that this orientation is independent of the choice of

the analytic chart. Applying the result of Exercise 2 once more, we can see that
an orientation of the triangulation is obtained.

5. Let $X = A_1 \cup \ldots \cup A_n$ be a finite closed covering of a topological space and let
   $f : X \to Y$ be a map into another topological space Y. Show that f is continuous
   iff all restrictions $f_i = f|A_i$ are continuous.

5. Gluing of Boundary Edges

A topological model of the torus can be obtained by gluing opposite sides
of a tetragon together. In a similar way, we shall construct for any connected
compact surface with an oriented triangulation a topological model which arises
from an n-gon by gluing boundary edges in a suitable way.

We start with an arbitrary oriented combinatorial scheme $\mathcal{S} = (\mathcal{P}, \mathcal{D})$.
Those edges which belong to only one triangle are called *boundary edges*. Vertices which belong to a boundary edge are called *exterior vertices*.

5.1 Definition. *A **boundary gluing** Σ of an oriented combinatorial scheme
S consists of an even number $2n$ of boundary edges, which are divided into n
unordered pairs $\{K, L\}$.*

*A boundary gluing of a polyhedron is a boundary gluing of the underlying
combinatorial scheme.*

Each of the $2n$ edges K has exactly one complementary edge L. We want to
glue complementary edges in a combinatorial scheme to obtain a combinatorial
counterpart of the gluing of vertices of polyhedra in the geometrical sense,
which is what we are interested in. We start with the latter construction.

So, let $(X, \mathcal{M})$ be a polyhedron with a boundary gluing. For each edge from
K with a complementary edge L, we choose a topological map $K \xrightarrow{\sim} L$ which
reverses the direction. Then we consider the generated equivalence relation $\sim$
and take the quotient space $X' = X/\sim$. As in the case of Proposition 4.7, we
can show the following.

5.2 Remark. *The topological space $X' = X/\sim$ which is associated with
the polyhedron $(X, \mathcal{M})$ with a boundary gluing, up to homeomorphism, does
not depend on the choice of the gluing map $K \xrightarrow{\sim} L$. In particular, a topological
space which is determined up to homeomorphism can be associated with each
combinatorial scheme with a boundary gluing (using Proposition 4.8).*

The space X' does not inherit a structure from X in the form of a polyhedron
in any case, since the composition of a triangle $\varphi \in \mathcal{M}$ with $X \to X'$ need not
to be injective. To get a better understanding of this phenomenon, we consider

the gluing on the level of combinatorial schemes. For this, we define a relation $\sim$ on the set of vertices. The relation $a \sim b$ means that a and b are both exterior vertices and that there exist complementary edges K and L such that a is the starting point (or endpoint) of K and b is the endpoint (or starting point, respectively) of L.

Example. We consider the "torus gluing" of a rectangle (gluing of opposite vertices).

Here, $\{a, b\}$ has to be glued to $\{c, d\}$ and $\{b, c\}$ to $\{a, d\}$. So we have

$$a \sim b, \quad c \sim d, \quad a \sim d, \quad b \sim c,$$

but not $a \sim c$. The relation $\sim$ is not an equivalence relation.

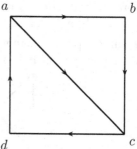

Hence we have to consider the equivalence relation which is generated by $\sim$. We denote the quotient set with respect to this equivalence relation by $\mathcal{P}' = \mathcal{P}/\Sigma$. There is a natural projection $\mathcal{P} \to \mathcal{P}'$. This means that we have identified certain exterior vertices in the set $\mathcal{P}$.

There is the obvious idea of equipping $\mathcal{P}'$ with a structure in the form of a combinatorial scheme. By definition, a triangle $D' \subset \mathcal{P}'$ should be the image of a triangle $D \subset \mathcal{P}$. But this does not always define a structure in the form of a combinatorial scheme, as the example of the torus gluing has shown, since here all four exterior vertices are identified. So, in this case, $\mathcal{P}'$ consists of one point only. The images of triangles are not triangles.

5.3 Definition.   *A boundary gluing Σ of $\mathcal{S} = (\mathcal{P}, \mathcal{D})$ is called **polyhedral** if the following conditions are satisfied:*

1) *If $D \in \mathcal{D}$ is a triangle in $\mathcal{P}$, then its image $D' \subset \mathcal{P}' = \mathcal{P}/\Sigma$ is a set of three elements.*
2) *If we denote the set of all these triangles by $\mathcal{D}'$, then the pair $\mathcal{S}' = (\mathcal{P}', \mathcal{D}')$ is a combinatorial scheme.*

A boundary gluing of a polyhedron is called polyhedral if this is the case for the underlying combinatorial scheme.

The map

$$\mathcal{D} \longrightarrow \mathcal{D}', \quad D \longmapsto D'$$

is then bijective. The images of the triangles $\varphi \in \mathcal{M}$, which means that their compositions with the canonical projection $X \to X'$ define a polyhedron $(X', \mathcal{M}')$. Obviously, its associated combinatorial scheme is isomorphic to $(\mathcal{P}', \mathcal{M}')$.

In this way, we can describe polyhedra and the gluing of boundary vertices by purely combinatorial data. This gives us the possibility of rigorous mathematical proofs of facts which, initially, are clear only intuitively. Nevertheless, for the purpose of visualization by figures, we prefer to draw polyhedra instead of finite sets. But one should bear in mind that only combinatorics plays a role.

As we have seen, boundary gluings need not be polyhedral. But this is not an essential restriction if we allow certain refinements which do not change the topological nature of the polyhedron and its quotient space. To be more precise, we allow two types of "elementary" refinements of a combinatorial scheme with a boundary gluing $(\mathcal{P}, \mathcal{D}, \Sigma)$. The case without any gluing is included ($\Sigma = \emptyset$).

The first type is that in which two adjacent refinements can be changed into four triangles as indicated in the following figure:

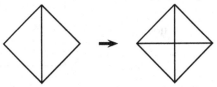

In the second type, there must be a boundary edge, drawn in bold in the following figure:

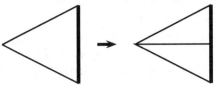

If this boundary edge belongs to Σ, the same construction has to be performed for the complementary edge.

Such an elementary refinement converts a combinatorial scheme with a boundary gluing $(\mathcal{P}_1, \mathcal{D}_1, \Sigma_1)$ into another combinatorial scheme $(\mathcal{P}_2, \mathcal{D}_2)$, which is well defined up to isomorphism. It should be clear how the gluing Σ_1 induces a gluing Σ_2. This has to be done in such a way that the quotient spaces X_1', X_2' in the sense of Remark 5.2 are homeomorphic. More generally, $(\tilde{\mathcal{P}}, \tilde{\mathcal{D}}, \tilde{\Sigma})$ is called a refinement of $(\mathcal{P}, \mathcal{D}, \Sigma)$ if there is a chain of elementary refinements

$$(\mathcal{P}, \mathcal{D}, \Sigma) \cong (\mathcal{P}_1, \mathcal{D}_1, \Sigma_1) \longmapsto \ldots \longmapsto (\mathcal{P}_n, \mathcal{D}_n, \Sigma_n) \cong (\tilde{\mathcal{P}}, \tilde{\mathcal{D}}, \tilde{\Sigma}).$$

A simple consideration shows that the following is true.

5.4 Remark.   *Any combinatorial scheme with a boundary gluing admits a polyhedral refinement.*

The same is true for polyhedra with a boundary gluing. As a consequence, the quotient space in the sense of Remark 5.2 can be equipped with a structure in the form of a polyhedron. We now illustrate by an example how boundary gluings can be described on the combinatorial level. We consider the gluing of adjacent edges in a tetragon:

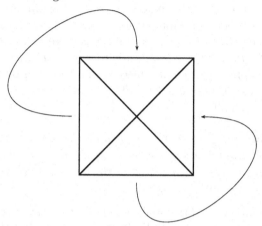

Since this is not polyhedral, we consider a modified tetragon:

Modified tetragon   *Tetrahedron from above*

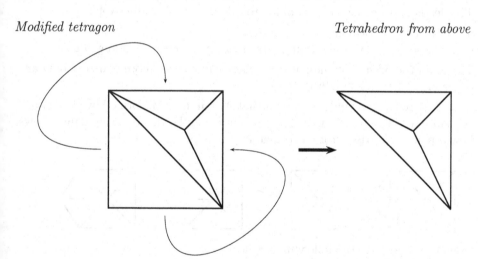

It is easy to see that the two tetragons have a common refinement. The boundary gluing of the second version is polyhedral. Its gluing is isomorphic to the tetrahedron triangulation of the sphere. Using Proposition 4.7, we see that they are homeomorphic. Hence we have obtained – just from inspection of combinatorial data – a rigorous proof of the following statement.

5.5 Remark. *If adjacent edges of a tetragon are glued together, then a space which is homeomorphic to the sphere is obtained.*

Following this pattern, we can obtain a topological classification of compact Riemann surfaces. This will be done in the next section.

6. The Normal Form of Compact Riemann Surfaces

All of the surfaces are assumed to be connected. In the middle of this section we shall describe a very simple combinatorial scheme, namely the n-gon, $n \geq 3$, with external vertices $P_1, P_2, \ldots, P_n$ and an inner vertex P_0 at the center. The triangles are $[P_0, P_i, P_{i+1}]$, $1 \leq i \leq n$, where we set $P_{n+1} := P_1$. We have already seen that a sphere can be obtained from a tetragon by gluing adjacent edges, and a torus can be obtained by gluing opposite edges. Now we shall show a more general result.

6.1 Lemma. *Any compact surface with an oriented triangulation, in particular any compact Riemann surface, is homeomorphic to the quotient of an n-gon by a boundary gluing, such that all exterior edges belong to this gluing. As a consequence, n has to be even.*

Proof. We consider an oriented triangulation of the given surface X. First of all, we make use of the connectedness of X. It is easy to see that the triangles can be ordered in such a way that each triangle shares an edge with its precursor. This leads us to construct a combinatorial scheme $\mathcal{X}_n$ inductively:

a) $\mathcal{X}_1$ is a single triangle.
b) $\mathcal{X}_i$ arises from $\mathcal{X}_{i-1}$ by gluing a triangle to the corresponding edge.

To obtain the original combinatorial scheme, of course, more pairs of boundary edges in $\mathcal{X}_n$ have to be glued.

It easy to show by induction on n that $\mathcal{X}_n$ can be transformed by elementary transformations into a standard n-gon. For example, the following figure shows how a pentagon arises from a tetragon by gluing on a triangle.

Now it is clear that the following is true:

If X is a surface with an oriented triangulation, then there exists an even natural number n and a boundary gluing Σ of a standard n-gon such that all exterior edges participate, and such that the quotient space $X(\Sigma)$ and X are homeomorphic.

We make the following assumption.

Assumption. *The number n has been chosen to be minimal for this property. We also exclude the case of a tetragon ($n = 4$) in which two adjacent vertices are glued. (This case leads to the sphere, as in Remark 5.5).*

Claim. *Under this assumption, it cannot happen that two adjacent exterior edges are glued.*

Proof. We assume that two adjacent edges in the n-gon have to be glued. We then have $n \geq 6$. We can show that X is then homeomorphic to a quotient of an $(n-2)$-gon. It is sufficient to illustrate this for a hexagon. In the following figure, the adjacent vertices which have to be glued together are drawn in bold.

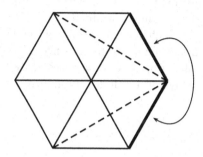

The two auxiliary dashed lines produce a refinement of the standard hexagon such that the gluing of the adjacent edges becomes polyhedral. The following figure shows the deformed hexagon, which indicates that the quotient becomes a tetragon.

Hexagon   *(refined, 10 triangles)*   *Tetragon*

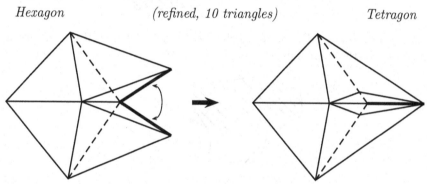

A simple transitivity property states that the total boundary gluing can be performed in two steps. We first glue two adjacent edges and then the rest. The gluing of the remaining edges of the hexagon induces a boundary gluing of the tetragon. The proof of the claim should be clear now.   □

A boundary gluing Σ of the n-gon $\mathcal{P}_n$ induces an equivalence relation on the set of the n exterior vertices. Two vertices are equivalent iff they define the same point in the quotient space X. For example, for the torus gluing of a tetragon, all four exterior vertices are equivalent.

Usually, the exterior vertices decompose into several equivalence classes. It is our next goal to construct a new boundary gluing Σ' of $\mathcal{P}_n$ such that the quotient spaces $X(\Sigma)$ on $X(\Sigma')$ are homeomorphic and that, for the new gluing, all exterior vertices become equivalent. For this, we assume that all exterior vertices are not yet equivalent. We choose an equivalence class of vertices with a maximal number m of elements. By assumption, $m < n$. The following construction will increase m. So, iterative application leads to a boundary gluing where all exterior vertices are equivalent.

In the maximal equivalence class $\mathcal{A}$, there exists a vertex a such that at least one of its two neighboring vertices is not contained in $\mathcal{A}$. Otherwise, all exterior vertices would be equivalent. So, there exists a boundary edge K which contains a but its other vertex b is not contained in $\mathcal{A}$. The vertex b belongs to another boundary edge L, which has, besides b, a vertex c. The edge L' which is complementary to L is different from K, since adjacent edges are not allowed to be glued. We cut the triangle with vertices a, b, c from the n-gon to obtain an $(n-1)$-gon as indicated in the following figure:

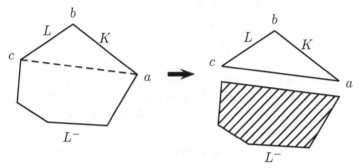

Now we glue the cut-off triangle to the $(n-1)$-gon by identifying L and L^-, as indicated in the next figure:

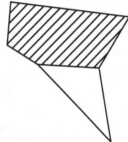

Again we obtain an n-gon with an obvious boundary gluing Σ' such that $X(\Sigma_n)$ and $X(\Sigma'_n)$ are homeomorphic. (One can argue on a topological and combinatorial level.) The equivalence class of a contains $m+1$ elements.

Our considerations so far show:

Every compact surface with an oriented triangulation which is not homeomorphic to a sphere is homeomorphic to the quotient space of an n-gon with respect to a boundary gluing Σ, such that adjacent edges are never glued together and such that all exterior vertices are glued to one point.

In the following constructions, these properties will be preserved.

Let K and K^- be two complementary edges.

Claim. There is a second pair L, L^- of complementary edges such that the edges K, L, K^-, L^- are oriented counterclockwise.

Proof of the claim. We consider one boundary edge L such that K, L, K' are oriented counterclockwise. Since K and K^- are not adjacent, such an edge exists. If, for all such L, the complementary edges L^- were such that K, L^-, K^- were also oriented counterclockwise, then the number of equivalence classes of exterior vertices would be greater than one. But this is not the case.   □

For the rest of the presentation, we need two types of constructions which will not change the topological nature of the quotient space. These can be described in the language of combinatorial schemes. We are satisfied to illustrate them with figures.

First construction. We want to arrange that there exists a counterclockwise ordered quadruple K, L, K^-, L^- such that K, L, K^- are then adjacent.

For this, we cut the polyhedron along a segment which lies between the endpoint of L and the starting point of L^-. We obtain two cut boundaries K_1, K_1^- (drawn in bold in the figure below). Now we glue K and K^- together. Instead of K, L, K^-, L^-, we now have the four boundary vertices L, K_1, L^-, K_1^-. These are oriented counterclockwise and the three K, L, K^- are adjacent. This is what we wanted to have.

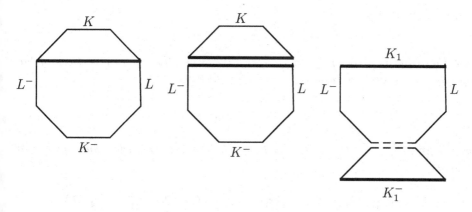

Second construction. We want to produce a quadruple K, L, K^-, L^- which is oriented counterclockwise and is such that all four K, L, K^-, L^- are then adjacent.

For this we start with two pairs of complementary edges such that K, L, K^-, L^- are oriented counterclockwise and that K, L, K^- are adjacent. Then we cut the polyhedron along the segment between the endpoint of L^- and the starting point of L. After that, we glue K and K^- together. If we denote the new cut edges by S and S^- in the correct ordering, then the edges L^-, S, L, S^- are oriented counterclockwise and they are adjacent. This is what we wanted to have.

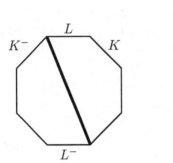 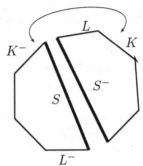

The two constructions together show that there exist two pairs of complementary edges K, L, K^-, L^- which are oriented counterclockwise and are
adjacent. If there are still boundary edges in the complement $(n > 4)$, then we
can find, by the same construction, a second quadruple of the prescribed type.
By repeating this construction, we finally obtain a situation where the set of
all boundary edges is the disjoint union of such quadruples. These quadruples
can be ordered counterclockwise, and we are led to the so-called *normal form.*

$\square$

The Normal Form

6.2 Definition. *Let $n = 4p$, $p \in \mathbb{N}$. We order the exterior vertices of the
$4p$-gon $\mathcal{P}_{4p}$ in their natural order (counterclockwise),*

$$P_1, \ldots, P_{4p},$$

*and define $P_{4p+1} := P_1$. We then divide the $4p$ boundary edges into groups of
four as follows:*

$$K_1 = (P_1, P_2), \ L_1 = (P_2, P_3), \quad K_1^- = (P_3, P_4), \ L_1^- = (P_4, P_5),$$

etc.; in general,

$$K_i = (P_{4i-3}, P_{4i-2}), \ L_i = (P_{4i-2}, P_{4i-1}),$$
$$K_i^- = (P_{4i-1}, P_{4i}), \ L_i^- = (P_{4i}, P_{4i+1}).$$

We consider the boundary gluing

$$K_i \leftrightarrow K_i^-, \qquad L_i \leftrightarrow L_i^-.$$

*The $4p$-gon, together with this boundary gluing, is called the **normal form**.*

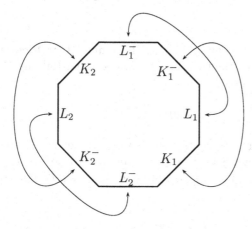

In the case $p = 1$ we obtain a torus. In general, we should think of a surface with handles, where each block K_i, L_i, K_i^-, L_i^- produces one handle. We illustrate this below for an octagon, which we draw in a modified form. The figure visualizes that the quotient space consists of two tori which have been linked by a tunnel.

Octagon   *Quotient space*

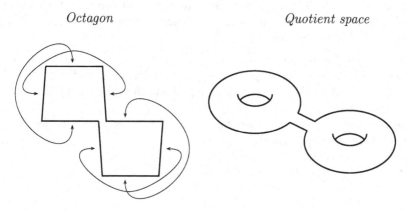

In this way, we obtain the *topological classification* of compact Riemann surfaces (more generally a classification of compact topological surfaces which admit an oriented triangulation).

6.3 Theorem. *A compact Riemann surface is homeomorphic either to a sphere or to the quotient space of a $4p$-gon $(p > 0)$ with respect to the normal form (Definition 6.2).*

We shall see later (Appendix to Sect. 7) that the number p is uniquely determined. It is called the *topological genus of the surface*. In the case of the sphere, we additionally set $p = 0$.

The Canonical Curves

We denote by $\mathcal{X}(p)$ the regular $4p$-gon in the plane. Our Riemann surface is homeomorphic to the quotient of $\mathcal{X}(p)$ by the standard gluing (the normal form). We have a natural projection

$$\pi : \mathcal{X}(p) \longrightarrow X.$$

The edges K_i of $\mathcal{X}(p)$ are segments, which we parametrize in their natural orientation (counterclockwise) by the unit interval. To be concrete, we take the affine parametrization. The image curves in X are closed curves,

$$\alpha_i : [0,1] \longrightarrow X,$$

and, correspondingly, the image curves of the edges L_i are closed curves,

$$\beta_i : [0,1] \longrightarrow X.$$

All of these $2p$ curves have the same starting point and endpoint q, namely the point which arises from gluing the $4p$ vertices together. Two different curves of the system $\alpha_1, \ldots, \alpha_p,\ \beta_1, \ldots, \beta_p$ have no common point other than this single point. We call this system of curves the *canonical system* (with respect to the representation of X in the normal form).

Let $R \subset X$ be the union of the images of these $2p$ curves. This is a compact subset. The interior of $\mathcal{X}(p)$ is mapped topologically under π onto the complement $X_0 = X - R$. As a consequence, X_0 is simply connected.

6.4 Proposition. *The fundamental group $\pi(X, q)$ is generated by the $2p$ curves of the canonical system*

$$\alpha_1, \ldots, \alpha_p,\ \beta_1, \ldots, \beta_p.$$

Proof. Let α be a closed curve in X which starts and ends at q.

Every point $x \in X$ has a suitable small simply connected neighborhood U which can be obtained – depending on its three possible positions – as the image of the shaded areas in the following figure:

$x \notin R$ $\qquad\qquad\qquad$ $x \in R,\ x \neq q$ $\qquad\qquad\qquad$ $x = q$

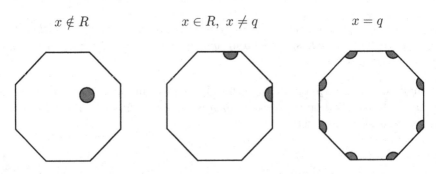

The union of these shaded areas is, in each of the three cases, the full inverse image of $U \subset X$. The topological nature of U is described in the following figure. (This could be performed in a mathematically rigorous way using the language of combinatorial schemes.)

The segments drawn here describe $U \cap R$ in each of the three cases. This figure makes it clear that any two points in U can be combined inside U by a curve which meets R only for finitely many parameter values. Now the proof of Proposition 6.4 can be given in two steps:

First step. There exists a curve β which is homotopic to α and is such that $\beta(t)$ is contained in R only for finitely many t.

We choose a partition

$$0 \le a_0 < a_1 < \ldots < a_n = 1$$

such that each piece $\alpha[a_{i-1}, a_i]$ $(1 \le i \le n)$ is contained in a small neighborhood of the kind described. Now the desired deformation is easy to perform.

Second step. From the first step, we can assume that α is the composition of finitely many curves $\alpha^{(\nu)} : [a_\nu, b_\nu] \to X$ such that $\alpha^{(\nu)}(t)$ is contained in R exactly for the boundary values $t = a_\nu$ and $t = b_\nu$. Hence, with the exception of these two points, the curve $\alpha^{(\nu)}$ runs in the interior, X_0. This part is topologically equivalent to the interior of the $4p$-gon. Hence the restriction $\alpha^{(\nu)}|(a_\nu, b_\nu)$ can be lifted to a curve $\tilde{\alpha}^{(\nu)} : (a_\nu, b_\nu) \to \mathcal{X}(p)$ which runs in the interior of the $4p$-gon. For a sufficiently small $\varepsilon > 0$, the starting piece $\alpha^{(\nu)}|[a_\nu, a_\nu + \varepsilon]$ (and analogously, the end piece) runs in a small neighborhood of the nature described. For reasons of connectedness, $\tilde{\alpha}^{(\nu)}|(a_\nu, a_\nu + \varepsilon)$ must then run in one of the shaded segments of circles in the figure at the start of the proof. This shows that $\tilde{\alpha}^{(\nu)}$ extends continuously to the closed interval. We denote this extension by $\tilde{\alpha}^{(\nu)} : [a_\nu, b_\mu] \to \mathcal{X}(p)$ again. The starting point and endpoint of the curve are on the boundary of the $4p$-gon. This is simply connected and the boundary is connected. Therefore $\tilde{\alpha}^{(\nu)}$ is homotopic to a curve whose image is contained in the boundary of the $4p$-gon. Hence $\alpha^{(\nu)}$ and, as a consequence, α are homotopic to a curve whose image is contained in R.

Third step. Now we can assume that:
1) The image of α is contained in R.
2) α passes through q only finitely often.

So, α is the composition of finitely many closed curves which, except at the starting point and endpoint, do not meet q. We can assume that $\alpha(t)$ equals the base point q only for $t = 0$ and $t = 1$. Obviously, $R - \{q\}$ consists of $2p$ connected components, namely the images of the members of the canonical system without q. Now it is clear that α runs in the image of *one* α_ν or β_ν. These images are homeomorphic to circles. Now the claim follows from the fact that the fundamental group of a circle is cyclic. $\square$

We still must clarify how the curves of the canonical system meet in the base point q. At this point, it is useful to use the notion of *free homotopy*, which is meaningful only for *closed* curves.

6.5 Definition.   *Two closed curves $\alpha, \beta : [0, 1] \longrightarrow X$ in a topological space X are called **freely homotopic** if there exists a family of closed curves*

$$\alpha_s : [0, 1] \longrightarrow X, \quad 0 \le s \le 1,$$

such that the map

$$H : [0, 1] \times [0, 1] \longrightarrow X, \quad H(s, t) = \alpha_s(t),$$

is continuous.

In contrast to the usual bounded homotopy, we do not demand that the base points of the curves are fixed. The notion of free homotopy is correlated with the notion of "fillable" (Definition III.4.6).

The curves of the canonical system exist in pairs (α_ν, β_ν). We call α_ν the *partner* of β_ν, and conversely.

6.6 Proposition.   *Any curve of the canonical system is freely homotopic to a curve whose image is disjoint with all images of the curves of the canonical system with one exception, namely the partner.*

Proof. As shown in the figure, we shift the boundary edge K in the $4p$-gon a little. $\square$

The statement Proposition 6.6 is enough for our purposes. Nevertheless, it is helpful to have in mind the following more precise statement about the position of the curves. This can also be seen from the $4p$-gon.

Two partners cross over in q:

In the other case, they bang together:

(In the above figures, one of the two curves is drawn dashed, and the other not.)

Exercises for Sect. IV.6

1. Study all possible boundary gluings of the octagon and determine p in each case.

2. Give an example of a boundary gluing of a hexagon such that the quotient space is a sphere.

7. Differentials of the First Kind

Let X be a (connected) compact Riemann surface. We fix a homeomorphism from X onto the quotient space of the $4p$-gon with respect to the standard boundary gluing and denote by

$$\pi : \mathcal{X}(p) \longrightarrow X$$

the canonical map. The canonical system $\alpha_1, \ldots, \beta_p$ is a system of closed curves on X, all with the same base point q.

We denote by $\Omega(X)$ the vector space of everywhere holomorphic differentials. (These are called "differentials of the first kind".)

In this section, we want to prove a precursor to the Riemann–Roch theorem, namely the following important theorem.

7.1 Theorem. *The set $\Omega(X)$ of holomorphic differentials is a complex vector space of dimension p:*

$$\dim_{\mathbb{C}} \Omega(X) = p.$$

Corollary 1. *The number p is determined by X.*

We call p the genus of X.

Corollary 2. *Topologically equivalent compact Riemann surfaces have the same genus.*

For the proof of Theorem 7.1, we have to consider line integrals of differentials. The notion of a differential $\omega \in A^1(X)$ was introduced in the appendix to Chap. II (Sect. 13). The line integral $\int_\alpha \omega$ along a piecewise smooth curve was also defined there. We recall that the total differential $df \in A^1(X)$ was defined for a C^∞-function $f : X \to \mathbb{C}$. We saw Stokes's theorem for curves (Theorem II.13.18),

$$\int_\alpha df = f(\alpha(1)) - f(\alpha(0)).$$

We are interested mainly in the case of *holomorphic* differentials ω. We recall the definition of the line integral in this special case. If $\varphi : U \to V$ is an analytic chart and α a smooth curve which runs in this chart, i.e. $\alpha : [a, b] \to U$, then

$$\int_\alpha \omega = \int_{\varphi \circ \alpha} f(\zeta)\,d\zeta \quad (\omega_\varphi = f(z)dz).$$

In the general case, we cut α into finitely many pieces which run inside analytic charts, and sum the individual integrals.

We need a homotopic version of the Cauchy integral theorem. In this connection, it is useful to define the line integral for *holomorphic* differentials along arbitrary continuous, not necessarily piecewise smooth, curves.

7.2 Lemma (homotopic version of the Cauchy integral theorem). *Let ω be a holomorphic differential on a Riemann surface X. Then the integral of ω along an arbitrary continuous curve can be defined in such a way that for piecewise smooth curves we obtain the old definition and that integration along homotopic curves leads to the same result.*

Supplement. *If α, β are two freely homotopic closed curves in X, then*

$$\int_\alpha \omega = \int_\beta \omega.$$

The simple proof is completely analogous to the case of the plane. We refer to [FB], Chap. IV, Proposition A2, for details. For the sake of completeness, we sketch the proof here. First of all, we define the integral along an arbitrary continuous curve by approximating by a piecewise smooth curve. For this, we decompose the parameter interval by means of a partition $0 = a_0 < \cdots a_n = 1$ into finitely many subintervals such that each piece of the curve is contained in a disk. Inside each disk, we replace the piece of the curve by a smooth curve (drawn in bold in the figure below). Using the Cauchy integral theorem for disks, it is easy to show that the integral along the approximating curve does not depend on the approximation.

For the proof of Lemma 7.2, the following has to be shown. Let $H : Q \to X$ be a continuous map of a rectangle into X. The image of the boundary of Q can be considered, in an obvious way, as a closed curve α in X. We have to show $\int_\alpha \omega = 0$. This is clear if the image of Q is contained in a disk. In general, we decompose Q as in the following figure into sufficiently small rectangles, and sum the corresponding integrals.

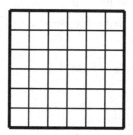

There is no need to take the rectangles so that they are of equal size. Hence we can avoid a finite number of points, and we obtain in this way the following generalization of Lemma 7.2.

7.3 Remark. *The homotopic version of the Cauchy integral theorem, Lemma 7.2, holds also for **meromorphic** differentials if the following conditions are satisfied:*

a) *The integration contour does not contain any poles.*
b) *The residues of all poles vanish.*

After these preparations, we come to the important notion of a *period* of a holomorphic differential. By definition, the set of periods is empty in the case $p = 0$. In the case $p > 0$, we define the periods as follows.

7.4 Definition. *Let $\omega \in \Omega(X)$ be an everywhere holomorphic differential. The numbers*

$$A_i = \int_{\alpha_i} \omega \quad and \quad B_i = \int_{\beta_i} \omega$$

*are called the **periods** of ω.*

Of course, the periods depend on the choice of the normal form. We shall se later how they transform if the normal form is changed.

7.5 Lemma. *An everywhere holomorphic differential $\omega \in \Omega(X)$ is determined by the **real part of periods**. In other words, the **period map***

$$\Omega(X) \longrightarrow \mathbb{R}^{2p}, \quad \omega \longmapsto \mathrm{Re}(A_1, \ldots, A_p, B_1, \ldots, B_p),$$

is injective.

Corollary 1. $\Omega(X)$ *is a real vector space of dimension* $\leq 2p$.

Corollary 2. $\Omega(X)$ *is a finite-dimensional complex vector space of dimension* $\leq p$.

Proof of the lemma. Let ω be an element of the kernel of the period map. The determination of the fundamental group and the homotopic version of the Cauchy integral theorem show the following.

If we join the base point q to an arbitrary point $x \in X$, then the integral

$$u(x) := \text{Re} \int_q^x \omega$$

does not depend on the choice of the curve.

The well-defined function u is locally the real part of an analytic function and hence harmonic. By the maximum principle for harmonic functions, u is constant. We obtain $\omega = 0$. $\square$

7.6 Proposition. *Let $A_1, \ldots, A_p, B_1, \ldots, B_p$ be a 2p-tuple of real numbers. There exists an everywhere holomorphic differential $\omega \in \Omega(X)$ whose periods have the following real parts:*

$$A_i = \text{Re} \int_{\alpha_i} \omega \quad and \quad B_i = \text{Re} \int_{\beta_i} \omega \qquad (1 \le i \le p).$$

Theorem 7.1 then follows from Lemma 7.5 and Proposition 7.6. For the proof of Proposition 7.6, we have to use the central existence theorem for harmonic functions (Theorem II.12.2). (Again we point out that we need this existence theorem only for compact, and not for arbitrary zero-bounded Riemann surfaces. In this case, the proof is simpler. We do not need the tedious Nevanlinna's lemma, (Lemma II.10.4), but only the simpler variant in Proposition II.10.2.)

  We recall that by Proposition 6.6, all curves α_i, β_i with the single exception α_1 (but including β_1) are freely homotopic to curves $\tilde{\alpha}_i, \tilde{\beta}_i$ which are disjoint with β_1. We want to replace the curve β_1 by a homotopic "polygonal path" β_1^*. This means the following in this connection: there exists a partition of the parameter interval $0 = a_0 < a_1 < \ldots < a_n = 1$ such that there exist disks

$$\varphi_i : U_i \xrightarrow{\sim} U_2(0) = \{z \in \mathbb{C}; \ |z| < 2\} \quad \text{with} \quad \gamma([a_{i-1}, a_i]) \subset U_i,$$

such that the pieces $\beta_i^* = \beta^*|[a_{i-1}, a_i]$ correspond in the disk $U_2(0)$ to the segment from -1 to $+1$. It is clear that this construction can be done in such a way that β_1^* like β_1, is in the complement of the curves $\tilde{\alpha}_2, \ldots, \tilde{\alpha}_p, \tilde{\beta}_1, \ldots, \tilde{\beta}_p$. After these preparations, we come to the following proof.

Proof of Proposition 7.6. We can assume that all real parts except for one vanish:

$$A_1 = 1; \ A_2 = \ldots = A_p = B_1 = \ldots = B_p = 0.$$

Now we use the existence theorem (Theorem II.12.4). In the complement of the image of β_i^* there exists a harmonic function u_i with the property that the function

$$u_i^{\varphi_i}(z) - \operatorname{Arg}\left(\frac{z+1}{z-1}\right)$$

extends to the full disk $U_2(0)$ as a harmonic function. It cannot be extended continuously to any point of $(-1,1)$.

We define u to be the sum of the $u_1, \ldots, u_n$. Then u is continuous in the complement of the image of β^*.

Now we consider the differentials associated with u_i and u in the sense of Remark 1.8 and denote them by ω_i, $1 \le i \le n$, and ω ($= \omega_1 + \cdots \omega_n$). Locally, a harmonic function is the real part of an analytic function. The associated differential is the total differential of this function. The differential associated with the harmonic function $\operatorname{Arg}((z+1)/(z-1))$ is the analytic function $-i/(z+1) + i/(z-1)$. Hence ω_i extends to a meromorphic differential on X. It has only two poles, and these are of order one with opposite residues. When these are summed, they cancel. Hence the differential ω which is associated with u is holomorphic on the whole of X. (We can think about this in the following way. The function u jumps when crossing β. But this disappears after differentiation.)

We know that each curve from the canonical system, with one exception α_1, is freely homotopic to one of the curves $\tilde{\alpha}_i, \tilde{\beta}_i$ which run completely in the complement of β_1^*. This implies that

$$\operatorname{Re}\int_{\alpha_i} \omega = \operatorname{Re}\int_{\tilde{\alpha}_i} \omega = \int_{\tilde{\alpha}_i} du = 0 \qquad (i > 1),$$

and correspondingly for all $\tilde{\beta}_i$. For the proof of Proposition 7.6, it is sufficient to show that the integral of ω along α_1 does not vanish. We argue indirectly and assume that the integral is zero. By Lemma 7.5, ω is then identically zero, and hence the function u is constant. This contradicts the fact that u does not extend as a continuous function to the whole of X. This completes the proof of Proposition 7.6. $\qquad\qquad\square$

We shall now illustrate Theorem 7.1 with an example which was of great importance during the historical development of the subject. We study the case of a hyperelliptic Riemann surface X. This is the surface which is attached to the algebraic function "$\sqrt{P(z)}$". Here P is a polynomial of degree $2p+1$ or $2p+2$ without multiple zeros. Recall that X is a compact Riemann surface with two distinguished meromorphic functions

$$f \qquad\qquad\qquad \text{``} = z\text{''},$$

$$g \qquad\qquad\qquad \text{``} = \sqrt{P(z)}\text{''}.$$

By means of f, the surface X is of degree 2 over the sphere. On this surface, $\sqrt{P(z)}$ appears as a (single-valued) meromorphic function g. The genus of X is p (Theorem 3.6). We can consider on X the meromorphic differentials

$$\frac{f^m\,dp}{g} \qquad\qquad \text{``} = \frac{z^m\,dz}{\sqrt{P(z)}} \text{''}.$$

It can be shown (Exercise 3) that these are holomorphic on the whole of X as long as $m < p$. Since these differentials are linearly independent, they have to generate $\Omega(X)$.

7.7 Remark. *Let X be the Riemann surface for "$\sqrt{P(z)}$", where P is a polynomial of degree $2p+1$ or $2p+2$ without multiple zeros. The differentials*

$$\text{``}\frac{z^m\,dz}{\sqrt{P(z)}}\text{''}, \qquad 0 \le m < p,$$

define a basis of $\Omega(X)$.

Exercises for Sect. IV.7

1. Show that the homotopic version of the Cauchy integral theorem holds for all differentials $\omega \in A^1(X)$ with the property $d\omega = 0$.

2. A C^∞-function is called a primitive of a differential $\omega \in A^1(X)$ if $df = \omega$. Show that ω admits a primitive if the integral of ω along every closed curve vanishes. Show that we obtain such a primitive if we integrate ω along a curve which starts at a fixed base point and ends at a variable point. Show that if ω is holomorphic, then this primitive is holomorphic too.

3. Let X be the Riemann surface with respect to "$\sqrt{P(z)}$", where P is a polynomial of degree $2p+1$ or $2p+2$ without multiple zeros. Show that

$$\text{``}\frac{z^m\,dz}{\sqrt{P(z)}}\text{''}, \qquad 0 \le m < p,$$

is holomorphic on the whole of X.

Appendix to Sect. 7. The Polyhedron Theorem

Let $(\mathcal{P}, \mathcal{D})$ be a combinatorial scheme together with a boundary gluing Σ, such that all boundary edges are included. As a consequence, the number of boundary edges is even. We define the following notation:

$$
\begin{aligned}
2R \quad &:= \quad \text{number of boundary edges;} \\
K \quad &:= \quad \text{number of "interior" vertices;} \\
A \quad &:= \quad \text{number of equivalence clasess of exterior vertices;} \\
E \quad &:= \quad \text{number of interior vertices;} \\
D \quad &:= \quad \text{number of triangles.}
\end{aligned}
$$

We call
$$ e := D - (K + R) + E + A $$
the Euler number of $(\mathcal{P}, \mathcal{D}, \Sigma)$.

First example. Let X be a compact surface with an oriented triangulation $\mathcal{M}$. The associated combinatorial scheme $(\mathcal{P}, \mathcal{D})$ has no boundary edges. In this case,
$$
\begin{aligned}
e(\mathcal{P}, \mathcal{D}) &= D - K + E \\
&= \#\text{triangles} - \#\text{edges} + \#\text{vertices}
\end{aligned}
$$
is the Euler number, which we have considered already.

Second example. Let $(\mathcal{P}, \mathcal{D})$ be the standard $4p$-gon with the normal gluing. In this case,
$$
2R = 4p, \quad K = 4p, \quad A = 1,
$$
$$
E = 1, \quad \text{and} \quad D = 4p.
$$

Hence
$$ e = 4p - (4p + 2p) + 2 = 2 - 2p. $$

7.8 Remark.   *If, from some oriented triangulation of a compact surface, we produce a normal form that is a $4p$-gon, then the Euler number e does not change in any of the constructions defined in Sects. 5 and 6.*

Corollary. *If a compact surface admits an oriented triangulation (for example if it admits a structure in the form of a Riemann surface), then the Euler number is independent of the choice of the oriented triangulation.*

Corollary (Euler, c. 1750; Legendre, 1794). *The Euler number of any triangulation of the sphere is 2.*

Corollary. *Let $f : X \to \bar{\mathbb{C}}$ be a nonconstant meromorphic function on a compact Riemann surface. Then the topological genus p can be computed by means of the Riemann–Hurwitz ramification formula (Theorem 3.5).*

Now the question arises of which compact topological surfaces admit a structure in the form of a Riemann surface. Actually, it can be shown that every orientable surface with a countable basis of topology admits such a structure. (Here, the notion of the orientability of a topological surface still has to be defined.) We shall prove only the following proposition.

7.9 Proposition.   *Every compact surface with an oriented triangulation admits a structure in the form of a Riemann surface.*

For the proof, it is enough to show that for each $p \geq 0$ there exists a compact Riemann surface. Take, for example, the hyperelliptic Riemann surface for

$$w^2 = z^{2p+1} - 1. \qquad\qquad \square$$

Exercises for Sect. IV.7

1. Describe a boundary gluing of a 12-gon which leads to $p = 1$.

2. Compute all boundary gluings of a decagon for $p = 0$.

8. Some Period Relations

Again we start with a compact Riemann surface X which is realized by a standard $4p$-gon $\mathcal{X}(p)$ in the normal form. We have described the canonical system α_i, β_i and defined the periods of a differential ω with respect to this system as

$$A_i = \int_{\alpha_i} \omega, \quad B_i = \int_{\beta_i} \omega \qquad (1 \leq i \leq p).$$

For a basis $\omega_1, \ldots, \omega_p$ of the $\mathbb{C}$-vector space of all holomorphic differentials, we can consider the *period matrix*

$$P := \begin{pmatrix} A_1^{(1)} & \cdots & A_1^{(p)} & B_1^{(1)} & \cdots & B_1^{(p)} \\ \vdots & & \vdots & \vdots & & \vdots \\ A_1^{(p)} & \cdots & A_p^{(p)} & B_p^{(1)} & \cdots & B_p^{(p)} \end{pmatrix}.$$

For an arbitrary holomorphic differential ω, the period vectors of ω are linear combinations, with integral coefficients, of the rows of P. Since a holomorphic differential vanishes if all its periods are zero, we have the following result.

8.1 Remark. *The period matrix P (with respect to the given canonical system and the given basis of $\Omega(X)$) has rank p.*

The periods underlie certain strong restrictions which are called "period relations". In this section, we shall prove only those period relations which are needed for the proof of the Riemann–Roch theorem. Further period relations will be derived in connection with Abel's theorem (Sect. 11).

We denote the canonical projection of the $4p$-gon onto X by

$$\pi : \mathcal{X}(p) \longrightarrow X.$$

Let ω be a differential which is holomorphic on the whole of X. We want to associate with this differential a continuous function

$$f : \mathcal{X}(p) \longrightarrow \mathbb{C}.$$

For this, we choose an arbitrary point a in the $4p$-gon. For every other point x of the $4p$-gon, we choose a curve α in $\mathcal{X}(p)$ from a to x. For example, we can take a straight segment. We then define

$$f(x) := \int_{\pi \circ \alpha} \omega.$$

By the homotopic version of the Cauchy integral theorem, the integral does not depend on the choice of α. For two arbitrary points x, y from $\mathcal{X}(p)$, we can consider the straight line β from x to y. We have

$$f(y) - f(x) = \int_{\pi \circ \beta} \omega.$$

We apply this in the special case where x lies on an edge K_i and y is the complementary point on the edge K_i^-. The image of the segment from x to y is a closed curve in X which is freely homotopic to β_i (the image of L_i in the correct orientation). We obtain

$$f(y) - f(x) = \int_{\beta_i} \omega$$

and a corresponding equation for the edges L_i, L_i^-.

8.2 Lemma. *Let ω be a holomorphic differential on X. The associated (continuous) function*

$$f : \mathcal{X}(p) \longrightarrow \mathbb{C}$$

has the following property:

Let x be a point on K_i and let x^- be the complementary point on K_i^-; then

$$f(x^-) - f(x) = \int_{\beta_i} \omega.$$

Correspondingly, if y is a point from L_i and if y^- is its complementary point on L_i^-, then

$$f(y^-) - f(y) = -\int_{\alpha_i} \omega.$$

The proof of the Riemann period relations becomes transparent if we assume that the map π is smooth in the following sense: a function on $\mathcal{X}(p)$ or, more generally, on an open subset (with respect to the induced topology) $U \subset \mathcal{X}(p)$ is called smooth if its restriction to the intersection of U with the interior of $\mathcal{X}(p)$ is (infinitely often) differentiable and if if all partial derivatives of f extend continuously to U. By means of charts, we then define the notion of a smooth map into an arbitrary differentiable surface.

For the following proof, we want to assume in an initial approach that $\pi :$ $\mathcal{X}(p) \to X$ is smooth in this sense.

This is not really what we need; actually, it is sufficient to know that π is *piecewise smooth*. In the appendix to this section, we shall explain what this means and show that piecewise smooth normal forms $\pi : \mathcal{X}(p) \to X$ exist. We recommend the reader not to worry about this and to ignore this problem. After reading the appendix, it will become clear that the proof works in the piecewise smooth case.

By a smooth differential $\omega = f\,dx + g\,dy$ on $\mathcal{X}(p)$, we understand a pair of smooth functions $f, g : \mathcal{X}(p) \to \mathbb{C}$. For such smooth differentials, Stokes's theorem*) holds:

$$\oint_{\partial \mathcal{X}(p)} \omega = \int_{\mathcal{X}(p)} d\omega.$$

If ω is a $\mathcal{C}^\infty$-differential on X, then, as usual, we define the pulled-back differential $\pi^*\omega$ as a smooth differential on $\mathcal{X}(p)$.

8.3 Lemma. *Let ω be a holomorphic differential and let ω' be a differential with $d\omega' = 0$. Then*

$$\int_X \omega \wedge \omega' = \sum_{i=1}^{p} \left[\int_{\alpha_i} \omega \int_{\beta_i} \omega' - \int_{\alpha_i} \omega' \int_{\beta_i} \omega \right].$$

---

*) Stokes's theorem was formulated for triangles in Remark II.13.21. For the n-gon, it follows by summing over the triangles.

Proof. We consider the function $f : \mathcal{X}(p) \to \mathbb{C}$ which is obtained by integrating ω. Clearly f is smooth. We also consider the pulled-back differentials $\tilde{\omega} = \pi^*\omega$ and $\tilde{\omega}' = \pi^*\omega'$. Because $d\omega' = 0$, we have $d(f\tilde{\omega}') = \tilde{\omega} \wedge \tilde{\omega}'$. We apply Stokes's theorem and obtain

$$\int_X \omega \wedge \omega' = \int_{\partial \mathcal{X}(p)} f\tilde{\omega}'.$$

The contribution of the pair of edges K_i, K_i^- in the integral is

$$\int_{K_i} f\tilde{\omega}' - \int_{K_i^-} f\tilde{\omega}'.$$

Here we have oriented K_i counterclockwise but K_i^- clockwise (which produces the sign). By Lemma 8.2, the values of f differ at corresponding points of the two edges only by an additive constant $\int_{\beta_i} \omega$. Hence the contribution of the pair of edges is

$$\int_{\beta_i} \omega \int_{K_i} \tilde{\omega}' = \int_{\beta_i} \omega \int_{\alpha_i} \omega'.$$

Correspondingly, we obtain for the contribution of the pair L_i, L_i^- the expression

$$-\int_{\alpha_i} \omega \int_{\beta_i} \omega'.$$

By summing, we obtain the claim. $\qquad\square$

Period Relations of the Fist Kind

Now let ω, ω' be two *holomorphic* differentials on X. Then $\omega \wedge \omega' = 0$, and we obtain the following proposition from Lemma 8.3.

8.4 Proposition. *Let ω and ω' be two holomorphic differentials on X. Then we have the **period relation***

$$\sum_{i=1}^p \int_{\alpha_i} \omega \int_{\beta_i} \omega' = \sum_{i=1}^p \int_{\beta_i} \omega \int_{\alpha_i} \omega'.$$

For the proof of the Riemann–Roch theorem, we need a slight generalization of Lemma 8.2 and Proposition 8.4. This corresponds to the fact that ω is allowed to be meromorphic, i.e. poles are allowed. These should be contained in the complement of the canonical system. We also require that all residues of ω vanish. Then we can apply the homotopic version of Cauchy's theorem (Remark 7.3) to produce, by integration, a meromorphic function f_0 on $X - R$

with the property $df_0 = \omega$. We pull it back by means of π to obtain a function f which is continuous in the interior of the $4p$-gon up to finitely many exceptional points. If α is a curve which runs inside the interior of the $4p$-gon and does not meet an exceptional point, then

$$f(\alpha(1)) - f(\alpha(0)) = \int_{\pi \circ \alpha} \omega.$$

By means of this integral representation, we can extend f to a continuous function on the closed $4p$-gon,

$$f : \mathcal{X}(p) - S \longrightarrow \mathbb{C}.$$

8.5 Lemma (variant of Lemma 8.2).   *Let ω be a meromorphic function on X whose poles are contained in the complement R of the canonical system and such that all residues vanish. The associated continuous function*

$$f : \mathcal{X}(p) - S \longrightarrow \mathbb{C}$$

has the same behavior on the boundary as that described in Lemma 8.2.

The proof is the same as that of Lemma 8.2.   □

8.6 Proposition (variant of Proposition 8.4).   *As in Proposition 8.4, ω' is assumed to be a holomorphic differential, but the differential ω is assumed only to be meromorphic, where the possible poles are in the complement of the canonical system. The residues are assumed to be zero. Furthermore, we assume that for every pole a of order $m(a)$ of ω, the differential ω' has a zero at a of order at least $m(a) - 1$. Then the period relation formulated in Proposition 8.4 remains true.*

It follows from the above assumption that the differential $f_0\omega'$ is holomorphic in the complement of the canonical system. This suffices to imitate the proof of Proposition 8.4.   □

Exercises for Sect. IV.8

1. How does the period matrix change if the basis of $\Omega(X)$ is changed?

2. Let $X = \mathbb{C}/L$ be a torus. Give a basis of $\Omega(X)$ and a concrete canonical system, and compute the corresponding period matrix.

Appendix to Sect. 8. Piecewise Smoothness

Let $\Delta \subset \mathbb{C}$ be a closed triangular area. For our purposes, we need a notion of a *piecewise smooth* map

$$f : \Delta \longrightarrow X$$

from Δ into a Riemann surface X which corresponds in some sense to the notion of a piecewise smooth curve

$$\alpha : [0, 1] \longrightarrow X.$$

We consider a point in the interior of each of the three edges of Δ and divide Δ by means of the segments between these points into four triangles. We call this a *subdivision of level one* of Δ.

By dividing each of the four triangles in the same way into four subtriangles, we obtain a subdivision of level two and so on.

8.7 Definition. *A map*

$$f : \Delta \longrightarrow X$$

from a triangular area into a Riemann surface X is called **piecewise smooth** *if there exists a subdivision of some level such that the restriction to each of the 4^n subtriangles is smooth.*

8.8 Proposition. *The normal form of a Riemann surface can be realized in such a way that the canonical projection*

$$\mathcal{X}(p) \longrightarrow X$$

(more precisely, its restriction to each of the 4p triangles) is piecewise smooth.

IV. Compact Riemann Surfaces

For the *proof,* we simply have to observe that all of the constructions which we performed during the construction of the normal form can be formulated in the piecewise smooth world.

The advantage of a piecewise smooth realization

$$p : \mathcal{X}(p) \longrightarrow X$$

is that differentials ω on X can be pulled back to $\mathcal{X}(p)$ and that one can apply the Stokes's theorem to the pullback $p^*\omega$. We now explain this briefly.

Let ω be a smooth differential on a triangular area $\Delta \subset \mathbb{C}$. The integral along a piecewise smooth curve $\alpha : [0,1] \to \Delta$ is given by

$$\int_\alpha \omega = \int_0^1 h(t)\, dt,$$

(∗) $$h(t) = f(\alpha(t))\alpha_1'(t) + g(\alpha(t))\alpha_2'(t).$$

Now we consider a piecewise smooth realization of the Riemann surface X by a $4p$-gon,

$$\pi : \mathcal{X}(p) \longrightarrow X.$$

Let ω be a smooth differential on X. We want to define the pulled-back differential $p^*\omega$ on $\mathcal{X}(p)$. We have a triangulation of $\mathcal{X}(p)$ which arises from a standard decomposition of the $4p$ triangles, such that the restriction of f to each triangle of the triangulation is smooth. Then $f^*\omega$ can be defined as a smooth differential on each of these triangles. We obtain a differential which is piecewise smooth in the following sense.

8.9 Definition. *A piecewise smooth differential ω on the $4p$-gon $\mathcal{X}(p)$ is a pair consisting of a triangulation of $\mathcal{X}(p)$ arising from standard decompositions of the $4p$ triangles and a map which assigns to each triangle Δ of the triangulation a smooth differential ω_Δ on this triangle. These differentials have to fit together in the following sense.*

Assume that Δ, Δ' are two triangles of the triangulation with a joint edge K and that $\alpha : [0,1] \to K$ is a smooth parametrization of the edge. Then the integrands of the line integrals of ω_Δ and ω'_Δ (see (∗) above) agree.

In particular, we have

$$\int_K \omega_\Delta = \int_K \omega'_\Delta,$$

where the integral has been taken in the same direction in both cases.

So, piecewise differentials can jump if we pass from one triangle to a neighboring one. But these jumps are not visible by integration along edges.

For piecewise smooth differentials ω on $\mathcal{X}(p)$, the surface integral

$$\int_{\mathcal{X}(p)} d\omega$$

and the boundary integral

$$\oint_{\partial\mathcal{X}(p)} \omega$$

can be defined in an obvious way. The following version of Stokes's theorem follows immediately from Stokes's theorem for triangles by summing.

8.10 Theorem (Stoke's theorem). *Let ω be a piecewise smooth differential on the 4p-gon; then*

$$\oint_{\partial\mathcal{X}(p)} \omega = \int_{\mathcal{X}(p)} d\omega.$$

In the following, we tacitly assume that $\pi : \mathcal{X}(p) \to X$ is piecewise smooth.

9. The Riemann–Roch Theorem

Let X be a set. We consider the free abelian group $\mathcal{D}(X)$ generated by X. The elements of $\mathcal{D}(X)$ are maps

$$D : X \longrightarrow \mathbb{Z}$$

such that

$$D(a) = 0 \text{ for almost all } a \in X$$

(i.e. for all up to finitely many). Obviously, $\mathcal{D}(X)$ is an abelian group by means of the usual addition of maps.

Special case. For $X = \{1, \ldots, n\}$, we have $\mathcal{D}(X) = \mathbb{Z}^n$.

We call the elements of $\mathcal{D}(X)$ *divisors*. With each point $a \in X$ we can associate a divisor

$$(a) \in \mathcal{D}(X),$$

namely

$$(a)(x) = \begin{cases} 1 & \text{for } x = a, \\ 0 & \text{else.} \end{cases}$$

This defines an injective map

$$X \longrightarrow \mathcal{D}(X), \quad a \longmapsto (a).$$

Sometimes we identify X with its image in $\mathcal{D}(X)$.

Caution. If X already carries a structure in the form of an abelian group, then $(a+b)$ and $(a)+(b)$ may be different.

Each divisor $D \in \mathcal{D}(X)$ can be written in the form

$$D = \sum_{a \in X} D(a)(a) \qquad \text{(finite sum)}.$$

So, each divisor can also be written in the form

$$D = (a_1) + \ldots + (a_n) - (b_1) - \ldots - (b_m),$$

where the $\{a_1, \ldots, a_n\}$ and $\{b_1, \ldots, b_m\}$ are disjoint. Finally, we define the *degree of a divisor* by

$$\deg(D) = \sum_{a \in X} D(a).$$

Obviously,

$$\deg : \mathcal{D}(X) \longrightarrow \mathbb{Z}$$

is a group homomorphism. And obviously,

$$\deg\left((a_1) + \cdots + (a_n) - (b_1) - \cdots - (b_m)\right) = n - m.$$

Now let X be a (connected) compact Riemann surface. With each meromorphic function

$$f : X \longrightarrow \bar{\mathbb{C}}$$

which is different from zero, we can associate a divisor (f) by

$$D(a) = \operatorname{Ord}(f; a)$$

So we have

$$D(a) > 0 \iff a \text{ is a zero of } f,$$

$$D(a) < 0 \iff a \text{ is a pole of } f.$$

If we denote the zeros by $a_1, \ldots, a_n$ (each written as often as the multiplicity prescribes) and, correspondingly, the set of poles by $b_1, \ldots, b_m$, then

$$(f) = (a_1) + \cdots + (a_n) - (b_1) - \cdots - (b_m).$$

As we know, meromorphic functions have the same numbers of zeros and poles. Hence we have the following result:

The degree of the divisor (f) of a nonvanishing meromorphic function is 0.

9.1 Remark.   *Let $f, g \in \mathcal{M}(X) - \{0\}$ be two meromorphic functions which are different from 0. Then*

$$(f) + (g) = (f \cdot g).$$

The divisor of a constant function is the zero element in $\mathcal{D}(X)$.

9.2 Definition.   *A divisor $D \in \mathcal{D}(X)$ is called a **principal divisor** if there exists a meromorphic function with*

$$D = (f).$$

As we know that f is determined by D up to a constant factor, then because of Remark 9.1, the set of principal divisors

$$\mathcal{H}(X) = \{\, D \in \mathcal{D}(X); \quad D = (f); \; f \in \mathcal{M}(X) \,\}$$

is a subgroup of $\mathcal{D}(X)$. The factor group

$$\mathrm{Pic}(X) := \mathcal{D}(X)/\mathcal{H}(X)$$

is called the *divisor class group of X*. Since the degree of a principal divisor is zero, the degree

$$\deg : \mathrm{Pic}(X) \longrightarrow \mathbb{Z}, \quad [D] \longmapsto \deg(D),$$

is well defined on the divisor class group.

The Riemann–Roch Space

Let D, D' be two divisors on X. We define

$$D \geq D' \quad \Longleftrightarrow \quad D(a) \geq D'(a) \text{ for all } a \in X.$$

The Riemann–Roch space of a divisor D is the set of all meromorphic functions $f \in \mathcal{M}(X)$ with the property

$$(f) \geq -D,$$

together with the zero function. This means, symbolically, that the inequality

$$(0) \geq -D$$

is always assumed to be true, so we can write

$$\mathcal{L}(D) = \{\, f \in \mathcal{M}(X); \quad (f) \geq -D \,\}.$$

9.3 Remark.   *The set $\mathcal{L}(D)$ is a $\mathbb{C}$-vector space.*

The proof is trivial.   □

We shall see that $\mathcal{L}(D)$ is of finite dimension. The space $\mathcal{L}(D)$ essentially depends only on the divisor class of D. More precisely, we have the following:

The map

$$\mathcal{L}(D) \longrightarrow \mathcal{L}(D + (f)), \quad g \longmapsto g \cdot f,$$

is an isomorphism.

The Riemann–Roch problem is the determination of the dimension

$$l(D) := \dim_{\mathbb{C}} \mathcal{L}(D).$$

The Riemann–Roch theorem is a partial answer to this question.

Remark. *Assume that*

$$D := n \cdot (a); \quad n \in \mathbb{N}.$$

Then $\mathcal{L}(D)$ consists of all meromorphic functions which have a possible pole only at a and are such that its order is $\leq n$.

Exercise. Let $X = \mathbb{C}/L$ be a torus. Show that

$$l(n(a)) = \begin{cases} 1 & \text{for } n = 1, \\ 2 & \text{for } n = 2. \end{cases}$$

9.4 Theorem (Riemann–Roch inequality, B. Riemann, 1857).   *We have*

$$l(D) \geq \deg D - p + 1, \qquad p = \text{genus of } X.$$

This is an existence theorem for meromorphic functions. To make this clear, we consider again the case $D = n(a)$, and obtain the following corollary.

Corollary. *For each point $a \in X$, there exists a meromorphic function f which is holomorphic outside a and has a pole of order $\leq p + 1$ at a.*

The Riemann–Roch theorem is stronger than the inequality in Theorem 9.4. It also gives information about the "defect"

$$h(D) := l(D) - \deg(D) + p - 1.$$

To formulate this, we need the *canonical class,* which is an element of $\text{Pic}(X)$ of fundamental importance.

Let ω be a meromorphic differential which is different from zero. We have seen that it makes sense to speak of zeros and poles and of their orders. This means that we can attach to ω a divisor

$$K := (\omega) \in \mathcal{D}(X).$$

Such a divisor K is called a *canonical divisor.* Let f be meromorphic function which is different from zero; then

$$(f \cdot \omega) = (f) + (\omega).$$

As we know, $f \cdot \omega$ runs through *all* meromorphic differentials for fixed ω and variable f. We obtain the following result.

9.5 Remark. *All canonical divisors belong to one divisor class.*

9.6 Definition. *The divisor class of a meromorphic differential is called the* ***canonical class*** *of X.*

A fundamental refinement of the inequality $h(D) \geq 0$ is the equation

$$h(D) = l(K - D),$$

which means the following.

9.7 Theorem (Riemann–Roch theorem, G. Roch 1876). *We have*

$$\dim \mathcal{L}(D) - \dim \mathcal{L}(K - D) = \deg(D) - p + 1.$$

Corollary. $\deg K = 2p - 2$.

Remark. In the case of a torus $X = \mathbb{C}/L$, we can consider the holomorphic differential "dz". The associated divisor is the zero divisor. It has degree 0 in accordance with Theorem 9.7. But on the Riemann sphere $\bar{\mathbb{C}}$ ($p = 0$), the differential dz has a pole of second order at ∞ (because $d(-1/z) = -z^2 dz$); so we have $\deg(dz) = -2$, again in accordance with Theorem 9.7.

Corollary. *In the case $\deg(D) \geq 2p - 2$, the Riemann–Roch inequality is an equality,*

$$\dim \mathcal{L}(D) = \deg(D) - p + 1.$$

The space $\mathcal{L}(D)$ is different from zero only if $\deg(D) \geq 0$. This means that there are finitely many "exceptional numbers"

$$0 \leq \deg(D) \leq 2p - 2,$$

for which the Riemann–Roch theorem gives "only" an inequality.

The proof of the Riemann–Roch Theorem

In addition to

$$\mathcal{L}(D) = \{f; \quad (f) \geq -D\},$$

we introduce the "complementary" space

$$\mathcal{K}(D) := \{\omega \text{ meromorphic differential}; \quad (\omega) \geq D\}.$$

Obviously,

$$\dim \mathcal{K}(D) = \dim \mathcal{L}(K - D).$$

Hence the Riemann–Roch theorem can be written as follows:

$$\dim \mathcal{L}(D) - \dim \mathcal{K}(D) = \deg(D) - p + 1.$$

One special case of the Riemann–Roch theorem is obvious.

First step. Special case: $D < 0$.

In this case, we have $\mathcal{L}(D) = 0$. The space $\mathcal{K}(D)$ consists of all meromorphic differentials which, for all points $a \in X$ with $D(a) < 0$, have poles of order at most $-D(a)$ and are holomorphic elsewhere. We know that the principal parts of a meromorphic differential (with respect to some analytic chart), up to the necessary condition on the residues (their sum must vanish), can be freely prescribed. On the other hand, a meromorphic differential is determined by its principal parts up to an everywhere holomorphic differential. This gives us

$$\dim \mathcal{K}(D) = -(\deg(D) + 1) + p,$$

and this is the Riemann–Roch theorem.

Second step. Special case: $D = 0$.

In this case, we have

$$\dim \mathcal{L}(D) = 1, \qquad \dim \mathcal{K}(D) = \dim \Omega(D) = p,$$

and hence the Riemann–Roch theorem is true again.

Third step. Let D be an arbitrary divisor and let $a \in X$ be a given point. We do not exclude the case in which a occurs in D. We want to compare the divisors

$$D \text{ and } D' = D + (a).$$

Clearly, we have $D \leq D'$ and hence

$$\mathcal{L}(D) \subset \mathcal{L}(D').$$

We claim that

$$\dim \left(\mathcal{L}(D') / \mathcal{L}(D) \right) \leq 1.$$

For the proof, we consider a linear map

$$\mathcal{L}(D') \longrightarrow \mathbb{C},$$

whose kernel is $\mathcal{L}(D)$. For the construction, we consider a disk around a,

$$\varphi : U \overset{\sim}{\longrightarrow} \mathbb{E}, \qquad a \longmapsto 0,$$

and consider the Laurent expansion of f with respect to this disk,

$$f_\varphi(z) = \sum_{n=-\infty}^{\infty} a_n z^n.$$

Let $e := -D(a)$. If f occurs in $\mathcal{L}(D')$, then

$$a_n \neq 0 \Longrightarrow n \geq e - 1.$$

A function $f \in \mathcal{L}(D')$ is contained in $\mathcal{L}(D)$ iff

$$a_n \neq 0 \Rightarrow n \geq e.$$

The map

$$\mathcal{L}(D') \longrightarrow \mathbb{C}, \quad f \longmapsto a_e,$$

has the desired property.

Corollary. The spaces $\mathcal{L}(D)$ (and hence $\mathcal{K}(D)$) are finite-dimensional.

An analogous consideration shows that

$$\mathcal{K}(D') \subset \mathcal{K}(D) \text{ and } \dim(\mathcal{K}(D)/\mathcal{K}(D')) \leq 1.$$

Fourth step. We introduce the defect

$$\delta(D) = \dim \mathcal{L}(D) - \dim \mathcal{K}(D) - \deg(D) + p - 1.$$

The Riemann–Roch theorem states that

$$\delta(D) = 0.$$

In this step of the proof, we show that

$$\delta(D') \leq \delta(D) \qquad (D' = D + (a)).$$

We have

$$\delta(D) - \delta(D') = 1 - \dim(\mathcal{L}(D')/\mathcal{L}(D)) - \dim(\mathcal{K}(D)/\mathcal{K}(D')).$$

If the claimed inequality is false, we must have, by the third step,

$$\dim(\mathcal{L}(D')/\mathcal{L}(D)) = \dim(\mathcal{K}(D)/\mathcal{K}(D')) = 1.$$

Therefore there exist elements

$$f \in \mathcal{L}(D'), \quad f \notin \mathcal{L}(D),$$
$$\omega \in \mathcal{K}(D), \quad \omega \notin \mathcal{K}(D').$$

It is easy to show that:

a) $f \cdot \omega$ is holomorphic outside of a;

b) $f \cdot \omega$ has a simple pole at a.

But this is incompatible with the residue theorem (the sum of residues of $f \cdot \omega$ is 0).

By induction, we now obtain the more general result

$$D \geq D' \implies \delta(D) \leq \delta(D').$$

For an arbitrary divisor D, we find divisors

$$D_1 > 0, \quad D_2 < 0$$

with the property

$$D_2 \leq D \leq D_1.$$

We obtain

$$0 = \delta(D_2) \geq \delta(D) \geq \delta(D_1).$$

So, for the proof of the Riemann–Roch theorem, it is sufficient to show that

$$\delta(D_1) \geq 0 \text{ for } D_1 > 0.$$

Fifth and last step. We have to show that

$$\dim \mathcal{L}(D) - \dim \mathcal{K}(D) \geq \deg(D) - p + 1 \text{ for } D > 0.$$

For the proof, we construct a suitable vector space $\mathcal{I}(D)$ of meromorphic differentials, in which the two spaces $\mathcal{L}(D)$ and $\mathcal{K}(D)$ are hidden.

Definition. The vector space $\mathcal{I}(D)$ consists of all meromorphic differentials ω with the following two properties:

a) Let $a \in X$, $D(a) > 0$. The pole order of ω at a is at most $D(a) + 1$.
b) All residues of ω vanish.

From the existence theorem for meromorphic differentials and from our knowledge of the dimension of the space of all holomorphic differentials, we immediately obtain

$$\dim \mathcal{I}(D) = \deg(D) + p.$$

With every $f \in \mathcal{L}(D)$ there is associated an element of $\mathcal{I}(D)$, namely df. The map

$$\mathcal{L}(D) \longrightarrow \mathcal{I}(D), \quad f \longmapsto df,$$

is $\mathbb{C}$-linear; its kernel consists of constant functions. We want to construct the image of $\mathcal{L}(D)$ as the kernel of a suitable map. For this, we have to assume that none of the points a, $D(a) > 0$, lies on the canonical system (i.e. on the union of the images of the curves $\alpha_1, \ldots, \alpha_p$, $\beta_1, \ldots, \beta_p$). This is possible by a slight modification of the canonical system. Now we can consider the map

$$\mathcal{I}(D) \longrightarrow \mathbb{C}^{2p}, \quad \omega \longmapsto (A_1, \ldots, A_p, B_1, \ldots, B_p),$$

where

$$A_j := \int_{\alpha_j} \omega; \quad B_j := \int_{\beta_j} \omega \qquad (1 \le j \le p).$$

The kernel of this map consists of all $\omega \in \mathcal{I}$ with vanishing periods $A_1, \ldots, B_p$. By the homotopic version of Cauchy's integral theorem (Remark 7.3), the integral of ω along each closed curve vanishes. Therefore ω admits a meromorphic primitive f. It lies in $\mathcal{L}(D)$:

The kernel of the map above is precisely the image of $\mathcal{L}(D)$.

Readers who are familiar with the idea of exact sequences can reformulate this as the exactness of

$$0 \longrightarrow \mathbb{C} \longrightarrow \mathcal{L}(D) \longrightarrow \mathcal{I}(D) \longrightarrow \mathbb{C}^{2p}.$$

For the central argument, we now have to make use of the *period relations* of Proposition 8.4 in the version stated in Proposition 8.6. They imply the following result.

Let $\omega' \in \mathcal{K}(D)$ be a differential with the periods

$$A_j' = \int_{\alpha_j} \omega, \quad B_j' = \int_{\beta_j} \omega.$$

Then, for all $\omega \in \mathcal{I}(D)$,

$$A_1 A_1' + \ldots + A_p A_p' + B_1 B_1' + \ldots + B_p B_p' = 0.$$

So, for each element $\omega' \in \mathcal{K}(D)$, we have obtained a linear equation which all elements of the image

$$\mathcal{I}(D) \longrightarrow \mathbb{C}^{2p}$$

satisfy. If $\omega_1', \ldots, \omega_d'$ is a basis of $\mathcal{K}(D)$, then we obtain d such equations. The matrix of this system of linear equations is the period matrix

$$\begin{pmatrix} A_1^{(1)'} & \cdots & A_p^{(1)'} & B_1^{(1)'} & \cdots & B_p^{(1)'} \\ \vdots & & & & & \vdots \\ A_1^{(d)'} & \cdots & A_p^{(d)'} & B_1^{(d)'} & \cdots & B_p^{(d)'} \end{pmatrix}.$$

By Remark 8.1, this matrix has the rank d. We obtain the result that the dimension of the image of $\mathcal{I}(D)$ under the period map cannot exceed

$$2p - \dim \mathcal{K}(D).$$

From the equation

$$\dim \mathcal{I}(D) = \dim \text{Kernel} + \dim \text{Image},$$

follows that

$$\deg(D) + p \le \dim \mathcal{L}(D) - 1 + 2p - \dim \mathcal{K}(D).$$

This is the claimed inequality

$$\delta(D) \ge 0.$$

This completes the proof of the Riemann–Roch theorem.   □

Exercises for Sect. IV.9

1. Consider, for an element $f \in \mathcal{L}(D)$, the Laurent expansions in all a with $D(a) > 0$ with respect to arbitrarily chosen charts. Since f is determined by finitely many Laurent coefficients, obtain an easy proof of the finiteness of the dimension of $\mathcal{L}(D)$.

2. Give a direct proof of the Riemann–Roch theorem for the Riemann sphere.

3. Prove the Riemann–Roch theorem for a torus $X = \mathbb{C}/L$ by means of Abel's theorem for elliptic functions.

10. More Period Relations

We now apply the formula of Lemma 8.3 to the complex conjugate differential $\omega' = \bar{\omega}$. In local coordinates, this is defined by

$$\overline{h(z)dz} := \overline{h(z)}\,\overline{dz} \text{ with } \overline{dz} = dx - i\,dy.$$

From the definition of the line integral along a curve α, we immediately obtain the formula

$$\int_\alpha \bar{\omega}' = \overline{\int_\alpha \omega'}.$$

In local coordinates, $\omega \wedge \bar{\omega}$ is calculated as

$$-2i|h(z)|^2 dx\,dy.$$

The next statement follows from Lemma 8.3.

10.1 Proposition. *Let ω be a holomorphic differential on X. Then*

$$\operatorname{Im} \sum_{i=1}^{p} \int_{\alpha_i} \omega \overline{\int_{\beta_i} \omega} \geq 0.$$

The equality sign holds iff $\omega = 0$.

As an important consequence of the inequality in Proposition 10.1, we prove a variant of the existence and uniqueness theorem for holomorphic differentials.

10.2 Theorem. *The map*

$$\Omega(X) \longrightarrow \mathbb{C}^p, \quad \omega \longmapsto \left(\int_{\alpha_1} \omega, \ldots, \int_{\alpha_p} \omega \right),$$

is an isomorphism.

(One can replace the α-periods by the β-periods analogously.)

    Hence a holomorphic differential is determined by:

1) the real parts of all $2p$ periods, or
2) the p α-periods, or
3) the p β-periods.

10.3 Proposition and Definition. *With respect to a given canonical system, there exists a basis*

$$\omega_1, \ldots, \omega_p$$

of $\Omega(X)$ with the property

$$\int_{\alpha_i} \omega_j = \delta_{ij} = \begin{cases} 1 & \text{if } i = j, \\ 0 & \text{if } i \neq j. \end{cases}$$

*We call $\omega_1, \ldots, \omega_p$ the **canonical basis** of $\Omega(X)$ which belongs to this canonical system.*

The period relations in Proposition 8.4 and 10.1 can be rewritten in the following form.

10.4 Proposition. *Let $\omega_1, \ldots, \omega_p$ be a canonical basis (with respect to a given canonical system). We consider the "period matrix"*

$$Z = (z_{ij})_{1 \leq i,j \leq p}, \quad z_{ij} := \int_{\beta_i} \omega_j.$$

We then have:

1) *The matrix Z is symmetric: $Z = Z'$.*
2) *The imaginary part $Y := \operatorname{Im} Z$ is positive definite.*

We recall that a symmetric real matrix Y is called positive definite if the following two equivalent conditions are satisfied:

1) For every n-tuple of real numbers $a_1, \ldots a_n$ which are not all zero, we have

$$\sum_{i=1}^{n} y_{ij} a_i a_j > 0.$$

2) For every n-tuple of complex numbers $a_1, \ldots a_n$ which are not all zero, we have

$$\sum_{i=1}^{n} y_{ij} \bar{a}_i a_j > 0.$$

Now we want to investigate how the matrix Z changes if the canonical system is changed. So, assume that a second canonical system

$$\tilde{\alpha}_1, \ldots, \tilde{\alpha}_p, \tilde{\beta}_1, \ldots, \tilde{\beta}_p$$

is given. We consider the associated canonical basis

$$\tilde{\omega}_1, \ldots, \tilde{\omega}_p$$

and the associated period matrix

$$\tilde{Z} = \left(\int_{\beta_i} \tilde{\omega}_j \right)_{1 \le i, j \le p}.$$

10.5 Definition. *A $2p \times 2p$ matrix M is called **symplectic** if it satisfies the relation*

$$M'IM = I \quad \text{with } I = \begin{pmatrix} 0 & E \\ -E & 0 \end{pmatrix}.$$

Here E denotes the $n \times n$ unit matrix and 0 the zero matrix.

In principle, the coefficients of symplectic matrices could be elements of an arbitrary commutative ring with unity. At the moment, only integral coefficients are of interest to us.

10.6 Remark. *The set of all symplectic matrices is a group $\mathrm{Sp}(p, \mathbb{Z})$. It is called the **symplectic modular group**. In the case $p = 1$, it agrees with the elliptic modular group $\mathrm{SL}(2, \mathbb{Z})$:*

$$\mathrm{Sp}(1, \mathbb{Z}) = \mathrm{SL}(2, \mathbb{Z}).$$

The Intersection Pairing

As a further application of the period relations Proposition 10.1, we construct the intersection pairing. For this, we choose a base point q on the Riemann surface. Then we consider the fundamental group $\pi(X, q)$ with respect to this base point.

For a holomorphic differential ω, the integral

$$\omega(\alpha) := \operatorname{Re} \int_\alpha \omega$$

depends only on the homotopy class of α. Hence we obtain a homomorphism of $\pi(X, q)$ into the additive group of real numbers. So, we have associated with each ω an element of

$$H^1(X, \mathbb{R}) := \operatorname{Hom}(\pi(X, q), \mathbb{R}).$$

10.7 Proposition.   *The natural map*

$$\Omega(X) \longrightarrow H^1(X, \mathbb{R})$$

is an isomorphism.

This proposition is a reformulation of the proved result that a holomorphic differential is determined by the real part of its periods and that these can be prescribed arbitrarily.

We obtain the result that $H^1(X, \mathbb{R})$ is a real vector space of dimension $2p$. An element of $H^1(X, \mathbb{R})$ is determined by its values on the curves $\alpha_1, \ldots, \alpha_p; \beta_1, \ldots, \beta_p$ of a canonical system. These values can be prescribed arbitrarily. Hence we can consider, in $H^1(X, \mathbb{R})$, the *dual system*

$$\alpha_1^*, \ldots, \alpha_p^*; \beta_1^*, \ldots, \beta_p^*.$$

This means that α_i^* takes the value 1 for α_i and the value 0 for all other $2p - 1$ elements (and similarly for β_i^*).

Now we use isomorphism of Proposition 10.7 to define the "intersection pairing" on $H^1(X, \mathbb{R})$. First we define it on $\Omega(X)$ by

$$\langle \omega, \omega' \rangle = \operatorname{Re} \int_X \omega \wedge \bar{\omega}',$$

where $\omega \wedge \bar{\omega}'$ denotes the alternating product.

By means of Proposition 10.7, we transport this pairing to $H^1(X, \mathbb{R})$.

10.8 Proposition (Frobenius's theorem). *Let*

$$\alpha_1, \ldots, \alpha_p; \; \beta_1, \ldots, \beta_p$$

be the canonical system with respect to an arbitrary normal form. We have

$$\langle \alpha_i^*, \beta_i^* \rangle = -\langle \beta_i^*, \alpha_i^* \rangle = 1 \quad (1 \leq i \leq p).$$

All other brackets are zero.

Proof. We have to compare two $\mathbb{R}$-bases of $H^1(X, \mathbb{R})$, namely

$$\alpha_1^*, \ldots, \alpha_p^*, \beta_1^*, \ldots, \beta_p^*$$

and the images of

$$\omega_1, \ldots, \omega_p, i\omega, \ldots, i\omega_p.$$

The transition matrix can be computed by means of the relations in Lemma 8.3 as

$$\begin{pmatrix} E & 0 \\ X & Y \end{pmatrix} \quad (Z = X + iY).$$

The matrix of the products of the second matrix is (again by Lemma 8.3)

$$\begin{pmatrix} 0 & Y \\ -Y & 0 \end{pmatrix}.$$

Transformation to the first basis gives Proposition 10.8.   □

Frobenius's theorem gives a fundamental relation between two canonical bases. So, let

$$\alpha_1'^*, \ldots, \alpha_p'^*; \; \beta_1'^*, \ldots, \beta_p'^*$$

be the dual basis of a second canonical system. To simplify the formulae, we define

$$\alpha_{n+i} = \beta_i, \; \alpha_{n+i}'^* = \beta_i'^* \quad (1 \leq i \leq p).$$

There exists an integral $2p \times 2p$ matrix $M = (m_{ij})$ which describes the relation between the two systems,

$$\alpha_i'^* = \sum_{j=1}^{2p} m_{ij} \alpha_j^*.$$

From Frobenius's theorem, we immediately obtain the following result:

The transition matrix $M = (m_{ij})$ is symplectic.

We denote the canonical bases of $\Omega(X)$ with respect to the two canonical systems by

$$\omega_1, \ldots, \omega_p; \; \omega_1', \ldots, \omega_p'.$$

The matrix which describes the change from the first to the second is denoted by A:

$$\omega_i' = \sum_{j=1}^{p} a_{ij} \omega_j \quad (1 \leq i \leq p).$$

A direct computation gives the following result.

10.9 Proposition. *There exist*

a) *a matrix* $A \in \mathrm{GL}(n, \mathbb{C})$ *and*
b) *an integral symplectic matrix* $M \in \mathrm{Sp}(2n, \mathbb{Z})$ *such that*

$$A \cdot (E, Z) \cdot M = (E, \tilde{Z}).$$

As we have seen in connection with the theory of the elliptic modular group ([FB], Chap. V), the equation above means that in the case $p = 1$, there exists an elliptic modular substitution

$$M \in \mathrm{SL}(2, \mathbb{Z})$$

which transforms Z into $\tilde{Z}$. In the case $p > 1$, we shall see that the integral symplectic group will do the same in general. For the moment, we shall only define the corresponding equivalence relation.

10.10 Definition. *Let*

$$\mathbb{H}_p := \left\{ Z \in \mathbb{C}^{(p,p)}; \quad Z = Z', \ Y > 0 \right\}$$

be the set of symmetric complex $p \times p$ matrices with positive imaginary part. Two points $Z, \tilde{Z} \in \mathbb{H}_p$ are called equivalent if there exist matrices $A \in \mathrm{GL}(n, \mathbb{C})$, $M \in \mathrm{Sp}(n, \mathbb{Z})$ with the property

$$A \cdot (E, Z) \cdot M = (E, \tilde{Z}).$$

Obviously, this is an equivalence relation (see Exercise 1).

Notation. Let

$$\mathcal{A}_p = \mathbb{H}_p / \sim$$

be the set of equivalence classes with respect to this equivalence relation.

In the case $p = 1$, the space $\mathbb{H}_p$ is the usual upper half-plane, and

$$\mathcal{A}_1 = \mathbb{H}_1 / \mathrm{SL}(2, \mathbb{Z})$$

is its quotient by the elliptic modular group.

So, we have associated a well-defined point

$$\tau(X) \in \mathcal{A}_p$$

with each compact Riemann surface. Biholomorphically equivalent surfaces lead to the same point.

Notation. Let $\mathcal{M}_p$ be the set of all biholomorphy classes of compact Riemann surfaces of genus p.

10.11 Remark. *The assignment* $X \to \tau(X)$ *gives a map, called the period map,*

$$\tau : \mathcal{M}_p \longrightarrow \mathcal{A}_p.$$

A fundamental theorem of Torelli states that the period map τ is injective. We shall not prove this theorem in the case $p > 1$. The case $p = 0$ is relatively simple. By the uniformization theorem, every Riemann surface of genus 0 is biholomorphically equivalent to the Riemann sphere. So, $\mathcal{M}_0$ consists of a single element. The case $p = 1$ is somewhat more involved. Complex tori $\mathbb{C}/L$ are surfaces of genus 1. From the theory of elliptic functions, we know that the subset $\mathcal{M}_1' \subset \mathcal{M}_1$ of all isomorphy classes given by tori is mapped bijectively to

$$\mathcal{A}_1 = \mathbb{H}/\operatorname{SL}(2, \mathbb{Z}).$$

Hence, in the case $p = 1$, Torelli's theorem is equivalent to the following statement:

Every Riemann surface of genus 1 is biholomorphic equivalent to a complex torus $\mathbb{C}/L$.

We shall prove this result later as an application of Abel's theorem. (There is another proof which uses the uniformization theorem.)

Hence, in the case $p = 1$, the period map is not only injective but also bijective. The situation for $p > 1$ is different. One can show that $\mathcal{A}_p$ is a complex space of dimension

$$\dim_{\mathbb{C}} \mathcal{A}_p = \frac{p(p+1)}{2} \qquad (= \text{number of "variables" in } \mathbb{H}_p)$$

and $\mathcal{M}_p$ is a complex subspace of dimension

$$\dim_{\mathbb{C}} \mathcal{M}_p = 3p - 3.$$

Hence, for $p > 2$, $\mathcal{M}_p$ is a thin subset of $\mathcal{A}_p$.

The so-called Schottky problem asks for a description of $\mathcal{M}_p$ inside $\mathcal{A}_p$ by equations and inequalities.

There is another essential difference between the cases $p = 1$ and $p > 1$. Let

$$P = P^{(p,2p)} \in \mathbb{C}^{(p,2p)}$$

be a complex $p \times 2p$ matrix whose columns are $\mathbb{R}$-linearly independent. We can ask whether there exist matrices $A \in \operatorname{GL}(n, \mathbb{C})$, $U \in \operatorname{Sp}(n, \mathbb{Z})$ with the property

$$A \cdot P \cdot U = (E, Z), \quad Z \in \mathbb{H}_p.$$

For $p = 1$, this is always case. But in the case $p > 1$, this is not true, as the following rough dimensional argument explains. The number of free complex parameters of P modulo A and U is

$$p \cdot 2p - p^2 = p^2,$$

but the complex dimension of $\mathbb{H}_p$ is only $p(p+1)/2$. For $p > 1$, the two numbers are different. More detail will be given in Chap. VI about abelian functions.

Exercises for Sect. IV.10

1. Verify that an equivalence relation has been defined in Definition 10.10, and show that in the case $p = 1$ it leads to the usual equivalence mod $\mathrm{SL}(2, \mathbb{Z})$.

2. The choice of a lattice basis defines a normal form on $X = \mathbb{C}/L$. Determine the associated canonical basis of $\Omega(X)$.

11. Abel's Theorem

The Riemann–Roch theorem does not tell us when a divisor of degree 0 is a principal divisor. The case of a torus $\mathbb{C}/L$ here is exceptional. Since the canonical class is trivial, we have

$$\dim \mathcal{L}(D) > 0 \text{ for } \deg D > 0.$$

From this result, it is easy to deduce the difficult direction in Abel's theorem for elliptic functions and, conversely, the Riemann–Roch theorem for tori follows easily from Abel's theorem for elliptic functions, as we already have pointed out.

It is more complicated to find an analogue for Abel's theorem for an arbitrary Riemann surface of genus $p > 1$.

The Jacobi Variety

In this section, we use the notion of a "period" in a slightly modified form. Let $\omega_1, \ldots, \omega_p$ be a basis of the space $\Omega(X)$ of everywhere holomorphic differentials, and let α be a closed curve X. The p-tuple

$$(A_1, \ldots, A_p) \text{ with } A_j := \int_\alpha \omega_j \quad (1 \le j \le p)$$

is called a *period* of X with respect to the given basis.

11.1 Definition. *A subset L of a finite-dimensional real vector space is called a **lattice** if there exists a basis $e_1, \ldots, e_n$ such that*

$$L = \mathbb{Z}e_1 + \cdots + \mathbb{Z}e_n.$$

By a lattice in a finite-dimensional complex vector space, we understand a lattice of the underlying real vector space.

11.2 Remark. *The set L of all periods*

$$L = L(\omega_1, \ldots, \omega_p) \subset \mathbb{C}^p$$

with respect to a given basis of $\Omega(X)$ is a lattice.

Proof. If α runs through all curves of a canonical system, then we obtain $2p$ periods, which, as we know, are linearly independent over $\mathbb{R}$. Any period can be written as an integral linear combination of these $2p$ periods.   □

We consider the $2p$-dimensional torus

$$\operatorname{Jac}(X) := \mathbb{C}^p / L,$$

and call $\operatorname{Jac}(X)$ the *Jacobi variety* of X. The Jacobi variety does not depend in an essential manner on the choice of the basis. If $\omega_1', \ldots, \omega_p'$ is a second basis and L' is the associated lattice, then we have

$$A(L) = L',$$

where A is the matrix which transforms $\omega_1, \ldots, \omega_p$ into $\omega_1', \ldots, \omega_p'$. The isomorphism

$$A : \mathbb{C}^p \longrightarrow \mathbb{C}^p$$

then induces a bijection of tori

$$A : \mathbb{C}^p / L \longrightarrow \mathbb{C}^p / L'.$$

If we prefer, we can describe the Jacobi variety in a basis-invariant form as follows. Let

$$\Omega(X)^* := \operatorname{Hom}_{\mathbb{C}}(\Omega(X), \mathbb{C})$$

be the dual space of $\Omega(X)$. With each closed curve α in X we associate an element of this dual space, namely the linear form

$$\omega \longmapsto \int_\alpha \omega.$$

The set of these linear forms is a lattice $L \subset \Omega(X)^*$, and the Jacobi variety is

$$\text{Jac}(X) = \Omega(X)^*/L.$$

We choose a point $q \in X$ and define a map

$$\lambda = \lambda_q : X \longrightarrow \text{Jac}(X)$$

as follows. First we associate with a point $a \in X$ the tuple

$$\left(\int_q^a \omega_1, \ldots, \int_q^a \omega_p \right),$$

where we have chosen a fixed curve from q to a. This tuple is determined up to a period from L. Hence the coset in $\text{Jac}(X)$ is well defined. We call λ the *period map*.

We shall make essential use of the fact that, as in the case $p = 1$, the torus $\mathbb{C}^p/L$ has a structure in the form of an abelian group. This is defined in such a way that the natural projection

$$\mathbb{C}^p \longrightarrow \mathbb{C}^p/L$$

is a homomorphism. This fact gives us the possibility to extend the map λ to a map

$$\Lambda : \mathcal{D}(X) \longrightarrow \text{Jac}(X),$$

which is defined on the set of *all divisors*. We simply define this map by

$$\Lambda(D) = \sum_{a \in X} D(a)\lambda(a) \qquad \text{(finite sum)}.$$

The diagram

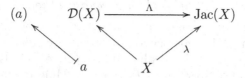

$$(a) \qquad \mathcal{D}(X) \xrightarrow{\quad \Lambda \quad} \text{Jac}(X)$$

commutes. The map Λ has been constructed in such a way that it is a homomorphism of groups.

11.3 Remark. *Let $\mathcal{D}^{(0)}(X)$ be the set of divisors of degree zero. The restriction of Λ to $\mathcal{D}^{(0)}(X)$,*

$$\Lambda : \mathcal{D}^{(0)}(X) \longrightarrow \text{Jac}(X),$$

is independent of the choice of the base point q.

The well-known Abel's theorem states the following.

11.4 Theorem. *A divisor $D \in \mathcal{D}(X)$ is the divisor of a meromorphic function if and only if the following two conditions are satisfied:*

1) $\deg D = 0$, *i.e.* $D \in \mathcal{D}^{(0)}(X)$.

2) $\Lambda(D) = 0$ *(in* $\mathrm{Jac}(X)$*)*.

This theorem can be formulated as a theorem about integrals of algebraic functions without talking about Riemann surfaces. The second condition was proved in this form by N.A. Abel in 1828. The converse was proved for the first time by A. Clebsch in 1865.

Before the proof, we formulate some corollaries.

Corollary 1. *The map*
$$\lambda : X \longrightarrow \mathrm{Jac}(X)$$
is injective in the case $p \geq 1$.

Proof of Corollary 1. Let $a, b \in X$, such that
$$a \neq b, \quad \lambda(a) = \lambda(b).$$

By Abel's theorem, there exists a meromorphic function f with
$$(f) = (a) - (b).$$

But this would be a function of degree one; it would give a biholomorphic map from X onto the Riemann sphere.   □

Corollary 2. *Let $p = 1$. The map*
$$\lambda : X \longrightarrow \mathrm{Jac}(X) = \mathbb{C}/L$$
is a biholomorphic map of Riemann surfaces. In particular, every Riemann surface of genus one is biholomorphically equivalent to a torus $\mathbb{C}/L$.

Abel's theorem for elliptic functions is a consequence of the general theorem stated in Theorem 11.4. For the proof of the second corollary, we simply observe that the map λ is holomorphic. It is injective, but also surjective, since its image is open and compact in $\mathbb{C}/L$. Hence λ is bijective, which implies that it is biholomorphic.

The Proof of Abel's Theorem

This proof rests on a certain period relation for the so-called normalized (abelian) differentials of the third kind,*) which we shall now derive.

---

*) One subdivides the meromorphic differentials in those of the first kind (everywhere holomorphic), those of the second kind (the residues of all poles vanish), and those of the third kind (all poles are simple).

Consider a normal form with a corresponding projection

$$\pi : \mathcal{X}(p) \longrightarrow X,$$

and denote by

$$\alpha_1, \ldots, \alpha_p, \ \beta_1, \ldots, \beta_p$$

the corresponding canonical system. Let a, b be two different points on X. By an abelian differential of the third kind with respect to these two points, we understand a differential $\omega = \omega_{ab}$ with the following properties:

1) ω is holomorphic outside $\{a, b\}$.
2) ω has poles of order one at a and b.
3) $\mathrm{Res}_a \, \omega = -\,\mathrm{Res}_b \, \omega = 1$.

We know that such a differential exists. Of course, ω is determined only up to an everywhere holomorphic differential. We assume that neither of the two points is contained in the canonical system. We can then normalize ω in such a way that

$$\int_{\alpha_k} \omega = 0 \qquad \text{for } k = 1, \ldots, p.$$

After this normalization, ω is uniquely determined. We call it the *normalized abelian differential of the third kind*. This normalized differential satisfies an important *period relation*.

11.5 Lemma. *Let*

$$\pi : \mathcal{X}(p) \longrightarrow X$$

be the natural projection with respect to a given normal form, let $\alpha_1, \ldots, \alpha_p$, $\beta_1, \ldots, \beta_p$ the corresponding canonical system, and let $\omega_1, \ldots, \omega_p$ be the corresponding canonical basis of $\Omega(X)$. Let a, b be two points which do not lie on the canonical system, and let ω_{ab} be the corresponding normalized abelian differential of the third kind. We then have

$$\int_{\beta_k} \omega_{ab} = 2\pi i \int_b^a \omega_k \qquad (1 \le k \le p),$$

where, on the right-hand side, the integral is taken along some path from b to a which does not meet the canonical system.

Note. This integral does not depend on the choice of the curve, since the complement of the canonical system is simply connected.

Proof. In the complement of the canonical system, ω_k admits a (holomorphic) primitive

$$f_k : X_0 \longrightarrow \mathbb{C}, \qquad \omega_k = df_k.$$

We denote the corresponding function in the interior of the $4p$-gon by

$$f_k : \mathcal{X}(p)^\circ \longrightarrow \mathbb{C},$$

and similarly for the pulled-back differential of ω_{ab}. We know that f_k extends to $\mathcal{X}(p)$ as a continuous function. From the behavior of f_k, we obtain

$$\oint_{\partial \mathcal{X}(p)} f_k \omega_{ab} = \int_{\beta_k} \omega_{ab}.$$

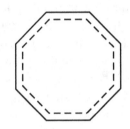

We want to compute the integral on the left-hand side by means of the residue theorem, and prefer for this purpose to argue on the Riemann surface and not on the $4p$-gon. For this, we shrink the $4p$-gon by a factor $0 < r < 1$, which can be arbitrarily close to 1. The shrunken $4p$-gon is denoted by $\mathcal{X}_r(p)$. We run through $\partial \mathcal{X}_r(p)$ in the usual orientation and denote the image curve in X by $\partial(r)$.

A simple continuity argument shows that

$$\oint_{\partial \mathcal{X}(p)} f_k \omega_{ab} = \lim_{r \longrightarrow 1} \oint_{\partial(r)} f_k \omega_{ab}.$$

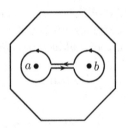

The curve $\partial(r)$ runs in the complement of the canonical system. We draw a small circle around each of the points a and b and join the circles by a segment. This gives a closed curve α in $X_0 - \{a, b\}$ which is freely homotopic to $\partial(r)$. It is easy to prove this in the $4p$-gon.

By the homotopic version of the Cauchy integral theorem for the Riemann surface $X_0 - \{a, b\}$, we obtain

$$\oint_{\partial(r)} f_k \omega_{ab} = \int_\alpha f_k \omega_{ab}.$$

The integral along α equals the sum over the residues of

$$f_k \omega_{ab},$$

so it equals $f(a) - f(b)$. Since f_k is a primitive of ω_k, we obtain

$$f(a) - f(b) = \int_a^b \omega_k.$$

This proves the lemma. $\square$

Besides Lemma 11.5, we need another lemma which generalizes the fact that the winding number (defined by the index integral) is an integer.

11.6 Lemma. *Let $f : X \to \mathbb{C}^*$ be a holomorphic function without zeros on a Riemann surface X. The integral*

$$\frac{1}{2\pi i} \int_\alpha \frac{df}{f}$$

is an integer for every closed curve α.

Proof. If f has a holomorphic logarithm F, we have

$$\frac{1}{2\pi i} \int_\alpha \frac{df}{f} = \frac{1}{2\pi i} \left(F(\alpha(1)) - F(\alpha(0)) \right)$$

for an arbitrary (not necessarily closed) curve α.

In general, each point of X admits an open neighborhood in which f admits a holomorphic logarithm. The construction of the *analytisches Gebilde* now shows that there exists a Riemann surface $\tilde{X}$ and a covering

$$p : \tilde{X} \longrightarrow X$$

such that the pullback $\tilde{f} = f \circ p$ admits a holomorphic logarithm $\tilde{F}$. We lift α to a curve $\tilde{\alpha}$ on $\tilde{X}$ and obtain, by means of the transformation invariance of the line integral,

$$\frac{1}{2\pi i} \int_\alpha \frac{df}{f} = \frac{1}{2\pi i} \int_{\tilde{\alpha}} \frac{d\tilde{f}}{\tilde{f}} = \frac{1}{2\pi i} \left(\tilde{F}(\tilde{\alpha}(1)) - \tilde{F}(\tilde{\alpha}(0)) \right).$$

Now $\tilde{F}(\tilde{\alpha}(1))$ and $\tilde{F}(\tilde{\alpha}(0))$ are both logarithms of $f(\alpha(0)) = f(\alpha(1))$. They differ by an integral multiple of $2\pi i$. $\qquad\square$

After these preparations, we now prove one direction of Abel's theorem, namely that the conditions 1) and 2) are necessary. For the condition 1), this is already known. It remains to show the following.

Claim. *If f is a nonzero meromorphic function, then*

$$\Lambda((f)) = 0 \quad (in \ \mathrm{Jac}(X)).$$

Proof. We choose a normal form of X:

$$\mathcal{X}(p) \longrightarrow X.$$

We may assume that none of the poles or zeros of f is contained in the canonical system. (Even though we have not mentioned it elsewhere, it should be clear that the construction of the normal form can be performed in such a way that finitely many points are outside the canonical system.)

We also choose some base point q in which f has neither a pole nor a zero and which also avoids the canonical system. We write the divisor of f in the form

$$(f) = (a_1) + \ldots + (a_n) - (b_1) - \ldots - (b_n),$$

where a_i are the zeros and b_i the poles of f. Then we consider the differential

$$\frac{df}{f} - \sum_{k=1}^{n} (\omega_{a_k q} - \omega_{b_k q}).$$

Here ω_{ab} denotes the normalized abelian differential of the third kind. This differential is holomorphic everywhere. Hence we have

$$\frac{df}{f} = \sum_{k=1}^{n} (\omega_{a_k q} - \omega_{b_k q}) + \sum_{k=1}^{n} c_k \omega_k,$$

with suitable constants c_k. Here $(\omega_1, \ldots, \omega_p)$ denotes the canonical basis (with respect to the given canonical system) of $\Omega(X)$. Now we compute the image of the principal divisor (f) in the Jacobi variety $\mathrm{Jac}(X)$. For this, we consider

$$\Lambda_j := \sum_{k=1}^{p} \left(\int_q^{a_k} \omega_j - \int_q^{b_k} \omega_j \right),$$

where the curve is chosen in the complement of the canonical system. The tuple

$$\Lambda := (\Lambda_1, \ldots, \Lambda_p) \in \mathbb{C}^p$$

is a well-defined representative of $\Lambda((f)) \in \mathrm{Jac}(X)$. Now the period relation for the abelian differentials of the third kind shows that

$$\Lambda_j = \frac{1}{2\pi i} \int_{\beta_j} \frac{df}{f} - \frac{1}{2\pi i} \sum_{k=1}^{p} c_k \int_{\beta_j} \omega_k.$$

Obviously, $\Lambda := (\Lambda_1, \ldots, \Lambda_p)$ is an element of the period lattice if the following two conditions are satisfied:

a)

$$\frac{1}{2\pi i} \int_{\beta_j} \frac{df}{f} \in \mathbb{Z},$$

b)

$$\frac{1}{2\pi i} c_k \in \mathbb{Z}.$$

Making use of

$$\int\limits_{\alpha_k} \frac{df}{f} = c_k,$$

then a) and b) follow from Lemma 11.6.   □

It remains to show that the following claim is true.

Claim. *The conditions 1) and 2) in Theorem 11.4 are sufficient.*
We start with a divisor

$$D = (a_1) + \ldots + (a_n) - (b_1) - \ldots - (b_n).$$

We choose a further base point b_0 which is different from $a_1, \ldots, a_n, b_1, \ldots, b_n$, and which is not contained in the canonical system. We consider the differential

$$\omega := \sum_{k=1}^{n} (\omega_{a_k b_0} - \omega_{b_k b_0}) + \sum_{k=1}^{n} c_k \omega_k,$$

with constants c_k which still have to be determined. Then we consider the function f, where

$$f(a) := \exp \int\limits_{b_0}^{a} \omega.$$

If this expression does not depend on the choice of the curve from b_0 to a, then f is meromorphic on X with a divisor $(f) = D$. What remains to be shown is the following:

Let $\Lambda(D) = 0$ in $\mathrm{Jac}(X)$. Then the constants c_k can be chosen in such a way that

$$\frac{1}{2\pi i} \int\limits_{\alpha_k} \omega, \quad \frac{1}{2\pi i} \int\limits_{\alpha_k} \omega \quad (1 \le k \le p),$$

are integral numbers.
We have

$$\int\limits_{\alpha_k} \omega = c_k \quad (1 \le k \le p)$$

and, because of the period relations in Lemma 11.5,

$$\int\limits_{\beta_j} \omega = 2\pi i \sum_{k=1}^{p} \left(\int\limits_{q}^{a_k} \omega_j - \int\limits_{q}^{b_k} \omega_j \right) + \sum_{k=1}^{p} c_k \int\limits_{\beta_j} \omega_k.$$

Now we use the assumption

$$\Lambda(D) = 0 \ \text{ in } \ \mathrm{Jac}(X).$$

This means that

$$\sum_{k=1}^{p} \left(\int_q^{a_k} \omega_j - \int_q^{b_k} \omega_j \right) = n_j + \sum_{k=1}^{p} m_k \int_{\beta_k} \omega_j$$

with integers n_j, m_k. If we define

$$c_k := -2\pi i m_k,$$

everything has been shown and Abel's theorem is proved.   □

Exercise for Sect. IV.11

1. How does the map $\Lambda : \mathcal{D}(X) \to \mathrm{Jac}(X)$ change if the base point is changed?

12. The Jacobi Inversion Problem

Let $\mathrm{Pic}^{(0)}(X)$ be the group of divisor classes of degree 0. From Abel's theorem, it follows that the natural map

$$\Lambda : \mathcal{D}^{(0)} \longrightarrow \mathrm{Jac}(X)$$

induces an injective map

$$\mathrm{Pic}^{(0)}(X) \longrightarrow \mathrm{Jac}(X).$$

It is natural to ask whether this map is also surjective. We shall see that the answers is yes. The map

$$\mathrm{Pic}^{(0)}(X) \longrightarrow \mathrm{Jac}(X)$$

will turn out to be an isomorphism. One can interpret this as follows: the bijection equips $\mathrm{Pic}^{(0)}(X)$ with a structure in the form of a complex torus.

Actually, we shall prove more. For this, we introduce the *symmetric power* of a set X. The nth Cartesian power is defined by

$$X^n = X \times \ldots \times X \qquad (n \text{ times}).$$

The symmetric group (the permutation group of the digits $1, \ldots, n$) S_n acts on X^n by

$$\sigma(x_1, \ldots, x_n) := (x_{\sigma^{-1}(1)}, \ldots, x_{\sigma^{-1}(n)}).$$

If we identify two n-tuples iff they differ by such a permutation, we obtain the so-called nth symmetric power,

$$X^{(n)} := X^n / S_n.$$

12.1 Remark. *The assignment*

$$X^{(n)} \longrightarrow \mathcal{D}(X),$$

$$(a_1, \dots, a_n) \longmapsto (a_1) + \dots + (a_n),$$

defines a bijection of the nth symmetric power $X^{(n)}$ with the set of divisors with the properties

$$D \geq 0, \quad \deg D = n.$$

Now, let X be a Riemann surface with base point q. We consider the map

$$\Lambda = \Lambda_n : X^{(n)} \longrightarrow \mathrm{Jac}(X),$$

which has been defined by

$$\Lambda_n(a_1, \dots, a_n) = \left(\int_q^{a_1} \omega_j + \dots + \int_q^{a_n} \omega_j \right)_{1 \leq j \leq p}.$$

The Jacobi inversion theorem*) states the following.

12.2 Theorem. *Let X be a compact Riemann surface of genus p.*

1) *The map*

$$\Lambda_p : X^{(p)} \longrightarrow \mathrm{Jac}(X)$$

is surjective. Its fibers are connected.

2) *There exists an open, dense subset $U \subset X^{(p)}$ such that the restriction of Λ_p to U defines a topological map from U onto an open, dense subset of $\mathrm{Jac}(X)$.*

Of course, $X^{(p)} = X^p / S_p$ has been equipped here with the quotient topology of the p-fold Cartesian product (equipped with the product topology).

(Why the pth and not another symmetric power? This is explained by a dimensional consideration: $\mathrm{Jac}(X)$ is a torus of complex dimension p.)

Jacobi's theorem tells us that Λ_p is "nearly bijective". But only in the case $p = 1$ is it really bijective. The analysis of the "degeneration locus" of Λ_p is very interesting but difficult.

Before the proof of the inversion theorem we give an important corollary.

---

*) This name is not correct historically. In Theorem VI.13.13, we will see that the map Λ_p defines a bijection between the meromorphic functions on $X^{(p)}$ and on $\mathrm{Jac}(X)$. This is the true solution of the Jacobi inversion problem. We shall treat this question and will give more historical background in connection with the proof of Theorem VI.13.13.

12.3 Corollary. *The map*

$$\Lambda : \mathrm{Pic}^{(0)}(X) \overset{\sim}{\longrightarrow} \mathrm{Jac}(X)$$

is bijective.

Proof of Theorem 12.2.

First step. *There exists an open, dense subset U of the p-fold Cartesian product X^p such that the restriction of Λ_p (more precisely, its composition with the natural projection $X^p \to X^{(p)}$) is locally topological.*

For the proof, we use an argument from calculus. Let $U, V \subset \mathbb{C}^n$ be open subsets and let $\varphi : U \longrightarrow V$ be a map with the following properties:

1) The functions

$$z_\nu \longmapsto \varphi_\mu(z_1, \ldots, z_n) \qquad (1 \le \nu, \mu \le n)$$

are holomorphic for fixed $z_1, \ldots, z_{\nu-1}, z_{\nu+1}, \ldots, z_n$; the (complex) partial derivatives are continuous.

2) The matrix of the complex derivatives

$$J_{\mathbb{C}}(\varphi, z) := \left(\frac{\partial \varphi_\mu}{\partial z_\nu} \right)_{1 \le \mu, \nu \le n}$$

is invertible for all $z \in U$.

Claim. Then φ is locally topological.

For the proof, we observe that it follows from the existence and continuity of the complex partial derivatives that φ is continuously partially differentiable in the sense of real analysis. Let $J_{\mathbb{R}}(\varphi, z)$ be the real $2n \times 2n$ Jacobi matrix. From the Cauchy–Riemann differential equations, we obtain a formula to compute $J_{\mathbb{R}}$ from $J_{\mathbb{C}}$. We can use this to show that

$$\det J_{\mathbb{R}}(\varphi, z) = |\det J_{\mathbb{C}}(\varphi, z)|^2 .$$

Hence the claim follows from the real theorem of invertible functions. We used the same method in [FB], Chap. I, to reduce the theorem of invertible functions to its real analogue, and also used this kind of argument in the appendix B of Sect. I.6 ("A Theorem of Implicit Functions").

For the proof of the first step, it is sufficient to construct for each nonempty open subset $U \subset X^p$ a nonempty open subset $U_0 \subset U$ such that the restriction of Λ_p to U_0 is locally topological. We can assume that U is of the form $U = U_1 \times \ldots \times U_p$ with disks

$$\varphi_i : U_i \longrightarrow \mathbb{E}.$$

Our holomorphic differentials ω_ν correspond, in these disks, to holomorphic functions

$$f_\nu : \mathbb{E} \longrightarrow \mathbb{C}.$$

The period map is changed by a translation only if we change the base point q. Hence we can assume that the base point corresponds to $(0, \ldots, 0) \in \mathbb{E}^p$. In these coordinates, the period map is described by

$$\Lambda_p : \mathbb{E}^p \longrightarrow \mathbb{C}^p,$$
$$(z_1, \ldots, z_p) \longmapsto (A_1(z_1, \ldots, z_p), \ldots, A_p(z_1, \ldots, z_p)),$$
$$A_\nu(z_1, \ldots, z_p) := \sum_{\mu=1}^{p} \int_0^{z_\mu} f_\nu(\zeta) \, d\zeta.$$

The complex Jacobian of this map is

$$(f_\nu(z_\mu)).$$

The functions f_ν are linearly independent, as are the ω_ν. We have to show that they are invertible for at least one tuple $(z_1, \ldots, z_p) \in U$. This is an easy consequence of the linear independence of the functions $f_1, \ldots, f_p$ (see Exercise 1).

Second step. *Surjectivity of Λ_p.*
Consider a point

$$C = (C_1, \ldots, C_p) \in \mathbb{C}^p / L = \mathrm{Jac}(X).$$

We have to construct a divisor

$$D \geq 0, \quad \deg D = p,$$

such that

$$\Lambda_p(D) = C \quad \text{in} \quad \mathbb{C}^p / L.$$

For the proof, we consider a tuple $(a_1, \ldots, a_p) \in X^p$ such that a full neighborhood of this tuple is mapped topologically onto an open subset of $\mathrm{Jac}(X)$ by Λ_p (see step 1). Consider

$$C' = \Lambda_p(a_1, \ldots, a_p) \quad (= \Lambda_p((a_1) + \cdots + (a_p))).$$

For sufficiently large n, the point

$$C' + C/n$$

is contained in this open subset of $\mathrm{Jac}(X)$ and hence in the image of Λ_p. Hence there exists a p-tuple

$$(b_1, \ldots, b_p) \in X^p$$

with

$$\Lambda_p(b_1, \ldots, b_p) = C' + C/n = \Lambda_p(a_1, \ldots, a_p) + C/n$$

or
$$C = n(\Lambda_p((b_1) + \cdots (b_p) - (a_1) - \cdots - (a_p))).$$

Now we consider the divisor

$$D := n(b_1) + \cdots + n(b_p) - n(a_1) - \cdots - n(a_p) + p(q).$$

Its degree is p. From the Riemann–Roch theorem, we obtain the existence of a meromorphic function
$$f \in \mathcal{L}(D), \quad f \neq 0.$$

The divisor

$$\tilde{D} = (f) + D = (f) + (b_1) + \cdots + (b_p) - (a_1) - \cdots - (a_p) + p(q)$$

is ≥ 0 and has degree p. By Abel's theorem (the "necessary" part), we have $\Lambda((f)) = 0$ and hence
$$\Lambda(\tilde{D}) = C.$$

Remark. We have used the Riemann–Roch theorem for the proof of the surjectivity of Λ_p. By using fundamental results of complex analysis for several variables, we could have avoided the Riemann–Roch theorem. Those who are already familiar with the following argument will understand it immediately. $\Lambda_p(X^p)$ is an analytic subset of $\mathrm{Jac}(X)$ which has dimension p, by the first step. Hence it must agree with $\mathrm{Jac}(X)$.

Third Step. *The fibers $\Lambda_p^{-1}(C)$ of the map*

$$\Lambda_p : X^{(p)} \longrightarrow \mathrm{Jac}(X)$$

are connected.

This will complete the proof of the Jacobi inversion theorem, since by step 1 there exists an open, discrete subset in X^p, and hence also an open, dense subset in $U \subset X^{(p)}$, such that each point $a \in U$ is isolated in its fiber

$$a \in \Lambda_p^{-1}\left(\Lambda_p(a)\right).$$

Since the fiber is connected, it can consist only of the point a. This means that the restriction of Λ_p to U is injective. By the first step, we can choose U in such a way that the map $U \to \mathrm{Jac}(X)$ is open. Then the image $V \subset \mathrm{Jac}(X)$ is open and $U \to V$ is topological.

Proof of the third step. Let $D \in X^{(p)}$. We consider D as a divisor:

$$D \geq 0, \quad \deg D = p.$$

Let

$$f \in \mathcal{L}(D) - \{0\}.$$

Then $D' = D + (f)$ also has the property

$$D' \geq 0, \ \deg D' = p.$$

By Abel's theorem, D and D' have the same image in $\mathrm{Jac}(X)$. We obtain the following result:

The map

$$H : \mathcal{L}(D) - \{0\} \longrightarrow X^{(p)}, \quad f \longmapsto D + (f),$$

defines a surjective map from $\mathcal{L}(D) - \{0\}$ onto the fiber $\Lambda_p^{-1}(\Lambda_p(D))$ which contains D.

More precisely, we have the following. If we identify two elements of $\mathcal{L}(D) - \{0\}$ which differ only by a constant factor, we obtain the associated projective space $P(\mathcal{L}(D))$. The above map induces a bijection

$$P(\mathcal{L}(D)) \xrightarrow{\sim} \Lambda_p^{-1}(\Lambda_p(D)).$$

The space $\mathcal{L}(D)$ is a finite-dimensional vector space and hence carries a well-defined topology. (We choose a basis, identify $\mathcal{L}(D)$ with $\mathbb{C}^d$, and transport the usual topology from $\mathbb{C}^d$. This topology on $\mathcal{L}(D)$ is independent of the choice of the basis.)

For the proof of the connectedness of the fiber, it is sufficient to show the following:

The map

$$\mathcal{L}(D) - \{0\} \longrightarrow X^{(p)}, \quad f \longmapsto D + (f),$$

is continuous.

The proof of the continuity rests on the *continuity of the roots by variation of an analytic function.*

We want to prove the continuity at a given element $f \in \mathcal{L}(D) - \{0\}$. For this, we take f as the first element of some basis $f = f_1, \ldots, f_d$. Now we consider, on $\mathcal{L}(D)$, the maximum norm with respect to this basis. The topology on $\mathcal{L}(D)$ can be defined by this norm. From the assumption, we have

$$(f) + D = (a_1) + \cdots + (a_n)$$

with certain (not necessarily pairwise different) points $a_i \in X$. Let $U \subset X^{(p)}$ be a neighborhood of the image point. We have to construct $\varepsilon > 0$ such that

$$\|g - f\| < \varepsilon \Longrightarrow (g) + D \in U.$$

Since the map $X^p \to X^{(p)}$ is open and continuous, we can assume that U is the image of a set of the form $W_1 \times \cdots \times W_p$ with open neighborhoods $a_i \in W_i \subset X$. Hence we have to show that

$$\|g - f\| < \varepsilon \Longrightarrow (g) + D = (b_1) + \cdots + (b_n) \quad \text{with} \quad b_i \in W_i.$$

Since we can shrink the neighborhoods W_i, we can realize them in the following way:

$$a_i = a_j \implies W_i = W_j, \qquad a_i \neq a_j \implies W_i \cap W_j = \emptyset.$$

We can assume that W_i is biholomorphically equivalent to a disk and (after shrinking this disk) that it has a "good" boundary. A further assumption which we can make is that f has neither poles nor zeros on the boundary of W_i. A simple compactness argument now shows that $\varepsilon > 0$ can be chosen so small that every g with $\|g - f\| < \varepsilon$ also has no zeros or poles on the boundary of W_i.

On an open neighborhood W of $\bar{W}_1 \cup \ldots \cup \bar{W}_n$, we find a meromorphic function h which fits D there. This means $\mathrm{Ord}(h, a) = D(a)$ for all $a \in W$. The functions $F := fh$ and, more generally $G := gh$ for arbitrary $g \in \mathcal{L}(D)$ are holomorphic on W. We know the numbers of the zeros of the function in U_i. These numbers are n_i, where n_i describes how often a_i occurs among the $a_1, \ldots, a_n$. On the other hand, the number of zeros is given by the integral

$$\frac{1}{2\pi i} \int\limits_{\partial U_i} \frac{dF}{F}.$$

Since the zero-counting integral is always an integer, we obtain by a continuity argument

$$\frac{1}{2\pi i} \int\limits_{\partial U_i} \frac{dG}{G} = \frac{1}{2\pi i} \int\limits_{\partial U_i} \frac{dF}{F} = n_i.$$

Hence the function has precisely n_i zeros in U_i (counted with multiplicity). If we use the index j with $a_j = a_i$ for them, we obtain in total n zeros $b_1, \ldots, b_n$, with $b_i \in W_i$, which represent the divisor of G inside $W_1 \cup \ldots \cup W_n$. Hence the divisor $(g) + D$ inside $W_1 \cup \ldots \cup W_n$ agrees with the divisor $(b_1) + \ldots + (b_n)$. Since we know $(g) + D \geq 0$ everywhere, we obtain generally

$$(g) + D \geq (b_1) + \cdots + (b_n).$$

Equality must hold, since the degrees are equal. This proves the claimed continuity.   □

Functions with Several Periods

Let X be a compact Riemann surface of genus $p > 0$. We consider a nonconstant meromorphic function

$$f : X \longrightarrow \bar{\mathbb{C}}.$$

This is holomorphic on the complement of a finite subset $\mathcal{S} = f^{-1}(\infty)$,

$$f_0 : X_0 \longrightarrow \mathbb{C}, \quad X_0 := X - \mathcal{S}.$$

The function f_0 induces in an obvious way a map

$$f_0 : X_0^{(p)} \longrightarrow \mathbb{C}^{(p)}.$$

In the appendix to this section, we define by means of elementary symmetric functions a topological map $E : \mathbb{C}^{(n)} \xrightarrow{\sim} \mathbb{C}^n$. We compose this map with f_0 to obtain a map $F : X^{(p)} \to \mathbb{C}^p$. This is nothing but a p-tuple of functions

$$F_1, \ldots, F_p : X_0^{(p)} \longrightarrow \mathbb{C}.$$

We compose this with the "inverse map" of

$$X^{(p)} \longrightarrow \mathrm{Jac}(X)$$

to obtain a p-tuple of functions $A_1, \ldots, A_p$ which are defined on an open, dense subset of $\mathrm{Jac}(X) = \mathbb{C}^p/L$. Pulling them back to $\mathbb{C}^p$, we obtain

a) an open subset $U \subset \mathbb{C}^p$ with the property

$$a \in U \implies a + g \in U \text{ for all } g \in L;$$

b) a p-tuple of functions

$$A_1, \ldots, A_p : U \longrightarrow \mathbb{C}$$

with the property

$$A_\nu(z + g) = A_\nu(z) \text{ for } z \in L.$$

In this way, we obtain functions with $2p$ periods.

In the case $p = 1$ it is clear that we are dealing with elliptic functions. So, for $p > 1$, we are forced to deal with the following problems:

1) Develop the notion of a meromorphic function of several complex variables.

2) Show that the functions $A_1, \ldots, A_p$ are meromorphic on $\mathbb{C}^p$.

3) Develop a theory of meromorphic functions on $\mathbb{C}^p$ which have L as their period lattice.

These questions determine the subjects of the rest of this book.

Appendix to Sect. 12. Continuity of Roots

We denote by

$$E_\nu = \sum_{k_1 + \ldots + k_n = \nu} z_1^{k_1} \ldots z_n^{k_n}, \qquad 1 \le \nu \le n,$$

the elementary symmetric polynomials. The induced map

$$E : \mathbb{C}^n \longrightarrow \mathbb{C}^n, \quad z \longmapsto (E_1(z), \ldots, E_n(z)),$$

factorizes for trivial reasons through the symmetric power

$$\mathbb{C}^{(n)} = \mathbb{C}^n / S_n :$$

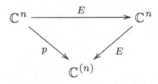

12.4 Proposition. *The map which is induced by the elementary symmetric functions*

$$\mathbb{C}^{(n)} \overset{E}{\underset{\sim}{\longrightarrow}} \mathbb{C}^n$$

is topological.

Remark. Instead of elementary symmetric functions, we could take the power sums

$$T_\nu(z) = \sum_{j=1}^n z_j^\nu \qquad (1 \le \nu \le n)$$

(since it can be shown by elementary algebra that the T_ν can be written as polynomials in the E_ν and conversely).

Proof.

1) The continuity of E follows from the definition of the quotient topology on $\mathbb{C}^n / S_n$.

2) $E : \mathbb{C}^{(n)} \to \mathbb{C}^n$ is bijective.

3) We define the inverse map as follows. Let $(\alpha_0, \ldots, \alpha_{n-1}) \in \mathbb{C}^n$. We consider the normalized polynomial with these coefficients and factorize it:

$$X^n + \alpha_{n-1} + \cdots \alpha_0 = (X - a_1) \cdot \ldots \cdot (X - a_n).$$

The zeros $a_1, \ldots, a_n$ are determined up to their ordering. So, they define a point in $\mathbb{C}^{(n)}$. It is well known that the coefficients α_i up to the sign, are the elementary symmetric expressions in the zeros. This proves the bijectivity.

4) $E : \mathbb{C}^n \to \mathbb{C}^n$ is proper, i.e. the inverse image of a compact subset is compact. This follows from a version of the continuity of roots, which we know already (Lemma I.3.5).

A continuous, bijective, and proper map is topological, since the images of closed subsets are closed (and hence the inverse images of closed subsets under E^{-1} are closed, which shows that E^{-1} is continuous.) $\qquad\square$

Exercise for Sect. IV.12

1. Let
$$g_1, \ldots, g_n : \mathcal{M} \longrightarrow \mathbb{C}$$
be n linearly independent functions on some set $\mathcal{M}$. Show that there exists a finite subset $\mathcal{X} \subset \mathcal{M}$, consisting of n elements, such that the restrictions
$$g_1|\mathcal{X}, \ldots, g_n|\mathcal{X}$$
are linearly independent.

Appendices to Chapter IV.
Dimension Formulae for Spaces of Modular Forms

13. Multicanonical Forms

Let $X = (X, \mathcal{A})$ be a Riemann surface. We recall the notion of a holomorphic differential.

By definition, a holomorphic differential is given by a family
$$\omega = (\omega_\varphi)_{\varphi \in \mathcal{A}} \qquad (\varphi : U_\varphi \longrightarrow V_\varphi)$$
of holomorphic functions $\omega_\varphi : V_\varphi \to \mathbb{C}$ such that, in the intersection $U_\varphi \cap U_\psi$ of two charts $\varphi, \psi \in \mathcal{A}$, the transformation formula
$$\gamma^* \omega_\varphi = \omega_\psi \text{ with } \gamma := \psi \circ \varphi^{-1}$$
is valid. This means that if $a \in U_\varphi \cap U_\psi$ is a point in the intersection and if $z = \varphi(a)$, $w = \psi(a)$ are the corresponding chart points, then
$$\omega_\psi(w) = \gamma'(z) \omega_\varphi(z).$$

This notion can be generalized if the derivative $\gamma'(z)$ is replaced by a power. In this way, we arrive at the notion of a *higher differential* or a *multicanonical form*. We prefer the latter terminology.

13.1 Definition. *A (holomorphic) multicanonical form of weight $m \in \mathbb{Z}$ on an analytic atlas $\mathcal{A}$ is a family*

$$\omega = (\omega_\varphi)_{\varphi \in \mathcal{A}} \qquad (\varphi : U_\varphi \longrightarrow V_\varphi)$$

of holomorphic functions $\omega_\varphi : V_\varphi \to \mathbb{C}$ such that for any two charts $\varphi, \psi \in \mathcal{A}$ the transformation formula

$$\omega_\psi(w) = \gamma'(z)^m \omega_\varphi(z)$$
$$(z = \varphi(a), \ w = \psi(a) \text{ with } a \in U_\varphi \cap V_\psi)$$

is valid.

Multi-canonical forms of weight 0 are 0-forms which can be identified with holomorphic functions on X. Multi-canonical forms of degree 1 are holomorphic differentials.

13.2 Supplement (supplement to Definition 13.1). *We can replace "holomorphic" by "meromorphic" in the definition and obtain in this way the notion of a meromorphic multicanonical form.*

Some of the rules for differentials immediately carry over to multicanonical forms. We have to observe that the powers of the derivative satisfy the same chain rule as the derivative itself. We collect together the most basic rules below:

1) Every holomorphic (or meromorphic) multicanonical form extends in a unique manner to a multicanonical form on the maximal atlas. Hence one can talk about multicanonical forms on a Riemann surface.

2) If $U \subset X$ is an open Riemann subsurface, we can define in a natural way the restriction $\omega|U$ of a multicanonical form on X to U. If $X = \bigcup_i U_i$ is an open covering and if, on each U_i, a multicanonical form ω_i of some fixed weight k is given, then we have the following.

   There exists a multicanonical form ω on X with $\omega_i = \omega|U_i$ for all i iff

$$\omega_i|(U_i \cap U_j) = \omega_j|(U_i \cap U_j) \text{ for all } i, j$$

   holds.

3) If $f : X \to Y$ is a holomorphic map of Riemann surfaces, then for a multicanonical form ω on Y, we can define a pullback $f^*\omega$ on X. This is a multicanonical form of the same type. This pullback has the following properties (and can be characterized by these properties):

   a) In the case of the canonical inclusion $f : U \hookrightarrow X$ of an open subsurface, f^* is the restriction in the sense of 2).

b) If $f : X \to Y$ and $g : Y \to Z$ are two holomorphic maps of Riemann surfaces, then

$$f^* \circ g^* = (g \circ f)^*.$$

c) Let $U \subset \mathbb{C}$ be an open subset of the plane, considered as a Riemann surface. Then the holomorphic (or meromomorphic) multicanonical forms on U correspond to the holomorphic (or meromorphic) functions on U. If f is a meromorphic function on U, then we may write

$$\omega = f(z)(dz)^m$$

for the corresponding multicanonical form of weight m.

Let $\varphi : U \to V$ be an analytic map between open subsets of the complex plane, let

$$\omega = g(w)(dw)^m$$

be a multicanonical form of weight m on V, and let

$$\varphi^*\omega = f(w)(dw)^m$$

be the pulled-back form; then

$$f(z) = \varphi'(z)^m g(\varphi(z)).$$

Algebraic Computation Rules for Multicanonical Forms

1) Multicanonical forms ω, ω' *of the same weight* can be added:

$$(\omega + \omega')_\varphi := \omega_\varphi + \omega'_\varphi.$$

A multicanonical form of the same type is obtained.

2) A holomorphic (or meromorphic) multicanonical form ω can be multiplied by a holomorphic (or meromorphic) function:

$$(f\omega)_\varphi := f_\varphi \omega_\varphi.$$

A multicanonical form of the same type is obtained.

The latter operation admits an important generalization:

3) Let ω, ω' be multicanonical forms of weight m, m'. We can define the product $\omega\omega'$ by

$$(\omega\omega')_\varphi := \omega_\varphi \omega'_\varphi,$$

and we obtain a multicanonical from of weight $m + m'$.

Since this product is different form the alternating product of differential forms, we sometimes write

$$\omega \otimes \omega' = \omega\omega'$$

and call $\omega \otimes \omega'$ the tensor product of the two forms. This tensor product is commutative, associative, and distributive for trivial reasons. So, we can define the powers

$$\omega^{\otimes n} = \omega^n := \omega \cdots \omega \quad (n \text{ times}).$$

This notation is compatible with the notation $f(z)(dz)^m$ already introduced for open subsets of the plane.

Let X be a connected Riemann surface and let ω be a meromorphic multicanonical form which is not identically zero. As in the case of functions and differentials, it can be shown that none of the components ω_φ vanishes. This observation allows us to define the multicanonical form ω^{-1} of weight $-m$ by

$$\left(\omega^{-1}\right)_\varphi = \left(\omega_\varphi\right)^{-1}.$$

Invariant Multicanoncal Forms

Let $\gamma : X \to X$ be a biholomorphic self-map of a Riemann surface X and let ω a multicanonical form on X. We call ω invariant under γ if $\gamma^*\omega = \omega$. More generally, let Γ be a group of biholomorphic transformations of X; we then call ω invariant under Γ if it is invariant under all $\gamma \in \Gamma$:

$$\gamma^*\omega = \omega \text{ for all } \gamma \in \Gamma.$$

As for functions and differentials, we have the following lemma.

13.3 Lemma. *Let Γ be a group of biholomorphic transformations of a Riemann surface X which acts freely on X, let $Y = X/\Gamma$ be the quotient surface, and let $\pi : X \to Y$ be the natural projection. The assignment*

$$\omega \longmapsto \pi^*\omega$$

defines a one-to-one correspondence between the set of holomorphic (or meromorphic) multicanonical forms on Y and the set of Γ-invariant holomorphic (or meromorphic) multicanonical forms on X.

In the special case $X = D \subset \mathbb{C}$ of an open subset of the plane, the invariance property $\gamma^*\omega = \omega$ for a multicanonical form $\omega = f(z)(dz)^m$ means nothing more than

$$\boxed{f(\gamma z)\left(\gamma'(z)\right)^m = f(z).}$$

In particular, if $D = \mathbb{H}$ is the upper half-plane and γ is the Möbius transformation

$$\gamma(z) = Mz = \frac{az + b}{cz + d}, \quad M = \begin{pmatrix} a & b \\ c & d \end{pmatrix} \in \mathrm{SL}(2, \mathbb{R}),$$

then the invariance means

$$\boxed{f(Mz)(cz + d)^{-2m} = f(z).}$$

Functions with such a transformation property are familiar to us from the theory of elliptic modular forms.

13.4 Proposition. *Let $\Gamma \subset \mathrm{SL}(2, \mathbb{R})$ be a subgroup whose image in* $\mathrm{Bihol}\,\mathbb{H}$ *acts freely. The holomorphic (or meromorphic) forms of weight m on the Riemann surface*

$$X = \mathbb{H}/\Gamma$$

are in one-to-one correspondence with the holomorphic (or meromorphic) functions on $\mathbb{H}$ with the transformation property

$$f(Mz) = (cz + d)^{2m} f(z) \text{ for all } M \in \Gamma.$$

Functions which have a transformation property of this type are called *automorphic forms* with respect to Γ. Modular forms are nothing more than automorphic forms with respect to special groups Γ, namely congruence subgroups of the modular group, where, in addition certain conditions usually have to be required.

The question arises of whether the theory of Riemann surface helps with the theory of modular forms. In fact, we can determine the dimensions of vector spaces of modular forms in many cases.

In the first step, we associate a divisor with a multicanonical form. Let ω be a meromorphic multicanonical form on the Riemann surface X, and let $a \in X$ be a given point. We choose an analytic chart $\varphi : U_\varphi \to V_\varphi$ at a, i.e. $a \in U_\varphi$. We assume that ω_φ does not vanish identically. If X is connected, this means that ω is not identically zero. The order of the function ω_φ at the point $z := \varphi(a)$ is independent of the choice of φ, since the order of a meromorphic function does not change if one multiplies it by a holomorphic function without zeros. Hence we can define the order of ω at a by

$$\mathrm{Ord}(\omega; a) := \mathrm{Ord}(\omega_\varphi; z).$$

If X is a connected compact Riemann surface and ω does not vanish identically, then the order is defined at all points. It is different from zero only on a discrete subset. Hence it is finite if X is compact. So we can associate with ω a divisor, which we denote by (ω). We obviously have

$$(\omega\omega') = (\omega) + (\omega').$$

The following remark justifies the terminology "multicanonical form".

13.5 Remark. *Let X be a (connected) compact Riemann surface and let K be a canonical divisor on X (the divisor of a meromorphic differential). The divisor of an arbitrary nonzero multicanonical form ω of weight m is equivalent to mK. As a consequence, we have*

$$\deg(\omega) = m(2p - 2) \quad (p = \text{genus of } X).$$

Proof. We choose a meromorphic differential ω_0 which is different from zero. Then ω/ω_0^m is a multicanonical form of weight 0, which corresponds to a meromorphic function which has a degree 0. □

We can also see that the divisors of two multicanonical forms ω, ω' of the same weight which are different from 0 are equivalent, since we have $\omega' = f\omega$, with a meromorphic function f. Then ω' is holomorphic if and only if

$$(f) \geq -(\omega').$$

Hence the vector space of all multicanonical forms of weight m, which we denote by $\Omega^{\otimes m}(X)$, is isomorphic to the Riemann–Roch space $\mathcal{L}((\omega'))$ and hence to the space $\mathcal{L}(mK)$, where K is some canonical divisor. This gives us the possibility to compute the dimension of $\Omega^{\otimes m}(X)$ by means of the Riemann–Roch theorem. For the moment we shall restrict ourselves to the case where the degree of a canonical divisor is positive, i.e. the case $p > 1$.

13.6 Theorem. *Let X be a compact Riemann surface of genus $p \geq 2$. The dimension of the vector space of all holomorphic multicanonical forms of weight m is*

$$\dim \Omega^{\otimes m}(X) = \begin{cases} 0 & \text{for } m < 0, \\ p & \text{for } m = 1, \\ (p-1)(2m-1) & \text{for } m > 1. \end{cases}$$

We translate this result into the language of automorphic forms as follows.

13.7 Corollary. *Let $\Gamma \subset \mathrm{SL}(2, \mathbb{R})$ be a subgroup whose image in $\mathrm{Bihol}\,\mathbb{H}$ acts freely, and is such that $\mathbb{H}/\Gamma$ is compact. The dimension of the vector space of all automorphic forms of weight k, i.e. of all holomorphic functions $f : \mathbb{H} \to \mathbb{C}$ with the transformation behavior*

$$f(Mz) = (cz + d)^k f(z) \quad \text{for all } M \in \Gamma,$$

equals

$$(p - 1)(k - 1) \quad (p = \text{genus of } \mathbb{H}/\Gamma).$$

for even $k > 2$. This dimension is equal to 0 for $k < 0$, 1 for $k = 0$, and p for $k = 2$.

(Uniformization theory implies $p \geq 2$.)

14. Dimensions of Vector Spaces of Modular Forms

We want to use the Riemann-Roch theorem to compute the dimensions of spaces of modular forms with respect to congruence subsroups of the modular group. All that we need about modular forms can be found in [FB], Chap. VI. The main difficulty comes from the fact that $\mathbb{H}/\Gamma$ is not compact. To overcome this difficulty, we compactify this space. Before this, we have to verify that this space is Hausdorff (compare Sect. III.2, Exercises 6 and 7):

14.1 Remark. *Let Γ be a subgroup of the modular group $\mathrm{SL}(2,\mathbb{Z})$. The quotient space $\mathbb{H}/\Gamma$ is Hausdorff.*

Proof. Let $a, b \in \mathbb{H}$ be two points which are inequivalent mod Γ. We have to show that there are two neighborhoods $U(a), U(b)$ such that no point from $U(a)$ is equivalent to a point from $U(b)$. (The images of $U(a), U(b)$ are then disjoint neighborhoods of the images of a, b in the quotient.) We give an indirect argument and assume the contrary. There then exist sequences $a_n \to a$, $b_n \to b$ such that a_n and b_n are equivalent, i.e. $M_n a_n = b_n$, $M_n \in \Gamma$. There exists a number $\delta > 0$ such that both sequences are contained in the set defined by

$$|x| \le \delta^{-1}, \quad y \ge \delta.$$

Because of [FB], Lemma VI.1.2, the sequence M_n is contained in a finite set. Taking a subsequence, we can assume that it is a constant M. Taking limits, we get $Ma = b$, in contradiction to the assumption that a and b are inequivalent. $\square$

This proof shows a little more: for every point a in the upper half-plane, there exists a small neighborhood $U(a)$ such that $M(U(a)) \cap U(a) \ne \emptyset$ implies that M is contained in the stabilizer of a ($M(a) = a$). This shows the following.

14.2 Lemma. *Let Γ be a subgroup of the modular group which, besides the unit matrix E and possibly $-E$, does not contain elements of finite order. Its image in $\mathrm{Bihol}\,\mathbb{H}$ acts freely. As a consequence, $\mathbb{H}/\Gamma$ carries a structure in the form of a Riemann surface. The natural map $\mathbb{H} \to \mathbb{H}/\Gamma$ is locally biholomorphic.*

We give some examples of such groups below.

14.3 Remark. *The so-called principal congruence subgroup of level $q \in \mathbb{N}$,*

$$\Gamma[q] := \mathrm{Kernel}\big(\mathrm{SL}(2,\mathbb{Z}) \longrightarrow \mathrm{SL}(2,\mathbb{Z}/q\mathbb{Z})\big),$$

is a subgroup of finite index, which for $q \ge 2$ does not contain elements of finite order which are different from $\pm E$.

Proof. The kernel of a homomorphism into a *finite* group always has a finite index. The elements of finite order of $SL(2, \mathbb{Z})$ are known ([FB], Proposition VI.1.8). This description proves the remark. $\square$

By a *congruence subgroup*, we understand a subgroup of $SL(2, \mathbb{Z})$ which contains $\Gamma[q]$ for suitable q. Congruence subgroups have a finite index in $SL(2, \mathbb{Z})$. The space $\mathbb{H}/\Gamma$ is not compact, as the construction of the fundamental domain of the modular group and its basic properties show. We want to compactify it by adding a finite number of points. For this, we extend the upper half-plane by cusps:

$$\mathbb{H}^* = \mathbb{H} \cup \bar{\mathbb{Q}}, \quad \bar{\mathbb{Q}} = \mathbb{Q} \cup \{\infty\}.$$

We recall that the modular group also acts on $\mathbb{H}^*$ by means of the usual formulae. Hence we can consider, for an arbitrary subgroup Γ of the modular group, the set

$$X_\Gamma := \mathbb{H}^*/\Gamma.$$

The points which have been added are the cusp classes, i.e. the elements of the set

$$S_\Gamma := \bar{\mathbb{Q}}/\Gamma.$$

We know ([FB], Lemma VI.5.3, Corollary) that this set is finite. In the case of the full modular group, it consists of one element. We want to introduce a topology on X_Γ. It will be defined as the quotient topology of a certain topology on $\mathbb{H}^*$. This topology will have the property that $\mathbb{H}$ is an open subset and that the topology induced by $\mathbb{H}^*$ is the usual topology on $\mathbb{H}$. We also want to have that the full modular group acts topologically on $\mathbb{H}^*$.

The essential part of the construction is to define the neighborhoods of the cusp $i\infty$. In particular, we must define when a sequence $z_n \in \mathbb{H}$ converges to $i\infty$. From the theory of modular forms, we can expect that this means $\operatorname{Im} z_n \to \infty$ (and not $|z|_n \to \infty$; the topology to be constructed will not be the topology induced by the Riemann sphere, which is also the reason why we prefer the notation $i\infty$ instead of ∞). So, the sets

$$U_C^* = U_C \cup \{i\infty\}$$

with

$$U_C := \{z, \quad \operatorname{Im} z > C\} \quad (C > 0)$$

should be typical neighborhoods of $i\infty$. Since we want the modular group to act topologically, the typical neighborhoods of a cusp $\kappa = M(i\infty)$, $M \in SL(2, \mathbb{Z})$, should be the transformed sets $M(U_C^*) = M(U_C) \cup \{\kappa\}$. These sets, called *horocycles*, can easily be described.

14.4 Remark. *Let $M \in SL(2, \mathbb{Z})$, $M(i\infty) = \kappa \in \mathbb{Q}$. Each set of the family $M(U_C)$, $C > 0$, is an open disk in the upper half-plane which touches the real axis at κ:*

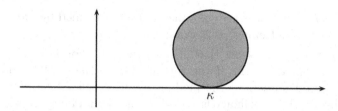

After these preparations, it should be clear how the topology has to be defined. A set $U \subset \mathbb{H}^*$ is called open if its intersection with $\mathbb{H}$ is open in the usual sense, and, if the cusp $\kappa = M(\mathrm{i}\infty)$ is contained in U, a horocycle $M(U_C^*)$ is contained in U for suitable C. It is easy to prove that this is a topology. We can also characterize this topology as follows.

14.5 Remark. *On $\mathbb{H}^* = \mathbb{H} \cup \bar{\mathbb{Q}}$, there exists a unique topology with the following properties:*

1) *The upper half-plane is an open part; the topology induced by $\mathbb{H}^*$ gives the usual topology on $\mathbb{H}$.*

2) *The modular group acts topolologically on $\mathbb{H}^*$.*

3) *A subset $U \subset \mathbb{H}$ is a neighborhood of $\mathrm{i}\infty$ iff it contains a set U_C^* (for suitable C).*

Again we point out that the topology on $\mathbb{H}^*$ is an unusual one, since it is not induced by the topology of the Riemann sphere. The set of cusps is discrete in $\mathbb{H}^*$. Its importance follows from the following statement.

14.6 Proposition. *Let Γ be a subgroup of finite index of the modular group, for example a congruence subgroup; then the quotient*

$$X_\Gamma := \mathbb{H}^*/\Gamma$$

is compact.

Proof. In the first step, we have to show that X_Γ is Hausdorff. For this, we have to show that two Γ-inequivalent points $a, b \in \mathbb{H}^*$ admit neighborhoods such that no point of one of the neighborhoods can be equivalent to a point of the other. We can assume that one of the points is a cusp; without loss of generality, we assume $b = \mathrm{i}\infty$. Again we use an indirect argument and assume that there exist sequences $a_n \to a$ and $b_n \to b = \mathrm{i}\infty$ such that a_n and b_n are equivalent, i.e. $M_n a_n = b_n$, $M_n \in \Gamma$. We have to distinguish the two cases of whether a is a cusp or not. We shall treat only the somewhat more difficult case, where a is a cusp. We choose $N \in \mathrm{SL}(2, \mathbb{Z})$, with $Na = \mathrm{i}\infty$. There exists $\delta > 0$ such that both of the sequences b_n and Na_n are contained in the set defined by $y \geq \delta$. We can modify the elements a_n and b_n by applying elements of the stabilizers Γ_a and Γ_b, respectively. This shows that it is possible to assume that both sequences are contained in the set defined by $|x| \leq \delta^{-1}$. Now the same proof as for Remark 14.1 works.

In the next step, we show that X_Γ is compact. For $\Gamma = \mathrm{SL}(2, \mathbb{Z})$, this follows easily from the form of the fundamental domain. If $\mathcal{F}$ denotes the usual fundamental domain of the modular group, then $\mathcal{F}^* := \mathcal{F} \cup \{i\infty\}$ is a compact subset of $\mathbb{H}^*$. Its image is the whole of X_Γ and hence this is compact. In the general case, we use the fact that the natural projection

$$X_\Gamma \longrightarrow X_{\mathrm{SL}(2,\mathbb{Z})}$$

is proper. This proves the compactness. $\square$

We shall show more, namely that X_Γ is a compact surface. (The existence of a mere compactification is not very remarkable, since we always have the possibility of a one-point compactification.) To prove this, we have to investigate the structure of X_Γ close to a cusp class. Here we can restrict ourselves to the cusp $i\infty$, since an arbitrary modular substitution $M \in \mathrm{SL}(2, \mathbb{Z})$ induces a topological map

$$X_\Gamma \longrightarrow X_{M\Gamma M^{-1}}, \quad [a] \longmapsto [Ma].$$

If we choose the constant C large enough, two points of U_C are Γ-equivalent if they can obtained from each other by translation. In other words, the natural map

$$U_C/\Gamma_\infty \longrightarrow \mathbb{H}/\Gamma$$

is then injective. Here Γ_∞ denotes the stabilizer of $i\infty$, i.e. all matrices $M \in \Gamma$ with the property $c = 0$. These are of the form

$$M = \pm \begin{pmatrix} 1 & b \\ 0 & 1 \end{pmatrix}.$$

The closure of U_C in $\mathbb{H}^*$ is

$$U_C^* := U_C \cup \{i\infty\}.$$

The natural map

$$U_C^*/\Gamma_\infty \longrightarrow X_\Gamma = \mathbb{H}^*/\Gamma$$

is injective too. It follows from the definition of the quotient topology that it is continuous and open. So, it defines a topological map from U_C^*/Γ_∞ onto an open neighborhood of the cusp class $[\infty]$ in X_Γ. Let R be the smallest positive number R such the translation $z \mapsto z + R$ is contained in the group Γ. We consider the disk $U_r(0)$ of radius $r := \exp(-2\pi i/R)$ in the complex plane. The function $z \to \exp(2\pi i z/R)$ defines a bijective map

$$U_C/\Gamma \longrightarrow U_r(0)^{\cdot}.$$

If we associate the origin with the cusp, we obtain an extension to a bijective map

$$U_C^*/\Gamma_\infty \longrightarrow U_r(0).$$

Now the topology of $\mathbb{H}^*$ has been defined in such a way that this map is topological. By composing it with the embedding into X_Γ, we obtain a topological map of the disk onto a neighborhood of $[i\infty]$ in X_Γ. Its inverse map is a chart on X_Γ. It is obvious that this chart is analytically equivalent to the charts which define the analytic structure of $\mathbb{H}/\Gamma$. The reason is based simply on the fact that the exponential map is locally biholomorphic. Collecting the above considerations together, we obtain the following result.

14.7 Proposition.   *For every congruence group Γ which, besides E and possibly $-E$, contains no elements of finite order, there can be constructed a structure in the form of a compact Riemann surface on $X_\Gamma = \mathbb{H}^*/\Gamma$, such that the following conditions are satisfied:*

1) $\mathbb{H}/\Gamma$ *is an open Riemann subsurface, and the natural map $\mathbb{H} \to X_\Gamma$ is locally biholomorphic.*

2) *Let $C > 0$ be large enough that the natural map $U_C/\Gamma_\infty \to \mathbb{H}/\Gamma$ is injective. Let $U_r(0) \to U_C^*/\Gamma_\infty$ be the inverse map of the map induced by $z \to \exp(2\pi i z/R)$. Then the composition*

$$U_r(0) \longrightarrow X_\Gamma$$

defines a biholomorphic map from the disk $U(0)$ onto an open neighborhood of the cusp class $i\infty$ in X_Γ.

3) *The map*

$$X_\Gamma \longrightarrow X_{M\Gamma M^{-1}}$$

which is induced by an arbitrary modular substitution $M \in \mathrm{SL}(2,\mathbb{Z})$ is biholomorphic.

It is clear that the analytic structure of the surfaces X_Γ is determined by the conditions 1)–3).

Let $f \in \{\Gamma, k\}$ be a meromorphic modular form of even weight $k = 2m$ (in the sense of [FB], Definition VI.2.1) and let ω_f be the associated multicanonical form on $\mathbb{H}/\Gamma$. We want to show that ω_f extends meromorphically to X_Γ. For this, we investigate ω_f at the cusp $i\infty$. We pull back ω_f to the chart domain $U_r(0)^{\cdot}$ and obtain a multicanonical form $g(q)(dq)^m$. Here $g(q)$ denotes the component of ω_f with respect to this chart. So, we have to show that $g(q)$ has an inessential singularity at the origin. We pull $g(q)(dq)^m$ back to U_C by means of the function $q := \exp(2\pi i z/R)$. This gives $f(z)(dz)^m$. The rule for the pullback of a multicanonical form gives

$$f(z) = \left(\frac{dq}{dz}\right)^m g(q) = \left(\frac{2\pi i}{R}\right)^m q^m g(q).$$

Now we see that g is meromorphic at the origin. The factor q^m induces only a shift of the orders. For example, let $m = 1$. We see that g has a removable singularity at the origin iff f vanishes at the cusp $i\infty$.

14.8 Proposition.   *The vector space $\Omega(X_\Gamma)$ of the holomorphic differentials on X_Γ is isomorphic to the space of cusp forms $[\Gamma, 2]_0$ of weight two. As a consequence, we have*

$$\dim[\Gamma, 2]_0 = p \quad (= genus\ of\ X_\Gamma).$$

Let f be a meromorphic modular form different from zero. In [FB], Sect. VI.2, we defined the order $\mathrm{Ord}(f; a)$ for an arbitrary point $a \in \mathbb{H}^*$. We recall that if a is a point in the upper half-plane, then $\mathrm{Ord}(f; a)$ is the usual order. If a is the cusp $i\infty$, then the order is defined by means of the Fourier expansion. For this, let us consider the smallest positive number $R > 0$ such that

$$\begin{pmatrix} 1 & R \\ 0 & 1 \end{pmatrix} \quad \text{or} \quad \begin{pmatrix} -1 & R \\ 0 & -1 \end{pmatrix}$$

is contained in Γ. Since the weight of f is even, we have $f(z + R) = f(z)$ and we have an expansion

$$f(z) = \sum_{n=-\infty}^{\infty} a_n q^n \quad (q = e^{2\pi i z / R}).$$

We define

$$\mathrm{Ord}(f; i\infty) := \min\{n, \quad a_n \neq 0\}.$$

If $\kappa = M(i\infty)$, $M \in \mathrm{SL}(2, \mathbb{Z})$, is an arbitrary cusp, we can replace Γ by $M\Gamma M^{-1}$ and f by $f|M$ to define

$$\mathrm{Ord}(f; \kappa) = \mathrm{Ord}(f|M; i\infty).$$

It is easy to see that this definition is independent of the choice of M. So, $\mathrm{Ord}(f; a)$ is defined for all $a \in \mathbb{H}^*$. It is trivial to show that this definition depends only on the Γ-equivalence class of a. So, we can define

$$\mathrm{Ord}(f; x) := \mathrm{Ord}(f; a), \quad x = [a] \in \mathbb{H}/\Gamma.$$

Only for finitely many $x \in X_\Gamma$ is the order different from zero. So, we have associated with a modular form $f \neq 0$ a divisor, which we denote by (f). Let us collect these results together.

14.9 Remark. *With any meromorphic modular form $f \neq 0$ of even weight $k = 2m$ we can associate a divisor (f) on the compact Riemann surface X_Γ, such that the following conditions are satisfied:*

1) *The modular form is entire iff $(f) \geq 0$. It is a cusp form iff $(f) \geq \sum_{s \in S_\Gamma}(s)$.*
2) *If ω_f is the multicanonical form on X_Γ which is associated with f, then*

$$(f) = (\omega_f) + m \sum_{s \in S_\Gamma} s.$$

In particular,

$$\deg((f)) = m(2p - 2 + h).$$

Furthermore, we have

$$(fg) = (f) + (g).$$

Now we choose, for each weight $k = 2m$, a nonzero meromorphic modular form. We can take, for example, $f_0 = (G_6/G_4)^m$. Then any other f of the same weight is of the form $f = h f_0$, with a modular function (= modular form of weight zero) h. The form f is entire if $(h) \geq -(f_0)$, i.e. f is contained in the Riemann–Roch space $\mathcal{L}((f_0))$. We obtain the following result.

14.10 Proposition. *Let K be a canonical divisor on X_Γ; then*

$$[\Gamma, k] \cong \mathcal{L}\left(mK + m \sum_{s \in S_\Gamma} s\right)$$

and, correspondingly,

$$[\Gamma, k]_0 \cong \mathcal{L}\left(mK + (m-1) \sum_{s \in S_\Gamma} s\right).$$

From the Riemann–Roch theorem, we obtain the following proposition.

14.11 Proposition. *Let Γ be a congruence subgroup of the modular group which, besides E and possibly $-E$, does not contain an element of finite order. For even $k > 0$, we have*

$$\dim[\Gamma, k] = \frac{k}{2}(2p - 2 + h) + 1 - p$$

and

$$\dim[\Gamma, k]_0 = \begin{cases} \dim[\Gamma, k] - h & \text{for } k > 2, \\ \dim[\Gamma, k] - h + 1 & \text{for } k = 2. \end{cases}$$

Here h denotes the number of cusp classes and p the genus of X_Γ.

Next we want to determine, for many groups, the genus $p =: p(\Gamma)$ and the number of cusp classes $h =: h(\Gamma)$.

The Topological Genus

In the definition of the order of a modular form, a natural number R occurred, namely the smallest positive number such that the translation $z \mapsto z + R$ belongs to Γ. Every other translation is then a multiple of R. If the negative unit matrix is contained in Γ, then the stabilizer of the cusp $i\infty$ equals

$$\Gamma_\infty := \{M \in \Gamma; \quad M(i\infty) = i\infty\} = \left\{ \pm \begin{pmatrix} 1 & xR \\ 0 & 1 \end{pmatrix}; \quad x \in \mathbb{Z} \right\}.$$

This a subgroup of index R of the stabilizer $i\infty$ in the full modular form. We call R the *width of the cusp* ∞. For an arbitrary cusp $\kappa = M(i\infty)$, $M \in \mathrm{SL}(2, \mathbb{Z})$, we define the width of κ as the width of $i\infty$ with respect to the conjugate group $M\Gamma M^{-1}$. It should be clear that this definition is independent of the choice of M and, moreover, depends only on the Γ-equivalence class. We use the notation

$$R(\kappa) = R_\Gamma(\kappa) = \text{width of } \kappa.$$

It is clear that this definition is independent of the choice of M and that it depends only on the Γ-equivalence class. If the negative unit matrix is contained in Γ, then the cusp width $R(\kappa)$ equals the index of the stabilizers of the cusp,

$$[\mathrm{SL}(2, \mathbb{Z})_\kappa : \Gamma_\kappa].$$

14.12 Remark. *Let $\kappa = M(i\infty)$, $M \in \mathrm{SL}(2, \mathbb{Z})$, be an arbitrary cusp of Γ. The width of the cusp ∞ of $M\Gamma M^{-1}$ depends only on κ and not on the choice of M. It is called the width of the cusp κ of Γ. It depends only on the Γ-equivalence class of κ.*

The sum of all cusp widths

$$R(\Gamma) := \sum_{x \in S_\Gamma} R_\Gamma(x)$$

is of great importance.

14.13 Remark. *Let Γ be a congruence group which contains the negative unit matrix. The sum of all cusp widths equals the index of Γ in the full modular group:*

$$R(\Gamma) = [\mathrm{SL}(2, \mathbb{Z}) : \Gamma].$$

Proof. Let

$$\kappa_1 = M_1(i\infty), \dots, \kappa_h = M_h(i\infty)$$

be a system of representatives of the cusp classes. For each representative κ_ν, we choose a system of representatives $N_{\nu,1}, \dots N_{\nu,R_\nu}$ of the cosets of Γ_κ in

$\mathrm{SL}(2,\mathbb{Z})_\kappa$. Obviously, $M_\nu N_{\nu,\mu}$ runs through a system of representatives of Γ in $\mathrm{SL}(2,\mathbb{Z})$. □

We recall that there exists a modular form of weight 12 with respect to the full modular group which has no zeros in the upper half-plane and vanishes at the cusp ∞ of first order. We consider it as a modular form with respect to a congruence subgroup Γ; the vanishing order at the cusp ∞ then equals the width of the cusp $i\infty$. We obtain the following result.

14.14 Proposition. *Let Γ be a congruence subgroup which contains the negative unit matrix and is such that its image in Bihol $\mathbb{H}$ acts freely. Let $R(\Gamma)$ be the sum of all cusp widths of Γ. We have*

$$R(\Gamma) = [\mathrm{SL}(2,\mathbb{Z}):\Gamma] = 6(2p-2+h).$$

14.15 Corollary. *For even $k > 0$, we have*

$$\dim[\Gamma,k] = \frac{k}{12}[\mathrm{SL}(2,\mathbb{Z}):\Gamma]+1-p.$$

The assumption $-E \in \Gamma$ is harmless, since we can replace Γ by the group $\Gamma \cup -\Gamma$. The two groups have the same modular forms of even weight.

The basic numbers p and h can easily be determined for the principal congruence group of level two. This group has index of 6 in the full modular group. The width of the cusp $i\infty$ is 2. Since $\Gamma[2]$ is a normal subgroup, all widths are 2. We obtain the result that $\Gamma[2]$ has three cusp classes. The genus can be determined using Proposition 14.8 by means of the structure theorem ([FB], Theorem VI.6.3). It follows that every cusp form of weight two vanishes for $\Gamma[2]$. This result can also be derived in a purely topological manner by means of the polyhedron theorem, since we can use a fundamental domain to construct a triangulation. We see the following.

14.16 Lemma. *We have*

$$p(\Gamma[2]) = 0, \quad h(\Gamma[2]) = 3.$$

The index and number of cusp classes can be computed if we have a coset system of Γ in the full modular group. We perform this computation here for the principal congruence subgroup $\Gamma[q]$, where we use the fact that

$$\mathrm{SL}(2,\mathbb{Z}) \longrightarrow \mathrm{SL}(2,\mathbb{Z}/q\mathbb{Z})$$

is surjective and that, as a consequence,

$$[\mathrm{SL}(2,\mathbb{Z}):\Gamma[q]] = \#\,\mathrm{SL}(2,\mathbb{Z}/q\mathbb{Z}) = q^3 \prod_{l \text{ prim},\, l|q}\left(1-\frac{1}{l^2}\right).$$

This is a special case of a result which will be proved later (Proposition VII.6.5). We shall use this result here.

14.17 Proposition. *For the principal congruence subgroup of level $q > 2$, we have:*

$$[\mathrm{SL}(2, \mathbb{Z}) : \Gamma[q]] = q^3 \prod_{l \mid q} \left(1 - \frac{1}{l^2} \right),$$

$$h = \frac{1}{2} q^2 \prod_{l \mid q} \left(1 - \frac{1}{l^2} \right),$$

$$p = 1 + \frac{q - 6}{12} h.$$

The dimension formulae

$$\dim[\Gamma[q], k] = \frac{1}{24} q^3 \prod_{l \mid q} \left(1 - \frac{1}{l^2} \right) + 1 - p$$

hold for even $k > 2$.

Corollary. *In the cases $q \le 5$, the genus is zero.*

Proof. It is better to work with the group $\tilde{\Gamma}[q] = \Gamma[q] \cup -\Gamma[q]$. The number of cusp classes is the same, but the index has to be divided by two. Since the cusp widths are all q, the formula for the number of cusp classes follows from the formula for the index.

For the determination of the genus, we apply the Riemann–Hurwitz ramification formula. First we consider, for an even level q, the natural map

$$X_{\Gamma[q]} \longrightarrow X_{\Gamma[2]}.$$

We determine its degree. It is easy to see that each point of $\mathbb{H}/\Gamma[2]$ has

$$[\Gamma[2] : \tilde{\Gamma}[q]] = \frac{[\Gamma[1] : \tilde{\Gamma}[q]]}{[\Gamma[1] : \Gamma[2]]}$$

inverse images. So this is the degree of the map. The only ramification points are the cusps. The ramification order at each cusp is q.

If q is odd, we apply the Riemann–Hurwitz ramification formula to

$$X_{\Gamma[2q]} \longrightarrow X_{\Gamma[q]}. \qquad \square$$

15. Dimensions of Vector Spaces of Modular Forms with Multiplier Systems

In this section, we want to give up the restriction that the weights of the modular forms are even. We also want to admit multiplier systems. Let v be a multiplier system of weight $r/2$, $r \in \mathbb{N}$, with respect to a congruence subgroup Γ. We use the notation of [FB], Sect. VI.5. Again we have to define a divisor on X_Γ for a meromorphic modular form $f \in \{\Gamma, r/2, v\}$. For this, we have to define the order of f at the cusp $i\infty$. Again let R be the smallest positive number such that

$$\begin{pmatrix} 1 & R \\ 0 & 1 \end{pmatrix} \quad \text{or} \quad \begin{pmatrix} -1 & R \\ 0 & -1 \end{pmatrix}$$

is contained in Γ. A difficulty arises, since f need not have a period R. We only have $f(z + R) = \varepsilon f(z)$, with a certain root of unity ε. This difficulty arises even for the trivial multiplier system when the weight k is odd and when $\begin{pmatrix} -1 & R \\ 0 & -1 \end{pmatrix}$ is contained in Γ. We then have

$$\varepsilon = \begin{cases} v \begin{pmatrix} 1 & R \\ 0 & 1 \end{pmatrix} & \text{if } \begin{pmatrix} 1 & R \\ 0 & 1 \end{pmatrix} \in \Gamma, \\[2ex] (-1)^k v \begin{pmatrix} -1 & R \\ 0 & 1 \end{pmatrix} & \text{if } \begin{pmatrix} -1 & R \\ 0 & -1 \end{pmatrix} \in \Gamma. \end{cases}$$

In general, we call the root of unity ε the *irregularity* of the cusp $i\infty$. It depends only on the triple $(\Gamma, r/2, v)$. The occurrence of the irregularity is responsible for the following extra considerations.

We write the irregularity in the form

$$\varepsilon = e^{2\pi i a}, \quad 0 \le a < 1.$$

The function $z \mapsto \exp(2\pi i a)$ has the same transformation property as f under the translation $z \mapsto z + R$. Hence the function $z \mapsto f(z)\exp(-2\pi i a z)$ has period R and admits a Fourier expansion

$$f(z)e^{-2\pi i a z} = \sum_{n=-\infty}^{\infty} a_n e^{2\pi i n z/R}.$$

Now we define

$$\mathrm{Ord}(f, i\infty) = \min\{n; \quad a_n \ne 0\}.$$

We should point out that this definition is artificial in some sense, because it depends on the representation of ε in the form $\exp(2\pi i a)$. Another normalization of a would lead to another order. One should bear in mind that in our general setting, modular forms no longer have an interpretation as multicanonical forms as in the case of even weight and the trivial multiplier system.

Analogously to the case of even weight and the trivial multiplier system, we now define $\mathrm{Ord}(f; a)$ for an arbitrary cusp a by transforming it to $i\infty$ and considering a conjugate group and the conjugate form. This definition is independent of the choice of the substitution which is needed to transform a to $i\infty$. For a point a in the upper half-plane, we again use the usual order of the meromorphic function f at a. In this way, we again obtain, for any modular form f, a divisor (f).

But, in contrast to the case of even weight and the trivial multiplier system, there is no requirement that the formula $(fg) = (f) + (g)$ remains true. Obviously, however, this formula is true if one of the two forms has no irregularity. In particular, we have the following fact.

15.1 Lemma.   *Let $f \in \{\Gamma, k, v\}$ be a meromorphic modular form which is different from 0, and let $h \in \{\Gamma, 0\}$ be a fully invariant nonzero modular function. Then*

$$(hf) = (h) + (f).$$

From this we obtain the following result.

15.2 Lemma.   *If there exists a nonzero meromorphic modular form $f \in \{\Gamma, k, v\}$, then*

$$[\Gamma, k, v] \cong \mathcal{L}((f)).$$

We shall not discuss the existence of an (only) meromorphic modular form here, since this will obvious in all our applications. It remains to determine the degree of the divisor (f). An obvious idea is to take a natural number N such that Nk is even and f^N has a trivial multiplier system. From the results of the previous section, we get

$$\deg((f^N)) = \frac{Nk}{2}(2p - 2 + h).$$

So, we need a relation between $\mathrm{Ord}(f; i\infty)$ and $\mathrm{Ord}(f^N, i\infty)$. The expansion of f is

$$f(z) = e^{2\pi i a z} \sum a_n e^{2\pi i n z/R}.$$

We obtain

$$f(z)^N = e^{2\pi i N a z} \left(\sum a_n e^{2\pi i n z/R} \right)^N.$$

The number Na is integral. We see that

$$\mathrm{Ord}(f^N; i\infty) = aN + N\,\mathrm{Ord}(f, i\infty).$$

15.3 Remark.   *Let $\kappa = M(i\infty)$, $M \in \mathrm{SL}(2, \mathbb{Z})$, be an arbitrary cusp of Γ. The irregularity at $i\infty$ of $(M\Gamma M^{-1}, v^M)$ is independent of the choice of M. We call it the irregularity of $(\Gamma, r/2, v)$ at the cusp κ. Moreover, the irregularity depends only on the Γ-equivalence class of κ.*

Now we obtain the following result.

15.4 Proposition. *Let* Γ *be a congruence subgroup whose image in* Bihol $\mathbb{H}$ *acts freely. Let* $f \in \{\Gamma, k, v\}$ *be a meromorphic modular form and let* N *be a natural number such that* Nk *is even and such that all occurring cusp widths divide* N. *Then*

$$(f^N) = N(f) + \sum_{s \in S_\Gamma} Na(s).$$

In particular,

$$\deg(f) = \frac{k}{2}(2p - 2 + h) - \sum_{s \in S_\Gamma} a(s).$$

Here $R(s)$ *denotes the cusp width and* $\varepsilon(s) = \exp(2\pi i a(s))$ *the irregularity (in the standard representation* $0 \le a(s) < 1$*).*

15.5 Corollary. *In the case* $k \ge 2$, *we have*

$$\dim[\Gamma, k, v] = \frac{k}{2}(2p - 2 + h) + 1 - p - \sum_{s \in S_\Gamma} a(s).$$

Exercises for the appendices to Chap. IV

1. In [FB], at the end of Sect. VI.6, we proved the following formula for $r \ge 2$:
    $$\dim[\Gamma[4, 8], r/2, v_\vartheta^r] = 4r - 2.$$
    Reprove this formula in the case $r \ge 4$ by means of Corollary 15.5.

2. Let $\Gamma \subset \mathrm{SL}(2, \mathbb{Z})$ be a congruence subgroup whose image in Bihol $\mathbb{H}$ does not necessarily act freely. As we have seen in Exercise 4 for Sect. III.2, the quotient $\mathbb{H}/\Gamma$ also carries a natural structure as Riemann surface in this case.

    Show that X_Γ for an arbitrary congruence group admits a unique structure in the form of a Riemann surface, such the natural projection
    $$\mathbb{H} \longrightarrow X_\Gamma$$
    is holomorphic.

3. In the case of the full modular group $\Gamma = \mathrm{SL}(2, \mathbb{Z})$, the Riemann surface X_Γ is biholomorphically equivalent to the Riemann sphere.

    Give three different proofs:

1) Use the j-function (compare Exercise 7 for Sect. III.2).

2) Using the uniformization theorem, it is sufficient to show that the genus of X_Γ is zero. This can be done by means of Euler's polyhedron formula, using a suitable triangulation of the fundamental domain.

3) Apply the ramification formula to $X_{\Gamma[2]} \to X_{\Gamma[1]}$.

4. Compute the genus $p(\Gamma)$ of X_Γ for an arbitrary congruence subgroup by means of the Riemann–Hurwitz ramification formula using the natural projection

$$X_\Gamma \longrightarrow X_{\mathrm{SL}(2,\mathbb{Z})} = \bar{\mathbb{C}}.$$

If $-E$ is not contained in Γ, then the result is

$$p(\Gamma) = 1 + \frac{[\mathrm{SL}(2,\mathbb{Z}) : \Gamma]}{12} - \frac{a}{4} - \frac{b}{3} - \frac{h}{2}.$$

Here $a = a(\Gamma)$ and $b = b(\Gamma)$ denote the numbers of Γ-equivalence classes of fixed points of order two and three, respectively. By definition, the order $e(a)$ of a point $a \in \mathbb{H}$ is the order of the image of Γ_a in Bihol $\mathbb{H}$. This is the order of Γ_a if $-E$ is not contained in Γ, and half of it otherwise. Of course, it depends only on the Γ-equivalence class.

5. Show that the formula

$$\dim[\Gamma, 2]_0 = p(\Gamma)$$

holds for all congruence subgroups.

6. Show that, for even k, the formula

$$\dim[\Gamma, k] = \begin{cases} \dim[\Gamma, k]_0 + h & \text{for } k > 2, \\ \dim[\Gamma, k]_0 + h - 1 & \text{for } k = 2 \end{cases}$$

holds for all congruence subgroups.

7. Show that, for even $k > 0$ and arbitrary congruence groups, we have

$$\dim[\Gamma, k] = (k-1)(p-1) + \frac{kh}{2} + \sum_{a \in \mathbb{H}/\Gamma} \left[\frac{k}{2}\left(1 - \frac{1}{e(a)}\right) \right].$$

Here $[x]$ means the greatest integer $\leq x$. Of course, the sum is finite, since $e(a)$ is 1 almost always.

8. Use the result of the previous exercise to give a new proof of the structure theorem (Theorem VI.3.4 in [FB]). (This states that the ring of modular forms with respect to the full modular group is generated by two forms of weight 4 and 6.)

9. To obtain a dimension formula for a arbitrary weight $k \in \frac{1}{2}\mathbb{Z}$ and for an arbitrary multiplier system v, we have to associate an *irregularity* with the elliptic fixed

points $a \in \mathbb{H}$ in analogy to the case of cusps. Again we assume the existence of a meromorphic modular form f. The function

$$g(w) = (w - 1)^{-k} f\left(\frac{\bar{a}w - a}{w - 1}\right)$$

is defined in the unit circle. If $e(a)$ is the order of an fixed point, then g transforms as

$$g(e^{2\pi i/e(a)} w) = \eta g(w),$$

where

$$\eta = e^{2\pi i \alpha}, \quad 0 \leq \alpha = \alpha(a) < 1,$$

is a root of unity, which now plays the role of the irregularity. This depends only on Γ, v, k, and the Γ-equivivalence class of a.

Show that for $k > 2$, we have

$$\dim[\Gamma, k, v] = (k - 1)(p - 1) + \frac{kh}{2} + \sum_{a \in \mathbb{H}/\Gamma} \left(\frac{k}{2}\left(1 - \frac{1}{e(a)}\right) - \alpha(a)\right).$$

V. Analytic Functions of Several Complex Variables

The Jacobi inversion theorem leads to functions of several variables with many periods. So, we are led to the problem of developing a theory of them which is analogous to the theory of elliptic functions. First of all, we need to give an introduction to the theory of functions of several complex variables. In this chapter, we give an elementary introduction which essentially follows Weierstrass. One of the main topics will be the proof of the theorem that any meromorphic function on $\mathbb{C}^n$ can be written as a quotient of two entire functions. Weierstrass called this a very difficult problem. The first proof was given by Poincaré. In the case $n = 1$, it is not difficult to show this by means of the theory of Weierstrass products, even for arbitrary domains $D \subset \mathbb{C}$ instead of $\mathbb{C}$. The case $n > 1$ is more involved, since the zero sets and pole sets of analytic functions of several complex variables are not discrete.

The investigation of the zero set is related to the division theory of the ring of convergent power series. This theory is governed by two central theorems, the Weierstrass preparation theorem and the division theorem. The two theorems are closely related. They are equivalent in the sense that it is rather easy to derive one from the other. The Weierstrass preparation theorem appeared in print in 1886, but it already appeared in 1860 in Weierstrass's lectures. The division theorem is frequently called the preparation theorem, but this is historically false. Historical comments and amendments can be found in Siegel's paper "Zu den Beweisen des Vorbereitungssatzes von Weierstrass" (Collected Papers, Vol. IV, No. 83) [Si2]. There, it was pointed out that the division theorem was proved for the first time in 1887 by Stickelberger and was rediscovered by Späth in 1929. Siegel gave, in the above paper, a simple proof of the preparation theorem which rests on a calculation with power series. Here, we shall present a different proof which uses the Cauchy integral.

At the end of this chapter, we shall also give a short introduction to the local calculus of alternating differential forms, which extends the two-dimensional-case considered in the appendix to Chap. II.

1. Elementary Properties of Analytic Functions of Several Variables

We are familiar with the notion of an *analytic (= holomorphic) function* of one complex variable and now want to use it to develop a notion of an analytic function of several variables.

1.1 Definition. *A function*

$$f : D \to \mathbb{C}$$

*on an open subset $D \subset \mathbb{C}^n$ is called **analytic** if it is continuous, and if it is analytic in each of the n variables if the rest of the variables are fixed.*

Remark. *A nontrivial result of Hartogs states that the assumption of continuity in Definition 1.1 is superfluous.*

The following properties of analytic functions can easily be reduced to the onevariable case.

1. The sum and product of two analytic functions are analytic. The function $1/f$ is analytic if f is an analytic function without zeros.

Notation.

$$\mathcal{O}(D) = \{f : D \to \mathbb{C}, \ f \text{ analytic}\}.$$

So, $\mathcal{O}(D)$ is a $\mathbb{C}$-algebra.

2. *Maximum principle. Assume that D is connected. If $|f(z)|$ attains its maximum in D, then f is constant.*

Let $a \in D$ be a point at which $|f(z)|$ attains its maximum. For the proof, we consider the set of all points such that $f(z) = f(a)$. This set is closed, by a continuity argument. Using the maximum principle of complex analysis for one variable, we can easily show that this set is also open in D. Since D is connected, it coincides with D.

3. If $f_n : D \to \mathbb{C}$ is a locally uniformly convergent sequence of analytic functions, then the limit function is analytic too.

4. *Identity theorem for analytic functions of several complex variables:*

An analytic function $f : D \to \mathbb{C}$ on a domain D which vanishes on an open nonempty subset is identically zero.

Proof. There exists a largest open subset $U \subset D$ on which f vanishes identically. If U is different from D, there exists a boundary point $a \in D$ of U. For the proof, we can replace D by a small open neighborhood of a. Hence we can assume that $D = D_1 \times \ldots \times D_n$, where the D_ν are domains in $\mathbb{C}$. By assumption, there exist nonempty open sets $U_\nu \subset D_\nu$ such that f vanishes on $U_1 \times \ldots \times U_n$. Now we can apply inductively the identity theorem for the onevariable case. $\qquad \square$

Another proof will follow from the local expansibility into power series (Proposition 2.2).

As in the case $n = 1$, we need the power series expansion of an analytic function. This will be the subject of the next section.

Exercises for Sect. V.1

1. Let $D \subset \mathbb{C}^n$ be a domain which has a nonempty intersection with $\mathbb{R}^n$. Show that an analytic function on D which vanishes on $D \cap \mathbb{R}^n$ is identically zero.

2. A function $f : U \to \mathbb{R}$ on an open subset $U \subset \mathbb{R}^n$ is called a *real analytic* function if each point $a \in U$ admits an open neighborhood $U(a) \subset \mathbb{C}^n$ and a complex analytic function $f_a : U(a) \to \mathbb{C}^n$ with the property

$$f(x) = f_a(x) \text{ for } x \in U(a) \cap U.$$

Show that the identity theorem in the form given in Sect. V.1 is true for real analytic functions. That is, if U is connected and f vanishes on an open nonempty subset of U, then f is zero on the whole of U.

3. Let $f : U \to \mathbb{C}$ be a nonconstant analytic function on a domain $U \subset \mathbb{C}$ and let $P : \mathbb{C} \times \mathbb{C} \to \mathbb{C}$ be an analytic function with the property

$$P(\operatorname{Re} f, \operatorname{Im} f) \equiv 0.$$

Show that P vanishes identically.

4. Show that the image $f(D)$ of a nonconstant analytic function which is defined on a domain $D \subset \mathbb{C}^n$ is open in $\mathbb{C}$.

2. Power Series in Several Variables

When we are studying power series of several variables, it is useful to separate the algebraic computational rules from questions of convergence. Many of the algebraic properties can be formulated for *formal power series*. These are power series without any assumption of convergence. The coefficients of power series are usually complex numbers for us. For the definition of formal power series, we can take as coefficients elements of arbitrary *commutative rings with unity* $1 = 1_R \in R$. Sometimes we require that R is an integral domain, i.e.

$$ab = 0 \Longrightarrow a = 0 \text{ or } b = 0.$$

For the moment, R can be an arbitrary commutative ring with unity.

A (formal) *power series in n variables* over R is a map

$$P : \mathbb{N}_0^n \longrightarrow R, \qquad (\nu_1, \ldots, \nu_n) \longmapsto a_{\nu_1, \ldots, \nu_n},$$

which, owing to its later use, is written in the form

$$P = \sum a_{\nu_1, \ldots, \nu_n} X_1^{\nu_1} \ldots X_n^{\nu_n}.$$

Here $X_1, \ldots, X_n$ are merely symbols.

It is useful to make use of the calculus of multi-indices:

$$
\begin{aligned}
\nu &:= (\nu_1, \ldots, \nu_n), & X &:= (X_1, \ldots, X_n), \\
\nu! &:= \nu_1! \ldots \nu_n!, & X^\nu &:= X_1^{\nu_1} \ldots X_n^{\nu_n}.
\end{aligned}
$$

The power series is then of the form

$$
P = \sum_{\nu \in \mathbb{N}_0^n} a_\nu X^\nu.
$$

The summation is taken over all multi-indices $\nu \in \mathbb{N}_0^n$.

For two power series

$$
P = \sum_{\nu \in \mathbb{N}_0^n} a_\nu X^\nu, \quad Q = \sum_{\nu \in \mathbb{N}_0^n} b_\nu X^\nu,
$$

we *define* the sum and product as follows:

(1) $$P + Q := \sum_{\nu \in \mathbb{N}_0^n} (a_\nu + b_\nu) X^\nu,$$

(2) $$P \cdot Q := \sum_{\nu \in \mathbb{N}_0^n} c_\nu X^\nu, \quad c_\nu := \sum_{\alpha + \beta = \nu} a_\alpha b_\beta \quad \text{(finite sum!)}.$$

It is easy to check that the set of all formal power series with this addition and multiplication becomes an associative and commutative ring with unity

$$
1 = \sum_{\nu \in \mathbb{N}_0^n} a_\nu X^\nu, \quad a_\nu = \begin{cases} 0 & \text{for } \nu \neq (0, \ldots, 0), \\ 1 & \text{for } \nu = (0, \ldots, 0). \end{cases}
$$

This ring is denoted by

$$
R[\![X_1, \ldots, X_n]\!].
$$

We call this the ring of formal power series over R in n "variables". This ring contains the polynomial ring $R[X_1, \ldots, X_n]$ as a subring. A polynomial is nothing but a formal power series which has only finitely many coefficients which are different from zero. The ground ring R can be embedded into $R[X_1, \ldots, X_n]$ and hence also into $R[\![X_1, \ldots, X_n]\!]$ in a natural way, $r \mapsto r \cdot 1$. We shall usually identify r and $r \cdot 1$.

Other Notations for Power Series

Let P be a power series and let $d \geq 0$ be a nonnegative integer. We then define

$$P_d := \sum_{\nu_1 + \cdots + \nu_n = d} a_\nu X^\nu.$$

This is a polynomial. It collects together all monomials of degree d which occur in P. If P is a polynomial, we have

$$P = \sum_{d=0}^{\infty} P_d,$$

where the sum is finite in reality. A power series is determined by the totality of its homogeneous terms P_d. Hence we can use the notation

$$P = \sum_{d=0}^{\infty} P_d.$$

If P and Q are power series, we can immediately verify that

$$(PQ)_d = \sum_{d_1 + d_2 = d} P_{d_1} Q_{d_2} \qquad \text{(finite sum)}.$$

By the *order* of a power series P different from zero*), we understand the smallest d such that P_d is different from 0. If d_1, d_2 are the orders of P, Q, then

$$(PQ)_{d_1 + d_2} = P_{d_1} Q_{d_2}.$$

We assume that it is known that the polynomial ring over an integral domain is an integral domain too. We obtain the following result.

The ring of power series over an integral domain is an integral domain too.

Power Series as Coefficients of Power Series

Let P be a power series in n variables. We fix one variable, say X_n. For a nonnegative integer k, we consider the power series $P^{(k)}$ in $n-1$ variables

$$P^{(k)} := \sum_{\nu_n = k} a_\nu X_1^{\nu_1} \cdots X_{n-1}^{\nu_{n-1}}.$$

It is easy to check that the map

$$P \longmapsto \sum_{k=0}^{\infty} P^{(k)} X_n^k$$

---

*) A power series is different from 0 if not all of its coefficients are zero. This means that it is not the zero element in the ring of power series.

defines an isomorphism

$$R[\![X_1, \ldots, X_n]\!] \longrightarrow R[\![X_1, \ldots, X_{n-1}]\!][\![X_n]\!].$$

We identify the two rings and write

$$P = \sum_{k=0}^{\infty} P^{(k)} X_n^k.$$

Next, we introduce convergent power series. For this, we assume that the ground ring is the field of complex numbers.

2.1 Definition. *A formal power series*

$$P = \sum_{\nu \in \mathbb{N}_0^n} a_\nu X^\nu$$

is called **convergent** *if there exists an n-tuple of complex numbers* $(z_1, \ldots, z_n)$, *which all are different from zero, such that*

$$P(z_1, \ldots, z_n) = \sum_{\nu \in \mathbb{N}_0^n} a_{\nu_1, \ldots, \nu_n} z_1^{\nu_1} \cdots z_n^{\nu_n}$$

converges absolutely.

We recall some well-known facts about convergent series.

Let $(a_s)_{s \in S}$ be a family of complex numbers which is parametrized by a countable set S. The "series" $\sum_{s \in S} a_s$ is called *absolutely convergent* if there exists a number $C > 0$ such that

$$\sum_{s \in S_0} |a_s| \leq C$$

for each finite subset $S_0 \subset S$. In this case the value of the series can be defined, for example by ordering the elements of S somehow;

$$S = \{s_1, s_2, s_3, \ldots\}.$$

The number

$$\sum_{s \in S} a_s := a_{s_1} + a_{s_2} + a_{s_3} + \ldots$$

is independent of the choice of this ordering.

More generally, we consider a family

$$f_s : X \longrightarrow \mathbb{C}$$

of functions on a topological space. The series

$$\sum_{s \in S} f_s$$

is called *normally convergent* if each point $a \in X$ admits a neighborhood $U = U(a)$, and if there exist constants m_s such that

$$|f(x)| \le m_s \text{ for all } x \in U \text{ and } s \in S$$

and such that

$$\sum_{s \in S} m_s$$

converges. The series then converges in any ordering, absolutely and locally uniformly (the Weierstrass majorant criterion).

If the power series P converges absolutely at a point $(w_1, \ldots, w_n)$ with $w_k \ne 0$ for $k = 1, \ldots, n$, then, by the majorant criterion, it also converges at every point

$$(z_1, \ldots, z_n), \quad |z_k| < |w_k|.$$

*An n-tuple $r = (r_1, \ldots, r_n)$ of positive real numbers is called a **multiradius of convergence** of P if P converges at all points*

$$(z_1, \ldots, z_n), \quad |z_k| < r_k \text{ for } k = 1, \ldots, n.$$

Notation. Let $b \in \mathbb{C}^n$, $r \in \mathbb{R}_{>0}^n$. The set

$$U_r(b) := \left\{ z \in \mathbb{C}^n; \ |z_k - b_k| < r_k \text{ for } k = 1, \ldots, n \right\}$$
$$= U_{r_1}(b_1) \times \ldots \times U_{r_n}(b_n)$$

is called the *polydisk* with center b and multiradius r.

2.2 Proposition. *Let $b \in \mathbb{C}^n$ be a given point and let be r be an n-tuple of positive real numbers.*

1) *Let P be a power series and let r be a multiradius of convergence. Then the series*

$$P(z - b) = \sum_{\nu \in \mathbb{N}_0^n} a_\nu (z - b)^\nu$$

   converges normally in $U_r(b)$ and defines an analytic function there.

2) *An analytic function*

$$f : U_r(b) \longrightarrow \mathbb{C}$$

   can be expanded in the whole of $U_r(b)$ as a power series P which converges normally there, so that

$$f(z) = P(z - b) = \sum_{\nu \in \mathbb{N}_0^n} a_\nu (z - b)^\nu,$$

and we have

$$a_\nu = \frac{f^{(\nu)}(b)}{\nu!}.$$

Proof. 1) The known proof in the case $n = 1$ works in the general case. The validity of the Taylor formula in 2) also follows from the standard stability theorems in the case $n = 1$.

2) Since we have seen already the uniqueness of the expansion, we can enlarge the polydisk slightly: we can assume that f is analytic in an open neighborhood of the closure of $U_r(b)$. We also can assume $b = 0$.

The idea is to generalize the well-known proof of the case $n = 1$ ([FB], Theorem III.2.2). For this, we need a suitable generalization of the Cauchy integral formula. Again, we shall reduce it by induction to the case $n = 1$.

First we apply the usual Cauchy integral formula to the analytic function in the single variable z_n

$$z_n \longmapsto f(z_1, \ldots, z_n),$$

keeping $z_1, \ldots, z_{n-1}$ fixed. We obtain

$$f(z_1, \ldots, z_n) = \frac{1}{2\pi i} \oint_{|\zeta_n| = r_n} \frac{f(z_1, \ldots, z_{n-1}, \zeta_n)}{\zeta_n - z_n} d\zeta_n.$$

Now we apply the Cauchy integral formula step by step for the variables $z_1, \ldots, z_n$. We obtain the following formula:

The Cauchy integral formula in several variables.

$$f(z_1, \ldots, z_n) = \frac{1}{(2\pi i)^n} \oint_{|\zeta_1| = r_1} \cdots \oint_{|\zeta_n| = r_n} \frac{f(\zeta_1, \ldots, \zeta_n)}{(\zeta_1 - z_1) \cdots (\zeta_n - z_n)} d\zeta_1 \ldots d\zeta_n.$$

Now the power series expansion of f can be obtained as in the case $n = 1$. We expand the integrand into a *geometric series* and interchanges integration and summation. The geometric series in several variables can be obtained from the usual series

$$\frac{1}{\zeta - z} = \frac{1}{\zeta} \frac{1}{1 - z/\zeta} = \frac{1}{\zeta} \sum_{\nu=0}^{\infty} \left(\frac{z}{\zeta}\right)^\nu$$

by termwise multiplication:

$$\frac{1}{\zeta_1 - z_1} \cdot \ldots \cdot \frac{1}{\zeta_n - z_n} = \frac{1}{\zeta_1 \cdot \ldots \cdot \zeta_n} \sum_{\nu \in \mathbb{N}_0^n} \left(\frac{z_1}{\zeta_1}\right)^{\nu_1} \cdot \ldots \cdot \left(\frac{z_n}{\zeta_n}\right)^{\nu_n}.$$

(This holds for $|z_k| < |\zeta_k| = r_k, \quad k = 1, \ldots, n$.)

Computation Rules for Power Series

We collect together some rules for computations with convergent power series here. The proofs are the same as in the case $n = 1$.

1) *Addition and multiplication of convergent power series.* We assume that the power series

$$\sum a_\nu (z-a)^\nu, \quad \sum b_\nu (z-a)^\nu$$

converge in a polydisk U around a. In U, we then have

$$\sum a_\nu (z-a)^\nu + \sum b_\nu (z-a)^\nu = \sum (a_\nu + b_\nu)(z-a)^\nu,$$

$$\left(\sum a_\nu (z-a)^\nu \right) \cdot \left(\sum b_\nu (z-a))^\nu \right) = \sum_n \left(\sum_{\nu + \mu = n} a_\nu b_\nu \right)(z-a)^n.$$

This can be expressed as follows: the map

$$\mathcal{O}(U) \longrightarrow \mathbb{C}[[X_1, \ldots, X_n]],$$

which associates the formal power series $\sum a_\nu X^\nu$ with a function f with power series $f(z) = \sum a_\nu (z-a)^\nu$, is a ring homomorphism.

2) *Reordering of power series.* We assume that the power series

$$\sum a_\nu (z-a)^\nu$$

converges in the polydisk $U_r(a)$. Let

$$U_\rho(b) \subset U_r(a)$$

be another polydisk which is contained in it. The expansion of the analytic function which is defined by the original power series around b can be obtained by formal reordering using

$$(z-a)^\nu = [(z-b) + (b-a)]^\nu = \sum \binom{\nu}{k}(z-b)^k (b-a)^{\nu-k},$$

$$\binom{\nu}{k} := \binom{\nu_1}{k_1} \cdots \binom{\nu_n}{k_n}.$$

So, in $U_\rho(b)$, we have

$$\sum a_\nu (z-a)^\nu = \sum b_\nu (z-b)^\nu$$

with

$$b_\nu = \sum_k a_\nu \binom{\nu}{k}(b-a)^{\nu-k}.$$

3) *Composition of power series.* Let

$$U \xrightarrow{f} V \xrightarrow{g} \mathbb{C},$$
$$U \subset \mathbb{C}^n,\ V \subset \mathbb{C}^m \text{ open,}$$

be maps whose components are analytic functions, and let

$$a \in U,\ b = f(a).$$

We consider the expansions of the components of f

$$f_j(z) = \sum_\nu a_\nu^{(j)}(z-a)^\nu$$

in a neighborhood of a, and the components of g in a neighborhood of b,

$$g(z) = \sum_\mu b_\mu(z-b)^\mu.$$

The power series expansion of $g(f(z))$ in a small neighborhood of a can be obtained by formal replacement and reordering.

In contrast to the case $n = 1$, in the case $n > 1$ there exists no *largest* radius of convergence which could be called *the* radius of convergence. This is shown by the following example:

$$\sum_{n=0}^{\infty} z_1^n z_2^n.$$

The domain

$$\{(z_1, z_2) \in \mathbb{C} \times \mathbb{C};\quad |z_1 z_2| < 1\}$$

is the largest open set in which this series converges. It is not a polydisk. In Exercise 4, we shall obtain the shape of the precise domains of convergence.

Exercises for Sect. V.2

1. Show that every , analytic and bounded function in $\mathbb{C}^n$ is constant ("Liouville's theorem").

2. Expand the function
$$\frac{1}{(z_1 - 1)^2(z_2 - 2)^3}$$
into a power series around the origin.

3. Let $\sum_{s \in S} a_s$ be an absolutely convergent series. Show that its limit A can be characterized without ordering S, as follows:

   For any $\varepsilon > 0$, there exists a finite subset $S_0 \subset S$ such that for any finite intermediate set $S_0 \subset T \subset S$ we have
$$\left| \sum_{s \in T} a_s - A \right| < \varepsilon.$$

4. By a Reinhardt domain $D \subset \mathbb{C}^n$, we understand a domain which is invariant under "rotations" of the kind
$$(z_1, \ldots, z_n) \longmapsto (\zeta_1 z_1, \ldots, \zeta_n z_n), \quad |\zeta_\nu| = 1 \ (1 \leq \nu \leq n).$$
It is called *complete* if this true for all $|\zeta_\nu| \leq 1$.

   Show the following:
   a) In the case $n = 1$, the Reinhardt domains are precisely the annuli with center 0, and the complete Reinhardt domains are the disks.

   b) The largest open set in which a convergent power series $\sum a_\nu z^\nu$ is absolutely convergent is a complete Reinhardt domain. It converges normally there and defines an analytic function there.

   c) Any function which is analytic on a complete Reinhardt domain can be expanded in it into a power series.

   Hint. Use the obvious fact that every complete Reinhardt domain is a union of polydisks with center 0.

5. By a Laurent series of several variables, we understand a series
$$\sum_{\nu \in \mathbb{Z}^n} a_{\nu_1, \ldots \nu_n} z_1^{\nu_1} \cdot \ldots \cdot z_n^{\nu_n}.$$

   The sum is taken over all n-tuples of integers (including the negative integers).

   Show the following. We assume that the Laurent series converges absolutely in at least one point $(w_1, \ldots, w_n)$ with $w_\nu \neq 0$ for $1 \leq \nu \leq n$. The largest open set in which $\sum a_\nu z^\nu$ converges absolutely is a Reinhardt domain. The expansion converges normally there and defines an analytic function there. Conversely, any analytic function on a Reinhardt domain can be expanded into a Laurent series in the whole domain.

Hint. Every Reinhardt domain can be written as a union of "polyannuli" $D_1 \times \ldots \times D_n$. Here the D_ν are annuli with center 0 in the complex plane.

3. Analytic Maps

A map
$$f : U \longrightarrow V, \quad U \subset \mathbb{C}^n, \quad V \subset \mathbb{C}^m \text{ open},$$
is called *analytic* (or *holomorphic*) if its components
$$f_k = p_k \circ f, \quad p_k : V \longrightarrow \mathbb{C}, \quad k\text{th projection} \quad (1 \le k \le m)$$
are analytic functions.

3.1 Remark. *A map*
$$f : U \longrightarrow V, \quad U \subset \mathbb{C}^n, \quad V \subset \mathbb{C}^m \text{ open},$$

is analytic iff it is **totally complex differentiable** *at each point $a \in U$. This means that*
$$f(z) - f(a) = A(z - a) + r(z)$$
with a $\mathbb{C}$-linear map
$$A : \mathbb{C}^n \longrightarrow \mathbb{C}^m$$
and a remainder term r with the property

$$\frac{r(z)}{\|z - a\|} \longrightarrow 0 \text{ for } z \longrightarrow a$$

(where $\|\cdot\|$ means the Euclidean norm).

Proof. It follows from Proposition 2.2 that analytic maps are totally complex differentiable. Conversely, as in the real case, totally complex differentiable functions are continuous and partially complex differentiable. $\square$

The linear map A is described by an $m \times n$ matrix,

$$A(z_1, \ldots, z_n) = (w_1, \ldots, w_m),$$
$$w_i = \sum_{j=1}^{n} a_{ij} z_j, \quad 1 \le i \le m.$$

This matrix is the complex Jacobian

$$A = J(f; a) = \begin{pmatrix} \frac{\partial f_1}{\partial z_1} & \cdots & \frac{\partial f_1}{\partial z_n}, \\ \vdots & \ddots & \vdots \\ \frac{\partial f_n}{\partial z_1} & \cdots & \frac{\partial f_n}{\partial z_n} \end{pmatrix}(a).$$

Each $\mathbb{C}$-linear map is $\mathbb{R}$-linear as well. Therefore we can also describe A by a *real $2m \times 2n$ matrix!*

From the real chain rule and the fact that a composition of $\mathbb{C}$-linear maps is again $\mathbb{C}$-linear, we obtain the complex chain rule as below.

3.2 Remark. *Let*

$$f : U \longrightarrow V, \quad g : V \longrightarrow W$$

($U \subset \mathbb{C}^n$, $V \subset \mathbb{C}^m$, $W \subset \mathbb{C}^p$ open subsets)

be analytic maps. Then the composition $g \circ f$ is analytic too, and we have

$$J(g \circ f; a) = J(g; f(a)) \cdot J(f; a).$$

In the same manner, the real theorem of invertible functions implies the complex version. We simply have to observe that the inverse of an invertible $\mathbb{C}$-linear map is $\mathbb{C}$-linear as well.

3.3 Remark (theorem of invertible functions). *Let*

$$f : U \longrightarrow V, \quad U, V \subset \mathbb{C}^n \ open,$$

be an analytic map and let a be a point from U. The following statements are equivalent:

1) *f maps a suitable open neighborhood $U(a)$ biholomorphically onto an open neighborhood $V(f(a))$.*

2) *$J(f; a)$ is invertible.*

(A map f is called biholomorphic if it is bijective and if both f and f^{-1} are analytic.)

Remark. Let

$$A : \mathbb{C}^n \longrightarrow \mathbb{C}^n$$

be a $\mathbb{C}$-linear map. It can be described by a complex $n \times n$ matrix. The determinant of this matrix is denoted by $\det(A)$. We can also consider A as an $\mathbb{R}$-linear map and describe this by a *real $2n \times 2n$ matrix*. The determinant of this real matrix is denoted by $\det_{\mathbb{R}}(A)$. We have (see Exercise 1)

$$\det_{\mathbb{R}}(A) = |\det(A)|^2.$$

As in real analysis, the theorem of invertible functions implies the theorem of implicit functions. We formulate a special case as follows.

3.4 Remark.   *Let $f(w, z_1, \ldots, z_n)$ be analytic on some open subset of $\mathbb{C}^{n+1}$. Let $(b, a_1, \ldots, a_n)$ be a point with*

$$f(b, a_1, \ldots, a_n) = 0 \quad and \quad \frac{\partial f}{\partial w}(b, a_1, \ldots, a_n) \neq 0.$$

Then there exists a holomorphic function φ in a small open neighborhood of $(a_1, \ldots, a_n)$ with the properties

$$b = \varphi(a_1, \ldots, a_n), \quad f(\varphi(z_1, \ldots, z_n), z_1, \ldots, z_n) \equiv 0.$$

We can compare this with the treatment of the case $n = 1$ in Sect. I.3, Appendix B.

Zeros of Analytic Functions

In Sect. 1, we formulated the identity theorem for analytic functions and showed how to reduce it to the case $n = 1$. Another proof can be obtained by means of the power series expansion. A reformulation of the identity theorem says the following.

3.5 Remark.   *Let $f : D \to \mathbb{C}$ be an analytic function on a domain $D \subset \mathbb{C}^n$ which does not vanish identically. Then the set of zeros contains no inner points.*

Corollary. *The ring $\mathcal{O}(D)$ of holomorphic functions on a domain $D \subset \mathbb{C}^n$ is an integral domain.*

In contrast to the case $n = 1$, in the case $n > 1$ the set of zeros of an analytic function is never discrete if it is nonempty. We shall see this in Sect. 4. Here we consider only a simple example,

$$f(z_1, z_2) = z_1 \cdot z_2.$$

The zero set of this function is the union of the "axes" $\mathbb{C} \times \{0\} \cup \{0\} \times \mathbb{C}$. This is one of the reasons why complex analysis of several variables is much more difficult than the one-variable case. One has to be careful with the notion of a meromorphic function. A meaningful definition should imply that rational functions are meromorphic. For example, z_1/z_2 should be meromorphic on $\mathbb{C}^2$. Its zero set ($z_1 = 0$) and its pole set ($z_2 = 0$) (whatever that means) cross at the origin. Hence there is no meaningful way to assign a value to the origin and it does not help to allow ∞ as a value. This consideration shows that in the case of several variables we need a different approach compared with $n = 1$ for the introduction of the notion of a meromorphic function.

Meromorphic Functions

3.6 Definition. *Let $D \subset \mathbb{C}^n$ be a domain (a connected and open nonempty subset of $\mathbb{C}^n$) and let $D_0 \subset D$ be an open and dense subset of D. An analytic function*

$$f : D_0 \longrightarrow \mathbb{C}$$

*is called **meromorphic on** D if the following conditions are satisfied:*

For each point $a \in D$ there exist an open connected neighborhood U and two analytic functions $g, h : U \to \mathbb{C}$, where h is not identically zero, such that

$$f(z) = \frac{g(z)}{h(z)} \text{ for all } z \in U \cap D_0 \text{ with the property } h(z) \neq 0.$$

In this definition we have to observe two fine points:

1) We did not require that h has no zeros on $U \cap D_0$. Hence the representation $f(z) = g(z)/h(z)$ is valid only in the set

$$\{z \in U \cap D_0; \quad h(z) \neq 0\}.$$

By the identity theorem, this set is still open and dense in $U \cap D_0$. Independently of this, one can raise the question of whether this local representation of f can be chosen in such a way that h has no zeros in the whole of $U \cap D_0$. It can be shown that this is true, but for a proof we need deeper insight into the structure of the zero set, which is not available at the moment.

2) We required that D is connected, but not that D_0 is connected. The reason for this is as follows. Let g be an analytic function on D which does not vanish identically. We can then consider $D_0 := \{z \in D; \quad g(z) \neq 0\}$ and, on D_0, the analytic function $f(z) = 1/g(z)$. The notion of meromorphy should imply that this function is meromorphic on D. But, at the moment, we do not know that D_0 is connected. Actually, this is true, but the proof will come later. For this reason, we did not require D_0 to be connected. Nevertheless, the principle of analytic continuation also holds for meromorphic functions, as we now show.

3.7 Lemma. *Let $D \in \mathbb{C}^n$ be a domain and let $f : D_0 \to \mathbb{C}$ be an analytic function on an open and dense subset of D which is meromorphic on D. If f vanishes on an open and dense subset of D_0, then it is identically zero.*

Proof. The problem is that there might be connected components of D_0 where f vanishes identically, and others where this is not the case. We shall give an indirect proof and assume that this actually happens. So, let A be the union of all connected components of D_0 on which f vanishes, and let B be the union of the other components. The sets A and B are open, nonempty, and disjoint. Their union is D_0. The function f vanishes identically on A, but its set of zeros is thin in B. (This means that it contains no inner points.) Since D_0 is dense in D, we have $D = \bar{A} \cup \bar{B}$, where $\bar{A}$ and $\bar{B}$ denote the closure of A and B in

D. Both of these are closed in D. Since D is connected, the intersection of $\bar{A}$ and $\bar{B}$ is not empty. As a consequence, there exists a joint boundary point $a \in D$, $a \in \partial A \cap \partial B$. We use the definition of meromorphy at this point; f can be written in a small connected open neighborhood as a quotient of analytic functions in the above sense, $f = g/h$. The function g vanishes on $U \cap A$. By the identity theorem, it vanishes on U and hence on an open nonempty subset of B. This contradicts the fact that the zero set of f is thin in B. $\qquad\square$

The Field of Meromorphic Functions

When a function $f : D_0 \to \mathbb{C}$ is meromorphic on a larger domain D, it can of course happen that f extends holomorphically to a larger open set $D_0 \subset D' \subset D$. For this reason, one has to be careful with the definition of a "meromorphic function".

We consider pairs (D_ν, f_ν), $\nu = 1, 2$,

$$f_\nu : D_\nu \longrightarrow \mathbb{C} \text{ analytic,}$$

where the D_ν are open and dense in the domain D and the f_ν are meromorphic on D. The intersection $D_1 \cap D_2$ is again open and dense in D. The two pairs are said to be equivalent,

$$(D_1, f_1) \sim (D_2, f_2),$$

if f_1 and f_2 agree on an open nonempty subset of $D_1 \cap D_2$. By the above identity theorem (Lemma 3.7), they agree on the whole intersection $D_1 \cap D_2$. It should be clear that this relation is really an equivalence relation.

3.8 Definition. *Let $D \subset \mathbb{C}^n$ be a domain. A meromorphic function on D is a full equivalence class of analytic functions $f : D_0 \to \mathbb{C}$ which are meromorphic on D with respect to the equivalence relation described above. We denote the equivalence class of (D_0, f) by $[D_0, f]$. (Later, we shall write simply f instead of $[D_0, f]$.)*

If (D_0, f) represents a meromorphic function on D, we call D_0 a *domain of holomorphy* of $[D_0, f]$. A union of such domains of holomorphy is also a domain of holomorphy. Hence every meromorphic function has a unique maximal domain of holomorphy. This maximal domain of holomorphy is called *the domain of holomorphy* of the given meromorphic function.

As we have already mentioned, the intersection of two open and dense subsets is open and dense again. If (D_1, f_1) and (D_2, f_2) represent two meromorphic functions, we can define the sum and product of f_1 and f_2 as analytic functions on the intersection $D_1 \cap D_2$. They are meromorphic on D. The definition

$$[D_1, f_1] \dotplus [D_2, f_2] := [D_1 \cap D_2, f_1 \dotplus f_2]$$

is independent of the choice of the representatives.

So, the set of all meromorphic functions is an associative and commutative ring with unity. But we can state more, as follows.

3.9 Proposition. *The set $\mathcal{M}(D)$ of all meromorphic functions on a domain D is a field.*

Proof. If $[D_0, f]$ is a meromorphic function which does not vanish identically, then
$$D_0' := \{\, a \in D_0; \quad f(a) \neq 0 \,\}$$
is open and, by the identity theorem of Lemma 3.7, also dense in D_0, and hence dense in D. The function $g(z) = 1/f(z)$ is analytic on D_0' and meromorphic on D. We have
$$[D_0, f]^{-1} = [D_0', g]. \qquad \qquad \square$$

Every function which is analytic on D is meromorphic on D ($f = f/1$). If we associate the meromorphic function $[D, f]$ with f, we obtain an embedding
$$\mathcal{O}(D) \hookrightarrow \mathcal{M}(D), \qquad f \mapsto [D, f].$$

We shall identify $\mathcal{O}(D)$ with its image in $\mathcal{M}(D)$. $\mathcal{O}(D)$ consists of all meromorphic functions whose domain of holomorphy is D. $\mathcal{M}(D)$ contains the quotients of functions from $\mathcal{O}(D)$:
$$\mathcal{M}(D) \supset \left\{ \frac{f}{g}; \quad f, g \in \mathcal{O}(D),\, g \neq 0 \right\}.$$

It is an important problem whether $\mathcal{M}(D)$ agrees with the quotient field of $\mathcal{O}(D)$. This is true in the case $n = 1$. In [FB], this was proved – but only for the case $D = \mathbb{C}$ – by means of the Weierstrass product theorem.

The case $n > 1$ is more involved. It will take us some effort to show that in the case $D = \mathbb{C}^n$ every meromorphic function is the quotient of two entire functions.

Exercises for Sect. V.3

1. Let A be a complex $m \times n$ matrix and let $\mathbb{C}^m \to \mathbb{C}^n$ the associated linear map. Identify $\mathbb{C}^m$ (similarly $\mathbb{C}^n$) with $\mathbb{R}^{2m}$ via
    $$(z_1, \ldots, z_m) \longmapsto (x_1, \ldots, x_m, y_1, \ldots, y_m)$$
    to obtain a linear map $\mathbb{R}^{2m} \to \mathbb{R}^{2n}$. What is the associated real $(2m) \times (2n)$ matrix? What are the conditions for a real $(2m) \times (2n)$ matrix to be derived from a complex $m \times n$-matrix?

2. Let V a vector space of finite dimension over some field K, and let $A : V \to V$ be a K-linear map. Then the determinant $\det_K A$ is well-defined. It is the determinant

of the matrix of A with respect to some basis. Now consider $K = \mathbb{C}$. Since V can be considered as a vector space over $\mathbb{R}$, and since a $\mathbb{C}$-linear map is $\mathbb{R}$-linear, one can consider $\det_{\mathbb{C}} A$ and $\det_{\mathbb{R}} A$. Show that

$$\det_{\mathbb{R}} A = |\det_{\mathbb{C}} A|^2.$$

Hint. Reduce this to the case where a $\mathbb{C}$-basis exists, such that A can be represented by a diagonal matrix. Then reduce the statement to the one-dimensional case $V = \mathbb{C}$.

3. Let $D \subset \mathbb{C}^m$ be an open set and let $f : D \to \mathbb{C}^n$ be an analytic map such that the (complex) Jacobian has rank n at all points. Show that the image $f(D)$ is open in $\mathbb{C}^n$.

4. The Weierstrass Preparation Theorem

The complex analysis of one variable is distinguished by the fact that the zero sets of nonvanishing analytic functions are discrete. This is related to the fact that the divisibility properties of the ring of convergent power series of one variable are very simple. Every nonzero convergent power series of one variable is the product of a power of z and a power series Q which does not vanish at the origin. Such power series are invertible in the ring of power series. Hence, in the ring of power series of one variable, there is essentially only one prime element, namely z, and $P = Q \cdot z^n$ is the decomposition of P into primes. Hence the ring of power series is simpler than the the the ring of polynomials $\mathbb{C}[z]$. The prime elements of this ring are the nonconstant linear polynomials, and the decomposition into primes is just the decomposition which results from the fundamental theorem of algebra,

$$P(z) = C(z - a_1) \cdots (z - a_n).$$

The polynomials $z - a$, $a \neq 0$, are prime elements in the polynomial ring, but in the ring of power series they are units. For example, the inverse of $1 - z$ is given by the geometric series.

In the complex analysis of several variables, the situation is much more involved. This is already visible in the case of the ring of polynomials. In the case $n > 1$, this ring is not a Euclidean ring. Nevertheless, the theorem of unique decomposition into prime elements holds. In the algebraic appendix at the end of this volume, we shall treat division theory and obtain a proof of this fundamental result of Gauss in Corollary VIII.2.4.

There are two fundamental theorems for the shape of the zero sets and for the division theory in the ring of power series of several variables, namely the *preparation theorem* of Weierstrass and the *division theorem*. Both theorems play fundamental roles in the complex analysis of several variables. For example, we shall deduce from them that unique prime factorization holds also in the ring of power series.

There is a close relation between the division theory of the ring of convergent power series and the study of the *zero set* of a power series, since we have

$$P|Q \quad \Longrightarrow \quad (P(z) = 0 \Rightarrow Q(z) = 0 \text{ in a neighborhood of } z = 0).$$

First we notice that every power series P with $P(0) = 0$ can be written as a product of finitely many indecomposable elements. For this, we recall from Sect. 2 the notion of the order

$$o(P) := \min\{\,\nu_1 + \nu_2 + \cdots + \nu_n;\quad a_{\nu_1,\ldots,\nu_n} \neq 0\,\}.$$

Of course, we have to assume that $P \not\equiv 0$. Additionally, we define

$$o(0) := \infty.$$

If Q, $Q(0) = 0$, is a second nonunit, then we know that

$$o(P \cdot Q) = o(P) + o(Q).$$

For the proof, it is convenient to characterize the order in a different manner. Recall that a power series P can be written in the form

$$P = P_1 + P_2 + \cdots,$$

where P_m is a homogeneous polynomial of degree m (Sect. 2). We have

$$o(P) = \min\{m;\quad P_m \neq 0\} \quad (P \neq 0).$$

The claimed relation $o(PQ) = o(P) + o(Q)$ follows.

Now the decomposition into a product of finitely many indecomposable elements follows by induction on $o(P)$.

4.1 Definition.   *A power series $P \in \mathcal{O}_n := \mathbb{C}\{z_1, \ldots, z_n\}$ is called z_n-general if*

$$P(0, \ldots, 0, z_n) \not\equiv 0.$$

A power series is z_n-general if it contains a monomial which is independent of $z_1, \ldots z_{n-1}$. For example, $z_1 + z_2$ is z_2-general but $z_1 z_2$ is not.

Let $A = (a_{\mu\nu})_{1 \leq \mu,\nu \leq n}$ be an invertible complex $n \times n$ matrix. We consider A as a linear map

$$A : \mathbb{C}^n \longrightarrow \mathbb{C}^n \quad z \longmapsto w, \qquad w_\mu = \sum_{\nu=1}^{n} a_{\mu\nu} z_\nu.$$

For a power series $P \in \mathcal{O}_n$, we obtain, by substitution and reordering, the power series

$$P^A(z) := P(A^{-1}z).$$

Obviously, the map

$$\mathcal{O}_n \xrightarrow{\sim} \mathcal{O}_n, \qquad P \longmapsto P^A$$

is an ring automorphism, i.e.

$$(P + Q)^A = P^A + Q^A.$$

The inverse map is given by A^{-1}.

4.2 Remark. *For every finite set of convergent power series $P \in \mathcal{O}_n$, $P \neq 0$, there exists an invertible $n \times n$ matrix A such that all P^A are z_n-general.*

Proof. There exists a point $a \neq 0$ in a joint convergence polydisk such that $P(a) \neq 0$ for all P. After a suitable coordinate transformation (a choice of A) has been chosen, we can assume $A(0, \ldots, 0, 1) = a$. Then all P^A are z_n-general. $\square$

Zeros of Power Series

For the local behavior of the zeros of analytic functions, it is sufficient to investigate the zeros of z_n-general power series. The following lemma gives a rough description.

4.3 Lemma. *Let P, $P(0) = 0$, be a z_n-general power series, and let d be the zero order of $P(0, \ldots, 0, z_n)$ at $z_n = 0$. Choose the number $r > 0$ such that P converges absolutely for $|z_\nu| \leq r$ and such that $P(0, \ldots, 0, z_n)$ has no zeros in the disk $|z_n| \leq r$ besides 0. Then there exists a number ε, $0 < \varepsilon < r$, with the following properties:*

1. $P(z_1, \ldots, z_{n-1}, z_n) \neq 0$ *for* $|z_n| = r$ *and* $|z_\nu| < \varepsilon$ $(1 \leq \nu \leq n-1)$.
2. *For fixed* $(z_1, \ldots, z_{n-1})$ *with* $|z_\nu| < \varepsilon$, *the function* $z_n \mapsto P(z_1, \ldots, z_n)$ *has precisely d zeros (counted with multiplicity) for* $|z_n| < r$.

Proof. The first statement is clearly true for any fixed chosen z_n, by a continuity argument. For the general case, one has to use a simple compactness argument.

The second statement follows by means of the zero-counting integral of the usual complex analysis ([FB], Proposition III.7.1). This integral shows that the number of zeros depends continuously on $z_1, \ldots, z_{n-1}$. Since it is an integer, it must be constant and hence be equal to the value for $z_1 = \ldots = z_{n-1} = 0$. $\square$

The fact that the parameters $z_1, \ldots, z_{n-1}$ can be chosen arbitrarily can be expressed as follows:

The zero set of an analytic function on a domain D either is empty, the whole D, or a complex $(n-1)$-dimensional set.

We shall not give a precise explanation for this, since we do not want to introduce dimension theory here.

Let $P_0, \ldots, P_m$ be convergent power series in $(n-1)$ variables, say elements of $\mathcal{O}_{n-1} = \mathbb{C}\{z_1, \ldots, z_{n-1}\}$; then

$$P_0 + P_1 z_n + \ldots + P_m z_n^m$$

can be considered as a convergent power series in $\mathcal{O}_n$. In other words, the polynomial ring $\mathcal{O}_{n-1}[z_n]$ in one variable over $\mathcal{O}_{n-1}$ is embedded into $\mathcal{O}_n$:

$$\mathcal{O}_{n-1}[z_n] \hookrightarrow \mathcal{O}_n.$$

4.4 Definition.   *An element $P \in \mathcal{O}_{n-1}[z_n]$ is called a* **Weierstrass polynomial** *if the highest coefficient is one and if all other coefficients are nonunits:*

$$P = z_n^d + P_{d-1} z_n^{d-1} + \ldots + P_0, \ d \geq 1,$$
$$P_\nu \in \mathcal{O}_{n-1}, \quad P_\nu(0) = 0 \ for \ 0 \leq \nu \leq d-1.$$

A normalized polynomial from $\mathcal{O}_{n-1}[z_n]$ is a Weierstrass polynomial if

$$P(0, \ldots, z_n) = z_n^d \quad (d \geq 1).$$

Weierstrass polynomials are z_n-general.

4.5 Lemma.   *Let $Q \in \mathcal{O}_{n-1}[z_n]$ be a Weierstrass polynomial and let $A \in \mathcal{O}_n$ be a power series with the property*

$$P = AQ \in \mathcal{O}_{n-1}[z_n].$$

Then $A \in \mathcal{O}_{n-1}[z_n]$ also.

For arbitrary polynomials $Q \in \mathcal{O}_{n-1}[z_n]$ instead of Weierstrass polynomials, this statement is false, as the example

$$(1 - z_n)(1 + z_n + z_n^2 + \cdots) = 1$$

shows.

Proof of Lemma 4.5. First step. In addition, we assume that the degree of P (as a polynomial over $\mathcal{O}_{n-1}$) is smaller than the degree of Q. In this case, we show $A = 0$. We choose $r > 0$ small enough that all occurring power series converge for $|z_\nu| < r$. Then we choose $\varepsilon > 0$ small enough that $\varepsilon < r$ and that each zero

$$Q(z_1, \ldots, z_n) = 0, \quad |z_\nu| < \varepsilon \ \text{for} \ \nu = 1, \ldots, n-1,$$

automatically has the property $|z_n| < r$. Then the polynomial

$$z_n \longmapsto P(z_1, \ldots, z_n)$$

has, for each $(n-1)$-tuple $(z_1, \ldots, z_{n-1})$, $|z_\nu| < \varepsilon$, at least $d = \deg Q$ zeros, as Q has, counted with multiplicity. Because $\deg P < \deg Q$, we get

$$P(z_1, \ldots, z_n) \equiv 0 \ \text{for} \ |z_\nu| < \varepsilon, \ \nu = 1, \ldots, n-1.$$

We obtain $P = 0$ and $A = 0$.

Second step. We use a simple but basic fact about division with a remainder in the polynomial ring in one variable over a commutative ring R with unity. (In our application, $R = \mathcal{O}_{n-1}$).

Let $P \in R[X]$ be an arbitrary polynomial and $Q \in R[X]$ a normalized polynomial, i.e. the highest coefficient of Q is assumed to be one. We then have

$$P = AQ + B, \quad \deg B < \deg Q \quad (or \ B = 0))$$

with unique polynomials A, B. (In this context, we define the degree of the zero polynomial to be $-\infty$.)

We apply this simple fact to $R = \mathcal{O}_{n-1}$ and to the two polynomials P and Q given in Lemma 4.5. Division with a remainder gives

$$P = CQ + D, \quad \deg D < \deg Q.$$

Using the equation $P = AQ$, we get

$$(A - C)Q = D, \quad \deg D < \deg Q.$$

Now it follows from the first step that $A = C$, and A, like C, is a polynomial over $\mathcal{O}_{n-1}$. □

We mention already that Weierstrass polynomials are z_n-general. Units in $\mathcal{O}_n$ are z_n-general too. So, the product of a Weierstrass polynomial and a unit is z_n-general. The fundamental preparation theorem of Weierstrass states that *every* z_n-general power series is the product of a unit and a Weierstrass polynomial. Since z_n-generality is not a restrictive property (Remark 4.2), the preparation theorem gives a link between the rings $\mathcal{O}_{n-1}[z_n]$ and $\mathcal{O}_n$.

4.6 Theorem (Weierstrass preparation theorem) (Weierstrass, 1886).
Let $P \in \mathcal{O}_n = \mathbb{C}\{z_1, \ldots, z_n\}$ be a z_n-general power series. Then there exists a unique decomposition

$$P = U \cdot Q,$$

where Q is a Weierstrass polynomial and U is a unit $(U(0) \neq 0)$.

The preparation theorem is related to the division theorem, which is sometimes incorrectly also called the preparation theorem.

4.7 Theorem (Division theorem) (Stickelberger, 1887).
Let Q be a z_n-general power series with $Q(0) = 0$. Let d be the zero order of the power series $Q(0, \ldots, 0, z_n)$ at $z_n = 0$ $(0 < d < \infty)$. Every power series $P \in \mathcal{O}_n$ admits a unique decomposition of the form

$$P = RQ + S,$$

where
   a) $R \in \mathcal{O}_n$;
   b) $S \in \mathcal{O}_{n-1}[z_n]$, $\deg_{z_n}(S) < d$ (or $S = 0$).

Before the proofs of the two theorems, we treat their basic applications to the division theory of $\mathcal{O}_n$.

Division Theory for the Ring of Power Series

We compare the division theory of the rings $\mathcal{O}_{n-1}[z_n]$ and $\mathcal{O}_n$ here.

An application of Lemma 4.5 and the preparation theorem is the following lemma.

4.8 Lemma. *A Weierstrass polynomial $P \in \mathcal{O}_{n-1}[z_n]$ is a prime element in $\mathcal{O}_n$ if and only if it is a prime element in $\mathcal{O}_{n-1}[z_n]$.*

On the other hand, $1 - z_n$ is a unit in $\mathcal{O}_n$ but a prime element in $\mathcal{O}_{n-1}[z_n]$.

Proof of Lemma 4.8. 1) If P is a prime element in $\mathcal{O}_n$, then Lemma 4.5 shows that P is a prime element in $\mathcal{O}_{n-1}[z_n]$ too.

2) Conversely, let P be a prime element in $\mathcal{O}_{n-1}[z_n]$. We have to show that P is prime in $\mathcal{O}_n$. So, we assume

$$P | AB, \quad A, B \in \mathcal{O}_n.$$

After a suitable change of coordinates, we can assume that A, B are both z_n-general. By means of the preparation theorem, we obtain

$$A = A_0 \cdot U, \quad B = B_0 \cdot V,$$

where A_0, B_0 are Weierstrass polynomials and U, V are units. From Lemma 4.5, we get $P | A_0 B_0$ in $\mathcal{O}_{n-1}[z_n]$ and therefore

$$P | A_0 \text{ or } P | B_0 \text{ in } \mathcal{O}_{n-1}[z_n],$$

since P is prime in this ring. So we get

$$P | A \text{ or } P | B \text{ in } \mathcal{O}_n. \qquad \square$$

Now we prove the following proposition as fundamental consequence of the preparation theorem.

4.9 Proposition. *The ring $\mathcal{O}_n$ of convergent power series is a UFD ring.*

As we have seen already, every power series P can be written as a product of finitely many indecomposable elements. So, it remains to show that any indecomposable element of $\mathcal{O}_n$ is prime.

Proof by induction on n. The beginning of the induction is trivial, so we assume that the proposition has been proved for $(n-1)$ instead of n. We have to show that it holds for n. Of course, we can assume that P is z_n-general and then, by the preparation theorem, that it is a Weierstrass polynomial. We show first that P is indecomposable in $\mathcal{O}_{n-1}[z_n]$. So, let $P = AB$ be a decomposition in $\mathcal{O}_{n-1}[z_n]$. Since P is indecomposable in $\mathcal{O}_n$, we can assume that A is a unit in $\mathcal{O}_n$, i.e. $A(0) \neq 0$. It follows from the equation $P(0, \ldots, 0, z_n) = z_n^d =$

$A(0,\ldots,0,z_n)B(0,\ldots,0,z_n)$ that, up to a constant factor, $A(0,\ldots,0,z_n)$ is a power of z_n, i.e. $A(0,\ldots,0,z_n) = Cz_n^\delta$. Since A does not vanish at the origin, we have $\delta = 0$. We obtain the result that B has a degree of at least d as a polynomial in z_n. But then it has a degree of precisely d and A must have degree 0. Hence A is contained in $\mathcal{O}_{n-1}$ and is a unit there.

Now we make use of the induction hypothesis. The ring $\mathcal{O}_{n-1}$ is factorial. By the theorem of Gauss mentioned earlier (Theorem VIII.2.2), $\mathcal{O}_{n-1}[z_n]$ is factorial as well. So, P is in $\mathcal{O}_{n-1}[z_n]$ and, because of Lemma 4.8, it is also a prime element in $\mathcal{O}_n$. $\qquad\square$

Proof of the Preparation and Division Theorems

First step. We start with a special case of the division theorem: *the division theorem (Theorem 4.7) is true if Q is a Weierstrass polynomial (and not only a z_n-general power series).*

For an arbitrary power series $P \in \mathcal{O}_n$, we have to construct a decomposition

$$P = AQ + B, \quad B \in \mathcal{O}_{n-1}[z_n], \quad \deg B < \deg Q,$$

and to show that it is unique.

Uniqueness. From $AQ + B = 0$, we get $A \in \mathcal{O}_{n-1}[z_n]$ because of Lemma 4.5. Comparing degrees, we get $A = B = 0$.

Existence. We want to define

$$A(z_1,\ldots,z_n) := \frac{1}{2\pi i} \oint_{|\zeta|=r} \frac{P(z_1,\ldots,z_{n-1},\zeta)}{Q(z_1,\ldots,z_{n-1},\zeta)} \frac{d\zeta}{\zeta - z_n}.$$

For this, we have to explain how $r > 0$ has to be chosen. It has to be so small that the power series P and Q converge in

$$U = \{z; \quad \|z\| < r\}, \quad \|z\| := \max\{|z_\nu|, \ \nu = 1,\ldots,n\}.$$

Then there exists a number ε, $0 < \varepsilon < r$, such that

$$Q(z_1,\ldots,z_n) \neq 0 \text{ for } |z_n| \geq r, \quad |z_\nu| < \varepsilon \text{ for } 1 \leq \nu \leq n-1.$$

The function A is analytic in $\|z\| < \varepsilon$ and can be expanded into a power series there. We denote this power series by A again. What we have to show now is that

$$B := P - AQ$$

is a polynomial in z_n, and that its degree is smaller than that of Q. By means of the Cauchy integral formula for P, we obtain (with $z := (z_1,\ldots,z_{n-1})$)

$$B(z, z_n) = \frac{1}{2\pi i} \oint_{|\zeta|=r} \frac{P(z,\zeta)}{\zeta - z_n} d\zeta - \frac{1}{2\pi i} \oint_{|\zeta|=r} Q(z, z_n) \frac{P(z,\zeta)}{Q(z,\zeta)} \frac{d\zeta}{\zeta - z_n}$$

$$= \frac{1}{2\pi i} \oint_{|\zeta|=r} \frac{P(z,\zeta)}{Q(z,\zeta)} \left[\frac{Q(z,\zeta) - Q(z,z_n)}{\zeta - z_n} \right] d\zeta.$$

The variable z_n occurs only inside the large brackets. For fixed $z_1, \ldots, z_{n-1}, \zeta$, we know that $Q(z_1, \ldots, z_{n-1}, \zeta) - Q(z_1, \ldots, z_n)$ is a polynomial of degree $d = \deg Q$ in z_n. This has a zero at $z_n = \zeta$ and hence is divisible by $z_n - \zeta$, such that the quotient is a polynomial of degree $d - 1$. Hence B is a polynomial of degree $< d$ in z_n.

Second step. Now let $P \in \mathcal{O}_n$ be an arbitrary z_n-general power series and let Q be a Weierstrass polynomial. Both are assumed to converge in $\|z\| \leq r$. We also assume that there exist a number ε, $0 < \varepsilon < r$, such that for each fixed $(z_1, \ldots, z_{n-1})$ with $|z_\nu| < \varepsilon$ for $(1 \leq \nu \leq n - 1)$ the functions

$$z_n \longmapsto Q(z_1, \ldots, z_n), \quad z_n \longmapsto P(z_1, \ldots, z_n)$$

have the same zeros (counted with multiplicity) in the disk $|z_n| < r$. Then

$$P = UQ,$$

with a unit U.

Proof. We can choose ε so small that all d zeros of Q are contained in $|z_n| < r$. By the special case of the division theorem, we have

$$P = AQ + B, \quad B \in \mathcal{O}_{n-1}[z_n], \quad \deg B < \deg Q.$$

We can assume that A and B both converge in $|z_n| < r$. The polynomial $z_n \longmapsto B(z_1, \ldots, z_n)$ has, for each $(z_1, \ldots, z_{n-1})$, $|z_\nu| < \varepsilon$, $1 \leq \nu \leq n - 1$, more zeros than its degree predicts. Hence it is identically zero. The same consideration shows that A is a unit. $\quad\square$

Third step. Proof of the preparation theorem. Let P be a z_n-general power series, and let d, $0 < d < \infty$, be the zero order of $P(0, \ldots, 0, z_n)$ at $z_n = 0$. The numbers $0 < \varepsilon < r$ are chosen as in Lemma 4.3. We consider the functions

$$\sigma_k(z_1, \ldots, z_{n-1}) = \frac{1}{2\pi i} \oint_{|\zeta|=r} \zeta^k \frac{\partial P(z, \zeta)}{\partial \zeta} \frac{d\zeta}{P(z, \zeta)}, \quad k = 0, 1, 2, \ldots.$$

These functions are analytic in the domain

$$z \in \mathbb{C}^{n-1}, \quad \|z\| < \varepsilon.$$

By the residue theorem of complex analysis (in relation to the zero-counting integral) in one variable, we know that $\sigma_0(z_1, \ldots, z_{n-1})$ is the number of zeros

$$z_n \longmapsto P(z_1, \ldots, z_n)$$

in $|z_n| < r$ (counted with multiplicity). As a consequence, σ_0 is an integer, and hence constant. We order the $d = \sigma_0(z_1, \ldots, z_{n-1})$ zeros arbitrarily,

$$t_1(z), \ldots, t_d(z).$$

Of course, we can not expect that the $t_\nu(z)$ will be analytic functions in z. But a simple generalization of the zero-counting integral gives

$$\sigma_k(z) = t_1(z)^k + \ldots + t_d(z)^k.$$

Therefore the symmetric expressions $t_1(z)^k + \ldots + t_d(z)^k$ are analytic functions in z. By a result of elementary algebra, which we shall use without proof, we have the following result.

The νth elementary symmetric polynomial $(1 \le \nu \le d)$,

$$E_\nu(X_1, \ldots, X_d) = (-1)^\nu \sum_{1 \le j_1 < \ldots < j_\nu \le d} X_{j_1} \ldots X_{j_\nu}$$

can be written as a polynomial (with rational coefficients) in the

$$\sigma_k(X_1, \ldots, X_d) = \sum_{j=1}^{d} X_j^k$$

$(1 \le k \le d$ is enough).

Example.

$$E_2(X_1, X_2) = X_1 X_2 = \frac{1}{2}\left[(X_1 + X_2)^2 - (X_1^2 + X_2^2)\right] = \frac{1}{2}[\sigma_1^2 - \sigma_2].$$

In particular, the elementary symmetric functions $t_1(z), \ldots, t_d(z)$ are analytic. We use them to define the Weierstrass polynomial

$$Q(z_1, \ldots, z_{n-1}, z_n) = z_n^d + E_1\left(t_1(z), \ldots, t_d(z)\right) z_n^{d-1} + \ldots + E_d\left(t_1(z), \ldots, t_d(z)\right).$$

For fixed $\|z\| < \varepsilon$, the zeros of these polynomials are $t_1(z), \ldots, t_d(z)$ by the (trivial) "Vieta theorem". By the second step, P and Q differ only by a unit. This proves the preparation theorem. ☐

Fourth step. Proof of the division theorem. This now follows immediately from the special cases proved above (in the first step) together with the preparation theorem. ☐

We now give another application of the preparation theorem which again shows the close connection between zeros and the divisibility of power series.

4.10 Proposition. *Let P, Q be two power series which are different from zero. Then the following two statements are equivalent:*

a) *In a small neighborhood of the origin,*

$$P(z) = 0 \Longrightarrow Q(z) = 0.$$

b) *There exists a natural number such that $P | Q^m$.*

Proof. We can assume that P and Q are Weierstrass polynomials. Let d be the degree of P. We want to show that $P|Q^d$, and perform polynomial division with a remainder for this purpose:

$$Q^d = AP + B, \quad \deg(B) < d.$$

For any $(z_1, \ldots, z_{n-1})$ in a small neighborhood of the origin, B has the same zeros as P. The multiplicities of the zeros of B are at least as large as those of P. This gives $B = 0$. $\square$

Exercises for Sect. V.4

1. The power series

$$z_2 + \sum_{\nu=1}^{\infty} z_1^{\nu}$$

    is z_2-general. Determine the associated Weierstrass polynomial.

2. Write the elementary symmetric polynomials

$$z_1 z_2 + z_1 z_3 + z_2 z_3, \quad z_1 z_2 z_3$$

    explicitly as polynomials in

$$z_1 + z_2 + z_3, \quad z_1^2 + z_2^2 + z_3^2, \quad z_1^3 + z_2^3 + z_3^3.$$

3. Show that $\mathbb{C}\{z\}$ contains only one nonzero prime ideal

4. Show that an analytic function of more than one variable never has an isolated zero.

5. Representation of Meromorphic Functions as Quotients of Analytic Functions

A meromorphic function on a domain D can be represented locally as a quotient of two analytic functions. Hence there exists a covering

$$D = \bigcup_{i \in I} U_i, \quad U_i \text{ open and connected},$$

and analytic functions

$$f_i, g_i : U_i \longrightarrow \mathbb{C}, \quad g_i \neq 0,$$

such that
$$f|_{U_i} = \frac{f_i}{g_i}.$$
Such a representation is not unique! We can try to enforce some kind of uniqueness by demanding that f_i and g_i are coprime. Here, elements σ, ϱ of a commutative ring with unity are called coprime if
$$x|\varrho \text{ and } x|\sigma \implies x \in R^*.$$
There arises a difficulty: the ring of analytic functions on a domain is not factorial (i.e. a UFD domain). Only the ring $\mathcal{O}_n = \mathbb{C}\{z_1,\ldots,z_n\}$ of convergent power series is factorial. To overcome the difficulty, we need the following statement.

5.1 Proposition. *Let*
$$f, g : D \longrightarrow \mathbb{C}, \quad D \subset \mathbb{C}^n \text{ open,}$$
be analytic functions and let $a \in D$ be a point such that the the power series of f and g at a are coprime elements of the ring of convergent power series
$$\mathbb{C}\{X_1,\ldots,X_n\} \qquad (\text{``}X_\nu = z_\nu - a_\nu\text{''}).$$
Then there exists a neighborhood of a such that the power series expansions of f and g at all points b in this neighborhood are coprime.

We shall now prove another proposition (Proposition 5.2) which will imply Proposition 5.1. For this, we denote the power series expansion of f (and analogously that of g) at a point $a \in D$ by
$$[f]_a \in \mathbb{C}\{X_1,\ldots,X_n\} \qquad (\text{``}X_\nu = z_\nu - a_\nu\text{''}).$$
Now we factorize $[f]_a$ and $[g]_a$ into primes,
$$[f]_a = \prod_{j=1}^{r}[f_j]_a^{\nu_j} \text{ and } [g]_a = \prod_{j=1}^{s}[g_j]_a^{\nu_j},$$
where $[f_j]_a$ (and analogously $[g_j]_a$) are pairwise not associated. Here two elements ϱ, σ of a ring R are called *associated* if they differ by a unit, i.e.
$$\sigma = \varepsilon\varrho, \quad \varepsilon \in R^*.$$
Since we can shrink D, we can assume that the representatives f_j, g_j are analytic in D. Now we consider
$$F := f_1 \cdot \ldots \cdot f_r \text{ and } G := g_1 \cdot \ldots \cdot g_s$$
instead of f and g. It is clear that $[F]_a$ and $[G]_a$ are coprime if this is the case for $[f]_a$ and $[g]_a$. The elements $[F]_a$ and $[G]_a$ are coprime if and only if $[F]_a \cdot [G]_a$ is a square-free element of the ring of convergent power series. An element ϱ of a ring R is called *square-free* if
$$x^2|\varrho \implies x \in R^*.$$
Conversely, for an arbitrary element $b \in D$, the power series $[F]_b$ and $[G]_b$ will be coprime if $[F]_b \cdot [G]_b$ is square-free. Hence Proposition 5.1 follows from the next proposition.

5.2 Proposition. *Let*

$$f : D \longrightarrow \mathbb{C}, \quad D \subset \mathbb{C}^n \ open,$$

be an analytic function. The set of all points $a \in D$ for which the power series expansion $[f]_a$ is a square-free element of the ring of power series is open in D.

Proof. Let $a \in D$ be a point such that the power series expansion $[f]_a$ is square-free. We have to show that $[f]_b$ is square-free in a full neighborhood of a. Without loss of generality, we can assume that $[f]_a$ is X_n-general, which means

$$f(a_1, \ldots, a_{n-1}, z_n - a_n) \not\equiv 0.$$

This condition remains valid in a full neighborhood of a. So, we can assume that $[f]_b$ is X_n-general ("$X_n = z_n - b_n$") at all points $b \in D$. We want to prove a criterion for the square-freeness of an X_n-general power series which implies Proposition 5.2. To exclude trivial cases, we assume that the power series under consideration neither vanishes identically nor is a unit.

5.3 Criterion (Criterion for the square-freeness of a z_n-general power series). *A z_n-general power series*

$$P \in \mathbb{C}\{z_1, \ldots, z_n\}, \quad P(0) = 0, \quad P \neq 0,$$

is not square-free iff

$$\prod_{i<j} [t_i(z) - t_j(z)]^2 = 0 \ for \ z \in \mathbb{C}^{n-1}, \quad \|z\| < \varepsilon.$$

(We use the same notation as in the proof of the preparation theorem.)

Proof of the criterion. The condition of the criterion does not change if P is multiplied by a unit U ($U(0) \neq 0$). Because of the preparation theorem, we can assume that P is a Weierstrass polynomial. It follows easily from the preparation theorem and Lemma 4.5 that *a Weierstrass polynomial is square-free in $\mathcal{O}_n$ iff it is square-free in $\mathcal{O}_{n-1}[z_n]$.*

The latter is true (by a criterion which is valid for normalized polynomials over factorial domains; see Proposition VIII.3.2) if the *discriminant* Δ does not vanish. The discriminant can be computed as follows:

$$\Delta = \prod_{\mu < \nu} (t_\nu - t_\mu)^2 \in \mathcal{O}_{n-1}.$$

This proves Criterion 5.3.   □

Now we come back to our problem of representing a meromorphic function on a domain D globally as a quotient of analytic functions. For each point $a \in D$, there exist a connected open neighborhood $U(a)$ and analytic functions

$$g_a, h_a : U(a) \longrightarrow \mathbb{C}$$

such that

$$f|U(a) = \frac{g_a}{h_a}.$$

Here we can assume – at least after shrinking of $U(a)$ – that the power series expansions of g_a and h_a at a are coprime. By Proposition 5.1, this remains true in a full neighborhood of a. So, we obtain the following result.

5.4 Proposition. *Let f be a meromorphic function on a domain $D \subset \mathbb{C}^n$. Then there exist an open covering*

$$D = \bigcup_{i \in I} U_i, \quad U_i \subset D \text{ open and connected,}$$

and a family of analytic functions

$$g_i, h_i : U_i \longrightarrow \mathbb{C} \qquad (h_i \neq 0)$$

such that the following two conditions are satisfied:

a) $f|_{U_i} = \dfrac{g_i}{h_i}.$

b) *The power series expansions $[g_i]_a$ and $[h_i]_a$ are coprime at all points $a \in U_i$.*

Supplement. *In the intersection of two members U_i, U_j of the covering, we have*

$$g_i|_{U_i \cap U_j} = \varphi_{ij} \cdot g_j|_{U_i \cap U_j}$$

with a function $\varphi_{ij} \in \mathcal{O}(U_i \cap U_j)^*$.

The supplement is a consequence of the following simple algebraic consideration.

Remark. *Let (ϱ, σ), (ϱ', σ') be two pairs of coprime elements of a factorial domain. From the equation*

$$\varrho \sigma' = \sigma \varrho' \qquad (\textit{i.e. "}\varrho/\sigma = \varrho'/\sigma' \textit{ ")},$$

it follows that

$$\varrho = \varepsilon \varrho' \quad \text{and} \quad \sigma = \varepsilon \sigma',$$

*with a **unit** $\varepsilon \in R^*$.*

It is worthwhile to formulate the condition in the Supplement of Proposition 5.4 in the form of a definition.

5.5 Definition.   *A (multiplicative)* **Cousin distribution** *on a domain* $D \subset \mathbb{C}^n$ *is a family* $(U_i, f_i)_{i \in I}$ *consisting of*
   a) *an open covering*

$$X = \bigcup_{i \in I} U_i, \quad U_i \subset D \text{ open and connected,}$$

   b) *a family of analytic functions* $f_i : U_i \to \mathbb{C}$ *with the property*

$$f_i = \varphi_{ij} \cdot f_j \text{ in } U_i \cap U_j, \quad \varphi_{ij} \in \mathcal{O}(U_i \cap U_j)^*.$$

We say that an analytic function $f : D \to \mathbb{C}$ *fits the Cousin distribution* if for all i,
$$f|_{U_i} = \varphi_i \cdot f_i, \quad \varphi_i \in \mathcal{O}(U_i)^*.$$
Then the zero sets of f inside U_i agree with that of f_i.

Since the sets of zeros of f_i and f_j agree in $U_i \cap U_j$, we should think of a Cousin distribution as "prescription of zeros with multiplicities".

The construction of f will be done along the following lines.

First, we determine a family of *invertible* analytic functions

$$g_i \in \mathcal{O}(U_i)^*$$

with the property
$$g_i = \varphi_{ij} \cdot g_j \text{ in } U_i \cap U_j,$$

where the φ_{ij} are the transition functions which occur in Definition 5.5. Then we have, for all i, j,
$$\frac{f_i}{g_i} = \frac{f_j}{g_j} \text{ in } U_i \cap U_j,$$

As a consequence, there exists an analytic function f with $f = f/g_i$ on U_i. This function solves the problem.

The construction of the invertible functions g_i rests on a certain *gluing lemma*, which we shall now formulate and prove in a situation which is topologically very simple. We assume that the U_i are rectangles, parallel to the axes, in a very special position. The rectangles which we consider are Cartesian products of open intervals in $\mathbb{R}$. As usual, we identify $\mathbb{C}^n$ and $\mathbb{R}^{2n}$ by means of

$$(z_1, \ldots, z_n) \longleftrightarrow (x_1, y_1, \ldots, x_n, y_n).$$

Now, let the following be given:

   a) an open rectangle $Q' \subset \mathbb{R}^{2n-1}$;
   b) real numbers $a < b < c < d$.

Then we build
$$Q_1 := (a, c) \times Q' \subset \mathbb{C}^n$$
and
$$Q_2 := (b, d) \times Q' \subset \mathbb{C}^n.$$
So, we have
$$Q_1 \cap Q_2 := (b, c) \times Q'.$$

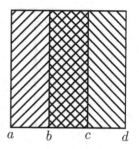

$a \qquad b \qquad c \qquad d$

5.6 Lemma. *Let Q_1, Q_2 be two open rectangles in $\mathbb{C}^n$ ($= \mathbb{R}^{2n}$) in the special positions*

$$Q_1 := (a, c) \times Q' \subset \mathbb{C}^n, \quad Q_2 := (b, d) \times Q' \subset \mathbb{C}^n \qquad (a < b < c < d).$$

Furthermore, let f be an analytic function on an open set U which contains the closure $\overline{Q_1 \cap Q_2}$. We then have the following two lemmas.

1) *Additive gluing lemma. There exist analytic functions*

$$f_\nu : Q_\nu \longrightarrow \mathbb{C}, \quad \nu = 1, 2,$$

  with the property

$$f(z) = f_1(z) - f_2(z) \text{ for } z \in Q_1 \cap Q_2.$$

2) *Multiplicative gluing lemma. Assume that f is invertible ($f \in \mathcal{O}(U)^*$); then there exist invertible analytic functions*

$$f_\nu : Q_\nu \longrightarrow \mathbb{C}, \quad f_\nu \in \mathcal{O}(U)^*, \quad \nu = 1, 2,$$

  such that

$$f(z) = \frac{f_1(z)}{f_2(z)}.$$

Proof of the additive gluing lemma.
The proof uses the Cauchy integral formula applied to f as a function of z_1. In the proof, $z_2, \ldots, z_n$ will be kept fixed. The integrals under consideration will depend analytically on $z_2, \ldots, z_n$ by Leibniz's criterion. Hence it is sufficient to restrict ourselves to the case $n = 1$. The Cauchy integral formula gives

$$f(z) = \frac{1}{2\pi i} \oint_{\partial(Q_1 \cap Q_2)} \frac{f(\zeta)}{\zeta - z} d\zeta \quad \text{for } z \in Q_1 \cap Q_2.$$

It is clear that the boundary $\partial(Q_1 \cap Q_2)$ is the composition of two paths W_1 and W_2, where W_1 is contained in the boundary of Q_1 and W_2 in the boundary of Q_2:

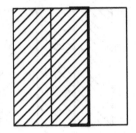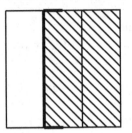

We then have

$$f(z) = f_1(z) - f_2(z) \text{ for } z \in Q_1 \cap Q_2,$$

with

$$f_\nu(z) := \oint_{W_\nu} \frac{f(\zeta)}{\zeta - z}\, d\zeta, \quad \nu = 1, 2.$$

The functions f_ν, $\nu = 1, 2$, are analytic in the complements of W_ν, and hence in the whole of Q_ν (and actually in a much bigger domain!).   □

Proof of the multiplicative gluing lemma.

This can be reduced to the additive lemma if f can be written in the form

$$f(z) = e^{F(z)}, \quad F : D \longrightarrow \mathbb{C} \text{ analytic},$$

since then we have to solve the additive problem for F:

$$F = F_1 - F_2, \quad F_\nu \in \mathcal{O}(Q_\nu),$$

to obtain

$$f = \frac{f_1}{f_2} \text{ with } f_\nu = e^{F_\nu}.$$

Hence it remains to verify the existence of holomorphic logarithms on suitable domains. This can be done similarly to the case $n = 1$ as follows.

5.7 Lemma.   *Let*

$$f : D \longrightarrow \mathbb{C}, \quad D \subset \mathbb{C}^n \text{ open and convex},$$

be an analytic function without zeros. Then f admits an analytic logarithm F, i.e. an analytic function $F : D \to \mathbb{C}$ with

$$e^F = f.$$

Proof. We choose a point $a \in D$. For arbitrary $z \in D$, the segment between a and z is contained in D. Hence the function

$$\alpha(t) := f\left(a + t(z - a)\right), \quad 0 \le t \le 1,$$

is well defined and continuous. We set

$$F(z) := \int_0^1 \frac{\alpha'(t)}{\alpha(t)} \, dt + \mathrm{Log}\, f(a).$$

This function depends (because of the Leibniz rule) analytically on z. We have

$$e^F = f \qquad \left(\text{since } \text{``} \frac{\alpha'(t)}{\alpha(t)} = \frac{d}{dt} \mathrm{Log}\, \alpha(t)\text{''}\right). \qquad \square$$

Why is Lemma 5.6 called the "gluing lemma"?
Consider analytic functions

$$g_\nu : Q_\nu \longrightarrow \mathbb{C}, \quad \nu = 1, 2.$$

Assume that there exists an invertible analytic function

$$\varphi_{12} \in \mathcal{O}(U)^* \qquad (U \supset \overline{Q_1 \cap Q_2})$$

with the property

$$g_1 = \varphi_{12} \cdot g_2 \text{ in } Q_1 \cap Q_2$$

There then exists an analytic function

$$g : Q_1 \cup Q_2 \longrightarrow \mathbb{C}$$

such that

$$g = \varphi_\nu g_\nu \text{ in } Q_\nu, \quad \varphi_\nu \in \mathcal{O}(Q_\nu)^*$$

(g is the "glued function").

For the proof, we apply the multiplicative gluing lemma and write φ_{12} in the form

$$\varphi_{12} = \frac{f_1}{f_2}, \quad f_\nu \in \mathcal{O}(Q_\nu)^*.$$

The functions

$$\frac{g_2}{f_1} \text{ and } \frac{g_1}{f_2}$$

agree in $Q_1 \cap Q_2$ and "glue" to a function defined on $Q_1 \cup Q_2$.

Now we are in the position to prove the following important proposition.

5.8 Proposition (P. Cousin, 1895). *Let (U_i, f_i) be a Cousin distribution on $\mathbb{C}^n$ $\left(= \bigcup_{i \in I} U_i\right)$. Then there exists an analytic function*

$$f : \mathbb{C}^n \longrightarrow \mathbb{C}$$

such that

$$f|_{U_i} = \varphi_i \cdot f_i, \quad \varphi_i \in \mathcal{O}(U_i)^*.$$

Remark. In the case $n = 1$, this proposition remains true for arbitrary domains $D \subset \mathbb{C}^n = \mathbb{C}$, but not in the case $n > 1$!

Proof of Proposition 5.8. In the first step, we prove that on an open neighborhood U of the closed unit cube

$$W := \left\{ z \in \mathbb{C}^n; \quad 0 \leq x_\nu, y_\nu \leq 1 \text{ for } 1 \leq \nu \leq n \right\}$$

there exists an analytic function $f : U \to \mathbb{C}$ which fits the given Cousin distribution,

$$f = \varphi_i \cdot f_i \text{ in } U \cap U_i, \qquad \varphi_i \in \mathcal{O}(U \cap U_i)^*.$$

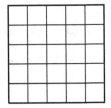

To show this, we decompose the cube W into N^{2n} closed small subcubes by dividing each edge into N equidistant pieces, as illustrated in the figure. We denote the small cubes by $W_{\nu_1, \ldots, \nu_{2n}}$, $1 \leq \nu_j \leq N$ $\quad (1 \leq j \leq 2n)$.

We choose N large enough that each $W_{\nu_1, \ldots, \nu_{2n}}$ is contained in a U_i, and we denote such a U_i by $U_{\nu_1, \ldots, \nu_{2n}}$. Correspondingly, we use the notation f_i for $f_{\nu_1, \ldots, \nu_{2n}}$.

For the rest of the proof, we restrict ourselves to the case $n = 1$, which can be presented more easily. But the case $n > 1$ can be treated in the same manner.

Claim. In a suitable small neighborhood of $W_{11} \cup W_{12}$, there exists an analytic function which fits the Cousin distribution.

This is a direct consequence of the gluing lemma. By induction, we obtain an analytic function in an open neighborhood of $W_{11} \cup \ldots \cup W_{1N}$ which fits the Cousin distribution. We denote this function by F_1. In the same manner, we construct functions F_i on an open neighborhood of $W_{i1} \cup \ldots \cup W_{iN}$. Now, using the gluing lemma again, we can glue the functions F_1 and F_2 and so on, to obtain finally a function which fits the Cousin distribution. Now we obtain the following result.

For each compact subset $K \subset \mathbb{C}^n$, there exist an open neighborhood $U \supset K$ and an analytic function $f : U \to \mathbb{C}$ which fits the Cousin distribution.

To obtain an analytic function f on the whole of $\mathbb{C}^n$, we use an approximation argument.

Let

$$V_k = \left\{ z \in \mathbb{C}^n \mid \|z\| < k \right\}, \quad k = 1, 2, \ldots.$$

We choose an arbitrary analytic function

$$h_k : V_k \longrightarrow \mathbb{C}$$

which fits the Cousin distribution. Then we have, for $k < l$,

$$h_l = h_k \, e^{\varphi_{kl}} \quad \text{in } V_k,$$

with a function φ_{kl} which is analytic on V_k.

$$\boxed{h_{k+1} = h_k \, e^{\varphi_k} \quad \text{in } V_k} \qquad (\varphi_k := \varphi_{k,k+1}).$$

The sequence (V_k, h_k) is again a Cousin distribution. Every analytic function which fits this Cousin distribution fits the original one. So we have succeeded (by means of the gluing lemma) in reducing an arbitrary covering $(U_i)_{i \in I}$ to the ascending chain $V_1 \subset V_2 \subset \ldots$.

We still have the freedom to replace h_k by $\tilde{h}_k = h_k \, e^{\psi_k}$, with analytic functions $\psi_k : V_k \to \mathbb{C}$.

Claim. After a suitable choice of the $\psi_k \in \mathcal{O}(V_k)$, we obtain

$$\tilde{h}_{k+1} = \tilde{h}_k \, e^{\tilde{\varphi}_k},$$

with

a) $\tilde{\varphi}_k \in \mathcal{O}(V_k)$;
b) $|\tilde{\varphi}_k(z)| \leq 2^{-k}$ *for $z \in V_{k-1}$.*

Proof. We shall see that one can find polynomials ψ_k with the desired properties. (They are analytic on the whole of $\mathbb{C}^n$.) It is easy to construct the ψ_k by induction on k. We simply use the fact that every analytic function on V_k (in our case φ_k) can be approximated on the relatively compact subset $\bar{U}_{k-1} \subset U_k$ arbitrarily closely by polynomials. We use the Taylor expansion on polydisks. $\square$

Now we can assume

$$|\varphi_k(z)| \leq 2^{-k} \quad \text{for } z \in V_{k-1}.$$

For each $z \in \mathbb{C}^n$, there exists k_0 with $z \in V_{k_0}$. The series

$$\sum_{k \geq k_0} \varphi_k(z)$$

converges. As a consequence, the sequence

$$(h_k(z))_{k \geq k_0}$$

converges, since we have

$$h_l = h_k \, e^{\varphi_k + \dots + \varphi_{l-1}} \quad \text{for } l > k.$$

We denote the limit by

$$f(z) := \lim_{k \geq k_0} h_k(z).$$

The convergence is clearly locally uniform. Hence f is an analytic function which obviously fits our Cousin distribution. This proves Proposition 5.4. □

From Theorems 5.4 and 5.8, we get the main result of this section, stated below.

5.9 Theorem (Poincaré 1883). *Every meromorphic function f on $\mathbb{C}^n$ can be written as a quotient of two analytic functions, i.e.*

$$f = \frac{g}{h}, \quad g, h : \mathbb{C}^n \longrightarrow \mathbb{C} \text{ analytic functions.}$$

One can achieve the result that the power series expansions of g and h are coprime for each point.

Supplement. *If*

$$f = \frac{g}{h} = \frac{\tilde{g}}{\tilde{h}}$$

are two "coprime" representations of this type, then

$$\tilde{g} = g \cdot e^{\varphi} \quad and \quad \tilde{h} = h \cdot e^{\varphi}$$

with an analytic function $\varphi : \mathbb{C}^n \to \mathbb{C}$.

For the proof of the supplement, we have to use the UFD property of the ring of power series and the existence of analytic logarithms on $\mathbb{C}^n$ (Lemma 5.7).

Exercises for Sect. V.5

1. Show that the Weierstrass product theorem in the form given in [FB], Theorem IV.2.1, is a special case of Proposition 5.8.

2. Let P be a prime element of the ring of power series $\mathcal{O}_n$ and let Q be an arbitrary power series. Assume that $P(z) = 0 \Rightarrow Q(z) = 0$ in a small neighborhood of the origin. Show that $P|Q$.

3. An analytic function $f : U \to \mathbb{C}$ on an open domain $U \subset \mathbb{C}^n$ is called *reduced* if its power series expansion at each point is reduced (square-free). Show that if f, g are two reduced analytic functions with the same zero set, then $g = \varphi f$, where φ is an analytic function without zeros.

4. Show that every nonzero analytic function $f : \mathbb{C}^n \to \mathbb{C}$ can be written in the form
$$f = f^{\text{red}} g.$$
Here f^{red} is a reduced function with the same zero set as f.

   Hint. Use Proposition 5.8 and the result of the previous exercise.

6. Alternating Differential Forms

We developed the calculus of differential forms for the two-dimensional case in the appendix to Chap. II (on Stokes's theorem). The local part of this theory will now be generalized to the case of arbitrary dimensions.

In the following, n denotes a fixed natural number. We denote by
$$\mathcal{M}_p := \mathcal{M}_p^{(n)} := \{a \subset \{1, \dots, n\}; \quad \#a = p\}$$
the set of all subsets of $\{1, \dots, n\}$ which contain p elements. Their number is
$$\binom{n}{p} \qquad (= 0 \text{ if } p < 0 \text{ or } p > n).$$

6.1 Definition. *An (alternating)* **differential form** ω *of degree* p *on an open subset* $D \subset \mathbb{R}^n$ *is a map which assigns to each* $a \in \mathcal{M}_p$ *a* C^∞*-function* $f_a : D \to \mathbb{C}$:
$$\omega = (f_a)_{a \in \mathcal{M}_p}.$$

We denote by $A^p(D)$ the set of all differential forms of degree p on D, and call them p-forms for short.

The set $\mathcal{M}_0^{(n)}$ consists of one element only (namely the empty set). Hence a zero form has only one component. So, we can identify zero forms and functions:

$$A^0(D) = C^\infty(D).$$

Calculus of Differential Forms

I. Algebraic Computation Rules

Since p-forms can be added componentwise and multiplied by functions, $A^p(D)$ is a *module* over $C^\infty(D)$:

$$(f_a) + (g_a) = (f_a + g_a),$$
$$f \cdot (f_a) = (f \cdot f_a).$$

By the way,

$$A^p(D) = 0 \text{ for } p < 0 \text{ or } p > n,$$

since $\mathcal{M}_p$ is empty in these cases.

II. The Total Differential of a Function

Since we can identify the one-element subsets of $\{1, \ldots, n\}$ with the elements of $\{1, \ldots, n\}$, a 1-form can be considered as an n-tuple of functions

$$A^1(D) = A^0(D) \times \ldots \times A^0(D),$$
$$A^1(D) = \underbrace{C^\infty(D) \times \ldots \times C^\infty(D)}_{n-\text{times}}.$$

The *total differential* of a C^∞-function f is defined by

$$df := \left(\frac{\partial f}{\partial x_1}, \ldots, \frac{\partial f}{\partial x_n} \right).$$

One can verify the rules

a) $d(f + g) = df + dg$;
b) $d(f \cdot g) = f \cdot dg + g \cdot df$;
c) $df = 0 \iff f$ is locally constant.

If we denote by

$$p_\nu : D \to \mathbb{C}, \quad p_\nu(x) = x_\nu,$$

the projection to the νth coordinate, we have

$$dp_\nu = (0,\ldots,0,1,0,\ldots,0)\,.$$
$$\uparrow$$
$$\nu\text{th component}$$

Notation: $dx_\nu := dp_\nu$.

So, we can write a 1-form $(f_1,\ldots,f_n)$ also as

$$(f_1,\ldots,f_n) = \sum_{\nu=1}^{n} f_\nu\,dx_\nu.$$

Now the total differential can be written as

$$df = \sum_{\nu=1}^{n} \frac{\partial f}{\partial x_\nu}\,dx_\nu.$$

III. The Alternating Product

Analogously to the case $p = 1$, we define for a p-element subset $a \subset \{1,\ldots,n\}$ the p-form

$$(dx_a)_b = \begin{cases} 1 & \text{for } a = b, \\ 0 & \text{for } a \neq b. \end{cases}$$

Each p-form $\omega = (f_a)_{a\in\mathcal{M}_p}$ can be written in the form

$$\omega = \sum_{a\in\mathcal{M}_p} f_a\,dx_a.$$

Let

$$a,b \subset \{1,\ldots,n\}, \quad \#a = p, \quad \#b = q.$$

We define a "sign factor" $\varepsilon(a,b)$.

First case. We set

$$\varepsilon(a,b) = 0 \text{ if } a \cap b \neq \emptyset.$$

Second case. Let $a \cap b = \emptyset$.

We order the elements of a naturally:

$$a = \{a_1,\ldots,a_p\}, \quad a_1 < a_2 < \ldots < a_p,$$

and correspondingly

$$b = \{b_1,\ldots,b_q\}, \quad b_1 < b_2 < \ldots < b_q.$$

We then have

$$a \cup b = \{a_1, \ldots, a_p, b_1, \ldots, b_q\}, \quad \#(a \cup b) = p + q.$$

But the elements in the brackets are not in their natural order, since there is no need for $a_p < b_1$.

We denote by $\varepsilon(a, b)$ the sign of the permutation (of $p + q$ elements) which is used to bring $(a_1, \ldots, a_p, b_1, \ldots, b_q)$ into their natural order.

Example.

$$a = \{1, 2\}, \quad b = \{2, 3\}, \quad \varepsilon(a, b) = 0;$$
$$a = \{1, 3\}, \quad b = \{2, 4\}, \quad \varepsilon(a, b) = -1;$$
$$a = \{2, 4\}, \quad b = \{3, 5\}, \quad \varepsilon(a, b) = -1.$$

Now we define the *alternating product*

$$A^p(D) \quad \times \quad A^q(D) \quad \longrightarrow \quad A^{p+q}(D),$$
$$\omega \quad , \quad \omega' \quad \longmapsto \quad \omega \wedge \omega',$$

by the formulae

$$\left(\sum_{a \in \mathcal{M}_p} f_a \, dx_a \right) \wedge \left(\sum_{b \in \mathcal{M}_q} g_b \, dx_b \right) := \left(\sum_{\substack{a \in \mathcal{M}_p \\ b \in \mathcal{M}_q}} f_a g_b \, dx_a \wedge dx_b \right)$$

$$\boxed{dx_a \wedge dx_b := \varepsilon(a, b) \, dx_{a \cup b}.}$$

It is a simple exercise to verify the following formulae.

1) In the case $p = 0$, the alternating product agrees with the usual (componentwise) product:

$$f \wedge \omega = f \cdot \omega \text{ for } f \in A^0(D).$$

2) The alternating product is *skew commutative:*

$$\omega \wedge \omega' = (-1)^{pq} \, \omega' \wedge \omega \text{ for } \omega \in A^p(D), \ \omega' \in A^q(D).$$

In particular,

$$\omega \wedge \omega = 0$$

if p is odd. The alternating product is *associative:*

$$(\omega \wedge \omega') \wedge \omega'' = \omega \wedge (\omega' \wedge \omega''),$$
$$\omega \in A^p(D), \quad \omega' \in A^q(D), \quad \omega'' \in A^r(D).$$

3) The alternating product is *bilinear:*

$$(\omega_1 + \omega_2) \wedge \omega = \omega_1 \wedge \omega + \omega_2 \wedge \omega,$$
$$\omega_1, \omega_2 \in A^p(D), \quad \omega \in A^q(D).$$

Because of the associativity, we can define the alternating product

$$\omega_1 \wedge \ldots \wedge \omega_m$$

of several alternating differential forms.

It is easy to show, by induction with respect to p, that for a p-element subset $a = \{a_1, \ldots, a_p\}$ of $\{1, \ldots n\}$ in its natural order, the formula

$$dx_a = dx_{a_1} \wedge \ldots \wedge dx_{a_p}$$

holds. If we write $f_{a_1,\ldots,a_p}$ instead of f_a, we obtain the standard representation of a p-form ω,

$$\omega = \sum_{a \in \mathcal{M}_p} f_a \, dx_a$$
$$= \sum_{1 \le a_1 < \ldots < a_p \le n} f_{a_1,\ldots,a_p} \, dx_{a_1} \wedge \ldots \wedge dx_{a_p}.$$

As an example, we can compute the alternating product of two 1-forms. Using

$$dx_\nu \wedge dx_\mu = -dx_\mu \wedge dx_\nu = 0, \text{ if } \mu = \nu,$$

we get

$$\left(\sum_{\nu=1}^{n} f_\nu \, dx_\nu\right) \wedge \left(\sum_{\mu=1}^{n} g_\mu \, dx_\mu\right) = \sum_{1 \le \nu < \mu \le n} (f_\nu g_\mu - f_\mu g_\nu) \, dx_\nu \wedge dx_\mu.$$

IV. The Exterior Derivative

Generalizing the total differential of a function, we define

$$d : A^p \to A^{p+1}$$

by the formula

$$d\left(\sum f_a\, dx_a\right) = \sum df_a \wedge dx_a.$$

The following formulae can easily be checked:

1. $d(\omega + \omega') = d\omega + d\omega'$.
2. $d(\omega \wedge \omega') = (d\omega) \wedge \omega' + (-1)^p\, \omega \wedge (d\omega')$,

   $\omega \in A^p(D), \quad \omega' \in A^q(D)$.

   In particular, $d(c\omega) = c\,d\omega$ for $c \in \mathbb{C}$.
3. $d(d\omega) = 0$.

An important special case is

$$d(\omega \wedge \omega') = d\omega \wedge \omega' \text{ if } d\omega' = 0.$$

By induction, we obtain

$$d(\omega_1 \wedge \ldots \wedge \omega_m) = 0 \text{ if } d\omega_1 = \ldots = d\omega_m = 0.$$

Hence we have

$$d(\omega \wedge df_1 \wedge \ldots \wedge df_m) = d\omega \wedge df_1 \wedge \ldots \wedge df_m.$$

V. Complex Coordinates

Now we consider an open subset $D \subset \mathbb{C}^n$. Since we can identify $\mathbb{C}^n$ with $\mathbb{R}^{2n}$ by

$$\mathbb{C}^n \longleftrightarrow \mathbb{R}^{2n},$$
$$(z_1, \ldots, z_n) \longleftrightarrow (x_1, y_1, \ldots, x_n, y_n),$$

all of what we have said for the real case holds in the complex case too. Nevertheless it is frequently useful to introduces "complex coordinates" in the complex case:

$$dz_\nu := dx_\nu + i\,dy_\nu, \quad d\bar{z}_\nu := dx_\nu - dy_\nu.$$

We have

$$dx_\nu = \frac{1}{2}(dz_\nu + d\bar{z}_\nu), \quad dy_\nu = \frac{1}{2i}(dz_\nu - d\bar{z}_\nu).$$

So, we see that each 1-form can be written as

$$\omega = \sum_{\nu=1}^{n} f_\nu\, dz_\nu + \sum_{\nu=1}^{n} g_\nu\, d\bar{z}_\nu.$$

If we set

$$A^{1,0}(D) = \left\{ \sum_{\nu=1}^{n} f_\nu \, dz_\nu \right\},$$

$$A^{0,1}(D) = \left\{ \sum_{\nu=1}^{n} g_\nu \, d\bar{z}_\nu \right\},$$

we obtain

$$A^1(D) = A^{1,0}(D) + A^{0,1}(D).$$

We want to generalize this decomposition to arbitrary degrees, and define for this purpose

$$A^{p,q}(D) := \left\{ \sum_{\substack{1 \le i_1 < \ldots < i_p \le n \\ 1 \le j_1 < \ldots < j_q \le n}} f\binom{i_1,\ldots,i_p}{j_1,\ldots,j_q} \, dz_{i_1} \wedge \ldots \wedge dz_{i_p} \wedge d\bar{z}_{j_1} \wedge \ldots \wedge d\bar{z}_{j_q} \right\}.$$

Then, obviously,

$$A^m(D) = \sum_{p+q=m} A^{p,q}(D).$$

This decomposition is *direct*, i.e. the decomposition of an m-form ω as

$$\omega = \sum \omega_{p,q}, \quad \omega_{p,q} \in A^{p,q}(D),$$

is unique.

Obviously, the alternating product respects this decomposition in the following sense: if

$$\omega \in A^{p,q}(D), \quad \omega' \in A^{p',q'}(D),$$

then

$$\omega \wedge \omega' \in A^{(p+p',q+q')}(D).$$

But the exterior derivative does not respect the decomposition of $A^m(D)$. However, it is possible to split it into a sum such that each summand respects this decomposition. For this, we set

$$\frac{\partial}{\partial z_\nu} = \frac{1}{2}\left(\frac{\partial}{\partial x_\nu} - \mathrm{i}\frac{\partial}{\partial y_\nu} \right),$$

$$\frac{\partial}{\partial \bar{z}_\nu} = \frac{1}{2}\left(\frac{\partial}{\partial x_\nu} + \mathrm{i}\frac{\partial}{\partial y_\nu} \right).$$

If we define the operators

$$\partial : C^\infty(D) \longrightarrow A^{1,0}(D),$$

$$\bar{\partial} : C^\infty(D) \longrightarrow A^{0,1}(D)$$

by

$$\partial f := \sum_{\nu=1}^{n} \frac{\partial f}{\partial z_\nu} \, dz_\nu,$$

$$\bar{\partial} f := \sum_{\nu=1}^{n} \frac{\partial f}{\partial \bar{z}_\nu} \, d\bar{z}_\nu,$$

then we have

$$df = \partial f + \bar{\partial} f.$$

Now we define the more general maps

$$\partial : A^{p,q}(D) \to A^{p+1,q}(D),$$

$$\bar{\partial} : A^{p,q}(D) \to A^{p,q+1}(D)$$

by the formulae

$$\partial(f \, dz_{i_1} \wedge \ldots \wedge dz_{i_p} \wedge d\bar{z}_{j_1} \wedge \ldots \wedge d\bar{z}_{j_q}) = \partial(f) \wedge dz_{i_1} \wedge \ldots \wedge dz_{i_p} \wedge d\bar{z}_{j_1} \wedge \ldots \wedge d\bar{z}_{j_q}$$

and

$$\bar{\partial}(f \, dz_{i_1} \wedge \ldots \wedge dz_{i_p} \wedge d\bar{z}_{j_1} \wedge \ldots \wedge d\bar{z}_{j_q}) = \bar{\partial}(f) \wedge dz_{i_1} \wedge \ldots \wedge dz_{i_p} \wedge d\bar{z}_{j_1} \wedge \ldots \wedge d\bar{z}_{j_q}$$

$$(1 \le i_1 < \ldots < i_p \le n, \quad 1 \le j_1 < \ldots < j_q \le n)$$

Then the rules

$$\partial \circ \partial = 0, \quad \bar{\partial} \circ \bar{\partial} = 0,$$

$$\partial \circ \bar{\partial} = -\bar{\partial} \circ \partial$$

hold. We now formulate the two most important theorems about alternating differential forms in the "local case". For both of them, we assume that $D \subset \mathbb{R}^n$ is a *convex domain* (i.e. D is open and the straight segment between two points of D is contained in D). First, we have to give a definition.

6.2 Definition. *A differential form ω is called **closed** if*

$$d\omega = 0.$$

Sometimes this is called "d-closed". The terms "∂-closed" ($\partial\omega = 0$) and "$\bar{\partial}$ closed" ($\bar{\partial}\omega = 0$) are defined analogously.

6.3 Lemma (Poincaré's lemma).
For every closed p-form ω, there exists a $(p-1)$-form ω' with

$$\omega = d\omega'.$$

In the case $n = p = 1$, this follows from the main theorem of differential and integral calculus.

6.4 Lemma (Dolbeault's lemma).

For every (p,q)-form ω which is $\bar{\partial}$-closed (i.e. $\bar{\partial}\omega = 0$), there exists a $(p, q-1)$-form ω' such that

$$\omega = \bar{\partial}\omega'.$$

(The same lemma holds for the ∂-complex as well.)

We shall not prove these two lemmas, since we shall not use them in the following.

VI. Analytic Differential Forms

A differential form ω on an open subset of $\mathbb{C}^n$ is called analytic if the following two conditions are satisfied:

a) ω is of the type $(p,0)$, i.e. of the form

$$\omega = \sum f_{i_1,\ldots,i_p} \, dz_{i_1} \wedge \ldots \wedge dz_{i_p}.$$

b) The components $f_{i_1,\ldots,i_p}$ are analytic functions.

Obviously, a p-form ω of type $(p,0)$ is analytic iff

$$\bar{\partial}\omega = 0.$$

The *exterior derivative* of an analytic p-form is also analytic. Moreover, the formula

$$d\omega = \partial\omega = \sum df_{i_1,\ldots,i_p} \wedge dz_{i_1} \wedge \ldots \wedge dz_{i_p}$$

holds. The exterior derivative of an analytic function is

$$df = \partial f = \sum \frac{\partial f}{\partial z_\nu} \, dz_\nu.$$

If we denote the set of all analytic p-forms on D by $\Omega^p(D)$, then the exterior derivative gives a map

$$\partial = d : \Omega^p(D) \to \Omega^{p+1}(D).$$

Of course, $\partial^2 = 0$.

6.5 Lemma (Poincaré's analytic lemma).

Let $D \subset \mathbb{C}^n$ be a convex domain. Then every closed analytic p-form ω ($d\omega = 0$) can be written in the form

$$\omega = d\omega', \quad \omega' \in \Omega^{p-1}(D).$$

In the case $n = p = 1$, this is the well-known theorem of complex analysis that any analytic function on a convex domain $D \subset \mathbb{C}$ admits a primitive.

We shall use Poincaré's analytic lemma only in the case $p = 1$ and $D = \mathbb{C}^n$. So, we shall prove it only in this special case. In this case it states:

Let $f_1, \ldots, f_n$ be analytic functions on $\mathbb{C}^n$. We assume that

$$\frac{\partial f_i}{\partial z_k} = \frac{\partial f_k}{\partial z_i} \text{ for } 1 \le i, k \le n \quad (\text{equivalent to } d\left(\sum f_\nu \, dz_\nu\right) = 0).$$

Then there exists an analytic function on $\mathbb{C}^n$ with

$$\frac{\partial f}{\partial z_i} = f_i \text{ for } i = 1, \ldots, n.$$

Proof by induction on n.
We have already considered the case $n = 1$. So, let $n > 1$. Termwise integration of the power series expansion of f_1 with respect to the first variable gives an analytic function f with

$$\frac{\partial f}{\partial z_1} = f_1.$$

We can replace f_i by $f_i - \partial f / \partial z_i$ without changing the assumptions. Hence we can assume

$$f_1 = 0.$$

But then,

$$\frac{\partial f_k}{\partial z_1} = 0 \text{ for all } k.$$

Hence the functions f_k do not depend on z_1. Now we can apply induction.

$\square$

Exercises for Sect. V.6

1. Determine an analytic function $f(z_1, z_2)$ with
$$df = z_2^2 \, dz_1 + 2 z_1 z_2 \, dz_2.$$

2. Prove the product rule
$$d(\omega \wedge \omega') = (d\omega) \wedge \omega' + (-1)^p \, \omega \wedge (d\omega').$$

3. Prove
$$d(\omega_1 \wedge \ldots \wedge \omega_k) = d\omega_1 \wedge \ldots \wedge \omega_k \quad \text{if} \quad d\omega_2 = \cdots = d\omega_k = 0.$$

VI. Abelian Functions

As the theory of elliptic integrals leads naturally to the theory of elliptic functions, the theory of general algebraic integrals leads to the theory of Riemann surfaces and then to the theory of abelian functions. These are meromorphic functions on a higher-dimensional complex torus. The case $n > 1$ is much more involved than the case of the elliptic functions ($n = 1$). The reason for this is that the Weierstrass approach using the $\wp$-function does not work, since the zeros and poles of meromorphic functions of several variables are no longer discrete. Hence the method of Mittag-Leffler partial fraction series and Weierstrass products is no longer available. In [FB], Sect. V.6, we touched on another approach to the theory of elliptic functions. This uses, instead of the Weierstrass σ-function, the Jacobi theta function for the proof of Abel's theorem. We shall follow this thread now. Many of the ideas of this chapter are taken from Igusa's fundamental book [Ig4].

1. Lattices and Tori

We start with the real theory.

1.1 Lemma. *Let $L \subset \mathbb{R}^n$ be a discrete additive subgroup. There exist linearly independent vectors $\omega^{(1)}, \dots \omega^{(k)}$ such that*

$$L = \mathbb{Z}\omega^{(1)} + \cdots + \mathbb{Z}\omega^{(k)}.$$

The number k is uniquely determined ($k \leq n$).

The proof is given by induction on n:

Beginning of the induction ($n = 1$). A discrete subgroup $L \subset \mathbb{R}$ which is different from zero has an element $a \in L$, $a \neq 0$, with minimal modulus. It is easy to show that $L = a\mathbb{Z}$.

Induction step. We assume that the proposition has been proved for $n - 1$ instead of n. The discreteness of L means that each compact subset contains only finitely many elements of L. We can assume that $L \neq 0$. Then we can choose, from all vectors in L which are different from zero, one with minimal Euclidean norm

$$|\omega^{(1)}| = \sqrt{\sum_{\nu=1}^{n} |\omega_\nu^{(1)}|^2}.$$

After a suitable coordinate transformation, we can assume that

$$\omega^{(1)} = (1, 0, \dots, 0).$$

It follows from the beginning of the induction that

$$\{x \in \mathbb{R}; \quad (x, 0, \ldots, 0) \in L\} = \mathbb{Z}.$$

Now we consider the projection

$$p : \mathbb{R}^n \longrightarrow \mathbb{R}^{n-1}, \quad (x_1, \ldots, x_n) \longmapsto (x_2, \ldots, x_n),$$

and claim that *the image $L' = p(L)$ is discrete in $\mathbb{R}^{n-1}$.*

Otherwise, there would exist infinitely many vectors $p(\omega)$, $\omega \in L$, such that

$$|\omega_\nu| \leq C \text{ for } \nu = 2, \ldots, n \quad (C \text{ suitable}).$$

We could add an integral multiple of $\omega^{(1)}$ to obtain also $|\omega_1| \leq C$, which gives a contradiction to the discreteness of L.

By the induction assumption, there exist vectors $\omega^{(2)}, \ldots, \omega^{(k)} \in \mathbb{R}^n$ whose images in $\mathbb{R}^{n-1}$ are linearly independent and generate L' as a $\mathbb{Z}$-module. Then $\omega^{(1)}, \ldots, \omega^{(k)}$ are of course linearly independent and we have

$$L = \sum_{\nu=1}^{k} \mathbb{Z}\omega^{(\nu)}.$$

The uniqueness of k is clear, since k is the dimension of the vector space spanned by L.   □

1.2 Definition.   *If the number k which occurs in Lemma 1.1 equals the dimension n, we call L a **lattice**. The point set*

$$P = \left\{ \sum_{\nu=1}^{n} t_\nu \omega^{(\nu)}; \quad 0 \leq t_\nu \leq 1 \right\}$$

*is called a **fundamental parallelogram** of L.*

Obviously,

$$\mathbb{R}^n = \bigcup_{a \in L} P_a, \quad P_a = \{a + x; \ x \in P\},$$

i.e. $\mathbb{R}^n$ is covered by L-translates of the fundamental parallelogram. Up to boundary points, this decomposition is disjoint.

1.3 Remark.   *Let $L \subset \mathbb{R}^n$ be a lattice. Then*

$$L^\circ := \{x \in \mathbb{R}^n; \quad \langle a, x \rangle \in \mathbb{Z} \text{ for all } a \in L\} \qquad \left(\langle a, x \rangle = \sum_{\nu=1}^{n} a_\nu x_\nu \right)$$

is a lattice as well. We have $(L^\circ)^\circ = L$.

We call $L°$ the *dual lattice* of L.

Proof. Obviously,

$$(\mathbb{Z}^n)° = \mathbb{Z}^n.$$

Each lattice is of the form

$$L = A\,\mathbb{Z}^n,$$

with a certain $n \times n$ matrix A. The columns of A are a lattice basis. Obviously,

$$L° = A^{t-1}\mathbb{Z}^n. \qquad \square$$

As usual, we associate with the lattice $L \subset \mathbb{R}^n$ a torus

$$X = \mathbb{R}^n/L.$$

The elements of X are equivalence classes

$$[x] = \{x + a;\ a \in L\}$$

with respect to the equivalence relation

$$x \sim y \quad \Longleftrightarrow \quad x - y \in L.$$

There is a natural projection

$$p : \mathbb{R}^n \longrightarrow X, \quad x \longmapsto [x].$$

Functions on X are in one-to-one correspondence with functions on $\mathbb{R}^n$ that are periodic under L, i.e.

$$f(x + a) = f(x) \text{ for all } a \in L.$$

Fourier Series

We denote by $\mathcal{C}^\infty(X)$ the set of all L-periodic $\mathcal{C}^\infty$-functions on $\mathbb{R}^n$. Some examples of periodic functions under L are

$$f(x) = e^{2\pi i\langle a,x\rangle}, \quad a \in L°.$$

The theory of Fourier series says that these are the basic functions for constructing all periodic functions.

1.4 Lemma. *Every periodic C^∞-function*

$$f \in C^\infty(X), \quad X = \mathbb{R}^n/L,$$

admits on the whole of $\mathbb{R}^n$ an expansion

$$f(x) = \sum_{g \in L^\circ} a_g e^{2\pi i \langle x, g \rangle}$$

which converges absolutely and uniformly. The coefficients a_g are unique, and

$$a_g = \frac{1}{\mathrm{vol}(P)} \int_P f(x) e^{-2\pi i \langle x, g \rangle} dx_1 \dots dx_n.$$

Here P is a fundamental parallelogram of L.

The case $L = \mathbb{Z}^n$ is usually treated in basic courses on analysis. In this case the Fourier series is of the form

$$f(x) = \sum_{g \text{ integral}} a_g e^{2\pi i (g_1 x_1 + \dots + g_n x_n)},$$

$$a_g = \int_0^1 \cdots \int_0^1 f(x) e^{-2\pi i (g_1 x_1 + \dots + g_n x_n)}.$$

We assume this to be known. In the general case, we write the lattice in the form $L = A\mathbb{Z}^n$. By means of

$$\langle Ax, A^{t^{-1}} y \rangle = \langle x, y \rangle,$$

we obtain the result that the Fourier series of f with respect to L corresponds to the Fourier series of

$$g(x) = f(Ax)$$

with respect to $\mathbb{Z}^n$. $\qquad \square$

What are the conditions for the coefficients a_g such that the corresponding Fourier series converges and represents a C^∞-function?

1.5 Lemma. *The series*

$$f(x) = \sum_{g \in L^\circ} a_g e^{2\pi i \langle x, g \rangle}$$

is the Fourier series of a C^∞-function iff for each polynomial $P(g_1, \dots, g_n)$ we have

$$|a_g P(g_1, \dots, g_n)| \longrightarrow 0 \text{ for } g_1^2 + \cdots + g_n^2 \longrightarrow \infty.$$

Proof. We can assume that $L = \mathbb{Z}^n$.

1) Assume that the coefficients a_g satisfy the decay behavior formulated in Lemma 1.5. We have to show that the Fourier series and all series which are obtained by repeated termwise partial differentiation converge absolutely and uniformly. Termwise differentiation with respect to x_ν means that a_g is replaced by $2\pi i g_\nu a_g$. This does not affect the decay behavior. Hence it is sufficient to show that the Fourier series itself,

$$\sum_{g \in L^\circ} |a_g|,$$

converges. For this, it is sufficient to show that the majorants

$$\sideset{}{'}\sum (g_1^2 + \ldots + g_n^2)^{-k} \qquad \left(\sideset{}{'}\sum_\nu := \sum_{\nu \neq 0} \right)$$

converge for sufficiently large k. This series converges iff the integral

$$\int\limits_{x_1^2 + \ldots + x_n^2 \geq 1} (x_1^2 + \ldots + x_n^2)^{-k}$$

converges. By means of polar coordinates, one one can easily show that this is the case for $2k > n$. (compare [FB], Lemma V.2.1).

2) Assume that f is continuously differentiable arbitrarily often. By partial integration, one can show that

$$g_\nu \cdot a_g = \frac{1}{2\pi i} \int\limits_P \frac{\partial f}{\partial x_\nu} e^{-2\pi i \langle g, x \rangle}.$$

The integrand is bounded on the compact set P uniformly in g, since the exponential term has modulus one. Iterated application of this observation shows that

$$a_g P(g_1, \ldots, g_n)$$

is bounded for every polynomial. But this expression must in addition tend to 0, since we can multiply it by $g_1^2 + \ldots + g_n^2$. $\qquad \square$

Exercises for Sect. VI.1

1. Let L be a lattice. Show that a subgroup $L' \subset L$ is a lattice iff it has a finite index.

2. Show that the volume of a fundamental parallelogram of a lattice is independent of the choice of the lattice basis.

3. Let $L' \subset L$ be a subgroup of finite index $[L : L']$ of a lattice L, and let P', P be fundamental parallelograms. Show that

$$\mathrm{vol}(P') = [L : L']\mathrm{vol}(P).$$

2. Hodge Theory of the Real Torus

We denote by $A^p(X)$ the set of all p-forms on $\mathbb{R}^n$ whose components are periodic. We consider the de Rham complex of the torus,

$$\cdots \xrightarrow{d} A^p(X) \xrightarrow{d} A^{p+1}(X) \xrightarrow{d} \cdots.$$

The periodicity will not be destroyed by differentiation. We set

$$C^p(X) = \mathrm{Kernel}\big(A^p(X) \xrightarrow{d} A^{p+1}(X)\big),$$

$$B^p(X) = \mathrm{Image}\big(A^{p-1}(X) \xrightarrow{d} A^p(X)\big),$$

and

$$H^p(X) = C^p(X)/B^p(X) \quad \text{(we have } B^p(X) \subset C^p(X)),$$
$$h^p(X) = \dim_{\mathbb{C}} H^p(X).$$

We also set

$$\mathcal{H}^p(X) = \big\{ \omega \in A^p(X); \quad \text{the components of } \omega \text{ are constant} \big\}.$$

We have

$$\dim \mathcal{H}^p(X) = \binom{n}{p}.$$

Differentiation of a Fourier series gives another one without a constant Fourier coefficient. Hence d cannot be surjective.

2.1 Proposition. *Let $L \subset \mathbb{R}^n$ be a lattice. We have*

$$C^p(X) = B^p(X) \oplus \mathcal{H}^p(X).$$

The composition of $\mathcal{H}^p(X) \hookrightarrow C^p(X)$ with the projection $C^p(X) \to H^p(X)$ induces an isomorphism

$$H^p(X) \cong \mathcal{H}^p(X), \quad \text{and hence } h^p(X) = \binom{n}{p}.$$

Proof. We have already noticed that

$$B^p(X) \cap \mathcal{H}^p(X) = 0.$$

Hence it suffices to show that the image $B^p(X)$ under d contains all closed p-forms ω with vanishing constant Fourier coefficient. It is useful to consider first the special case $n = p = 1$. Here we have to show that *if*

$$f(x) = \sum_{k=-\infty}^{\infty} a_k e^{2\pi i k x}$$

is a C^∞-Fourier series whose constant Fourier coefficient vanishes, then $f(x)$ is the derivative of a C^∞-Fourier series.

It is clear what we have to do. Because of Lemma 1.5, the series

$$g(x) = \sum_{k \neq 0} \left(\frac{a_k}{2\pi i k} \right) e^{2\pi i k x}$$

is a C^∞-function like f.

In principle, the general case (n, p arbitrary) rests on the same effect. But working it out leads to certain combinatorial difficulties. The following formalism – which works also in the complex case – gets around these difficulties.

The Laplace operator is defined by the formula

$$\Delta f = \sum_{\nu=1}^{n} \frac{\partial^2 f}{\partial x_\nu^2}.$$

It transforms periodic functions into periodic functions. More generally, we define the Laplace operator for differential forms componentwise:

$$\Delta \left(\sum f_{i_1,\dots,i_p}\, dx_{i_1} \wedge \dots \wedge dx_{i_p} \right) = \sum \Delta f_{i_1,\dots,i_p}\, dx_{i_1} \wedge \dots \wedge dx_{i_p}.$$

This is an operator

$$\Delta : A^p(X) \to A^p(X).$$

Now we define, for arbitrary p, an operator called the *co-differentiation operator*,

$$\delta : A^p(X) \to A^{p-1}(X),$$

by means of the formula

$$\delta \left(\sum_{1 \leq i_1 < \dots < i_p \leq n} f_{i_1,\dots,i_p}\, dx_{i_1} \wedge \dots \wedge dx_{i_p} \right)$$

$$= \sum_{\nu=1}^{p} (-1)^\nu \sum_{1 \leq i_1 < \dots < i_p \leq n} \frac{\partial f_{i_1,\dots,i_p}}{\partial x_{i_\nu}}\, dx_{i_1} \wedge \dots \wedge d\hat{x}_{i_\nu} \wedge \dots \wedge dx_{i_p}.$$

Here the symbol "^" means that the term beyond it has to be canceled.

By means of co-differentiation, we obtain an important splitting of the Laplace operator. A simple calculation that can be left to the reader shows that

$$\boxed{-\Delta = d\delta + \delta\, d} \qquad\qquad \text{(on } A^p(X)\text{)}.$$

We consider two more operators which can be applied to Fourier series

$$f(x) = \sum_{g\in L^\circ} a_g e^{2\pi i\langle x,g\rangle},$$

namely

a) $\qquad\qquad (Hf)(x) = a_0,$

b) $\qquad\qquad (Gf)(x) = -\dfrac{1}{4\pi^2}\sum_{g\neq 0}\dfrac{a_g}{g_1^2 + \ldots + g_n^2}\,e^{2\pi i\langle x,g\rangle}.$

Again these operators can be generalized to differential forms by applying them componentwise.

The key to the proof of Proposition 2.1 is the following simple formula:

$$\boxed{(\Delta G)(\omega) = \omega - H\omega} \qquad\qquad (\omega \in A^p(X)).$$

Now let ω be a closed form without a constant Fourier coefficient;

$$d\omega = 0 \text{ and } H\omega = 0.$$

Obviously,
$$d\omega = 0 \Longrightarrow d(G\omega) = 0.$$

The above formula now says

$$-d(\delta G\omega) = \omega,$$

which proves Proposition 2.1.   □

Exercises for Sect. VI.2

1. Verify the formulae
$$-\Delta = d\delta + \delta d, \qquad (\Delta G)(\omega) = \omega - H\omega$$

in the case $n = 2$.

2. Show that in the case $n = 2$, every harmonic L-periodic function is constant.

Hint. Assume that the function is real-valued. Then you can apply the theory given in Chap. II.

In the following exercise, it can be assumed that this has been proved for all n.

3. Show that for a p-form $\omega \in A^p(X)$, we have
$$\Delta\omega = 0 \iff d\omega = \partial\omega = 0.$$

3. Hodge Theory of a Complex Torus

We now come to the case of a complex torus
$$X = \mathbb{C}^n/L, \quad L \subset \mathbb{C}^n \ (= \mathbb{R}^{2n}) \text{ a lattice.}$$

In Sect. V.6, we introduced the Dolbeault complex. Now we introduce its periodic version
$$A^{p,q}(X) := A^{p,q}(\mathbb{C}^n) \cap A^{p+q}(X).$$

The operators ∂, $\bar{\partial}$ preserve periodicity. In analogy to the real case, we can introduce the following vector spaces:

$$C^{p,q}(X) = \text{Kernel}\left(A^{p,q}(X) \overset{\bar{\partial}}{\longrightarrow} A^{p,q+1} \right),$$

$$B^{p,q}(X) = \text{Image}\left(A^{p,q-1}(X) \overset{\bar{\partial}}{\longrightarrow} A^{p,q}(X) \right),$$

$$H^{p,q}(X) = C^{p,q}(X)/B^{p,q}(X),$$

$$\mathcal{H}^{p,q}(X) = A^{p,q}(X) \cap \mathcal{H}^{p+q}(X)$$

$$= \{ \omega \in A^{p,q}(X), \quad \text{the components of } \omega \text{ are constant} \}.$$

The so-called Hodge numbers are
$$h^{p,q} = \dim_{\mathbb{C}} H^{p,q}(X).$$

3.1 Proposition. *Let $L \subset \mathbb{C}^n$ be a lattice. We have*
$$C^{p,q}(X) = B^{p,q}(X) \oplus \mathcal{H}^{p,q}(X);$$

in particular,

$$H^{p,q}(X) \cong \mathcal{H}^{p,q}(X) \text{ and } h^{p,q}(X) = \binom{n}{p} \cdot \binom{n}{q}.$$

The proof is analogous to the real case: we now give a brief indication of it.

First we verify that the operators Δ, G, H which we introduced in the real case respect the *bi-grading*

$$A^m(X) = \bigoplus_{p+q=m} A^{p,q}(X),$$

since the formulae

$$\Delta(f\omega_0) = (\Delta f)\omega_0, \quad G(f\omega_0) = (Gf)\omega_0, \quad \text{and } H(f\omega_0) = (Hf)\omega_0$$

are also valid in complex coordinates

$$\omega_0 = dz_{i_1} \wedge \ldots \wedge dz_{i_p} \wedge d\bar{z}_{j_1} \wedge \ldots \wedge d\bar{z}_{j_q}.$$

The reason for this is that ω_0 can be written as a linear combination with constant coefficients in the corresponding real basis elements.

We also have, in analogy to the real case,

$$\bar{\partial}\omega = 0 \Longrightarrow \bar{\partial}G\omega = 0.$$

So, the proof of the real version will work in the complex case as well if we can find an operator

$$\delta : A^{p,q}(X) \to A^{p,q-1}$$

such that

$$-\Delta = \bar{\partial}\delta + \delta\bar{\partial} \quad \text{on } A^{p,q}(X).$$

The formula

$$\delta(f\, dz_{i_1} \wedge \ldots \wedge dz_{i_p} \wedge d\bar{z}_{j_1} \wedge \ldots \wedge d\bar{z}_{j_q})$$
$$= \sum_{\nu=1}^{q}(-1)^\nu \frac{\partial f}{\partial \bar{z}_{j_\nu}}\, dz_{i_1} \wedge \ldots \wedge dz_{i_p} \wedge d\bar{z}_{j_1} \wedge d\hat{\bar{z}}_{j_\nu} \wedge \ldots \wedge d\bar{z}_{j_q}$$

defines such an operator. $\square$

Although the proofs of the real and complex Hodge decompositions are very similar, there is a fundamental difference between the real and the complex cases. In the real case, there exists in each dimension essentially only one torus and only one de Rham complex. The reason is that exterior differentiation d commutes with $\mathbb{R}$-linear maps and that every lattice in $\mathbb{R}^n$ can be transformed by an $\mathbb{R}$-linear isomorphism into another given lattice. In the complex case, the situation is different.

The operators $\partial/\partial z_\nu$ and $\partial/\partial \bar{z}_\nu$ commute only with $\mathbb{C}$-linear maps – not with arbitrary $\mathbb{R}$-linear ones.

Usually, two lattices $L, L' \subset \mathbb{C}^n$ cannot be transformed into each other by means of a $\mathbb{C}$-linear isomorphism. (In the case $n = 1$, the $\mathbb{C}$-linear automorphisms are the *similarity transformations*.)

Exercises for Sect. VI.3

1. Verify the formula
$$-\Delta = \bar{\partial}\delta + \delta\bar{\partial} \qquad \text{on } A^{p,q}(X))$$
   in the case $n = 1$.

2. Show that a differential form $\omega \in A^{p,q}(X)$ has constant coefficients if and only if
$$\bar{\partial}\omega = \delta\omega = 0.$$

4. Automorphy Summands

Let f be a meromorphic function on $\mathbb{C}^n$. For a given vector $\omega \in \mathbb{C}^n$, we can define the meromorphic function $g(z) = f(z + \omega)$ in an obvious way. We say that f has period ω if $g = f$.

In the following, $L \subset \mathbb{C}^n$ denotes a lattice. An *abelian function* f is a meromorphic function on $\mathbb{C}^n$ that is periodic with respect to L:

$$f(z + \omega) = f(z) \text{ for all } \omega \in L.$$

So, an abelian function has $2n$ periods which are linearly independent over $\mathbb{R}$. In the case $n = 1$, an abelian function is nothing but an elliptic function. In analogy to the first Liouville theorem in the theory of elliptic functions, we can make the following statement.

4.1 Remark. *Every analytic abelian function is constant.*

Proof. An abelian function takes all of its values in a fundamental parallelogram. Since this is compact, an everywhere analytic abelian function has a maximum and hence is constant by the maximum principle. □

We have seen (Theorem V.5.9) that every meromorphic function on $\mathbb{C}^n$, and, in particular, every abelian function f, can be written as a quotient of two analytic functions g and h, i.e. $f = h/g$. We can achieve the result that the power series expansions of h and g at each point are coprime. For a period

a, we have two coprime representations as quotients $f(z) = g(z)/h(z) = g(z + a)/h(z + a)$. This gives

$$h(z + a) = e^{2\pi i H_a(z)} h(z)$$

(and similarly for g), with certain analytic functions H_a. If f is not identically zero, then H_a has to satisfy the following functional equation:

$$\boxed{H_{a+b}(z) \equiv H_b(z + a) + H_a(z) \text{ mod } 1.}$$

(By definition, the congruence

$$a \equiv b \text{ mod } 1$$

means that $a - b$ is integral.)

4.2 Definition. *An **automorphy summand** (with respect to L) is a map*

$$H : L \times \mathbb{C}^n \longrightarrow \mathbb{C}, \quad (a, z) \longmapsto H_a(z),$$

with the following properties:

1) *$H_a(z)$ is analytic in z for each $a \in L$.*
2) *$H_{a+b}(z) \equiv H_b(z + a) + H_a(z) \text{ mod } 1$.*

Owing to the theorem concerning the representation of meromorphic functions as quotients of analytic functions, we can associate automorphy summands with abelian functions. A big part of the theory of abelian functions deals with the classification of these summands. We shall treat this classification in this and the following section (see Proposition 5.6):

Trivial Automorphy Summands

For any analytic function $\varphi : \mathbb{C}^n \to \mathbb{C}$,

$$H_a(z) := \varphi(z + a) - \varphi(z)$$

is an automorphy summand. For reasons which will immediately be clear, we call such summands *trivial automorphy summands*.

4.3 Definition. *Two automorphy summands $H_a, \tilde{H}_a$ are called equivalent if they differ only by a trivial summand:*

$$\tilde{H}_a(z) = H_a(z) + \varphi(z + a) - \varphi(z), \quad \varphi \text{ analytic on } \mathbb{C}^n.$$

Equivalent automorphy summands should be considered as "essentially equal": if h is a solution of the functional equation

$$h(z+a) = e^{2\pi i H_a(z)} h(z),$$

then

$$\tilde{h}(z) := e^{2\pi i \varphi(z)} h(z)$$

is a solution of

$$\tilde{h}(z+a) = e^{2\pi i \tilde{H}_a(z)} \tilde{h}(z).$$

This gives a one-to-one correspondence between the solution spaces. If g is a second solution and $\tilde{g}$ is the corresponding transformed function, we have

$$\frac{g(z)}{h(z)} = \frac{\tilde{g}(z)}{\tilde{h}(z)}.$$

Equivalent automorphy summands perform in the same way for the construction of abelian functions.

In the following, we shall choose a suitable representative from each equivalence class. We shall show in this section (Theorem 4.5) that $H_a(z)$ is a polynomial of degree ≤ 1 in z for each a. For the proof, we shall use the Hodge decomposition on a complex torus.

It is useful (but not necessary for our purposes) to equip the set of equivalence classes of automorphy summands with a structure in the form of a group. First, we can see that the sum of automorphy summands is an automorphy summand as well. So, the set of all automorphy summands is an abelian group. The set of trivial automorphy summands is a subgroup. We have to deal with the factor group. We shall determine the structure of this factor group. There is a connection between this factor group and the Picard group of a compact Riemann surface (see Exercise 4).

We want to change the congruence 2) in Definition 4.2 into an equality. For this, we consider the *imaginary part* $h_a(z) = \operatorname{Im} H_a(z)$ of the automorphy summand. We have

$$h_{a+b}(z) = h_b(z+a) + h_a(z).$$

We accept that h_a is not analytic. The goal of the following construction is to associate with the system of functions h_a a closed periodic differential form of type $(p,q) = (1,1)$. We shall apply the Hodge decomposition (Proposition 3.1) to this differential.

4.4 Lemma. *Let*

$$h : L \times \mathbb{C}^n \longrightarrow \mathbb{R}, \quad (a,z) \longmapsto h_a(z),$$

be a map with the following properties:

1) $h_a(z)$ *is a (real)* C^∞*-function for fixed* $a \in L$;
2) $h_{a+b}(z) = h_b(z+a) + h_a(z)$.

Then there exists a real C^∞*-function* $h : \mathbb{C}^n \to \mathbb{R}$ *with the property*

$$h_a(z) = h(z+a) - h(z).$$

(In the real theory, we can consider h_a as a trivial $\mathcal{C}^\infty$-automorphy summand.)

Proof of Lemma 4.4. We choose a real C^∞-function

$$\varphi : \mathbb{C}^n \longrightarrow \mathbb{R}, \quad \varphi \geq 0,$$

with the following properties:

1) The support of φ is compact.
2) The set of points $z \in \mathbb{C}^n$ with $\varphi(z) > 0$ contains a fundamental parallelogram of L.

It is not difficult to prove the existence of such a function. We shall skip the proof. Because of 1) and 2), the series

$$\sum_{a \in L} \varphi(z - a) \qquad \text{(the sum is finite for fixed } z)$$

is an everywhere (!) positive C^∞-function. If we set

$$\psi_b(z) := \frac{\varphi(z - b)}{\sum_{a \in L} \varphi(z - a)},$$

we obtain for each $b \in L$ a C^∞-*function with compact support* with the property

$$\psi_{a+b}(z + b) = \psi_a(z).$$

It is easy to check that

$$h(z) := \sum_{a \in L} \psi_a(z) h_a(z - a)$$

has the desired property. $\square$

We come back to our automorphy summand $H_a(z)$. By Lemma 4.4, there exists a real C^∞-function h with the property

$$\boxed{h(z + a) - h(z) = \operatorname{Im} H_a(z).}$$

We apply the operator $\partial\bar{\partial}$ to this equation (see Sect. V.6). Analytic functions are annihilated by $\bar{\partial}$, and antianalytic functions by ∂. Because $\partial\bar{\partial} = -\bar{\partial}\partial$, the operator $\partial\bar{\partial}$ annihilates the sum of an analytic and an antianalytic function. This shows that

$$\partial\bar{\partial} \operatorname{Im} H_a(z) = 0.$$

We obtain the result that *the (1,1)-form $\partial\bar{\partial}h$ is periodic under L,*

$$\omega := \partial\bar{\partial}h \in A^{1,1}(X), \quad X := \mathbb{C}^n/L.$$

Of course,

$$\bar\partial\omega = 0 \qquad \text{(since } \partial\bar\partial = -\bar\partial\partial \text{ and } \bar\partial^2 = 0\text{)}.$$

As announced, we have associated a closed differential form of type $(1,1)$ with the automorphy summand. We now apply the Hodge decomposition. By Proposition 3.1, we have

$$\partial\bar\partial h = \sum_{1\le i,k\le n} a_{ik}\, dz_i \wedge d\bar z_k + \bar\partial\phi$$

with a periodic $(1,0)$-form

$$\phi = \sum \phi_i\, dz_i \ \in A^{1,0}$$

and a matrix (a_{ik}) of complex numbers.

We point out again that h itself need not to be periodic.

Obviously, we have

$$\sum_{1\le i,k\le n} a_{ik}\, dz_i \wedge d\bar z_k = \partial\bar\partial \sum a_{ik} z_i \bar z_k$$

and hence

$$\bar\partial\Big[\partial\Big(-h + \sum a_{ik} z_i \bar z_k\Big) - \phi\Big] = 0.$$

The differential form in the square brackets,

$$\omega_0 := \partial\Big(-h + \sum a_{ik} z_i \bar z_k\Big) - \phi,$$

is of type $(1,0)$. Since it is annihilated by $\bar\partial$, it is an *analytic differential form* of type $(p,q) = (1,0)$. From $\partial^2 = 0$, we get $\partial\omega_0 = -\partial\phi$. Hence $\partial\omega_0$ is periodic, like φ. By Liouville's theorem (Remark 4.1), the components of $\partial\omega_0$ and hence of $\partial\phi$ are constant! The constant terms in the Fourier expansion of ϕ are annihilated by ∂. For this reason, we have

$$\partial\omega_0 = -\partial\phi = 0.$$

We have shown that ω_0 is an analytic differential form of type $(1,0)$ which is annihilated by ∂.

From Poincaré's analytic lemma (Lemma V.6.5), we obtain $\omega_0 = \partial g$, with an analytic (but not necessarily) periodic function $g : \mathbb{C}^n \to \mathbb{C}$.

Now we use the fact that φ is a differential form of type $(1,0)$ which is annihilated by ∂. The Hodge decomposition (Proposition 3.1) holds for the ∂-complex as well as the $\bar\partial$-complex. This shows that

$$\phi = \sum C_i\, dz_i + \partial\psi,$$

with a *periodic* C^∞-function ψ and an n-tuple (C_i) of complex numbers.

Comparing the two representations for ω_0 and φ, we obtain

$$\partial g = \partial\left(-h + \sum a_{ik} z_i \bar{z}_k\right) - \sum C_i \, dz_i - \partial\psi.$$

If we replace g by the analytic function

$$g + \sum C_i z_i,$$

the above equation becomes

$$\partial\left(-h + \sum a_{ik} z_i \bar{z}_k - g - \psi\right) = 0.$$

Here g is an analytic (not necessarily periodic) function and φ is a periodic (not necessarily analytic) function. The expression in the brackets must be an analytic function. So, we obtain

$$-h = -\sum a_{ik} z_i \bar{z}_k + g + \tilde{g} + \varphi$$

with

a) g analytic,

b) $\tilde{g}$ antianalytic,

c) φ periodic.

We are interested only in the difference

$$\operatorname{Im} H_a(z) = h(z + a) - h(z) \qquad (a \in L).$$

Since this is independent of φ, we can assume $\varphi = 0$. We also know that h is real, so

$$-h = -\operatorname{Re}\left[\sum a_{ik} z_i \bar{z}_k\right] + \operatorname{Re} g + \operatorname{Re} \tilde{g}.$$

The real part of $\tilde{g}$ will not change if we replace $\tilde{g}$ by its complex conjugate. This function is also analytic and can be absorbed by g. This means that we can assume that

$$-h = -\operatorname{Re}\left[\sum a_{ik} z_i \bar{z}_k\right] + \operatorname{Re} g.$$

If we replace the automorphy summand H_a by the equivalent

$$\tilde{H}_a(z) := H_a(z) + i(g(z + a) - g(z)),$$

we get

$$\operatorname{Im} \tilde{H}_a(z) = \tilde{h}(z + a) - \tilde{h}(z)$$

with

$$\tilde{h}(z) = \mathrm{Re}\left[\sum a_{ik} z_i \bar{z}_k\right].$$

We obtain

$$\mathrm{Im}\big(\tilde{H}_a(z)\big) = \sum_\nu \alpha_\nu z_\nu + \sum_\nu \beta_\nu \bar{z}_\nu + K,$$

with certain complex constants α_ν, β_ν, K ($\alpha_\nu = \bar{\beta}_\nu$, $K \in \mathbb{R}$). Since an analytic function is determined by its real part up to an additive constant, we obtain from the last equation

$$\tilde{H}_a(z) = C + \sum_{\nu=1}^n C_\nu z_\nu,$$

with certain complex numbers C, C_ν which can depend on a. This gives the main result of this section, expressed in the following theorem.

4.5 Theorem. *Every equivalence class of automorphy summands contains an automorphy summand of the form*

$$H_a(z) = Q_a(z) + C_a.$$

Here Q_a is $\mathbb{C}$-linear in z and C_a is a constant.

(A function $Q : \mathbb{C}^n \to \mathbb{C}$ is called *linear* if it is of the form

$$Q(z) = \alpha_1 z_1 + \ldots + \alpha_n z_n.)$$

Now that we have proved this proposition, the Hodge decomposition has served its purpose and will no longer be used.

4.6 Definition. *Assume that the automorphy summand $H_a(z)$ has the normal form of Theorem 4.5. An analytic solution of the functional equation*

$$f(z + a) = e^{2\pi i H_a(z)} f(z) \text{ for } a \in L$$

is then called a **theta function.**

So far, we have proved the following proposition.

4.7 Proposition. *Every abelian function is the quotient of two theta function for a suitable automorphy summand.*

The classification of automorphy summands does not end with Theorem 4.5. Two problems remain to be investigated:

1) What are the conditions for the system (Q_a, C_a) such that H_a is an automorphy summand?

2) What does it mean that two such special automorphy summands are equivalent?

In the next section, we will attack both of these problems, with some effort in linear algebra.

Exercises for Sect. VI.4

1. Show that the Weierstrass σ-function ([FB], Section V.6) is a theta function.

2. Show that Jacobi's theta function ([FB], Sect. V.6) $f(z) := \vartheta(\tau, z)$ is a theta function with respect to the lattice $\mathbb{Z} + \tau\mathbb{Z}$.

3. Prove Proposition 4.7 in the case $n = 1$ by means of the theory of elliptic functions ([FB], Chap. V).

4. Consider a one-dimensional complex torus $X = \mathbb{C}/L$. Let D be a divisor on X. By the Weierstrass' product theorem, one can find a meromorphic function f on $\mathbb{C}$ which fits D in an obvious sense. Then, after the choice of a holomorphic logarithm, one can consider the automorphy summand $\log(f(z+\omega)/f(z))$. Show that this defines an isomorphism from $\mathrm{Pic}(X)$ onto the group of equivalence classes of automorphy summands.

5. Quasi-Hermitian Forms on Lattices

We are looking for a simple algebraic description of automorphy summands of the form
$$H_a(z) := Q_a(z) + C_a, \quad Q_a \text{ linear in } z.$$

Of course, this summand is determined if we know $H_a(z)$ for all a from a lattice basis $\omega_1, \ldots, \omega_{2n}$:

$$H_{\omega_\nu}(z) =: H^{(\nu)}(z) = Q^{(\nu)}(z) + C^{(\nu)}.$$

But we one cannot – and this is our problem – prescribe the linear forms $Q^{(\nu)}$ and the numbers $C^{(\nu)}$ arbitrarily. Because of the *commutativity* of the group L, we have to take care of the relations

$$H_b(z + a) + H_a(z) = H_a(z + b) + H_b(z).$$

The conditions which the linear forms $Q^{(\nu)}$ and constants $C^{(\nu)}$ have to satisfy need some effort in the field of linear algebra.

We now recall some ideas from linear algebra. Let $\mathcal{Z}$ be a finite-dimensional complex vector space. Without loss of generality, we could assume $\mathcal{Z} = \mathbb{C}^n$, but we prefer, for good reasons, a coordinate-free presentation. If we want, $\mathcal{Z}$ can be considered as a real vector space of doubled dimension.

We also recall that each finite-dimensional real vector space V has a natural topology. So, we can define the notion of a lattice $L \subset V$ in a natural way. If V, W are finite-dimensional complex vector spaces, then the notion of an analytic (= holomorphic) map from an open subset in V into an open subset of W is well defined.

A *symmetric bilinear form* S on $\mathcal{Z}$ is a map

$$S : \mathcal{Z} \times \mathcal{Z} \to \mathbb{C}$$

with the following properties:

a) $S(z, w)$ is linear in z for fixed w.
b) $S(z, w) = S(w, z)$.

A *Hermitian form* on $\mathcal{Z}$ is a map

$$H : \mathcal{Z} \times \mathcal{Z} \to \mathbb{C}$$

with the following properties:

a) $H(z, w)$ is linear in z for fixed w.
b) $H(z, w) = \overline{H(w, z)}$.

A *quasi-Hermitian form* Q on $\mathcal{Z}$ is a map

$$Q : \mathcal{Z} \times \mathcal{Z} \to \mathbb{C},$$

which can be written as the sum of a Hermitian form H and symmetric bilinear form S:

$$\boxed{Q = H + S.}$$

We associate with a quasi-Hermitian form Q the following $\mathbb{R}$-bilinear form:

$$A(z, w) := \frac{1}{2i} \left(Q(z, w) - Q(w, z) \right).$$

Obviously, A depends only on H:

$$\boxed{A(z, w) = \operatorname{Im} H(z, w).}$$

We have
$$A(z, w) = -A(w, z).$$

This means that A is an *alternating* $\mathbb{R}$-*bilinear form* on $\mathcal{Z}$.

One can reconstruct H from A. A simple calculation gives

$$\boxed{H(z, w) = A(\mathrm{i}z, w) + \mathrm{i}A(z, w), \quad A(\mathrm{i}z, w) = A(\mathrm{i}w, z).}$$

As a consequence, the decomposition of a quasi-Hermitian form into a sum of a Hermitian form and a symmetric bilinear form is unique!

5.1 Remark.   *The map*
$$Q : \mathcal{Z} \times \mathcal{Z} \to \mathbb{C}$$

is quasi-Hermitian if and only if the following conditions are satisfied:

a) $Q(z, w)$ *is* $\mathbb{C}$-*linear in* z *for fixed* w.
b) $Q(z, w)$ *is* $\mathbb{R}$-*linear in* w *for fixed* z.
c) $A(z, w) := (1/2\mathrm{i})\left(Q(z, w) - Q(w, z)\right)$ *is real.*

The proof is easy and will be omitted (see Exercise 1).

Representation by Matrices

Let $\mathcal{Z} := \mathbb{C}^n$, with the $\mathbb{C}$-standard basis $e_1, \ldots, e_n$. We shall denote the matrices associated with Hermitian or $\mathbb{C}$-bilinear forms by the same letter. There is no danger of confusion, because they determine each other.

1) The matrix S with entries

$$s_{\mu\nu} := S(e_\mu, e_\nu)$$

is symmetric.
2) The matrix H with entries

$$h_{\mu\nu} := H(e_\mu, e_\nu)$$

is Hermitian (i.e. $h_{\mu\nu} = \bar{h}_{\nu\mu}$).

Since A is only $\mathbb{R}$-bilinear, we should use an $\mathbb{R}$-basis to describe it as a matrix, for example
$$e_1, \ldots, e_n; \; e_{n+1} := \mathrm{i}e_1, \ldots, e_{2n} := \mathrm{i}e_n.$$

The matrix A with entries

$$a_{\mu\nu} = A(e_\mu, e_\nu) \qquad (1 \le \mu, \nu \le 2n)$$

is an *alternating* $2n \times 2n$-matrix ($a_{\mu\nu} = -a_{\nu\mu}$).

The connection between A and H is, in matrix notation,

$$A = \begin{pmatrix} \operatorname{Im} H & -\operatorname{Re} H \\ +\operatorname{Re} H & \operatorname{Im} H \end{pmatrix}.$$

What is the connection between quasi-Hermitian forms and affine automorphy summands? A function $H : \mathcal{Z} \to \mathbb{C}$ on a complex vector space $\mathcal{Z}$ is called *affine* if it can be written as a sum of a linear function Q and a constant C. By an (affine) automorphy summand in this somewhat more abstract context, we of course mean a map which associates with each lattice point $a \in L$ an affine function $H_a : \mathcal{Z} \to \mathbb{C}$, such that the relations

$$H_{a+b}(z) \equiv H_b(z + a) + H_a(z) \bmod 1$$

are valid. This means that

$$Q_{a+b}(z) + C_{a+b} \equiv Q_a(z + b) + C_a + Q_b(z) + C_b \bmod 1.$$

If we put $z = 0$, we obtain

$$C_{a+b} \equiv Q_a(b) + C_a + C_b \bmod 1$$

and then

$$Q_{a+b}(z) = Q_a(z) + Q_b(z).$$

Hence the map $a \mapsto Q_a(z)$ is a $\mathbb{Z}$-linear map in z for fixed a. We shall make use of the following trivial principle:

Every $\mathbb{Z}$-linear map of a lattice $L \subset \mathcal{Z}$ into an $\mathbb{R}$-vector space V is determined by its values on a lattice basis. Hence it can be extended to an $\mathbb{R}$-linear map $\mathcal{Z} \to V$.

When we apply this principle to the map $a \mapsto 2iQ_a$, we obtain the following result. There exists a unique map

$$Q : \mathcal{Z} \times \mathcal{Z} \to \mathbb{C}$$

with the properties

a) $Q(z, w) = 2iQ_w(z)$ for $w \in L$;
b) $Q(z, w)$ is $\mathbb{C}$-linear in the first variable, z;
c) $Q(z, w)$ is $\mathbb{R}$-linear in the second variable, w.

We claim that Q is quasi-Hermitian. Because of Remark 5.1, it is sufficient to show that

$$A(z, w) := \frac{1}{2i} \left(Q(z, w) - Q(w, z) \right)$$

takes only real values. It is enough to prove this for an $\mathbb{R}$-basis. Hence it is enough to prove this for L. But then we have more, namely

$$A(a, b) = Q_b(a) - Q_a(b) \in \mathbb{Z}.$$

So Q is quasi-Hermitian. The fact that A takes only integral values on $L \times L$ will turn out to be of high importance. Hence we fix this in a definition.

5.2 Definition. *Let $L \subset \mathcal{Z}$ be a lattice in a finite-dimensional $\mathbb{C}$-vector space $\mathcal{Z}$. A* **quasi-Hermitian form on the lattice** *L is a map*

$$Q : \mathcal{Z} \times \mathcal{Z} \to \mathbb{C}$$

with the following properties:

1) *Q is quasi-Hermitian, i.e. the sum of a symmetric bilinear form S and a Hermitian form H.*
2) *The alternating $\mathbb{R}$-bilinear form*

$$A(z, w) := \frac{1}{2\mathrm{i}} \left(Q(z, w) - Q(w, z) \right)$$

has the property

$$A(a, b) \in \mathbb{Z} \text{ for all } a, b \in L.$$

What we have proved is the following statement.

5.3 Remark. *Assume that*

$$H_a(z) := Q_a(z) + C_a$$

is an affine automorphy summand. There exists a unique quasi-Hermitian form Q on L with the property

$$Q(z, a) = 2\mathrm{i} Q_a(z) \text{ for } a \in L.$$

Now we can ask the following question. Let Q be a quasi-Hermitian form on L. What are the conditions for the constants C_a such that

$$H_a(z) := Q_a(z) + C_a, \quad Q_a(z) := \frac{1}{2\mathrm{i}} Q(z, a),$$

is an automorphy summand?

It is useful to replace the constants C_a by

$$D_a := C_a - \frac{1}{2} Q_a(a).$$

The characteristic equations are then

$$\boxed{D_{a+b} \equiv \frac{1}{2} A(a, b) + D_a + D_b \bmod 1.}$$

We shall describe all solutions of this system. In the first step, we shall show that it suffices to determine the *real* solutions. It follows from the equations for the D_a that

$$\operatorname{Im} D_{a+b} = \operatorname{Im} D_a + \operatorname{Im} D_b.$$

Hence the map $a \mapsto \operatorname{Im} D_a$ is $\mathbb{Z}$-linear, and can be extended to an $\mathbb{R}$-linear form $r : \mathcal{Z} \to \mathbb{R}$. Each $\mathbb{R}$-linear map r can be written as the imaginary part of a $\mathbb{C}$-linear form $l : \mathcal{Z} \to \mathbb{C}$, namely of

$$l(z) := r(\mathrm{i}z) + \mathrm{i}r(z).$$

So, we have shown that there exists a $\mathbb{C}$-linear form $l : \mathcal{Z} \to \mathbb{C}$ such that

$$E_a := D_a - l(a)$$

is real. Hence it is sufficient to analyze the *real* solutions of the equation

$$E_{a+b} \equiv \frac{1}{2}A(a,b) + E_a + E_b \bmod 1.$$

We fix them by the idea of an A-character.

5.4 Definition. *An A-character on L is a map*

$$E : L \longrightarrow \mathbb{R}$$

with the property

$$E_{a+b} \equiv \frac{1}{2}A(a,b) + E_a + E_b \bmod 1.$$

It is not difficult to classify all A-characters.

Since we are interested in E_a only mod 1, we compose it with the natural projection $\mathbb{R} \to \mathbb{R}/\mathbb{Z}$:

$$F_a : L \to \mathbb{R}/\mathbb{Z}.$$

First case. $A = 0$, and hence

$$F_{a+b} = F_a + F_b.$$

The solutions of this equation can be obtained as follows. Take a lattice basis of L. The values of E can be described arbitrarily on this basis. Hence the group of all characters is isomorphic to

$$(\mathbb{R}/\mathbb{Z})^{2n} \qquad (\cong \text{ group of } A\text{-characters in the case } A = 0).$$

Second case. A is arbitrary. The difference of two A-characters is a 0-character, which we have described in the context of the first case. So, we can obtain all A-characters from a single A-character by adding the 0-characters. Hence we have to clarify whether an A-character exists at all.

5.5 Remark. *Let A be an alternating bilinear form whose values on $L \times L$ are integral. Then there exists an A-character.*

Proof. Obviously, every integral alternating matrix $A = -A^t$ can be written in the form

$$A = B - B^t,$$

with an integral matrix. For example, we have in the case $n = 2$

$$\begin{pmatrix} 0 & a \\ -a & 0 \end{pmatrix} = \begin{pmatrix} 0 & a \\ 0 & 0 \end{pmatrix} - \begin{pmatrix} 0 & 0 \\ a & 0 \end{pmatrix}.$$

In the language of bilinear forms, this means that we can write A in the form

$$A(z, w) = B(z, w) - B(w, z),$$

with an $\mathbb{R}$-bilinear form which is integral on $L \times L$.

Obviously,

$$E(a) := \frac{1}{2} B(a, a)$$

is an A-character.   □

5.6 Proposition. *Let L be a lattice in the finite-dimensional complex vector space $\mathcal{Z}$. Assume that there are given*
a) *a quasi-Hermitian form Q on $\mathcal{Z}$;*
b) *a $\mathbb{C}$-linear form l on $\mathcal{Z}$;*
c) *an A-character E on L.*
Then

$$H_a(z) := \frac{1}{2i} Q(z, a) + \frac{1}{4i} Q(a, a) + l(a) + E_a$$

is an automorphy summand. Every affine automorphy summand is of this form.

Supplement. *The triple $(Q, l, E \bmod 1)$ is uniquely determined by the automorphy summand.*

Proof. It is easy to verify that H_a actually is an automorphy summand. The previous considerations show that every affine automorphy summand is of this form.   □

The automorphy summands in Proposition 5.6 can be trivial:

1) Let l be a $\mathbb{C}$-linear form on $\mathcal{Z}$; then the automorphy summand

$$H_a(z) := l(a) \qquad (= l(a + z) - l(z))$$

is trivial.

2) Let S be a *symmetric* $\mathbb{C}$-bilinear form on $\mathcal{Z}$; then the automorphy summand

$$H_a(z) := \frac{1}{2i} S(z, a) + \frac{1}{4i} S(a, a)$$

$$= \frac{1}{4i} (S(z + a, z + a) - S(z, z))$$

is trivial. This gives us the following statement.

5.7 Remark. *The automorphy summand which is described by the triple (Q, l, E) is equivalent to the automorphy summand which is described by*

$$(H, 0, E), \quad H := Q - S.$$

These are the only equivalences which can occur.

This completes the description of automorphy summands.

We can formalize Remark 5.7. For this, we denote by $\mathrm{Pic}(X)$ the group of equivalence classes of automorphy summands. Because of the result of Exercise 4 in Sect. 4, this notation is justified. Because of Remark 5.7, this group is isomorphic to the group of pairs (H, E). The group law is

$$(H_1, E_1) + (H_2, E_2) = (H_1 + H_2, E_1 + E_2).$$

We denote the subgroup of all $(0, E)$ by $\mathrm{Pic}^0(X)$. As we have seen, this subgroup is isomorphic to $\mathbb{R}^{2n}/\mathbb{Z}^{2n}$. The factor group $\mathrm{Pic}(X)/\mathrm{Pic}^0(X)$ is called the Neron Severi group. This can be written in the form of an exact sequence

$$0 \longrightarrow \mathrm{Pic}^0(X) \longrightarrow \mathrm{Pic}(X) \longrightarrow \mathrm{NS}(X) \longrightarrow 0.$$

The Neron Severi group is isomorphic to the additive group of all Hermitian forms on $\mathcal{Z}$ that are integral on $L \times L$.

Exercises for Sect. VI.5

1. Give the details of the proof of Remark 5.1.

2. Determine the triple $[Q, l, E]$ for Jacobi's theta function $\vartheta(\tau, z)$ ([FB], Sect. V.6).

3. Let $f(z)$ be a theta function which is different from zero. Show that
$$\frac{f(z + a)f(z - a)}{f(z)^2}$$
is, for every a, an abelian function for L.

4. Show that every theta function without zeros is constant.

5. The Neron Severi group is isomorphic to $\mathbb{Z}^m$ for suitable m. In the case $n = 1$, it is isomorphic to $\mathbb{Z}$. The map $\mathrm{Pic}(X) \to \mathrm{NS}(X)$ then corresponds to the degree of divisors (from the point of view of Exercise 4 in Sect. 4).

6. Riemannian Forms

As described in Proposition 5.6, we fix a triple (Q, l, E) and the associated automorphy summand H_a. We shall investigate theta functions

$$f : \mathcal{Z} := \mathbb{C}^n \to \mathbb{C} \text{ analytic}, \qquad f(a + z) = e^{2\pi i H_a(z)} f(z).$$

The space of these theta functions will be denoted by

$$[Q, l, E] \qquad (\cong [H, 0, E]).$$

Although it is possible, it will not always be useful to bring (Q, l, E) into the special form $(H, 0, E)$.

6.1 Lemma. *If $[Q, l, E]$ contains a theta function which does not vanish identically, then the Hermitian form H is semipositive, i.e. $H \geq 0$.*

"Semipositive" means
$$H(z, z) \geq 0 \text{ for all } z \in \mathcal{Z}.$$

Since H is Hermitian, the numbers $H(z, z)$ are real.

Proof of Lemma 6.1. Let $f \in [H, 0, E]$. We assume that there exists $z_0 \in \mathcal{Z}_0$ with $H(z_0, z_0) < 0$ and then show that $f \equiv 0$. The key to the proof is the claim that the function

$$g(z) := |f(z)| e^{-(\pi/2) H(z,z)}$$

is periodic under L.

This follows immediately from the equation

$$|f(z + a)| = e^{-2\pi \operatorname{Im} H_a(z)} |f(z)|$$

and from

$$- \operatorname{Im} H_a(z) = \frac{1}{2} \operatorname{Re} H(z, a) + \frac{1}{4} H(a, a).$$

Since the function g is continuous and periodic, it attains a maximum. Hence there exists a constant M such that

$$|f(z)| \leq M e^{(\pi/2)H(z,z)}.$$

We choose an arbitrary but fixed $z \in \mathcal{Z}$. For variable $t \in \mathbb{C}$, we have

$$H(z + tz_0, z + tz_0) = |t|^2 H(z_0, z_0) + \alpha t + \bar{\alpha} \bar{t} + \beta.$$

This expression tends to $-\infty$ for $t \to \infty$. Hence the function

$$h(t) := f(z + tz_0)$$

is bounded on $\mathbb{C}$ and is then constant by Liouville's theorem. The constant has to be 0. Since this is true for all $z \in \mathcal{Z}$, we obtain $f \equiv 0$. $\qquad\square$

6.2 Definition. *A **Riemannian form** on the lattice $L \subset \mathcal{Z}$ is a Hermitian form H on $\mathcal{Z}$ with the following properties:*

a) H is semipositive.
b) The alternating part A of H is integral on $L \times L$.

*The Riemannian form H is called **nondegenerate** if H is positive definite, i.e.*

$$H(z, z) > 0 \text{ for } z \neq 0.$$

The simplest example of a Riemannian form H is the zero matrix. But this is without any interest for us, since we have the following result:

Every function f from $[H = 0, 0, E]$ is constant.

Proof. In the case $H = 0$, $S = 0$, and $l = 0$, we have the result that H_a is real. But then

$$|f(z + a)| = |f(z)|.$$

It follows that f is bounded and hence constant. $\qquad\square$

The theory which we have developed so far has the following important consequence:

If there exists a nonconstant abelian function for L, then there exists a Riemannian form $H \neq 0$ on L.

Later, we shall see that in the case $n > 1$ there are lattices which do not admit a nonzero Riemannian form. Every abelian function for such a lattice must be constant!

Degenerate Abelian Functions.

We will show now that, in the theory of abelian functions, we can restrict ourselves to nondegenerate Riemannian forms

$$H > 0 \qquad \text{(and not only } H \geq 0\text{)}.$$

Let $l : \mathcal{Z} \to \mathcal{Z}'$ be a surjective linear map of finite-dimensional complex vector spaces. If $U' \subset \mathcal{Z}'$ is an open subset and f' an analytic function on U', then $f := f' \circ l$ is an analytic function on the inverse image $U := l^{-1}(U')$. If f' is meromorphic on the whole of $\mathcal{Z}'$, then f is meromorphic on the whole of $\mathcal{Z}$. A meromorphic function f on $\mathcal{Z}$ comes from a meromorphic function f' if all elements of the kernel of l are periods:

$$f(z + a) = f(z) \text{ for all } a \in \text{Kernel } l.$$

Now let $L \subset \mathcal{Z}$ be a lattice. It may then happen that

$$L' := l(L)$$

is a lattice in $\mathcal{Z}'$. (But it also can happen that L' is not discrete. Consider, for example, the projection

$$\mathbb{C} \times \mathbb{C} \longrightarrow \mathbb{C}, \quad (z, w) \longmapsto z + w.$$

By means of $L_1 = \mathbb{Z} + i\mathbb{Z}$, we can construct $L = L_1 \times L_1$. The image in $\mathcal{Z}'$ is L_1, and hence a lattice. But if we take $L = L_1 \times \sqrt{2}L_1$, the image L' is not discrete.) So, we make the assumption now that L' is a lattice. If f' is an abelian function on $\mathcal{Z}'$ with respect to L', then f is an abelian function on $\mathcal{Z}$ with respect to L.

We denote the set of all abelian functions on $\mathcal{Z}$ with respect to L by $K(L)$. This is a field which contains the constant functions. The map $g \mapsto f = l \circ g$ induces (under the assumption of the discreteness of L') an injective field homomorphism

$$K(L') \longrightarrow K(L).$$

6.3 Proposition. *For each lattice $L \subset \mathcal{Z}$, there exists a surjective $\mathbb{C}$-linear map*

$$l : \mathcal{Z} \to \mathcal{Z}'$$

onto a complex vector space of possibly smaller dimension, such that the following properties are satisfied:

1) $L' := l(L)$ *is a lattice in $\mathcal{Z}'$.*
2) *The map $g \mapsto g \circ l$ defines an isomorphism*

$$K(L') \to K(L).$$

3) *There exists a nondegenerate Riemannian form on L'.*

Proof. The vector space $\mathcal{Z}'$ will be constructed as the factor space by the degeneration locus. We first have to define this locus.

6.4 Lemma.   *Let H be a semipositive Hermitian form on $\mathcal{Z}$. For a vector $z_0 \in \mathcal{Z}$, the following conditions are equivalent:*

1) $H(z_0, z_0) = 0$;
2) $H(z_0, z) = 0$ *for all* $z \in \mathcal{Z}$;
3) $A(z_0, z) = 0$ *for all* $z \in \mathcal{Z}$.

The set $\mathcal{Z}_0$ of all vectors $z_0 \in \mathcal{Z}$ with the properties 1)–3) is a $\mathbb{C}$-subvector space of $\mathcal{Z}$ (because of 2)). We call $\mathcal{Z}_0$ the *degeneration locus* of H.

Proof of Lemma 6.4.

1) $\Rightarrow$ 2). For arbitrary $t \in \mathbb{C}$, we have

$$0 \le H(z + tz_0, z + tz_0) = H(z, z) + 2\operatorname{Re}\left(tH(z_0, z)\right).$$

From this, we get $H(z_0, z) = 0$.

2) $\Rightarrow$ 3). This is trivial.

3) $\Rightarrow$ 1). We use the formula

$$H(z, w) = A(\mathrm{i}z, w) + \mathrm{i}A(z, w), \quad A(\mathrm{i}z, w) = A(\mathrm{i}w, z). \qquad \square$$

6.5 Lemma.   *Let $f \in [H, 0, E]$ be a theta function. The elements of the degeneration locus of H are periods, i.e. f comes from a meromorphic function on $\mathcal{Z}/\mathcal{Z}_0$.*

Proof. The argument is similar to that in the proof of Lemma 6.1. The inequality $|f(z)| \le Me^{(\pi/2)H(z,z)}$ shows that the function

$$\mathcal{Z}_0 \longrightarrow \mathbb{C}, \quad z \longmapsto f(z + a),$$

is bounded for each $a \in \mathcal{Z}$. Hence it is constant. $\qquad \square$

Before we continue with the proof of Proposition 6.3, we derive a criterion for the discreteness of an additive subgroup of $\mathbb{R}^n$.

6.6 Lemma.   *Let $L \subset \mathbb{R}^n$ be an additive subgroup with the following properties:*

1) *It is finitely generated.*
2) *It generates $\mathbb{R}^n$ as a vector space (over $\mathbb{R}$).*
3) *The $\mathbb{Q}$-vector space which is generated by L has dimension $\le n$.*

Then L is a lattice.

Proof. We make use of the structure theorem for finitely generated abelian groups, which states that every torsion-free finitely generated abelian group is isomorphic to $\mathbb{Z}^m$ (for a suitable m). Because of 1), there exist vectors $\omega_1, \ldots, \omega_m$ such that each element of L can be represented as a unique integral linear combination of these vectors. These vectors are linearly independent over $\mathbb{Q}$, since a linear relation with rational coefficients produces, after multiplication by a joint denominator, a relation with integral coefficients. From 3), we obtain $m \le n$. Because of 2), the vectors $\omega_1, \ldots, \omega_m$ generate the $\mathbb{R}$-vector space $\mathbb{R}^n$. Hence these vectors must be a basis (and $m = n$). $\qquad \square$

6.7 Lemma. *Let H be a Riemannian form with respect to the lattice $L \subset \mathcal{Z}$ and let $\mathcal{Z}_0$ be the degeneration locus of H. We have*

1) $L_0 := L \cap \mathcal{Z}_0$ *is a lattice in* $\mathcal{Z}_0$.

2) *If*

$$p : \mathcal{Z} \to \mathcal{Z}' := \mathcal{Z}/\mathcal{Z}_0$$

denotes the canonical projection, then $L' := p(L)$ is a lattice in $\mathcal{Z}'$.

Proof. 1) Since L_0 is discrete, it is enough to show that the real vector space $\mathcal{Z}_0$ is generated by L_0. Let $\omega_1, \ldots, \omega_{2n}$ be a lattice basis of L. A vector

$$\omega := \sum x_\nu \omega_\nu, \quad x_\nu \in \mathbb{R} \ (1 \le \nu \le 2n),$$

is contained in the kernel $\mathcal{Z}_0$ if the real vector $(x_1, \ldots, x_{2n})$ solves the linear equations

$$\sum a_{ik} x_k = 0 \quad \text{with} \quad a_{ik} = A(\omega_i, \omega_k).$$

The elements of L_0 can be obtained from the integral solutions of this system. We have to show that each real solution can be written as a real linear combination of integral solutions. Of course, it is sufficient to show that each real solution is a real linear combination of rational (rather than integral) solutions. But this is in fact the case, since the matrix (a_{ik}) of the system is rational (and even integral). We use the well-known fact from linear algebra that the dimension of the space of solutions is governed by the rank of the matrix. But the rank does not depend on the field in which the matrix is considered ($\mathbb{Q}$ or $\mathbb{R}$).

2) The first part shows that the dimension of the $\mathbb{Q}$-vector space which is generated by L' does not exceed the dimension of $\mathcal{Z}'$. Hence we can apply Lemma 6.6 to conclude that L' is a lattice. $\qquad\square$

After these preparations, the proof of Proposition 6.3 is easy. The sum of two Riemannian forms is itself a Riemannian form. Its degeneration locus is the intersection of the degeneration loci. Hence there exists a Riemannian form H with the smallest degeneration locus $\mathcal{Z}_0$. The Hermitian form H factorizes through a Hermitian form on $\mathcal{Z}/\mathcal{Z}_0$. Its degeneration locus must be zero. It should be clear that the three properties in Proposition 6.3 hold. $\qquad\square$

Proposition 6.3 can be formulated very conveniently in a geometric form. Consider the tori $X := \mathcal{Z}/L$ and $X' = \mathcal{Z}'/L'$. We have a surjective homomorphism $X \to X'$ and can consider X' as a *factor torus* of X. Proposition 6.3 says that for each complex torus X there exists a factor torus X' such that fields of abelian functions are "equal", and such that X' admits a nondegenerate Riemannian form.

In any case, it is sufficient for the theory of abelian functions to consider lattices which admit a nondegenerate Riemannian form.

Examples of Riemannian Forms

Let P be a complex $n \times 2n$ matrix. If the columns of P are linearly independent over $\mathbb{R}$, they span a lattice $L_P \subset \mathbb{C}^n$:

$$L_P := \sum_{\nu=1}^{2n} \mathbb{Z}p_\nu, \quad P := (p_1, \ldots, p_{2n}).$$

If H is a Hermitian $n \times n$-matrix, we can consider the Hermitian form on $\mathbb{C}^n$

$$H(z, w) := z^t H \bar{w}.$$

We want to determine P and H such that H is a nondegenerate Riemannian form.

6.8 Remark. *Let T be an integral $n \times n$ matrix whose determinant is different from zero, and let Z be a symmetric complex $n \times n$ matrix whose imaginary part is positive definite. The columns of the matrix $P = (T, Z)$ generate a lattice L in $\mathbb{C}^n$. The Hermitian (and also real) matrix $H = (\mathrm{Im}\, Z)^{-1}$ defines a nondegenerate Riemannian form on L.*

We first show the $\mathbb{R}$-independence of the columns. Since the columns of T are $\mathbb{R}$-independent, it suffices to show that the columns of $\mathrm{Im}\, Z$ are $\mathbb{R}$-independent. But we know that a positive definite matrix has a positive determinant. Next, we show that $\mathrm{Im}\, H$ is integral on L. This means that the matrix $\mathrm{Im}\, \bar{P}^t H P$ is integral. We have

$$\bar{P}^t H P = \begin{pmatrix} T^t (\mathrm{Im}\, Z)^{-1} T & T^t (\mathrm{Im}\, Z)^{-1} Z \\ \bar{Z}^t (\mathrm{Im}\, Z)^{-1} T & \bar{Z}^t (\mathrm{Im}\, Z)^{-1} Z \end{pmatrix}.$$

We take the imaginary part. It is easy to show that the matrices

$$T^t (\mathrm{Im}\, Z)^{-1} T \text{ and } \bar{Z}^t (\mathrm{Im}\, Z)^{-1} Z$$

are real. Therefore we have

$$\mathrm{Im}\, \bar{P}^t H P = \begin{pmatrix} 0 & T^t \\ -T & 0 \end{pmatrix}.$$

By assumption, this is an integral matrix. $\square$

In the next section, we shall show that we can obtain all nondegenerate Riemannian forms in this way.

Exercises for Sect. VI.6

1. Show that each finitely generated subspace of $\mathbb{Q}^n$ is discrete.

2. Construct a finitely generated subgroup of $\mathbb{R}$ which is not discrete.

3. Two lattices $L, L' \subset V$ are called *commensurable* if their intersection has a finite index in L and in L'. Show that two lattices are commensurable iff they generate the same $\mathbb{Q}$-vector space.

4. Show that if L, L' are commensurable lattices, then $L \cap L'$ and $L + L'$ are lattices.

7. Canonical Lattice Bases

By an *elementary matrix*, we understand a diagonal matrix of the form

$$T := \begin{pmatrix} t_1 & & 0 \\ & \ddots & \\ 0 & & t_n \end{pmatrix}, \quad t_\nu \in \mathbb{N}, \quad t_\nu | t_{\nu+1} \qquad (1 \le \nu < n).$$

The significance of elementary matrices arises from the elementary divisor theorem:

For every integral $p \times q$ matrix, there exist matrices $U \in \mathrm{GL}(p, \mathbb{Z})$, $V \in \mathrm{GL}(q, \mathbb{Z})$ such that

$$UAV = \begin{pmatrix} T & 0 \\ 0 & 0 \end{pmatrix},$$

with a uniquely determined elementary matrix T.

This result is more or less equivalent to the following theorem

7.1 Theorem (main theorem for abelian groups).
For any finitely generated abelian group L, there exist a unique integer $m \ge 0$ and a unique elementary matrix T, $t_1 > 1$, with the property

$$L \cong \mathbb{Z}^m \oplus \mathbb{Z}/t_1 \oplus \ldots \oplus \mathbb{Z}/t_n.$$

In this context, we need to consider the classification theorem for integral alternating bilinear forms, as below.

7.2 Proposition. *Let*

$$A : L \times L \longrightarrow \mathbb{Z}, \quad L \cong \mathbb{Z}^m,$$

be a nondegenerate alternating bilinear form over $\mathbb{Z}$, *i.e.*
a) $A(a + b, c) = A(a, c) + A(b, c)$;
b) $A(a, b) = -A(b, a)$;
c) $A(a, x) = 0$ *for all* $x \in L \Longrightarrow a = 0$.

Then $m = 2n$ *is even. There exists a* $\mathbb{Z}$-*basis* $\omega_1, \ldots, \omega_m$ *of* $\mathbb{Z}^m$ *with the property*

$$(A(\omega_i, \omega_j))_{1 \le i, j \le 2n} = \begin{pmatrix} 0 & T \\ -T & 0 \end{pmatrix}.$$

Here T *is a unique elementary matrix.*

Another formulation of this result is as follows.

For every integral alternating matrix A *whose determinant is different from zero, there exists a unimodular matrix* $U \in \mathrm{GL}(n, \mathbb{Z})$ *and a uniquely determined elementary matrix* T *with the property*

$$U^t A U = \begin{pmatrix} 0 & T \\ -T & 0 \end{pmatrix}.$$

Proof of Proposition 7.2. We argue by induction on m. The strategy is as follows. In the case $m > 1$, we construct a splitting of L of the form

$$L = \mathbb{Z}\omega_1 \oplus \mathbb{Z}\omega_2 \oplus L', \quad L' \cong \mathbb{Z}^{m-2},$$

and such that the following conditions apply:

a) $(A(\omega_i, \omega_j)) = \begin{pmatrix} 0 & t_1 \\ -t_1 & 0 \end{pmatrix}$, $\quad t_1 \ne 0$.
b) The restriction of A to L' is nondegenerate and we have $A(\omega_i, L') = 0$.
c) $t_1 | A(x, y)$ for all $x, y \in L'$.

(After this splitting, Proposition 7.2 is proved, since the induction hypothesis can be applied to L'.)

We choose ω_1 and ω_2 such that

$$t_1 := |A(\omega_1, \omega_2)|$$

is different from zero and minimal with this property. Then we set

$$L' := \{x \in L; \quad A(\omega_1, x) = A(\omega_2, x) = 0\}.$$

We have to show a)–c).

a) is trivial (since A is alternating, we have $A(\omega_i, \omega_i) = 0$).

b) The image of the homomorphism

$$L \longrightarrow \mathbb{Z}, \qquad x \longmapsto A(\omega_1, x),$$

is a subgroup of $\mathbb{Z}$. Every such subgroup is cyclic. Since t_1 has the minimal absolute value, the image consists of the integral multiples of t_1. Therefore, for arbitrary $x \in L$, the element

$$x' := x - \frac{A(\omega_2, x)}{t_1}\omega_1 - \frac{A(\omega_1, x)}{t_1}\omega_2$$

is contained in L'. We also know that the restriction of A to L' is nondegenerate.

c) Let x, y be arbitrary elements of L' and let $m \in \mathbb{Z}$. We have

$$A(m\omega_1 + x, \omega_2 + y) = mt_1 + A(x, y).$$

If $A(x, y)$ were not an integral multiple of t_1, we could choose m in such a way that

$$|mt_1 + A(x, y)| < |t_1|.$$

This contradicts the minimality of t_1.

The uniqueness of the elementary matrix T follows from the fact that

$$t_1, t_1, t_2, t_2, \ldots, t_n, t_n$$

are *the* elementary divisors of the matrix A. □

7.3 Corollary.   *The determinant of a nondegenerate integral alternating matrix is the square of a natural number.*

We call

$$t_1 \cdot \ldots \cdot t_n =_+ \sqrt{\det A}$$

the *Pfaffian* of the alternating form A.

Now we consider triples $(\mathcal{Z}, L, H)$, where L is a lattice in a finite-dimensional complex vector space $\mathcal{Z}$ and H is a nondegenerate Riemannian form. Two such triples are called isomorphic, $(\mathcal{Z}, L, H) \cong (\mathcal{Z}', L', H')$, if there exists an isomorphism

$$\sigma : \mathcal{Z} \to \mathcal{Z}'$$

with the property

$$L' = \sigma(L), \qquad H'(\sigma(z), \sigma(w)) = H(z, w).$$

We want to select a simple representative from each isomorphy class. First we describe the representants which will be used (compare with Remark 6.8). We start with

a) T, an elementary matrix,
b) Z, a symmetric matrix with positive imaginary part.

We know that the columns of (T, Z) are linearly independent over $\mathbb{R}$. Hence they generate a lattice $L = L(T, Z)$. On this lattice L, we have a Riemannian form

$$H = H(T, Z),$$

namely

$$H(z, w) := z^t (\operatorname{Im} Z)^{-1} \bar{w},$$

as we have seen in Sect. 4. For the sake of completeness, we mention that a symmetric real invertible matrix Y is positive iff this is true for Y^{-1}. This follows from the formula

$$g^t Y g = (Y g)^t Y^{-1} (Y g).$$

7.4 Proposition. *For each nondegenerate Riemannian form H on a lattice $L \subset \mathcal{Z}$, there exist*

a) *an elementary matrix T,*
b) *a symmetric complex matrix Z with positive definite imaginary part,*
such that

$$(\mathcal{Z}, L, H) \cong (\mathbb{C}^n, L(T, Z), H(T, Z)).$$

The elementary matrix T is uniquely determined.

Proof. We choose a lattice basis $\omega_1, \ldots, \omega_{2n}$ such that $A := \operatorname{Im} H$ is of the form

$$(A(\omega_i, \omega_j)) = \begin{pmatrix} 0 & T \\ -T & 0 \end{pmatrix},$$

with an elementary matrix T. The lattice basis contains a $\mathbb{C}$-basis of $\mathcal{Z}$. Actually, we claim the following:

The vectors $\omega_1, \ldots, \omega_n$ define a $\mathbb{C}$-basis of $\mathcal{Z}$.

Since n is the complex dimension of $\mathcal{Z}$, we have to show that

$$\mathcal{Z} = \sum \mathbb{C}\omega_\nu \quad \left(= \sum \mathbb{R}\omega_\nu + \mathrm{i} \sum \mathbb{R}\omega_\nu \right).$$

Since the vectors $\omega_1, \ldots, \omega_n$ are $\mathbb{R}$-linearly independent, it suffices to show that

$$\left(\sum_{\nu=1}^{n} \mathbb{R}\omega_\nu \right) \cap \mathrm{i} \left(\sum_{\nu=1}^{n} \mathbb{R}\omega_\nu \right) = 0.$$

If $z = iw$ is an element in the intersection, we have

$$H(z, z) = A(iz, z) = A(-w, z) = 0 \quad \text{(because } A(\omega_i, \omega_j) = 0\text{).}$$

The definiteness of H shows that $z = 0$.

We can also use the vectors

$$t_1^{-1}\omega_1, \ldots, t_n^{-1}\omega_n$$

as a $\mathbb{C}$-basis of $\mathcal{Z}$. Now we consider the isomorphism $\mathcal{Z} \xrightarrow{\sim} \mathbb{C}^n$ which transforms this basis to the standard basis. Then we can assume that

$$\mathcal{Z} = \mathbb{C}^n \quad \text{and} \quad (\omega_1, \ldots, \omega_n) = T = \begin{pmatrix} t_1 & & 0 \\ & \ddots & \\ 0 & & t_n \end{pmatrix}.$$

We collect the remaining lattice vectors together in an $n \times n$ matrix

$$Z := (\omega_{n+1}, \ldots, \omega_{2n}).$$

For the proof of Proposition 7.4, we have to show:

a) Z is symmetric and $\operatorname{Im} Z$ is positive definite.
b) $H(z, w) = z^t (\operatorname{Im} Z)^{-1} \bar{w}$.

By the definition of Z and by the choice of the basis of $\mathcal{Z}$, we have

$$\omega_{n+\mu} = \sum_{\nu=1}^{n} Z_{\nu\mu} t_{n\nu}^{-1} \omega_\nu \qquad (1 \le \mu \le n).$$

b) is equivalent to

$$\text{b')} \qquad \left(H(t_\mu^{-1}\omega_\mu, t_\nu^{-1}\omega_\nu) \right) = (\operatorname{Im} Z)^{-1}.$$

We have the following information about A:

$$A(\omega_\mu, \omega_\nu) = A(\omega_{n+\mu}, \omega_{n+\nu}) = 0, \quad A(\omega_{n+\nu}, \omega_\mu) = \delta_{\mu\nu} t_\nu \ (1 \le \mu, \nu \le n).$$

We recall the relation between A and H:

a) $A(z, w) = \operatorname{Im} H(z, w)$;
b) $A(iz, w) = A(iw, z)$;
c) $H(z, w) = A(iz, w) + iA(z, w)$ (both sides have the same imaginary part and are $\mathbb{C}$-linear in z; this is true for the right-hand side because of b)).

Proof of b').

It follows from the equation

$$\omega_{n+\mu} = \sum_{\nu} \text{Re}(Z_{\nu\mu}) t_\nu^{-1} \omega_\nu + i \sum_{\nu} \text{Im}(Z_{\nu\mu}) t_\nu^{-1} \omega_\nu$$

that

$$\delta_{\mu\chi} t_\chi = A(\omega_{n+\mu}, \omega_\chi) = \sum_{\nu} \text{Im}(Z_{\nu\mu} t_\nu^{-1} A(i\omega_\nu, \omega_\chi).$$

The matrix

$$\left(A(\text{i} t_\mu^{-1}\omega_\mu, t_\chi^{-1}\omega_\chi)\right)_{1\leq\mu,\chi\leq n}$$

is inverse to $(\text{Im } Z)^t$. Because $A(\omega_\mu, \omega_\nu) = 0$ $(1 \leq \mu, \nu \leq n)$, this matrix equals

$$\left(H(t_\mu^{-1}\omega_\mu, t_\nu^{-1}\omega_\nu)\right)_{1\leq\mu,\nu\leq n}.$$

But this is a real matrix. Every real Hermitian matrix is symmetric. As we know, $(\text{Im } Z)^{-1}$ is positive definite if and only if this is the case for $\text{Im } Z$.

It remains to show that $\text{Re } Z$ is symmetric too. This follows from

$$A(\omega_{n+\mu}, \omega_{n+\nu}) = 0,$$

since because $H(z, w) = z^t (\text{Im } Z)^{-1} \bar{w})$ we have

$$\text{Im} \left(Z^t (\text{Im } Z)^{-1} \bar{Z}\right) = 0.$$

But the left-hand side equals $\text{Re } Z - \text{Re } Z'$. □

The following considerations should make it plausible that the *manifold of lattices* which admit a nondegenerate Riemannian form is a "thin subset of the manifold of all lattices".

Let $L_A \subset \mathbb{C}^n$ be a lattice with lattice basis

$$A := (\omega_1, \ldots, \omega_{2n}).$$

A is not unique, but it can easily be shown that

$$L_A = L_B \iff B = GAH, \qquad G \in \text{GL}(n, \mathbb{C}), \qquad H \in \text{GL}(2n, \mathbb{Z}).$$

(The matrix G changes the coordinate system in $\mathbb{C}^n$; the matrix H changes the lattice basis.) The number of free parameters (classically they are called "moduli") is

$$\begin{array}{ccccc} n \cdot (2n) & - & n^2 & = & n^2. \\ \uparrow & & \uparrow & & \\ \text{parameters of } A & & \text{parameters of } G & & \end{array}$$

Since lattices with nondegenerate Riemannian forms up to the discrete elementary divisors are determined by a *symmetric* matrix Z, the number of moduli is

$$\frac{n(n+1)}{2} \qquad (< n^2 \text{ if } n > 1).$$

These inexact dimensional considerations can at least be used to construct examples of lattices which admit no Riemannian form different from zero, so all abelian functions for these lattices are constant! An example is given in Exercise 2.

Exercises for Sect. VI.7

1. Determine, for the matrix $M = \begin{pmatrix} 2 & 3 \\ 5 & 7 \end{pmatrix}$, unimodular matrices U, V such that UMV is an elementary matrix.
2. Let a, b, c, d be four real numbers which are algebraically independent over $\mathbb{Q}$. This means that there exists no nonzero polynomial P in four variables and with rational coefficients that has the property $P(a, b, c, d) = 0$.

   Show that the lattice which is defined by the matrix

   $$\begin{pmatrix} 1 & 0 & ia & ib \\ 0 & 1 & ic & id \end{pmatrix}$$

   admits no nonzero Riemannian form.

3. Determine, for the matrix $A = \begin{pmatrix} 4 & 6 \\ -6 & 10 \end{pmatrix}$, a unimodular matrix U such that the diagonal of $U^t A U$ is zero.

8. Theta Series (Construction of the Spaces $[Q,l,E]$)

In the following, $L \subset \mathbb{C}^n$ denotes a lattice and

$$H : \mathbb{C}^n \times \mathbb{C}^n \longrightarrow \mathbb{C}$$

denotes a nondegenerate Riemannian form. We want to determine the dimensions of the spaces

$$[Q, l, E] \qquad (\cong [H, 0, E]).$$

Because of Proposition 7.4, we can assume that the lattice L is of the form $L = L(T, Z)$. Here T is an elementary matrix and Z is a symmetric matrix with positive definite imaginary part.

1) So, the columns of T and Z give a $\mathbb{Z}$-basis of L.

2) $H(z, w) = z^t (\operatorname{Im} Z)^{-1} \overline{w}$.

The automorphy summands are equivalent to

$$(H + S, l, E).$$

The dimensions do not depend on the choice of S and l. To get formulae which are as simple as possible, we prefer to take

$$l = 0$$

and

$$S(z, w) = -z^t (\operatorname{Im} Z)^{-1} w.$$

The latter is a symmetric $\mathbb{C}$-bilinear form. We have

$$\boxed{Q(z, w) = -2i z^t (\operatorname{Im} Z)^{-1} (\operatorname{Im} w).}$$

We still have to describe the A-characters

$$E : L \longrightarrow \mathbb{R}.$$

An arbitrary element $\omega \in L$ can be written in the form

$$\omega = T\alpha + Z\beta, \quad \alpha, \beta \in \mathbb{Z}^n \qquad \text{(columns)}.$$

A simple computation gives

$$A(\omega, \tilde{\omega}) = \operatorname{Im}[\omega^t (\operatorname{Im} Z)^{-1} \tilde{\omega}] = \beta^t T\tilde{\alpha} - \tilde{\beta}^t T\alpha \qquad (\tilde{\omega} = T\tilde{\alpha} + Z\tilde{\beta}).$$

Now we see that

$$E(\omega) = \frac{1}{2} \alpha^t T\beta$$

is an A-character:

$$E(\omega + \tilde{\omega}) - E(\omega) - E(\tilde{\omega}) =$$
$$\frac{1}{2}\alpha^t T\tilde{\beta} + \frac{1}{2}\tilde{\alpha}^t T\beta = \frac{1}{2} A(\omega, \tilde{\omega}) + \tilde{\beta}^t T\alpha \equiv \frac{1}{2} A(\omega, \tilde{\omega}) \bmod 1.$$

Since two A-characters differ by a character of the usual kind, we obtain the most general A-character in the form

$$\boxed{E(\omega) = E^{a,b}(\omega) = \frac{1}{2}\alpha^t T\beta + a^t\alpha - b^t\beta,}$$

where a and b are two arbitrary real columns. Since we are interested in E only mod 1, we have to consider the columns also only mod 1:

$$\text{``}a, b \in (\mathbb{R}/\mathbb{Z})^n\text{''}.$$

Now the automorphy summand can be written in the form

$$\frac{1}{2i}Q(z, \omega) + \frac{1}{4i}Q(\omega, \omega) + E(\omega) = -z^t\beta - \frac{1}{2}\beta^t Z\beta + a^t\alpha - b^t\beta \quad (\omega = T\alpha + Z\beta).$$

Hence a theta function $\theta \in [Q, 0, E]$ has the following transformation properties:

1)	$\theta(z + T\alpha)$	$= e^{2\pi i a^t \alpha}\theta(z);$
2)	$\theta(z + Z\beta)$	$= e^{-2\pi i[z^t\beta + \frac{1}{2}\beta^t Z\beta + b^t\beta]}\theta(z).$

It is our task to solve this functional equation and to compute the dimension of $[Q, 0, E]$.

For the sake of simplicity, we first treat the (typical) case

$$n = 1 \text{ and } a = b = 0.$$

We write $Z = (\tau)$ and $T = (t)$. Both are 1×1 matrices. In this case the characteristic functional equation says

1) $$\theta(z + t) = \theta(z);$$
2) $$\theta(z + \tau) = e^{-\pi i(2z + \tau)}\theta(z), \quad \beta \in \mathbb{Z}$$

Here $\theta : \mathbb{C} \to \mathbb{C}$ is an analytic function, t a natural number, and τ a point in the upper half-plane. Because of 1), the function θ admits a Fourier expansion

$$\theta(z) = \sum_{m=-\infty}^{\infty} a_m e^{(2\pi i/t)mz}.$$

The functional equation 2) gives a condition for the Fourier coefficients a_g:

a) $$\theta(z + \tau) = \sum a_m e^{(2\pi i/t)m\tau} e^{(2\pi i/t)mz};$$
b) $$e^{-\pi i[2z + \tau]}\theta(z) = \sum a_m e^{-\pi i\tau} e^{(2\pi i/t)(m - t\tau)}$$
$$= \sum a_{m+t} e^{-\pi i\tau} e^{(2\pi i/t)mz}.$$

The uniqueness of the Fourier expansion shows that

$$a_{m+t} e^{-\pi i\tau} = a_m e^{(2\pi i/t)m\tau}.$$

The equation can easily be solved.

The coefficients $a_{m+t\beta}$, $\beta \in \mathbb{Z}$, are determined by a_m. So, we can prescribe $a_0, \ldots, a_{t-1}$ arbitrarily and then compute the other coefficients from them. Slightly more abstractly this means that the linear map

$$[Q, 0, E] \to \mathbb{C}^t, \quad \theta \mapsto (a_0, \ldots, a_{t-1}),$$

has kernel zero and hence is injective. In particular,

$$\dim[Q, 0, E] \le t.$$

We want to show that the dimension equals t. For this, we have to prove the convergence of the Fourier series for given $a_0, \ldots, a_{t-1}$ and the computed a_m.

It is sufficient, for

$$r \in \{0, \ldots, t-1\},$$

to treat the case

$$a_m = \begin{cases} 1 & \text{for } m = r, \\ 0 & \text{for } m \ne r, \ m \in \{0, \ldots, t-1\}. \end{cases}$$

But then

$$\theta(z) = \sum_{\beta=-\infty}^{\infty} a_{r+t\beta} e^{2\pi i (r/t+\beta)z} \text{ with } a_{r+e\beta} = e^{\pi i \beta^2 \tau} e^{(2\pi i/t) r \beta \tau},$$

and hence

$$\theta(z) = e^{-\pi i \tau (r/e)^2} \sum_{\beta=-\infty}^{\infty} e^{\pi i \left(((\beta+r)/t)^2 \tau + 2((\beta+r)/t)z \right)}.$$

This series is called a *theta series*. It is closely related to the Jacobi theta series, introduced in [FB], Sect. V.6, in connection with elliptic functions. The simple proof of convergence is the same as for the Jacobi theta series. Hence we obtain the surjectivity of the map

$$[Q, 0, E] \longrightarrow \mathbb{C}^t.$$

Actually, we have obtained an explicit basis of $[Q, 0, E]$. We can rewrite it as

$$\sum_{\beta=-\infty}^{\infty} e^{\pi i [\tau(\beta+r)^2 + 2(\beta+r)z]}, \quad r \in \left\{ \frac{0}{e}, \frac{1}{e}, \ldots, \frac{e-1}{e} \right\}.$$

In the same way, we can construct a basis of $[Q, 0, E]$ for arbitrary n. We shall anticipate the result and immediately define the theta series occurring. We shall use the notation $Z[h] := h^t Z h$.

8.1 Definition.   *Let Z be a symmetric $n \times n$ matrix with positive definite imaginary part and let a, b be two columns from $\mathbb{R}^n$. We define*

$$\vartheta \begin{bmatrix} a \\ b \end{bmatrix} (Z, z) := \sum_{g \in \mathbb{Z}^n} e^{\pi \mathrm{i} \{ Z[g+a] + 2(g+a)^t (z+b) \}}.$$

We can reduce these series to the special case $a = b = 0$. This the Riemann theta function.

Riemann theta function

$$\vartheta(Z, z) := \sum_{g \in \mathbb{Z}^n} e^{\pi \mathrm{i} \{ Z[g] + 2g^t z \}}.$$

It generalizes the Jacobi theta function. A fundamental (but simple) result states the following.

8.2 Proposition.   *The theta series defined in Definition 8.1 converges as a function of z (for fixed Z, a, b) absolutely and locally uniformly in $\mathbb{C}^n$ and defines an analytic function there.*

Proof. We can assume $b = 0$. We have

$$\left| e^{\pi \mathrm{i} \{ Z[g+a] + 2(g+a)^t z \}} \right| = e^{\pi \{ (\operatorname{Im} Z)[g+a] + 2(g+a)^t y \}}, \quad y = \operatorname{Im} z.$$

Now we use a simple lemma about positive definite quadratic forms:

For every positive definite real symmetric matrix Y, there exists a positive number δ with the property

$$Y[g] \geq \delta g^t g = \delta \sum g_j^2 \ \textit{for } g \in \mathbb{R}^n.$$

It is sufficient to prove this for $g^t g = 1$. But this a compact set, and the continuous function $g \mapsto Y[g]$ has a minimum there.

Now the general term of the theta series can be estimated by

$$e^{-\pi \{ \delta (g+a)^t (g+a) + 2(g+a)^t y \}} = \prod_{\nu=1}^{n} e^{-\pi \{ \delta (g_\nu + a_\nu)^2 + 2(g_\nu + a_\nu) y_\nu \}}.$$

By Cauchy's multiplication theorem, it is enough to show the convergence of

$$\sum_{g_\nu = -\infty}^{\infty} e^{-\pi \{ \delta (g_\nu + a_\nu)^2 + 2(g_\nu + a_\nu) y_\nu \}}.$$

In other words, we have reduced the claim to the case $n = 1$. We now omit the index ν. Obviously,

$$\delta(g + a)^2 + 2(g + a)y \geq \frac{1}{2}\delta g^2$$

for all $g \in \mathbb{Z}$ up to finitely many exceptions, if y varies in a compact set. It remains to show that the series

$$\sum_{g=-\infty}^{\infty} e^{-\delta g^2} = 1 + 2\sum_{g=1}^{\infty} e^{-\delta g^2}$$

converges. This is clearly the case, since we can estimate the series with the geometric series because $e^{-\delta} < 1$ ($\delta > 0$). Now Proposition 8.2 has been proved.

$\square$

8.3 Lemma. *Up to a constant factor, the theta series* $\vartheta \begin{bmatrix} a \\ b \end{bmatrix} (Z, z)$ *depends only on* a, b *modulo* $\mathbb{Z}^n$. *More precisely, we have*

$$\vartheta \begin{bmatrix} a \\ b \end{bmatrix} = e^{2\pi i a^t (\tilde{b} - b)} \vartheta \begin{bmatrix} \tilde{a} \\ \tilde{b} \end{bmatrix} \qquad \text{if } a - \tilde{a}, \ b - \tilde{b} \in \mathbb{Z}^n.$$

The pair (a, b) is called the *characteristic* of the theta series. The proofs of this and the following lemma are trivial.

8.4 Lemma. *Let* $\alpha, \beta \in \mathbb{Z}^n$. *Then the transformation formulae*

$$\vartheta \begin{bmatrix} a \\ b \end{bmatrix} (Z, z + \alpha) = e^{2\pi i a^t \alpha} \vartheta \begin{bmatrix} a \\ b \end{bmatrix} (Z, z),$$

$$\vartheta \begin{bmatrix} a \\ b \end{bmatrix} (Z, z + Z\beta) = e^{-2\pi i \{z^t \beta + (1/2) Z[\beta] + \beta^t b\}} \vartheta \begin{bmatrix} a \\ b \end{bmatrix} (Z, z)$$

hold.

These formulae tell us that the theta series are theta functions with respect to the lattice

$$L(Z, E), \qquad E \text{ unit matrix.}$$

But we are interested in the sublattice $L(Z, T)$. From Lemma 8.4, we immediately obtain the following result.

8.5 Lemma. *Let* r *be a column such that* Tr *is integral* ($r \in T^{-1}\mathbb{Z}^n$). *Then the function*

$$z \mapsto \vartheta \begin{bmatrix} r + T^{-1}a \\ b \end{bmatrix} (Z, z)$$

is contained in $[Q, 0, E]$.

We recall that the automorphy summand which belongs to $(Q, 0, E)$ is given by

$$H_\omega(z) = -z^t\beta - \frac{1}{2}\beta^t Z\beta + a^t\alpha - b^t\beta \qquad (\omega = T\alpha + Z\beta).$$

If we change the vector r in Lemma 8.5 mod $\mathbb{Z}^n$, then the theta series changes only by a constant factor. Hence r should run through a system of representatives of $(T^{-1}\mathbb{Z}^n)/\mathbb{Z}^n$, for example

$$r_\nu \in \left\{ \frac{0}{t_\nu}, \frac{1}{t_\nu}, \ldots, \frac{t_\nu - 1}{t_\nu} \right\}.$$

This system consists of $t_1 \cdot \ldots \cdot t_n$ elements.

8.6 Theorem. *If r runs through a system of representatives $(T^{-1}\mathbb{Z}^n)/\mathbb{Z}^n$, the functions*

$$z \mapsto \vartheta \begin{bmatrix} r + T^{-1}a \\ b \end{bmatrix} (Z, z)$$

give a basis of $[Q, 0, E]$.

Corollary. *The space $[Q, l, E]$ has finite dimension and the dimension equals the Pfaffian of A.*

We shall see now that the *proof* of Theorem 8.6 is analogous to the case $n = 1$.

Let

$$\vartheta \in [Q, 0, E];$$

in particular,

$$\vartheta(z + T\alpha) = e^{2\pi i\alpha^t a}\vartheta(z).$$

It is useful to replace $\vartheta(z)$ by the function

$$\vartheta_0(z) := \vartheta(z) \cdot e^{-2\pi i a^t T^{-1} z},$$

since this function is periodic under $\mathbb{Z}^n$:

$$\vartheta_0(z + T\alpha) = \vartheta_0(z), \quad \alpha \in \mathbb{Z}^n.$$

As in the case $n = 1$ $(a = 0)$, we can expand such a function into a complex Fourier series:

$$\vartheta_0(z) = \sum_{g \in \mathbb{Z}^n} a_g e^{2\pi i g^t z}.$$

We shall prove this in the appendix to this section. Now we make use of the transformation formula under $z \mapsto z + Z\beta$. As in the case $n = 1$, the formula

$$\vartheta(z + Z\beta) = e^{-2\pi i[z^t\beta + \frac{1}{2}\beta^t Z\beta + b^t\beta]}\vartheta(z)$$

gives a recursion for the coefficients a_g. This recursion allows us to compute $a_{g+T\beta}$ from a_g. Hence ϑ is determined by finitely many a_g. The index g has to run through a system of representatives of $\mathbb{Z}^n/(T\mathbb{Z}^n)$. This gives us the estimate

$$\dim[Q, 0, E] \leq t_1 \cdot \ldots \cdot t_n.$$

For the proof of Theorem 8.6, it remains to show that the theta series in Lemma 8.5 are linearly independent or, equivalently, that the Fourier series

$$e^{-2\pi i a^t T^{-1} z} \vartheta \begin{bmatrix} r + T^{-1}a \\ b \end{bmatrix} (Z, z)$$

are linearly independent if r runs through a system of representatives of $(T^{-1}\mathbb{Z}^n) \bmod \mathbb{Z}^n$. This follows from the following simple criterion.

8.7 Remark. *Let*

$$f^{(\nu)}(z) = \sum_{g \in \mathbb{Z}^n} a_g^{(\nu)} e^{2\pi i g^t z}, \quad \nu = 1, \ldots, N,$$

be a Fourier series which converges in $\mathbb{C}^n$, and such the following conditions are satisfied:

1) *$f^{(\nu)} \neq 0$ for $\nu = 1, \ldots, N$.*
2) *If $g \in \mathbb{Z}^n$, then $a_g^{(\nu)}$ can be different from zero only for one ν.*

Then the functions $f^{(1)}, \ldots, f^{(N)}$ are linearly independent.

Proof. Let

$$\sum_{\nu=1}^{N} C_\nu f^{(\nu)}(g) = 0, \quad \text{and hence} \quad \sum_{\nu=1}^{N} C_\nu a_g^{(\nu)} = 0.$$

For a given ν_0, we choose an index

$$g \in \mathbb{Z}^n, \quad a_g^{(\nu_0)} \neq 0.$$

Then $a_g^{(\nu)} \neq 0$ for $\nu \neq \nu_0$, and hence $C_{\nu_0} = 0$. $\qquad\square$

This completes the proof of the fundamental *existence and finiteness theorem* stated in Theorem 8.6. $\qquad\square$

Exercises for Sect. VI.8

1. Assume that for $i = 1, 2$ there are given lattices L_i and two associated triples Q_i, l_i, E_i with a nondegenerate Hermitian form. Define a "Cartesian product" Q, l, E on $\mathbb{C}^n$ with $n = n_1 + n_2$, and show also that $L = L_1 \times L_2$ admits a nondegenerate Hermitian form. Show that each theta function $\theta \in [Q, l, E]$ can be written as a finite sum of functions of the form $\theta_1 \theta_2$, with $\theta_i \in [Q_i, l_i, E_i]$.*)

2. Construct, in the case $n = 1$, a triple Q, l, E such that the space of solutions $[Q, l, E]$ is spanned by Jacobi's theta function.

3. The theta function $\vartheta \begin{bmatrix} a \\ b \end{bmatrix} (Z, a, b)$ can be considered as a function in all of the variables Z, a, b, z. Show that the theta series converges normally on $A = \mathbb{H}_n \times \mathbb{C}^n \times \mathbb{C}^n \times \mathbb{C}^n$, where $\mathbb{H}_n$ denotes the space of all symmetric matrices with positive definite imaginary part.

Appendix to Sect. 8. Complex Fourier Series

Let the following be given:

1) *a domain $V \subset \mathbb{R}^n$;*
2) *a lattice $L \subset \mathbb{R}^n$ (not in $\mathbb{C}^n$);*
3) *a periodic analytic function*

$$ f : D \to \mathbb{C}, \quad f(z + a) = f(z) \text{ for all } a \in L, $$

where D denotes the domain

$$ D = \{ z \in \mathbb{C}^n; \quad z = x + iy, \ y \in V \}. $$

Claim. *The function f admits an expansion into an absolutely and locally uniformly convergent Fourier series of the kind*

$$ f(z) = \sum_{g \in L^\circ} a_g e^{2\pi i g^t z}. $$

The coefficients a_g are uniquely determined. For arbitrary $y \in V$, we have

$$ a_g = \int_P f(x + iy) e^{-2\pi i g^t (x + iy)} \, dx, $$

where P is a fundamental parallelogram of the lattice L.

---

*) The "correct" formula is $[Q, l, E] \cong [Q_1, l_1, E_1] \otimes [Q_2, l_2, E_2]$.

In the case $n = 1$, this is an easy consequence of the Laurent expansion. Since we have not developed Laurent expansions in several variables, however, we use a different approach. We reduce the complex case to the real case (which in fact is more fundamental).

From Lemma 1.4, we obtain for fixed y an expansion

$$f(x + iy) = \sum_{g \in L^\circ} b_g(y) e^{2\pi i g^t x}.$$

This can be written in the form

$$f(x + iy) = \sum_{g \in L^\circ} a_g(y) e^{2\pi i g^t z}$$

$(a_g(y) = e^{2\pi g^t y} b_g(y))$. We have to show that the coefficient

$$a_g(y) = \int_P f(x + iy) e^{-2\pi i g^t (x+iy)}\, dx$$

is independent of y. This can easily be reduced to the case $n = 1$, where the complex Fourier expansion is known. Another proof uses the Cauchy–Riemann differential equations: We have

$$\bar\partial f = 0.$$

It is easy to show that the operator $\partial/\partial\bar z$ can be applied termwise to the Fourier series. We obtain

$$\bar\partial \left(a_g(y) e^{2\pi i g^t z} \right) = 0$$

and then

$$\frac{\partial}{\partial y_\nu} a_g(y) = 0. \qquad \qquad \square$$

9. Graded Rings of Theta Series

As usual, $L \subset \mathbb{C}^n$ denotes a lattice here. Let H_1, H_2 be two Riemannian forms. Then $H_1 + H_2$ is also a Riemannian form. It is nondegenerate of one of the two is so.

9.1 Lemma. *Let $\tilde{H}, H$ be two Riemannian forms, H nondegenerate. Then there exists a natural number r such that*

$$rH = \tilde{H} + H_0,$$

with a nondegenerate Riemannian form H_0.

Proof. The imaginary part of H_0 is integral on $L \times L$ for any choice of r. Hence we have to show only that $rH - \tilde{H}$ is positive definite for sufficiently large r:

$$rH(z, z) - \tilde{H}(z, z) > 0 \text{ for } z \in \mathbb{C}^n - \{0\}.$$

It is sufficient to restrict ourselves to the compact set defined by $\|z\| = 1$. Now a simple compactness argument gives the proof.   □

The sum of two automorphy summands is again an automorphy summand. The product of two corresponding theta functions is a theta function with respect to this sum. So we have the following.

Assume that an automorphy summand is given by a triple (Q, l, E). Then the triple (rQ, rl, rE) $(r \in \mathbb{Z})$ is also an automorphy summand. We have

$$f \in [rQ, rl, rE], \; g \in [sQ, sl, sE] \Longrightarrow f \cdot g \in [(r+s)Q, (r+s)l, (r+s)E].$$

So, we are led to consider the set of all finite sums

$$\sum_{r \in \mathbb{Z}} f_r, \quad f_r \in [rQ, rl, rE], \quad f_r = 0 \text{ for almost all } r.$$

We denote this set by

$$A(Q, l, E) = \sum_{r \in \mathbb{Z}} A_r(Q, l, E), \text{ where } A_r(Q, l, E) := [rQ, rl, rE].$$

We have seen that this is a ring. We have

$$A_r(Q, l, E) = 0 \text{ for } r < 0 \text{ and } A_0(Q, l, E) = \mathbb{C}.$$

9.2 Lemma. *We have*

$$\sum_{r=0}^{\infty} \vartheta_r = 0 \quad (\vartheta_r \in [rQ, rl, rE], \text{ almost all } = 0)$$

if and only if

$$\vartheta_r = 0 \text{ for all } r.$$

Proof. Let $z \in \mathbb{C}^n$. We have

$$0 = \sum_{r=0} \vartheta_r(z+a) = \sum (e^{2\pi i H_a(z)})^r \vartheta_r(z).$$

Hence the polynomial

$$x \longmapsto \sum_{r=0}^{\infty} \vartheta_r(z) x^r$$

has infinitely many roots. Hence its coefficients are zero. $\qquad\square$

By Lemma 9.2, the sum decomposition of $A(Q, l, E)$ is direct. So, we can write

$$A(Q, l, E) = \bigoplus_{r=0}^{\infty} A_r(Q, l, E).$$

Instead of proving Lemma 9.2, one could take this direct sum for the definition of $A(Q, l, E)$. It should be clear how the ring structure and such an abstract direct sum have to be defined. Compare this with the final remarks in [FB], Sect. VI.3.

9.3 Lemma. *Let P be the Pfaffian of the alternating form which belongs to Q. Then*

$$\dim A_r(Q, l, E) = P \cdot r^n \text{ for } r \geq 0.$$

Proof. Let $e_1, \ldots, e_n$ be the elementary divisors of A. Then $rt_1, \ldots, rt_n$ are the elementary divisors of rA. The rest follows from Theorem 8.6. $\qquad\square$

We associate with the graded ring $A = A(Q, l, E)$ a subfield of the field of abelian functions, namely

$$K(A) := \left\{ \frac{f}{g}; \quad f, g \in A_r, \ r \in \mathbb{Z}, \ g \neq 0 \right\}.$$

Clearly, $K(A)$ is a field; for example,

$$\frac{f}{g} + \frac{\tilde{f}}{\tilde{g}} = \frac{f\tilde{g} + \tilde{f}g}{g\tilde{g}}$$

and

$$f\tilde{g} + \tilde{f}g, \ g\tilde{g} \in A_{r+\tilde{r}} \text{ for } f, g \in A_r; \ \tilde{f}, \tilde{g} \in A_{\tilde{r}}.$$

9.4 Proposition. *Let (Q, l, E) with a nondegenerate Riemannian form H be given. Then $K(A(Q, l, E))$ is the field of **all** abelian functions with respect to L.*

Proof. As we know, every abelian function can be written in the form

$$\frac{f}{g}, \quad f, g \in [\tilde{Q}, \tilde{l}, \tilde{E}],$$

for a suitable triple $(\tilde{Q}, \tilde{l}, \tilde{E})$.

We choose the natural number r as in Lemma 9.1. Then we choose an arbitrary theta function

$$h \in [rQ - \tilde{Q}, rl - \tilde{l}, rE - \tilde{E}], \quad h \neq 0.$$

The existence of h is ensured, since $rQ - \tilde{Q}$ leads to a nondegenerate Riemannian form. We have

$$\frac{f}{g} = \frac{fh}{gh} \text{ and } fh, gh \in [rQ, rl, rE]. \qquad \square$$

Exercises for Sect. VI.9

1. The polynomial ring $A = \mathbb{C}[X_1, \ldots, X_m]$ admits the grading

    $$A_r := \{P \in A; \quad P \text{ homogeneous of degree } r\}.$$

    Can there be an isomorphism from $A(Q, l, A)$ onto A (for suitable m) which is compatible with this grading?

2. Can there be an isomorphism from $A(Q, l, A)$ onto the graded ring of elliptic modular forms ([FB], Sect. V.3) which respects the gradings?

10. A Nondegenerateness Theorem

In principle, it could be possible that every abelian function for a lattice L is periodic with respect to a bigger lattice $\tilde{L}$, even if L admits a nondegenerate Riemannian form. Our next goal is to prove that such a pathological behavior is not possible at least for theta functions. We start with some notation.

Let A be an alternating nondegenerate bilinear form on $\mathbb{C}^n \times \mathbb{C}^n$ which takes only integral values on $L \times L$. We can define the dual lattice with respect to A (compare Remark 1.3):

$$L_* := \{z \in \mathbb{C}^n; \quad A(z, a) \in \mathbb{Z} \text{ for all } a \in L\}.$$

It is easy to show that L_* is a lattice and that $L \subset L_*$.

10.1 Lemma. *Let*

$$\theta \in [Q, l, E], \quad \theta \neq 0,$$

be a theta function with a nondegenerate Riemannian form. Let $a \in \mathbb{C}^n$ *be a vector such that*

$$\frac{\theta(z + a)}{\theta(z)}$$

is analytic on the whole of $\mathbb{C}^n$. *Then* a *is contained in the lattice* L_*.

Proof. A simple calculation shows that the function

$$\theta_0(z) := e^{-\pi H(z,a)} \frac{\theta(z + a)}{\theta(z)}$$

satisfies the transformation formula

$$\theta_0(z + b) = e^{2\pi i A(a,b)} \theta_0(z) \text{ for } b \in L.$$

It follows that $|\theta_0(z)|$ attains its maximum in $\mathbb{C}^n$. Hence it is constant. This implies that

$$A(a, b) \in \mathbb{Z}. \qquad \square$$

This proof shows more, as follows.

Let $\tilde{L}$ be the set of all vectors a which occur in Lemma 10.1. We have

$$L \subset \tilde{L} \subset L_*.$$

Obviously, $\tilde{L}$ is a lattice. The function θ which occurs in Lemma 10.1 is a theta function for $\tilde{L}$; more precisely,

$$\theta \in [Q, l, \tilde{E}],$$

where $\tilde{E}$ is some extension of the A-character E to $\tilde{L}$.

There are only finitely many extensions of an A-character on L to an A-character on $\tilde{L}$.

This is clear, since the extensions are determined by their values on a system of representatives of $\tilde{L}/L$ and this group is finite.

In the next step, we show that $[\tilde{Q}, \tilde{l}, \tilde{E}]$ are *proper subspaces* of $[Q, l, E]$ if L is a proper sublattice of $\tilde{L}$. Because of the dimension formulae (Theorem 8.6), this means that:

The Pfaffian $\tilde{P}$ of A with respect to $\tilde{L}$ is smaller than the Pfaffian P of A with respect to L.

For the proof, we consider normal bases

$$\omega_\nu \text{ and } \tilde{\omega}_\nu \qquad (1 \leq \nu \leq 2n)$$

of A with respect to L and $\tilde{L}$ respectively. Since L is contained in $\tilde{L}$, we have

$$\omega_\mu = \sum_\nu u_{\mu\nu}\tilde{\omega}_\nu,$$

with a suitable matrix

$$U = (u_{\mu\nu})_{1\leq\mu,\nu\leq2n}.$$

A simple computation shows that

$$\begin{pmatrix} 0 & T \\ -T & 0 \end{pmatrix} = U \begin{pmatrix} 0 & \tilde{T} \\ -\tilde{T} & 0 \end{pmatrix} U',$$

and hence

$$P = |\det U| \cdot \tilde{P}.$$

The determinant of U is integral. It cannot be ± 1, since otherwise U^{-1} would be integral as well and we would have $L = \tilde{L}$. But we have excluded this possibility.   □

Now we obtain the following result.

10.2 Lemma.   *We consider the automorphy summand defined by a triple (Q, l, E) with a nondegenerate Riemannian form H. If we take $\theta \in [Q, l, E]$ outside of the union of certain finitely many subspaces of smaller dimension, then the function $\theta(z + a)/\theta(z)$ is not analytic for any $a \in \mathbb{C}^n$, $a \notin L$.*

Note. Since a (real or complex) vector space can never be the union of finitely many subvector spaces of smaller dimension, there exists a θ with the properties formulated in Lemma 10.2.

10.3 Proposition (point separation theorem).   *Let $L \subset \mathbb{C}^n$ be a lattice on which an automorphy summand is given through a triple (Q, l, E) with a nondegenerate Riemannian form H.*
1) *Assume $m \geq 2$. Then, for each point $a \in \mathbb{C}^n$, there exists*

$$\theta \in [mQ, ml, mE] \text{ with } \theta(a) \neq 0.$$

2) *Assume $m \geq 3$. Then, for each pair $a, b \in \mathbb{C}^n$ of points which are not equivalent mod L, there exists*

$$\theta \in [mQ, ml, mE] \text{ with } \theta(a) = 0, \ \theta(b) \neq 0 \qquad \textbf{(point separation)}.$$

The proof of this fundamental theorem rests on a very simple but fundamental observation:

Let θ be an element of $[Q, l, E]$ and let

$$a_1, \ldots, a_m \in \mathbb{C}^n \text{ with } a_1 + \cdots + a_m = 0.$$

Then

$$\prod_{i=1}^{m} \theta(z + a_i) \in [mQ, ml, mE].$$

For the proof, we only have to observe that for "$a_1 + \cdots + a_m = 0$" and any automorphy summand

$$(a, z) \longmapsto Q_a(z) + C_a,$$

it follows that

$$m[Q_a(z) + C_a] = \sum_{i=1}^{m} [Q_a(z + a_i) + C_a].$$

Proof of Theorem 10.3.

1) We choose a theta function

$$\theta_0 \in [Q, l, E], \quad \theta_0 \neq 0,$$

and consider

$$\theta(z) := \theta_0(z + a)\theta_0 \left(z - \frac{a}{m-1} \right)^{m-1}.$$

We know that this function is contained in $[mQ, ml, mE]$. Now we show that for a given z, there exists an a such that $\theta(z) \neq 0$. For this, we consider θ for fixed z as an abelian function in a. In this case, this function is zero, and we have

$$\theta_0(z + a) \equiv 0 \text{ or } \theta_0 \left(z - \frac{a}{m-1} \right)^{m-1} \equiv 0 \qquad \text{(for all } a\text{).}$$

In both cases we would have $\theta_0 = 0$, which contradicts our assumption.

2) Using Lemma 10.2, we choose a theta function

$$\theta_0 \in [Q, l, E], \quad \frac{\theta_0(z + b - a)}{\theta_0(z)} \quad \text{is not analytic in } \mathbb{C}^n.$$

In our first approach, we make a restrictive assumption. Later we will see how to get rid of it.

Assumption. *The function θ_0 is reduced. This means that the power series expansion of θ_0 at an arbitrary point of $\mathbb{C}^n$ is square-free in the ring of power series.*

Claim. Under this assumption, there exists a c with

$$\theta_0(c) = 0, \quad \theta_0(c + b - a) \neq 0.$$

Proof. We choose c in such a way that $\theta_0(z + b - a)/\theta_0(z)$ is not analytic in any open neighborhood of c. The claim then follows from the theorem of the unique prime factor decomposition in $\mathbb{C}\{z_1 - c_1, \ldots, z_n - c_n\}$ and from the following remark. □

Remark. *Let P, Q be two prime elements in $\mathbb{C}\{z_1, \ldots, z_n\}$. Assume that*

$$P(z) = 0 \quad \Longrightarrow \quad Q(z) = 0$$

in a full neighborhood of $z = 0$. Then

$$P = UQ,$$

with a unit U $(U(0) \neq 0)$.

This remark follows from Proposition V.4.10.

Now we choose a further vector $c_1 \in \mathbb{C}^n$ and define c_2 by the equation

$$c - a + c_1 + (m - 2)c_2 = 0$$

(observe that $m > 2$). The function

$$\theta(z) := \theta_0(z + c - a)\theta_0(z + c_1)\theta_0(z + c_2)^{m-2}$$

is then contained in $[mQ, ml, mE]$. We have

$$\theta(a) = 0 \qquad (\text{because } \theta_0(c) = 0).$$

We claim that with a suitable choice of c_1,

$$\theta(b) \neq 0.$$

If this were not the case, we would have

$$\theta_0(b + c_1)\theta_0(b + c_2)^{m-2} = 0$$

(as a function of c_1) and this would imply $\theta_0 = 0$.

Finally, we want to get rid of the assumption of reducedness.

10.4 Lemma. *Each theta function*

$$\theta_0 \in [Q, l, E], \quad \theta_0 \neq 0 \qquad \text{(H nondegenerate)}$$

admits a decomposition

$$\theta_0 = \theta_0^{\mathrm{red}} \cdot \tilde{\theta}_0$$

as a product of two theta functions

$$\theta_0^{\mathrm{red}} \in [Q^{\mathrm{red}}, l^{\mathrm{red}}, E^{\mathrm{red}}], \quad \tilde{\theta}_0 \in [\tilde{Q}, \tilde{l}, \tilde{E}]$$

with the properties
1) θ_0^{red} *is reduced;*
2) H^{red} *is nondegenerate;*
3) $\tilde{\theta}_0(z) = 0 \implies \theta_0^{\mathrm{red}}(z) = 0.$

We assume that such a decomposition has been proved. We then find vectors $a_1, \ldots, a_m$, $a_1 + \cdots + a_m = 0$ such that

$$\theta^{\mathrm{red}}(z) := \prod_{\nu=1}^{m} \theta_0^{\mathrm{red}}(z + a_\nu)$$

has the desired properties

$$\theta^{\mathrm{red}}(a) = 0, \quad \theta^{\mathrm{red}}(b) \neq 0.$$

Then

$$\theta(z) := \prod_{\nu=1}^{m} \theta_0(z + a_\nu)$$

has the same properties, since because of 3) we have

$$\theta^{\mathrm{red}}(z) = 0 \iff \theta(z) = 0.$$

Proof of the existence of the decomposition $\theta_0 = \theta_0^{\mathrm{red}} \cdot \tilde{\theta}_0$ (Lemma 10.4).
As we know from Proposition V.5.2, the set of points at which the power series expansion of an analytic function is square-free is open. From this and the fact that for each Cousin distribution on $\mathbb{C}^n$ there exists an analytic function which fits it, we obtain the following result (compare Exercise 4 in Sect.V.5):
Every analytic function $f : \mathbb{C}^n \to \mathbb{C}$ ($f \neq 0$) admits a decomposition

$$f = f^{\mathrm{red}} \cdot \tilde{f},$$

where f^{red} and $\tilde{f}$ are both analytic functions in $\mathbb{C}^n$, such that the power series expansion of f^{red} at any point a is the square-free part of the power series expansion of f. So, we have

$$\tilde{f}(z) = 0 \implies f^{\mathrm{red}}(z) = 0.$$

Using the prime decomposition and a compactness argument, we obtain the result that *if $U \subset \mathbb{C}^n$ is a bounded open set, then there exists a natural number N such that*

$$\frac{(f^{\mathrm{red}})^N}{\tilde{f}}$$

is analytic in U.

We apply this to the theta function θ_0:

$$\theta_0 = \theta_0^{\mathrm{red}} \cdot \tilde{\theta}_0.$$

Since we can choose U in such a way that it contains a fundamental parallelogram U, we obtain

$$\frac{(\theta_0^{\mathrm{red}})^N}{\tilde{\theta}_0} \text{ is analytic in the whole of } \mathbb{C}^n.$$

We can multiply θ_0^{red} by a suitable function from $\mathcal{O}(\mathbb{C}^n)^*$. Hence we can assume that θ_0^{red} is a theta function. Then $\tilde{\theta}_0$ is also a theta function:

$$\theta_0^{\mathrm{red}} \in [Q^{\mathrm{red}}, l^{\mathrm{red}}, E^{\mathrm{red}}], \quad \tilde{\theta}_0 \in [\tilde{Q}, \tilde{l}, \tilde{E}].$$

The corresponding Riemannian forms are

$$H = H^{\mathrm{red}} + \tilde{H}.$$

We also know that

$$N H^{\mathrm{red}} - \tilde{H}$$

is a Riemannian form. From this we can deduce that H^{red} is nondegenerate. Assume that $H^{\mathrm{red}}(z_0, z_0) = 0$. Since $N H^{\mathrm{red}} - \tilde{H}$ is semipositive, we obtain $\tilde{H}(z_0, z_0) = 0$ and hence $H(z_0, z_0) = 0$. But, by assumption, H is nondegenerate. This completes the proof of Proposition 10.3.   □

Exercises for Sect. VI.10

1. Show that the index of L in L_* equals the Pfaffian.

2. Show that the Weierstrass σ-function and the Jacobi theta function $\vartheta(\tau, z)$ ([FB], Sect. V.6) are reduced.

11. The Field of Abelian Functions

We now have the tools to prove that the field of abelian functions with respect to a lattice $L \subset \mathbb{C}^n$ with a non-degenerated Riemannian form is an *algebraic function field of transcendental degree n*. The basic facts about algebraic function fields have been collected together in the algebraic appendix (Chap. VIII) to this volume. Once more, we arrange what is needed from the theory so far.

Let H be a nondegenerate Riemannian form and let (Q, l, E) be an associated triple. We consider

$$A_r(Q, l, E) = [rQ, rl, rE]$$

and

$$A(Q, l, E) = \bigoplus_{r=0}^{\infty} A_r(Q, l, E).$$

We then have:

1) $\dim A_r(Q, l, E) = Pr^n \quad (P > 0)$.
2) Each abelian function is the quotient of two elements from $A_r(Q, l, E)$ for a suitable r.
3) Let $a, b \in \mathbb{C}^n$ be two points which are inequivalent with respect to L, and assume $r \geq 3$. There then exists

$$\theta \in A_r(Q, l, E) \text{ with } \theta(a) = 0, \quad \theta(b) \neq 0.$$

11.1 Definition. *The analytic functions*

$$f_1, \ldots, f_m : U \to \mathbb{C}, \quad U \subset \mathbb{C}^n \text{ open, nonempty,}$$

are called analytically independent if there exists a point $a \in U$ at which the Jacobian

$$J(f, a) = \left(\frac{\partial f_i}{\partial z_k}(a) \right)_{\substack{i=1,\ldots,m \\ k=1,\ldots,n}}$$

has rank m.

In particular, $m \leq n$. It is well known that any $m \times n$ matrix of rank m admits an extension to an $n \times n$ matrix whose determinant is not zero. This shows the following.

11.2 Remark.   *Assume that the Jacobi matrix $J(f, a)$ of the analytic functions $f_1, \ldots, f_m$ has rank m. Then they can be extended to an n-tuple $f := (f_1, \ldots, f_n)$ of analytic functions such that the complex functional determinant of f is different from zero.*

(One can take linear functions for the $f_{m+1}, \ldots, f_n$.)

11.3 Remark.   *Analytically independent functions are algebraically independent. This means that there exists no nonzero polynomial P with*

$$P(f_1, \ldots, f_m) \equiv 0.$$

Because of Remark 11.3 we can assume $m = n$. The claim then follows from the theorem of invertible functions.   □

11.4 Lemma.   *Let*

$$f : U \to \mathbb{C}^m, \quad U \subset \mathbb{C}^n \text{ open and nonempty,}$$

*be an **injective** analytic map. Then the components of f contain an analytically independent subsystem consisting of n functions.*

Proof. Let d be the maximal number of analytically independent subsystems. We can assume that $f_1, \ldots, f_d$ are analytically independent. We extend $f_1, \ldots, f_d$ to an n-tuple

$$\varphi = (f_1, \ldots, f_d, g_{d+1}, \ldots, g_n)$$

of analytically independent functions (see Remark 11.3). We can assume that

$$\varphi : U \longrightarrow V, \quad V \subset \mathbb{C}^n,$$

is a biholomorphic map onto some open set V. Now we replace f by

$$F := f \circ \varphi^{-1} : V \to \mathbb{C}^m.$$

Obviously, the statement of Lemma 11.4 does not change if f is replaced by F.

From the equation $F = f \circ \varphi^{-1}$, it follows that $F \circ \varphi = f$, but

$$F_\nu(f_1, \ldots, f_d, g_{d+1}, \ldots, g_n) = f_\nu \qquad (1 \le \nu \le m)$$

and therefore

$$F_\nu(w_1, \ldots, w_n) = w_\nu \text{ for } 1 \le \nu \le d.$$

(We denote the coordinates of V by $w_1, \ldots, w_n$.)

The complex functional matrix of F is of the form

$$
\begin{pmatrix}
1 & & 0 & \Big| & 0 & \cdots & 0 \\
 & \ddots & & \Big| & \vdots & & \vdots \\
0 & & 1 & \Big| & 0 & \cdots & 0 \\
\hline
\dfrac{\partial F_{d+1}}{\partial w_1} & \cdots & \dfrac{\partial F_{d+1}}{\partial w_d} & \Big| & \dfrac{\partial F_{d+1}}{\partial w_{d+1}} & \cdots & \dfrac{\partial F_{d+1}}{\partial w_n} \\
\vdots & & \vdots & \Big| & \vdots & & \vdots \\
\dfrac{\partial F_m}{\partial w_1} & \cdots & \dfrac{\partial F_m}{\partial w_d} & \Big| & \dfrac{\partial F_m}{\partial w_{d+1}} & \cdots & \dfrac{\partial F_m}{\partial w_n}
\end{pmatrix}
$$

By assumption, the rank of this matrix must not exceed d. This gives

$$
\frac{\partial F_{d+1}}{\partial w_\nu} = 0 \text{ for } \nu > d.
$$

Since we can assume that V is connected, we obtain the result that F_{d+1} (and analogously $F_{d+2}, \ldots, F_m$) are independent of the variables $w_{d+1}, \ldots, w_n$. Since F, by assumption, is injective, such variables cannot exist. □

The next proposition now follows from the point separation theorem (Theorem 10.3).

11.5 Proposition. *Let $\theta_0, \ldots, \theta_N$ be a basis of the vector space $A_r(Q, l, E)$ with $r \geq 3$. We set*

$$
D = \{[z] \in \mathbb{C}^n/L; \quad \theta_0(z) \neq 0\}.
$$

The map

$$
D \longrightarrow \mathbb{C}^N, \quad [z] \longmapsto \left(\frac{\theta_1(z)}{\theta_0(z)}, \ldots, \frac{\theta_N(z)}{\theta_0(z)} \right),
$$

is injective. The abelian functions

$$
\frac{\theta_1}{\theta_0}, \ldots, \frac{\theta_N}{\theta_0}
$$

contain a subsystem of n analytically (and, in particular, algebraically) independent functions.

Proof. The injectivity follows from the point separation property. The analytic independence follows from Lemma 11.4 □

Algebraic Dependence

11.6 Lemma.   *There exists a natural number M, which depends only on $r \in \mathbb{N}$ and on the Pfaffian, such that the following holds:*

If

$$\theta_1, \ldots, \theta_n \in A_r(Q, l, E)$$

and

$$\theta \in A_{rs}(Q, l, E) \qquad\qquad (s \in \mathbb{N} \text{ arbitrary}),$$

then the monomials

$$\theta^\nu \theta_0^{\nu_0} \cdots \theta_n^{\nu_n}, \qquad \nu s + \nu_0 + \nu_1 + \nu_2 + \cdots n = Ms,$$

are linearly dependent.

Proof. The monomials are contained in the vector space $A_{rsM}(Q, l, E)$, which has the dimension

$$P(rsM)^n.$$

So we only have to take care that the number of monomials, and hence the number of solutions of

$$\nu s + \nu_0 + \nu_1 + \nu_2 + \cdots n = Ms,$$

is greater than $P(rsM)^n$. The number of solutions of

$$\nu_0 + \nu_1 + \nu_2 + \cdots n = s(M - \nu) \qquad\qquad (0 \le \nu \le M)$$

for fixed ν equals the binomial coefficient

$$\binom{s(M - \nu) + n}{n} \ge \frac{[s(M - \nu)]}{n!}.$$

Hence the number we are looking for is

$$\sum_{\nu=0}^{M} \binom{s(M - \nu) + n}{n} \ge s^n \frac{\sum_{\nu=1}^{M} \nu^n}{n!}.$$

We know that

$$Q(M) := \sum_{\nu=0}^{M} \nu^n$$

is a polynomial of degree $n + 1$ in M. We need the inequality

$$s^n Q(M) > P \cdot r^n s^n M^n \cdot n!.$$

It is clear that this inequality is valid for large enough $M = M(r, P)$, since on the left-hand side we have a polynomial of higher degree than on the right-hand size. $\qquad\square$

In the following, we choose elements $\theta_0, \ldots, \theta_n$ (by Proposition 11.5) from $A_r(Q, l, E)$ (for some r) such that the abelian functions

$$f_1 = \frac{\theta_1}{\theta_0}, \ldots, f_n = \frac{\theta_n}{\theta_0}$$

are algebraically independent. They generate a subfield $\mathbb{C}(f_1, \ldots, f_n)$ of the field of all abelian functions. Now let f be a further abelian function. We first assume that it is of the special form

$$f = \frac{\theta}{\theta_0^s}, \quad \theta \in A_{rs}(Q, l, E).$$

Because of Lemma 11.6, it satisfies an algebraic equation of degree $\leq M$ over the field $\mathbb{C}(f_1, \ldots, f_n)$. Now let f be arbitrary. It can be written in the form

$$f = \frac{\tilde{\theta}}{\theta}, \quad \theta, \tilde{\theta} \in A_t(Q, l, E), \quad t \text{ suitable.}$$

We can assume that t is divisible by r, i.e. $t = rs$, since we can write

$$f = \frac{\tilde{\theta}^r}{\theta \tilde{\theta}^{r-1}}.$$

Now we have

$$f = \frac{\tilde{\theta}/\theta_0^s}{\theta/\theta_0^s}.$$

This shows that an arbitrary f can written as a quotient of two functions which satisfy algebraic equations of degree $\leq M$. But then f satisfies an algebraic equation of degree $\leq M^2$. (This follows from Remark VIII.4.5.)

We have shown that there exist n algebraically independent abelian functions $f_1, \ldots, f_n$, and that each such abelian function satisfies an algebraic equation of bounded degree over $\mathbb{C}(f_1, \ldots, f_n)$. This says that the field of abelian functions is an algebraic function field of transcendental degree n (see Proposition VIII.4.7).

11.7 Theorem.   *Let $L \subset \mathbb{C}^n$ be a lattice which admits a nondegenerate Riemannian form. The field of all abelian functions is an **algebraic function field of transcendental degree n**.*

We know that algebraic function fields are finitely generated. (Actually, they can be generated by $n + 1$ elements.) This shows that there is an $r \in \mathbb{N}$ such that the field of abelian functions is generated by θ_ν/θ_0 $(1 \leq \nu \leq N)$, where $\theta_0, \ldots, \theta_N$ denotes a basis of $A_r(Q, l, E)$. This, in connection with Proposition 9.4 shows the following.

11.8 Theorem (theta theorem).   *For a suitable $r \geq 3$, we have the following. Let $\theta_0, \ldots, \theta_N$ be a basis of $A_r(Q, l, E)$. The field of all abelian functions is generated by the quotients*

$$\frac{\theta_1}{\theta_0}, \ldots, \frac{\theta_N}{\theta_0}.$$

Actually, a better result holds: one can take every $r \geq 3$. A proof can be found in [Co].

Exercise for Sect. VI.11

1. Let $L_i \subset \mathbb{C}^{n_i}$ be two lattices with a nondegenerate Riemannian form. Show that the field of abelian functions for $L_1 \times L_2$ is generated by the special function $f_1(z_1)f_2(z_2)$, where the f_i are abelian functions for L_i.

    Hint. Use the result of Exercise 1 in Sect. VI.8.

12. Polarized Abelian Manifolds

Similarly to the way in which the theory of elliptic functions leads to the theory of (elliptic) modular functions, the theory of abelian functions leads to the theory of modular functions of several variables. The link is achieved if we consider not only individual lattices but the set of all lattices $L \subset \mathbb{C}^n$. Of course, certain lattices have to be identified. How this has to be done becomes clear if we recall the construction of the canonical basis of a lattice. In that case, we had to consider triples $(\mathcal{Z}, L, H)$, where $L \subset \mathcal{Z}$ is a lattice and H is a nondegenerate Riemannian form. We saw that each triple is equivalent to one of the form $(\mathbb{C}^n, L(Z,T), H(Z,T))$. Here the elementary matrix T is unique, but not the matrix Z.

Change of the Canonical Basis

The dimension n and the $n \times n$ elementary-matrix T are fixed in the following.

Recall that the lattice is generated by the columns of the matrix $L = L(Z,T)$. The associated Riemannian form is $H(z,w) = z^t \operatorname{Im}(Z)^{-1} \bar{w}$.

Now we consider a second matrix $\tilde{Z}$ and seek an isomorphism

$$R : (\mathbb{C}^n, L(Z,T)), H(Z,T) \longrightarrow (\mathbb{C}^n, L(\tilde{Z},T), H(\tilde{Z},T)).$$

So, R is an automorphism of $\mathbb{C}^n$, which we shall identify with the corresponding matrix. The conditions for R are:

a) $RL(Z,T) = L(\tilde{Z},T)$;

b) $z^t \operatorname{Im}(Z)^{-1}\bar{w} = (Rz)^t \operatorname{Im}(\tilde{Z})^{-1}(\overline{Rw})$.

Condition a) can be formulated as follows:

There is a matrix $M \in \mathrm{GL}(2n, \mathbb{Z})$ with the property

$$(\tilde{Z},T) = R(T,Z)M^t.$$

It is known from linear algebra that condition b) can be rewritten as

$$M^t \begin{pmatrix} 0 & T \\ T & 0 \end{pmatrix} M = \begin{pmatrix} 0 & T \\ T & 0 \end{pmatrix}.$$

It is easy to verify that the set of all integral matrices with this property is a group.

12.1 Definition. *The paramodular group $\Gamma_0(T)$ of level T consists of all integral M with the property*

$$M^t \begin{pmatrix} 0 & T \\ T & 0 \end{pmatrix} M = \begin{pmatrix} 0 & T \\ T & 0 \end{pmatrix}.$$

If T is the unit matrix or a multiple of it, then $\Gamma_0(T)$ coincides with the integral symplectic group $\mathrm{Sp}(n, \mathbb{Z})$, which has already appeared during our investigations of the period relations of compact Riemann surfaces (see Remark IV.10.6). There, we saw that the period lattice of such a surface always admits a nondegenerate Riemannian form whose associated elementary matrix is the unit matrix.

Polarization

A Riemannian form is called minimal if the first elementary divisor t_1 equals 1. If H is an arbitrary Riemannian form, then $t_1^{-1}H$ is a minimal Riemannian form. For our purposes, two Riemannin forms can be considered as equal if they differ by a factor. Hence we can restrict ourselves to minimal Riemannian forms.

12.2 Definition. *A polarized abelian manifold is an isomorphy class of triples $(\mathcal{Z}, L, H)$, where L is a lattice in the vector space $\mathcal{Z}$ and H is a minimal Riemannian form on L.*

The corresponding elementary divisor matrix T, $t_1 = 1$, is called the *polarization type* of the polarized abelian manifold.

With each point $Z \in \mathbb{H}_n$ we associate a polarized abelian manifold of a prescribed polarization type T ($t_1 = 1$), namely the isomorphy class of $(\mathbb{C}^n, L(Z,T), H(Z,T))$. Hence the totality of all isomorphy classes is a set, which we denote by $\mathcal{A}(T)$. Two points Z, $\tilde{Z}$ have the same image in $\mathcal{A}(T)$ if there are matrices $R \in \mathrm{GL}(n, \mathbb{C})$ and $M \in \Gamma_0(T)$ such that

$$(\tilde{Z}, T) = R(T, Z)M^t.$$

If we decompose M into four $n \times n$ blocks

$$M = \begin{pmatrix} A & B \\ C & D \end{pmatrix},$$

then this equations reads as

$$\tilde{Z} = R(ZA^t + TB^t),$$
$$T = R(ZC^t + TD^t).$$

Hence the matrix R is determined:

$$R = T(ZC^t + TD^t)^{-1}.$$

This leads to the following definition.

12.3 Definition. *Two points $Z, \tilde{Z} \in \mathbb{H}_n$ are called equivalent mod $\Gamma_0(T)$ if there exists a matrix $M \in \Gamma_0(T)$ with the following two properties:*

1) *The matrix $ZC^t + TD^t$ is invertible.*
2) *We have*
$$T^{-1}\tilde{Z} = (ZC^t + TD^t)^{-1}(ZA^t + TB^t).$$

It is clear that this defines an equivalence relation. We denote the set of equivalence classes by

$$\mathbb{H}/\Gamma_0(T).$$

In the next chapter, we shall show that this equivalence relation comes from a group action. In fact, we shall see that condition 1) is automatically true for $Z \in \mathbb{H}_n$ and $M \in \Gamma_0(T)$, and that the matrix $\tilde{Z}$, which can now be defined by the equation 2), is automatically contained in $\mathbb{H}_n$. We should mention that we have already dealt with this problem in connection with period relations (compare Proposition IV.10.9).

In this section, we are satisfied to rewrite these relations in a standard form. If we use the fact that $\tilde{Z}$ is symmetric, then we can rewrite 2) of Definition 12.3 in the form

$$\tilde{Z} = (AZ + BT)(CZ + DT)^{-1}T.$$

If we introduce

$$\begin{pmatrix} \tilde{A} & \tilde{B} \\ \tilde{C} & \tilde{D} \end{pmatrix} = \begin{pmatrix} A & BT \\ T^{-1}C & T^{-1}DT \end{pmatrix} = \begin{pmatrix} E & 0 \\ 0 & T^{-1} \end{pmatrix} \begin{pmatrix} A & B \\ C & D \end{pmatrix} \begin{pmatrix} E & 0 \\ 0 & T \end{pmatrix},$$

we can rewrite the equation in the standard form

$$\tilde{Z} = (\tilde{A}Z + \tilde{B})(\tilde{C}Z + \tilde{D})^{-1}.$$

The matrix

$$N := \begin{pmatrix} \tilde{A} & \tilde{B} \\ \tilde{C} & \tilde{D} \end{pmatrix}$$

is symplectic, i.e.

$$N^t \begin{pmatrix} 0 & E \\ -E & 0 \end{pmatrix} N = \begin{pmatrix} 0 & E \\ -E & 0 \end{pmatrix}.$$

We recall that E denotes the unit matrix. The map

$$\begin{pmatrix} A & B \\ C & D \end{pmatrix} \longmapsto \begin{pmatrix} \tilde{A} & \tilde{B} \\ \tilde{C} & \tilde{D} \end{pmatrix}$$

is an injective homomorphism

$$\Gamma_0(T) \longrightarrow \mathrm{Sp}(n, \mathbb{Q}),$$

where $\mathrm{Sp}(n, \mathbb{Q})$ is the rational symplectic group. We denote the image of this homomorphism by $\Gamma(T)$ and call this group the *embedded paramodular group* of level T. The advantage is that all paramodular groups now occur as a subgroup of one group $\mathrm{Sp}(n, \mathbb{Q})$ and that the formula 2) of Definition 12.3 takes the unified form

$$\tilde{Z} = (AZ + B)(CZ + D)^{-1}.$$

We shall continue with this form in the next chapter.

Exercises for VI.12

1. Let t be a natural number. Show that

$$\Gamma_0(tT) = \Gamma_0(T), \quad \Gamma(tT) = M_T \Gamma(T) M_T^{-1},$$

with a rational symplectic matrix M_T. Describe this matrix explicitly.

2. In the next chapter, we shall study the principal congruence subgroup

$$\mathrm{Sp}(n, \mathbb{Z})[q] := \mathrm{Kernel}(\mathrm{Sp}(n, \mathbb{Z}) \longrightarrow \mathrm{Sp}(n, \mathbb{Z}/q\mathbb{Z})).$$

Show that

$$\Gamma(T) \supset \mathrm{Sp}(n, \mathbb{Z})[\det T].$$

3. Two subgroups of a group are called commensurable if their intersection has a finite index in both of them. Show that every two $\Gamma(T)$ in $\mathrm{Sp}(n, \mathbb{Q})$ are commensurable.

13. The Limits of Classical Complex Analysis

Complex tori $\mathbb{C}^n/L$ are examples of n-dimensional analytic manifolds. The notion of an analytic manifold is an obvious generalization of the notion of a Riemann surface for several variables.

Analytic Manifolds

A (complex) n-dimensional chart on a topological space X is a topological map $\varphi : U \to V$ of an open subset $U \subset X$ onto an open subset $V \subset \mathbb{C}^n$. Two n-dimensional charts φ, ψ are said to be analytically compatible if the chart transformation $\psi\varphi^{-1}$ is biholomorphic in the sense of complex analysis in several variables. An n-dimensional analytic atlas $\mathcal{A}$ is a set of n-dimensional analytic charts whose domains of definition cover X, such that any two charts from $\mathcal{A}$ are analytically compatible. Two analytic atlases are said to be equivalent if their union is an analytic atlas. A (complex) analytic manifold $(X, [\mathcal{A}])$ of dimension n is a pair consisting of a Hausdorff space X and a full equivalence class of analytic atlases. So, any analytic atlas $\mathcal{A}$ defines a structure in the form of an analytic manifold on X. We shall usually write simply $(X, \mathcal{A})$ instead of $(X, [\mathcal{A}])$. If it is clear which analytic structure is being considered at any particular moment, we shall simply write X. Riemann surfaces are nothing but one-dimensional analytic manifolds.

Some of the basic notions of the one-dimensional case carry over literally to the case of an arbitrary dimension. We briefly mention them here.

1) Any open subset $U \subset \mathbb{C}^n$ carries a natural structure in the form of an analytic manifold by means of the tautological atlas. This consists of only one chart, namely the identity $\mathrm{id} : U \to U$.

2) In analogy to the case $n = 1$, we can introduce the notion of an analytic ($=$ holomorphic) map $f : X \to Y$ of an n-dimensional analytic manifold $X = (X, \mathcal{A})$ into an m-dimensional analytic manifold $Y = (Y, \mathcal{B})$. The composition of analytic maps $X \to Y$, $Y \to Z$ is analytic. In the spacial case $Y = \mathbb{C}$ (equipped with the tautological structure), analytic maps are called analytic functions.

3) A map $f : U \to V$ between open sets $U \subset \mathbb{C}^n$, $V \subset \mathbb{C}^m$ is an analytic map of analytic manifolds iff it is analytic in the sense of Chap. V, which means that its components locally admit power series expansions.

4) A map $f : X \to Y$ between analytic manifolds is called biholomorphic if it is bijective and if both f and f^{-1} are holomorphic.

5) Every analytic atlas $\mathcal{A}$ is contained in a unique maximal analytic atlas $\mathcal{A}_{\max}$. This is the union of all atlases which are equivalent to $\mathcal{A}$. The elements of $\mathcal{A}_{\max}$ are called analytic charts.

6) If $U \subset X$ is an open subset of an analytic manifold $(X, \mathcal{A})$, we can define a restricted atlas $\mathcal{A}|U$. In this way, U is equipped with the structure of an analytic manifold. We call U (equipped with this structure) an open analytic submanifold. The elements of the maximal atlas $\mathcal{A}_{\max}$ are nothing but biholomorphic maps from open sub-manifolds $U \subset X$ onto open sets $V \subset \mathbb{C}^n$.

7. Finally, the Cartesian product $X \times Y$ of two analytic manifolds $(X, \mathcal{A})$ and $(Y, \mathcal{B})$ can be provided with a structure in the form of an analytic manifold. For two charts $\varphi \in \mathcal{A}$, $\psi \in \mathcal{B}$, we can define the product chart

$$\varphi \times \psi : U_\varphi \times U_\psi \longrightarrow V_\varphi \times V_\psi,$$
$$(x, y) \longmapsto (\varphi(x), \psi(x)).$$

The set

$$\mathcal{A} \times \mathcal{B} = \{\varphi \times \psi; \quad \varphi \in \mathcal{A}, \psi \in \mathcal{B}\}$$

is an analytic atlas on $X \times Y$. If X is an n-dimensional and Y an m-dimensional analytic manifold, then $X \times Y$ is an analytic manifold of dimension $n + m$.

More generally, we can define the Cartesian product $X_1 \times \cdots \times X_n$ of n analytic manifolds. The projections

$$p_\nu : X_1 \times \cdots \cdot X_n \longrightarrow X_\nu \quad (\nu = 1, \ldots, n)$$

are analytic. More generally, a map

$$f : X \longrightarrow X_1 \times \cdots \times X_n$$

of a further analytic manifold X into $X_1 \times \cdots \times X_n$ is analytic iff the compositions with the n projections are analytic.

An example of an analytic manifold is the complex torus $X_L = \mathbb{C}^n/L$, which can be associated with a lattice $L \subset \mathbb{C}^n$. It carries a natural structure in the form of an n-dimensional analytic manifold, such that the natural projection

$$\mathbb{C}^n \longrightarrow \mathbb{C}^n/L$$

is locally biholomorphic (and, in particular, analytic).

Meromorphic Functions

Let U be an open and dense subset of an analytic manifold X. An analytic function $f : U \to \mathbb{C}$ is called meromorphic on X if each point $a \in X$ admits an open connected neighborhood $U(a) \subset X$ and analytic functions

$$g : U(a) \longrightarrow \mathbb{C}, \quad h : U(a) \longrightarrow \mathbb{C}$$

with the following properties:

a) h does not vanish identically.

b) For all $x \in U(a) \cap U$ such that $h(x) \neq 0$, we have

$$f(x) = \frac{g(x)}{h(x)}.$$

Two pairs (U, f), (V, g) are said to be equivalent if $f|U \cap V = g|U \cap V$. It is of course sufficient to demand the equality of f and g on an open and dense subset of $U \cap V$. A *meromorphic function* is a full equivalence class $[U, f]$ of such pairs. If (U, f) is a representative, we call U a *domain of holomorphy* of the meromorphic function. The union of all domains of holomorphy is also a domain of holomorphy. We call it *the* domain of holomorphy. We shall frequently write f instead of $[U, f]$ and we denote *the* domain of holomorphy by D_f.

The set $\mathcal{M}(X)$ of all meromorphic functions is, in an obvious manner, a ring, furthermore, a field if X is connected. Any analytic function f on X can be considered as a meromorphic function ($[X, f]$). In this sense, we have $\mathcal{O}(X) \subset \mathcal{M}(X)$.

Caution. Let $f : X \to \bar{\mathbb{C}}$ be an analytic map of the analytic manifold X into the Riemann sphere. We assume that the fiber over ∞ is thin in X. Then f can be considered in an obvious way as a meromorphic function on X. But, in contrast to the case $n = 1$, not every meromorphic function needs to be of this form. A typical example is the meromorphic function on $\mathbb{C} \times \mathbb{C}$

$$f(z_1, z_2) = \frac{z_1}{z_2}, \quad D_f = \{z; \quad z_2 \neq 0\}.$$

It makes sense to define

$$f(z_1, z_2) = \infty \text{ for } z_2 = 0 \text{ and } z_1 \neq 0,$$

and we obtain in this way an analytic map

$$\mathbb{C} \times \mathbb{C} - \text{origin} \longrightarrow \bar{\mathbb{C}}.$$

But this map cannot be extended to the origin as a continuous function. The reason is that at the origin the pole set $(z_2 = 0)$ and the zero set $(z_1 = 0)$ intersect. By the way, for each $w \in \bar{\mathbb{C}}$ we can find a sequence $z_n \in \mathbb{C}^2 - \{0\}$ with the property $f(z_n) \to w$. So, if we wanted to associate function values with the origin, every point of $\bar{\mathbb{C}}$ would have the same right to be a value of f. Then f would be *multivalued*. Sometimes the notion "meromorphic" is understood in this way as "multishaped"*).

Of course, we shall insist on the rule that maps have to be single-valued. Hence there is no way for us to consider meromorphic functions as maps on the whole of X. (Complex analysis of several variables has another way to get around this problem. One constructs a "blow-up" $\tilde{X} \to X$ such that the meromorphic function f can be interpreted as an analytic map $f : \tilde{X} \to \bar{\mathbb{C}}$. We shall not need this here.)

In the following, we shall deal with some properties of compact analytic manifolds. These are closely related to the following two propositions, which we shall not prove here in full generality. But we shall obtain partial results in this direction which will be enough for our purposes.

13.1 Proposition. *Let X be a connected **compact** analytic manifold of dimension $n > 0$. The field $\mathcal{M}(X)$ of meromorphic functions is an algebraic function field of transcendental degree $\leq n$.*

The first complete proof seems to have been given by Remmert in 1956. Remmert mentions that this theorem had already been announced in 1953 by W.L. Chow without proof. There are important special cases due to Thimm (1954) and Siegel (1955). Interesting historical comments can be found in Siegel's paper "Meromorphe Funktionen auf kompakten analytischen Mannigfaltigkeiten" ([Si2], No 64).

Up to now, we have proved Proposition 13.1 for Riemann surfaces and complex tori. It is our goal to prove it also for the n-fold Cartesian product X^n of a compact Riemann surface and, as a consequence, also for the symmetric power $X^{(n)}$. This will imply an important consequence for the Jacobi inversion theorem.

---

*) But this is not correct historically. The notion of a "meromorphic function" has its origin in the complex analysis of one variable.

Removability Theorems

We need a higher-dimensional version of the Riemann removability theorem.

13.2 Lemma. *Let $U \subset \mathbb{C}^n$ an open subset and $g : U \to \mathbb{C}$ an analytic function with thin zero set $S = \{z \in U; \ g(z) = 0\}$. Any **bounded** analytic function $f : U - S \to \mathbb{C}$ extends to an analytic function on the whole of U.*

Proof. The function fg can be extended (by zero) to a continuous function on U. Moreover, the function fg^2 admits partial complex derivatives and hence is analytic. So, f is at least meromorphic on U. We assume that there exists a point $a \in S$ such that f cannot be extended as an analytic function to some open neighborhood $U(a) \subset U$. We can assume that in $U(a)$, f can be written as the quotient of two analytic functions whose power series expansions for all points of $U(a)$ are coprime. The zero set of the denominator cannot be contained in the zero set of the numerator. (Here we have to use the preparation theorem!) But then the function f cannot be bounded.   □

It us useful to reformulate the removability theorem in the following modified form.

13.3 Definition. *A closed subset $S \subset X$ of an analytic manifold X is called **analytically thin** if, for any point $a \in X$, there exist an open, connected neighborhood $U(a) \subset X$ and an analytic function $h_a : U(a) \to \mathbb{C}$, $h_a \neq 0$, with*

$$S \cap U(a) \subset \{x \in U(a); \ \ h_a(x) = 0\}.$$

If X is a Riemann surface, then the discrete subsets of X are analytically thin. The removability theorem can be reformulated as follows.

13.4 Proposition. *Let $S \subset X$ be a closed analytically thin subset of an analytic manifold X and let $f : X - S \to \mathbb{C}$ be an analytic function. Assume that for every point $a \in S$ there exists a neighborhood $U(a)$ such that f is bounded on $U(a) \cap (X - S)$. Then f extends to an analytic function on X.*

Corollary. *If X is connected, then $X - S$ is connected as well.*

Every locally constant function on $X - S$ extends to an analytic function on the whole of X and hence is constant.   □

The condition "analytically thin" can be weakened under additional assumptions for f, as follows.

13.5 Definition.   *A closed subset S of an analytic manifold X is called* ***nowhere decomposing*** *if it has no inner points and if, for each open, connected subset $U \subset X$, the complement*

$$U - S := \{x \in U; \ x \notin S\}$$

is connected.

Because of the corollary to Proposition 13.4, analytically thin subsets are nowhere decomposing. The union of two nowhere decomposing sets is nowhere decomposing. If $X_0 \subset X$ is an open submanifold, then $S \cap X_0 \subset X_0$ is nowhere decomposing if $S \subset X$ is so.

13.6 Proposition.   *Let X be a connected analytic manifold and let $S \subset X$ be a (closed) nowhere decomposing subset. Let f be a meromorphic function on $X - S$ which is algebraic over the field of meromorphic functions $\mathcal{M}(X)$ on X. Then f is meromorphic on the whole of X.*

Proof. The meromorphic function f is holomorphic outside an analytically thin subset. We can add this subset to S and hence assume that f is holomorphic on $X - S$. By assumption, there exist meromorphic functions $\varphi_0, \ldots, \varphi_n$ on X such that

$$\varphi_n f^n + \cdots + \varphi_0, \qquad \varphi_n \neq 0.$$

We can assume that the polynomial $P = \varphi_n t^n + \cdots + \varphi_0$ from $\mathcal{M}(X)[t]$ is irreducible. For the proof, we can replace X by a small open connected neighborhood of a given point. Hence we can assume that the functions φ_ν are quotients of functions which are holomorphic on the whole of X. Since we can multiply them by a common denominator, we can assume that they are all holomorphic. By multiplying the algebraic equation by φ_n^{n-1}, we obtain an algebraic equation for $\varphi_n f$ with a highest coefficient of one. Hence we can assume $\varphi_n = 1$. Now we shall show that f extends to a holomorphic function on X. We can assume that the φ_ν are bounded on X, since this is true in a small neighborhood of a given point. Now we obtain the result that f is bounded on $X - S$. In the case where S is analytically thin, the proof now follows from the Riemann removability theorem. The general case needs the following extra consideration.

It is enough to show that there exists an closed analytically thin subset $A \subset X$ such that f extends holomorphically to $X - A$, because then the Riemann removability theorem ensures holomorphic extendability to X. We apply this to the set A of all $x \in X$ such that the discriminant of the polynomial $P_x = \varphi_n(x) t^n + \cdots + \varphi_0(x)$ vanishes. Since our polynomial P is irreducible over the field $\mathcal{M}(X)$, its discriminant is a nonzero element of this field. It follows from the explicit formula for the discriminant that the discriminant of the specialized polynomial P_x depends holomorphically on x. Hence A is analytically thin. So, we can assume that the discriminant of P_x is different from zero for all x. We now choose some point a. The equation $P_a(w) = 0$ has n solutions $b_1, \ldots, b_n$. The derivative of P_a does not vanish at these points, since the discriminant is different from 0. From the theorem of implicit functions (Remark V.3.4), we obtain the result that there exist a small open neighborhood $U(a)$ and holomorphic functions w_ν on $U(a)$ with the properties $w_\nu(a) = b_\nu$ and $P_x(w_\nu(x)) = 0$ and such that the w_ν have disjoint images. Since we may replace X by $U(a)$, we can assume that *there exist holomorphic functions w_ν on the whole of X such that $P_x(w_\nu(x)) = 0$ and all zeros of P_x are exhausted by the $w_\nu(x)$.*

Since $f(x)$ is a zero of P_x for $x \in X - S$, we obtain

$$(w_1(x) - f(x)) \cdots (w_n(x) - f(x)) = 0 \quad \text{for} \quad x \in X - S.$$

Since $X - S$ is connected (!), we obtain $w_\nu(x) = f(x)$ for some ν. This shows that $f(x)$ can be extended to the whole of X. □

Galois Coverings

An important consequence of the removability theorem states the following.

13.7 Lemma. *Let $f : X \to Y$ be a surjective proper analytic map between analytic manifolds. Assume that there exists a closed analytically thin subset $S \subset Y$ such that $T = f^{-1}(S)$ is analytically thin in X. Assume that the restriction of f*

$$X - T \longrightarrow Y - S, \quad T = f^{-1}(S),$$

is locally biholomorphic. Let $g : Y - S \to \mathbb{C}$ be an analytic function whose pullback $g \circ f$ extends analytically to X. Then g extends analytically to Y.

Proof. The assumption of local biholomorphy implies that g is analytic on $Y - S$. For $s \in S$, we choose an open neighborhood $V(s)$ whose closure is compact. Since f is proper, the inverse image of $\overline{V(s)}$ is compact. Hence the analytic continuation of $g \circ f$ is bounded on the inverse image. Hence g is bounded on $V(s) - S$ and we can apply the Riemann removability theorem. □

We need a variant of Lemma 13.7 which includes meromorphic functions. Under the assumptions of Lemma 13.7, we have the following.

If $V \subset Y - S$ is an open and dense subset, then $U = f^{-1}(V)$ is (open and) dense in $X - T$. This follows from the fact that $X - T \to Y - S$ is proper and open. As a consequence, the inverse image of an open and dense subset of Y is open and dense in X. Let g be a meromorphic function on Y. We can then define the composition $g \circ f$ as a meromorphic function on X. In this way, we obtain an embedding (= injective homomorphism) of the fields of meromorphic functions,

$$\mathcal{M}(Y) \longrightarrow \mathcal{M}(X).$$

We are interested in the image. A very simple answer can be given if $f : X \to Y$ is Galois in the following sense.

13.8 Definition. *A surjective proper analytic map*

$$f : X \longrightarrow Y$$

of connected analytic manifolds is called a **ramified Galois covering** *if there exists a finite group G of biholomorphic automorphisms of X such that two points x, y from X have the same image under f iff there exists a $\gamma \in G$ with the property $y = \gamma(x)$. Then the map f factorizes over a bijective map*

$$X/G \longrightarrow Y.$$

We make two further assumptions (which in reality are always fulfilled; in our applications, their validity is immediately visible, so there is no need to prove them):

1) *The map is topological if X/G is equipped with the quotient topology or, equivalently, the map f is open.*

2) *There exists a closed analytically thin subset $S \subset Y$ such that the restriction*

$$X - T \longrightarrow X - S, \quad T = f^{-1}(S),$$

is locally biholomorphic.

The map $X - T \to X - S$ is proper and locally topological and hence a covering in sense of topological covering theory (Lemma I.3.13 and Definition I.3.14). It is Galois with deck transformation group G (see Definition III.5.15). We call G also simply the deck transformation group of $X \to Y$.

As we have mentioned already, there is a natural embedding $\mathcal{M}(Y) \hookrightarrow \mathcal{M}(X)$. For the sake of simplicity, we identify $\mathcal{M}(Y)$ with its image in $\mathcal{M}(X)$.

13.9 Proposition. *Let $f : X \to Y$ be a ramified Galois covering with a corresponding deck transformation group G. The image of $\mathcal{M}(Y)$ in $\mathcal{M}(X)$ consists precisely of the G-invariant function from $\mathcal{M}(X)$:*

$$\mathcal{M}(Y) = \mathcal{M}(X)^G \quad \text{(fixed field)}.$$

Proof. We know already from Lemma 13.7 that the analogous theorem is true for analytic instead of meromorphic maps. Let $g \in \mathcal{M}(X)^G$ be a G-invariant meromorphic function and let $U \subset X$ be an open and dense domain of holomorphy of g. The set $V = f(U)$ is open and dense in Y. There exists a holomorphic function $h : V \to \mathbb{C}$ with $h(x) = g(f(x))$ for $x \in U$. We have to show that h is meromorphic on the whole of Y. Let $b \in Y$ be an arbitrary point and let $a \in X$ be a preimage of b. We want to prove the meromorphy of h in a neighborhood of b. In a small open neighborhood $U(a)$ of a, we can write g as a quotient of analytic functions $g = g_1/g_2$. We can assume that $U(a)$ is G-invariant. We then have

$$g = \frac{g_1 g_3}{g_2 g_3}, \quad \text{with } g_3(x) = \prod_{\substack{\gamma \in G \ \gamma \neq \mathrm{id}}} g_2(\gamma x).$$

The denominator in this representation is G-invariant. Since f is G-invariant, the numerator must also be G-invariant. Now the denominator and numerator define analytic functions on the image $V(a) = f(U(a))$.   □

In the following, we shall make use of some simple facts of algebra, namely some rudiments of Galois theory. For simplicity, we assume that all of the fields considered have a characteristic of zero.

1) Let G be a finite group of automorphisms of a field L; then L is a finite algebraic extension of the fixed field

$$K = L^G = \{x \in L; \quad g(x) = x \text{ for all } g \in G\}.$$

We have

$$\#G = [L : K] \quad (= \dim_K L).$$

2) An extension of fields $K \subset L$ is said to be Galois if there exists a finite group G of automorphisms of L such that $K = L^G$ is the fixed field.

If $K \subset L$ is a finite field extension, then there exists an extension $L \subset \tilde{L}$ such that $\tilde{L}|L$ and $\tilde{L}|K$ are Galois.

The next statement follows from this and the closed relation between compact Riemann surfaces and function fields.

13.10 Remark. *Let X be a connected compact Riemann surface and let $f : X \to \bar{\mathbb{C}}$ a nonconstant meromorphic function. There exist a connected compact Riemann surface $\tilde{X}$ and an analytic map $g : \tilde{X} \to X$ such that*

$$g : \tilde{X} \to X, \quad g \circ f : \tilde{X} \to \bar{\mathbb{C}}$$

are ramified Galois coverings.

Supplement. *The maps f and g induce maps*

$$\tilde{X}^n \longrightarrow X^n \longrightarrow \mathbb{C}^n.$$

Here

$$g^n : \tilde{X}^n \longrightarrow X^n, \quad g^n \circ f^n : \tilde{X}^n \longrightarrow \bar{\mathbb{C}}^n$$

are ramified Galois coverings.

Now we are in a position to prove Proposition 13.1 for the Cartesian powers of Riemann surfaces.

13.11 Theorem.   *The field of meromorphic functions $\mathcal{M}(X^n)$ on the n-fold Cartesian power of a connected, compact Riemann surface is an algebraic function field of transcendental degree n.*

Proof. Let $\mathbb{C} \subset K \subset L$ be field extensions, $L|K$ finite (algebraic). If K is an algebraic function field of transcendental degree n, then this is the case for L for trivial reasons. But the converse is also true. Let $f_1, \ldots, f_m$ ($m \leq n$) be a maximal system of algebraically independent elements of K. Then K and hence also L are algebraic over $\mathbb{C}(f_1, \ldots, f_n)$. Since L is an algebraic function field of transcendental degree n, we obtain $m = n$ and the extension $L|\mathbb{C}(f_1, \ldots, f_n)$, and hence $K|\mathbb{C}(f_1, \ldots, f_n)$ is finite.

We obtain the result that Theorem 13.11 holds for a compact Riemann surface if and only if it holds for the Riemann sphere. Hence it holds if it is true for some compact Riemann surface. But we know that the proposition is true for powers of tori $\mathbb{C}/L$, since powers of tori are tori themselves. And we know from the theory of abelian functions that the field of meromorphic functions on a complex torus is an algebraic function field. $\qquad\square$

The Symmetric Power

Let X be a set and let $X^n = X \times \cdots \times X$ be the n-fold Cartesian power. The symmetric group S_n acts on X^n by permutation of the components. The quotient is the nth power,

$$X^{(n)} := X^n/S_n.$$

If X is a topological space, we equip X^n with the product topology and $X^{(n)}$ with the quotient topology. Now let $X = \mathbb{C}$ be the complex plane. The n elementary symmetric functions $E_1, \ldots, E_n$ define a map

$$E : \mathbb{C}^n \longrightarrow \mathbb{C}^n, \quad E(z) = (E_1(z), \ldots, E_n(z)).$$

This map factors through the nth symmetric power. For the sake of simplicity, we denote this map also by E:

$$E : \mathbb{C}^{(n)} \longrightarrow \mathbb{C}^n.$$

We can consider the statement that this map is topological (Proposition IV.12.4) as the *main theorem of elementary symmetric functions*. Hence we can equip $\mathbb{C}^{(n)}$ with a structure in the form of an analytic manifold such that this map is biholomorphic.

We want to extend this construction to Riemann surfaces X. We have to study the natural projection $X^n \longrightarrow X^{(n)}$. We study it locally, i.e. in a small open neighborhood U of a point $a \in X^n$. For this, we have to study the stabilizer

$$G := \{g \in S_n; \quad g(a) = a\}.$$

When the components of a are pairwise distinct, then G is trivial. In this case, $X^n \to X^{(n)}$ maps a small open neighborhood of a onto an open neighborhood of the image point in $X^{(n)}$. This map can be used to construct a chart.

Next we consider the extreme case in which all components of a are equal. In this case $G = S_n$. We can construct a small open neighborhood $U \subset X^n$ of a which is G-invariant and such that $U/G \to X^{(n)}$ defines a topological map onto some open subset of $X^{(n)}$. Since U can be identified with a subset of $\mathbb{C}^n$ and hence U/G with an open subset of $\mathbb{C}^{(n)}$, we can use this to construct a chart.

In the general case, the group G is isomorphic to a Cartesian product of groups S_d. We restrict our considerations to a typical case. Let $n = 3$ and $a_1 = a_2$, $a_3 \neq a_1$. Then G is isomorphic to S_2, where S_2 permutes the first two coordinates. Again it is easy to construct an open neighborhood U of a which is invariant under G and such that the natural map

$$U/G \longrightarrow X^{(n)}$$

defines a topological map onto an open neighborhood of the image of a. We can choose U in the form $U = U_1 \times U_2 \times U_3$ with $U_1 = U_2$. Here $U_i \subset \mathbb{C}$ is an open neighborhood of a_i. We have a natural identification

$$U/G = (U_1 \times U_2)/S_2 \times U_3 \qquad (U_1 = U_2).$$

We already know that U_1^2/S_2 carries a structure in the form of an analytic manifold. Hence U/G is an analytic manifold. The arguments indicated above easily lead to a proof of the following statement.

13.12 Remark. *Let X be a Riemann surface. Then the symmetric power $X^{(n)}$ admits a structure in the form of an analytic manifold such that the natural projection $X^n \to X^{(n)}$ is a ramified Galois covering $X^n \to X^{(n)}$ whose deck transformation group is isomorphic to S_n. The field of meromorphic functions on $X^{(n)}$ is an algebraic function field of transcendental degree n.*

We now return to the Jacobi inversion theorem. Recall that we have associated a compact Riemann surface with a complex torus

$$\mathrm{Jac}(X) = \mathbb{C}^n/L,$$

together with a map

$$X^{(n)} \longrightarrow \mathbb{C}^n/L.$$

It is clear from our construction that this map is analytic. We have seen that this map is "nearly bijective":

1) The fibers of this map are connected.

2) Let $U \subset X^{(n)}$ be the subset of all points such the map is locally biholomorphic in a small open neighborhood. The set U is open and dense. The same is true for the image $V \subset \mathrm{Jac}(X)$, and the induced map $U \to V$ is biholomorphic. (To be honest, we should note that the inversion theorem (Theorem IV.12.2) has only been formulated in such a way that one can find an open, dense subset U such that $U \to V$ is locally topological. But the explicit U which we constructed in the proof of Theorem IV.12.2 obviously has the property that $U \to V$ is biholomorphic. As a consequence, the set U which we have defined here is open and dense, and it follows from 1) that for this set, $U \to V$ is biholomorphic.)

3) If we set $T = X^{(n)} - U$ and $S = \mathrm{Jac}(X) - V$ then T is the full inverse image of S. (Let a, b be two different points which are mapped to the same image point in S. Then, because of 1), the map cannot be locally biholomorphic close to a or b. Hence both points are contained in T.)

The set T can be described locally as the zero set of a Jacobian determinant. Hence T is analytically thin and hence nowhere decomposing. We can show that its image S is nowhere decomposing. For this, we consider some open and connected subset $W \subset \mathrm{Jac}(X)$. We denote its inverse image by $\tilde{W}$. The map $\tilde{W} \to W$ is proper and has connected fibers. It is easy to deduce from this that $\tilde{W}$ is connected. We obtain the result that $\tilde{W} - T$ is connected. Hence the image $W - S$ is connected as well.

Now we consider the inclusion $\mathcal{M}(\mathrm{Jac}(X)) \hookrightarrow \mathcal{M}(X^{(n)})$. Both fields are algebraic function fields of transcendental degree n. Hence the field extension is algebraic. Let $g \in \mathcal{M}(X^{(n)})$. We can consider g as a meromorphic function on $\mathrm{Jac}(X) - S$. From Proposition 13.6, we obtain that g is meromorphic on $\mathrm{Jac}(X)$. Hence the two function fields agree. In this way, we obtain the following finer version of the Jacobi inversion theorem (compare Theorem IV.12.2):

13.13 Theorem. *Let $U \subset X^{(n)}$ be the set of all points such that the Jacobi map*

$$X^{(n)} \longrightarrow \mathbb{C}^n/L = \mathbb{C}^n/L$$

is locally biholomorphic close to them. This set is open and dense. The Jacobi map maps U biholomorphically onto an open and dense subset of $\mathrm{Jac}(X)$. It induces a bijection between the abelian functions with respect to L and the meromorphic functions on $X^{(n)}$.

In this sense, one can call the Jacobi map a *bimeromorphic map.*

As an application of the theory of elliptic functions, we have seen that the inverse function of an elliptic integral of the first kind is an elliptic function.

The solution of the Jacobi inversion problem yields a fantastic generalization for algebraic integrals. We shall formulate a special case of this inversion theorem and show how concrete these questions are.

We consider the example of the hyperelliptic integral of genus two which belongs to the algebraic function $\sqrt{1 + x^6}$ (see Remark IV.7.7). Following Jacobi,
we consider the integrals

$$
\begin{aligned}
y_1 = y_1(x_1, x_2) &= \int_{-\infty}^{x_1} \frac{dt}{\sqrt{1 + t^6}} + \int_{-\infty}^{x_2} \frac{dt}{\sqrt{1 + t^6}}, \\
y_2 = y_2(x_1, x_2) &= \int_{-\infty}^{x_1} \frac{t\, dt}{\sqrt{1 + t^6}} + \int_{-\infty}^{x_2} \frac{t\, dt}{\sqrt{1 + t^6}},
\end{aligned}
$$

and then – since both integrals are invariant when x_1, x_2 are permuted – take
the symmetric expressions (elementary symmetric functions)

$$ x_1 + x_2 \text{ and } x_1 \cdot x_2. $$

The Jacobi inversion problem asks for the inversion of

$$ (x_1 + x_2, x_1 x_2) \longmapsto (y_1, y_2). $$

Of course, we cannot expect that this map will be invertible globally. In an
initial step, we can consider local inversions and then study their analytic
continuation.

The following answer follows from Theorem 13.13.

13.14 Theorem. *There exists a lattice $L \subset \mathbb{C}^2$ and a pair of abelian functions
f, g with respect to L, such that the two following conditions are satisfied:*

1) *There exists an open and dense subset $U \subset \mathbb{R}^2$ which is contained in a
   common domain of holomorphy of f and g;*

2)

$$
\begin{aligned}
f(y_1, y_2) &= x_1 + x_2, \\
g(y_1, y_2) &= x_1 x_2.
\end{aligned}
$$

In this sense, Abel's discovery that the inversion of elliptic integrals of the
first kind leads to doubly periodic meromorphic functions has found a fantastic
generalization to hyperelliptic integrals. The triumph of the theory compact
Riemann surfaces was that it allowed to discover the true nature of the result
and to provide a proof of it.

Short Historical Note on the Jacobi Inversion Problem

Abel pointed out in 1825 that the inverse function of a hyperelliptic integral of genus two should have four independent periods. As Jacobi observed, more than two periods are not possible for *single-valued* meromorphic functions. This argument is no longer valid if the two hyperelliptic integrals are combined into a function of two independent variables. The formulation of the inversion problem that we gave just before Theorem 13.13 is roughly the same as that which Jacobi originally gave in 1834. This was a long time before the theory of Riemann surfaces was created. The inversion problem had a tremendous influence. Important mathematicians of that time worked on the solution of this problem, among them Riemann and Weierstrass. The latter called it a piece of good fortune that he could find, at the beginning of his scientific career, such a basic problem as the Jacobi inversion problem. The theory of Riemann surfaces was the basis for the general solution of this problem. But many famous mathematicians participated in the work of this problem and its partial solutions; we cannot mention them all here.

The inversion theorem also had another aspect in the second half of the nineteenth century. We have formulated it – as is nowadays usual – simply as a statement that the inversion of algebraic integrals leads to abelian functions. But, originally the question of an explicit inversion by means of theta functions was the central concern. The classical solutions of the inversion problem imply more than merely an existence theorem of the kind that we have formulated. The corresponding theta functions had already been introduced, in analogy to the Jacobi functions, before the theory of Riemann surfaces was developed. They lead to an explicit solution of the inversion problem in the hyperelliptic case $w^2 = P(z)$, where P has degree 5 or 6 ($p = 2$). The solution was found independently by A. Göpel in 1847 and J.G. Rosenhain in 1851. (Rosenhain's solution had already been submitted in 1846 because of a prize from the Paris Academy, and had even been mentioned in a letter to Jacobi in 1844).

The breakthrough for the most general case came with the theory of Riemann surfaces. In his fundamental paper of 1857, Riemann constructed the theta function for several variables directly from the Riemann surface of an algebraic function by means of the canonical dissection of the surface and the resulting period relations. This theta function is now called the Riemann theta function. Through a study of the theta function, Riemann was led to a solution of the inversion problem under certain restrictions. Since Riemann's paper contained only vague suggestions, it was necessary to work out the theory in special cases also. The case $p = 2$ was treated in 1862 by F.E. Prym.

The original proofs of the inversion theorem contain more than a mere existence result. The inverse function was constructed explicitly by means of theta functions.

The function-theoretic elaboration of the theory, which then also led to proofs of the inversion theorem without theta functions, was provided by

Clebsch and Gordan in 1866 in their work about abelian functions and by Weierstrass in his lectures on the theory of the *Abel'schen Transzendenten*.

Exercises for Sect. VI.13

1. Using the result of Exercise 2 in Sect. 11 and the theory of elliptic functions in [FB], Chap. V, show that the field of meromorphic functions on the nth power of a one-dimensional torus $(\mathbb{C}/L)^n$ is a finite extension of the rational function field

$$\mathbb{C}(\wp(z_1), \ldots, \wp(z_n)).$$

This extension has degree 2^n. A basis is given by the functions

$$\wp'(z_{\nu_1}) \cdots \wp'(z_{\nu_k}), \quad 1 \leq \nu_1 \leq \ldots \leq \nu_k \leq n.$$

2. Consider the map

$$\wp : \mathbb{C}/L \longrightarrow \bar{\mathbb{C}}.$$

This is a ramified Galois covering (ramified at four points). The group of deck transformations consists of the identity and the map $z \mapsto -z$. Correspondingly,

$$\wp^n : (\mathbb{C}/L)^n \longrightarrow \bar{\mathbb{C}}^n$$

is a Galois covering whose deck transformation group is isomorphic to $(\mathbb{Z}/2\mathbb{Z})^n$. The field of functions which is invariant under this group is precisely

$$\mathbb{C}(\wp(z_1), \ldots, \wp(z_n)).$$

Show that this implies the following theorem of Hurwitz:

Every meromorphic function on the product $\bar{\mathbb{C}}^n$ of n Riemann spheres is rational.

This theorem may be derived by a devious route from the theory of abelian functions. Actually, there is a direct elementary proof. We refer to the classic book by Osgood [Os], where this theorem is proved in Chap. 3, Sect. 23, rather laboriously.

3. Using the result of the previous exercise and Proposition 13.9 in conjunction with Remark 13.10, show the following.

Let X be a connected compact Riemann surface whose function field is realized as an algebraic extension of degree d of a rational function field,

$$\mathcal{M}(X) = \mathbb{C}(f)[g] = \bigoplus_{\nu=1}^{d} \mathbb{C}(f)g^\nu.$$

The field of meromorphic functions on the Cartesian power X^n is a finite algebraic extension of the rational function field

$$\mathbb{C}(f(x_1), \ldots, f(x_n))$$

of degree d^n. A possible basis is given by the monomials

$$g(x_1)^{\nu_1} \cdots g(x_n)^{\nu_n}, \quad 0 \leq \nu_1, \ldots, \nu_n \leq d.$$

VII. Modular Forms of Several Variables

In the same manner as the theory of elliptic functions leads to the theory of elliptic modular functions, the theory of compact Riemann surfaces and abelian functions leads to the theory of modular functions of several variables. We have tried here to give an introduction which is as simple as possible. In order to give complete proofs, we have restricted ourselves for the most part to the case $n = 2$. One of the main results is an elementary proof of a structure theorem of Igusa, which states that the ring of modular forms with respect to his group $\Gamma_2[4, 8]$ is generated by the ten classical theta nullwerte. The analogous result in the case $n = 1$ has been proved in [FB], Sect. VI.6 using a similar method.

The roots of the theory of modular functions of several variables lie in the nineteenth century. These functions occurred as theta functions in connection with the theory of compact Riemann surfaces and the theory of abelian functions. But at that time, modular functions came up as examples and there was no systematic theory of them. An epochmaking function-theoretic foundation of the theory of modular functions, with far reaching new results, was provided in 1935 by C.L. Siegel [Si2]. The modular group of degree n is called Siegel's modular group and modular functions (or forms) of severable variables are called Siegel modular functions (or forms) in his honor. We should mention that the theory of modular functions has now been generalized in many respects; for example, the symplectic group can be replaced by other Lie groups, and analyticity can be replaced by other conditions.

1. Siegel's Modular Group

We have already, in connection with the period relations of compact Riemann surfaces (Definition IV.10.10) and the canonical lattice bases (Remark VI.6.8), been led to the *generalized upper half-plane* $\mathbb{H}_n$. This consists of all complex symmetric $n \times n$ matrices $Z = X + iY$ whose imaginary part Y is positive (i.e. positive definite). We also have met the symplectic group $\mathrm{Sp}(n, \mathbb{R})$ (Definition IV.10.5 and Sect. VI.12). This consists of all real $2n \times 2n$ matrices M which leave invariant the standard alternating form I:

$$M^t I M = I, \quad I = \begin{pmatrix} 0 & E \\ -E & 0 \end{pmatrix}.$$

Frequently, a symplectic matrix M is decomposed into four $n \times n$ blocks,

$$M = \begin{pmatrix} A & B \\ C & D \end{pmatrix}.$$

A simple calculation shows the following.

1.1 Remark. 1) *A matrix* $M = \begin{pmatrix} A & B \\ C & D \end{pmatrix}$ *is symplectic if the relations*

$$A^t D - C^t B = E, \quad A^t C = C^t A, \quad B^t D = D^t B$$

are satisfied. In particular, we have

$$\mathrm{Sp}(1, \mathbb{R}) = \mathrm{SL}(2, \mathbb{R}).$$

2) *We have* $I^t = -I^{-1}$. *Hence* M^t *is symplectic if* M *is so, and*

$$AD^t - BC^t = E, \quad AB^t = BA^t, \quad CD^t = DC^t.$$

3) *The inverse of a symplectic matrix is*

$$M^{-1} = I^{-1} M^t I = \begin{pmatrix} D^t & -B^t \\ -C^t & A^t \end{pmatrix}.$$

4) *Some special examples of symplectic matrices are*

a) $\begin{pmatrix} E & S \\ 0 & E \end{pmatrix}, \quad S = S^t;$

b) $\begin{pmatrix} U^t & 0 \\ 0 & U^{-1} \end{pmatrix}, \quad U \in \mathrm{GL}(n, \mathbb{R});$

c) $I = \begin{pmatrix} 0 & E \\ -E & 0 \end{pmatrix}.$

1.2 Proposition. *Let* $M \in \mathrm{Sp}(n, \mathbb{R})$ *be a real symplectic matrix and let* $Z \in \mathbb{H}_n$ *be a point in the generalized upper half-plane. Then*

1) $\det(CZ + D) \neq 0$;
2) $MZ := (AZ + B)(CZ + D)^{-1} \in \mathbb{H}_n$.

The group $\mathrm{Sp}(n, \mathbb{R})$ *acts by means of* $(Z, M) \mapsto MZ$ *on* $\mathbb{H}_n$. *This means*

$$E^{(2n)} Z = Z, \quad (MN)Z = M(NZ).$$

Two symplectic matrices M, N *define the same symplectic substitution iff they differ by a sign.*

The substitutions obtained from a)–c) *in Remark 1.1 have the effect*

a) $Z \longmapsto Z + S$;
b) $Z \longmapsto Z[U] := U^t Z U$;
c) $Z \longmapsto -Z^{-1}$.

Proof. 1) We give an indirect argument and assume that the homogeneous system of linear equations $(CZ + D)^t \mathfrak{z} = 0$ has a solution $\mathfrak{z} \in \mathbb{C}^n$ which is different from zero. Multiplication by the rowvector $\bar{\mathfrak{z}}^t C$ from the left gives

$$\bar{\mathfrak{z}}^t C(CZ + D)^t \mathfrak{z} = \overline{C^t \mathfrak{z}}^t ZC\mathfrak{z} + \bar{\mathfrak{z}}^t CD^t \mathfrak{z} = 0.$$

If S is a *real symmetric matrix*, then

$$\bar{\mathfrak{z}}^t S\mathfrak{z} = S[\mathfrak{x}] + S[\mathfrak{y}] \quad (\mathfrak{z} = \mathfrak{x} + \mathfrak{y})$$

is real. Hence the imaginary part of the above expression is

$$Y[C^t \mathfrak{x}] + Y[C^t \mathfrak{y}].$$

Since Y is positive definite, we obtain $C^t \mathfrak{x} = C^t \mathfrak{y} = 0$. It follows that $D^t \mathfrak{x} = D^t \mathfrak{y} = 0$. This contradicts the fact that (C, D) has maximal rank n.

2) We can now take $(AZ + B)(CZ + D)^{-1}$ and show that this matrix is symmetric:

$$(CZ + D)^{t-1}(AZ + B)^t = (AZ + B)(CZ + D)^{-1}$$

or

$$(ZA^t + B^t)(CZ + D) = (ZC^t + D^t)(AZ + B).$$

This follows easily from the symplectic relations.

It remains to show that the imaginary part of MZ is positive. If $S = S^{(n)}$ is a real symmetric matrix and $A = A^{(n,m)}$ is a complex $n \times m$ matrix such that $\bar{A}^t SA$ is also real, then

$$\bar{A}^t SA = S[\operatorname{Re} A] + S[\operatorname{Re} A].$$

The claim that $\operatorname{Im} MZ$ is positive follows from this and from the following explicit formula.

1.3 Lemma.   *We have*

$$\operatorname{Im} MZ = (CZ + D)^{t-1}(\operatorname{Im} Z)\overline{(CZ + D)}^{-1}.$$

Proof. We multiply the expression

$$\operatorname{Im} MZ = \frac{1}{2i}\left[(AZ + B)(CZ + D)^{-1} + \overline{(AZ + B)(CZ + D)}^{-1}\right]$$
$$= \left[(CZ + D)^{t-1}(AZ + B)^t + \overline{(AZ + B)(CZ + D)}^{-1}\right]$$

from the left by $(CZ + D)^t$ and from the right by $\overline{(CZ + D)}$, and in this way remove the denominators. Now we can multiply term by term. Application of the symplectic relations gives the claim.   □

End of the proof of Proposition 1.2. It remains to show that only $\pm E^{(2n)}$ acts as an identity. From the fact that $MZ = Z$ for all Z, it follows that

$$AZ + B = Z(CZ + D).$$

By specializing to $Z = zE$, we obtain

$$C = 0, \quad B = 0 \quad \text{and} \quad A = D.$$

Now we get $AZ = ZA$ and, from this, $A = aE$. It follows from the symplectic relations that $a^2 = 1$. $\qquad\qquad\qquad\qquad\qquad\qquad\qquad\qquad\square$

The modular group of degree n consists of all *integral* symplectic matrices. We denote this group by

$$\Gamma_n = \mathrm{Sp}(n, \mathbb{Z}).$$

We have already encountered this group in connection with the period relations of compact Riemann surfaces (Remark IV.10.6) and later in connection with polarized abelian manifolds (Definition VI.12.1), as a special case of the so-called paramodular group $\Gamma(T)$. It would be possible to develop the theory of modular forms more generally for these paramodular groups. For the sake of simplicity, we restrict ourselves here to the important case $T = E$.

It is possible to define, for each *commutative ring R with unity*, the symplectic group $\mathrm{Sp}(n, R)$ with coefficients from R. This consists of all matrices $M \in R^{(2n,2n)}$ with the property $M^t I M = I$. We have $\det M^2 = 1$. By Cramer's rule, M is invertible in $R^{(2n,2n)}$. Hence $\mathrm{Sp}(n, R)$ is a group.

For natural numbers q, we can consider the ring of cosets $\mathbb{Z}/q\mathbb{Z}$. There is a natural group homomorphism

$$\mathrm{Sp}(n, \mathbb{Z}) \longrightarrow \mathrm{Sp}(n, \mathbb{Z}/q\mathbb{Z}).$$

The kernel of this homomorphism will be denoted by $\Gamma_n[q]$. We call this kernel the *principal congruence subgroup of level q*. This is a subgroup of finite index in Γ_n.

Analogously, we consider the groups

$$\mathrm{GL}(n, \mathbb{Z})[q] = \mathrm{Kernel}\big(\mathrm{GL}(n, \mathbb{Z}) \longrightarrow \mathrm{GL}(n, \mathbb{Z}/q\mathbb{Z})\big),$$
$$\mathrm{SL}(n, \mathbb{Z})[q] = \mathrm{Kernel}\big(\mathrm{SL}(n, \mathbb{Z}) \longrightarrow \mathrm{SL}(n, \mathbb{Z}/q\mathbb{Z})\big).$$

So, we have

$$\Gamma_n[q] = \Gamma_n \cap \mathrm{GL}(2n, \mathbb{Z})[q].$$

Exercises for Sect. VII.1

1. Show that M^t is symplectic if M is so.

2. Show that every symplectic matrix M with the property $C = 0$ can be uniquely written in the form
$$\begin{pmatrix} U^t & 0 \\ 0 & U^{-1} \end{pmatrix} \begin{pmatrix} E & S \\ 0 & E \end{pmatrix},$$
with a symmetric matrix S.

3. Describe all symplectic matrices M with $B = 0$.

4. Prove that the subgroups of the symplectic group which are defined by $C = 0$ and $B = 0$ are conjugated.

5. Prove that the subgroup which is defined by $C = 0$ inside $\mathrm{Sp}(n, \mathbb{R})$ acts transitively on $\mathbb{H}_n$.

2. The Notion of a Modular Form of Degree n

We denote by
$$\mathcal{Z}_n = \{Z = Z^{(n)} = Z^t\}$$
the vector space of all symmetric complex $n \times n$ matrices. This is a vector space of dimension $n(n+1)/2$. After the choice of an isomorphism
$$\mathcal{Z}_n \longrightarrow \mathbb{C}^N, \quad N = \frac{n(n+1)}{2},$$
we can define notions of open subsets, analytic functions, etc. All of these notions are independent of the choice of the isomorphism. If we want, we can order the pairs (i, j), $1 \le i \le j \le n$, lexicographically to produce a concrete isomorphism.

2.1 Remark. *The generalized upper half-plane $\mathbb{H}_n$ is an open and convex domain in $\mathcal{Z}_n$.*

Proof. A real symmetric matrix is positive definite if the n minors (principal subdeterminants) are positive. The convexity is clear; moreover, we have

a) $Z \in \mathbb{H}_n$, $t > 0 \Longrightarrow tZ \in \mathbb{H}_n$;
b) $Z, W \in \mathbb{H}_n \Longrightarrow Z + W \in \mathbb{H}_n$. $\qquad\square$

A simple generalization of the "chain rule" for $cz + d$ says the following.

2.2 Remark. *Let $I(M, Z) = CZ + D$. We have*
$$I(MN, Z) = I(M, NZ)I(N, Z).$$

Corollary. *Let $j(M, Z) = \det(CZ + D)$. We have*
$$j(MN, Z) = j(M, NZ)j(N, Z).$$

We want to admit *modular forms of half-integral weight*, and need for this a holomorphic square root of $j(M, Z)$. Its existence follows from the next lemma.

2.3 Lemma. *For every holomorphic function without zeros*
$$f : \mathbb{H}_n \longrightarrow \mathbb{C}^{\bullet},$$
there exists a holomorphic function
$$h : \mathbb{H}_n \longrightarrow \mathbb{C}^{\bullet}$$
with the property
$$h(Z)^2 = f(Z).$$
Like h, $-h$ is also a holomorphic square root. These are the only continuous square roots of f.

Proof. Uniqueness. If h, $\tilde{h}$ are two continuous square roots of f, then $h/\tilde{h}$ is a continuous function with values ± 1. Since $\mathbb{H}_n$ is connected, we obtain the result that $h/\tilde{h}$ is constant ± 1.

Existence of h. We can assume that $f(iE) = 1$. Since $\mathbb{H}_n$ is convex, the segment between iE and any given point $Z \in \mathbb{H}_n$ is contained in $\mathbb{H}_n$; in particular,
$$\alpha(t) = \alpha(Z; t) = f(iE + t(Z - iE)) \text{ for } 0 \le t \le 1$$
is defined and different from zero, and we can define
$$H(Z) := \int_0^1 \frac{\alpha'(t)}{\alpha(t)} \, dt.$$
Obviously, $H(Z)$ is analytic in $\mathbb{H}_n$, and we have
$$e^{H(Z)} = f(Z).$$
The function
$$h(Z) = E^{H(Z)/2}$$
has the desired property.   □

For every matrix $M \in \mathrm{Sp}(n, \mathbb{R})$, there exists a holomorphic square root of $\det(CZ + D)$. From now on, we choose one square root and denote it by
$$\sqrt{\det(CZ + D)} = \det(CZ + D)^{1/2}.$$
(This notation has to be used with some caution. It can happen that $\det(CZ_1 + D) = \det(CZ_2 + D)$ for two different points $Z_1, Z_2 \in \mathbb{H}_n$, but nevertheless $\sqrt{\det(CZ_1 + D)} = -\sqrt{\det(CZ_2 + D)}$.)

Since the square root is not unique, the chain rule only holds up to a sign (compare [FB], Remark VI.5.4):

2.4 Remark. *There exists a map*

$$w : \mathrm{Sp}(n, \mathbb{R}) \times \mathrm{Sp}(n, \mathbb{R}) \longrightarrow \{\pm 1\}$$

with the following property:

$$\sqrt{j(MN, Z)} = w(M, N)\sqrt{j(M, NZ)}\sqrt{j(N, Z)}.$$

As in the case $n = 1$ one defines the notion of a *multiplier system* on a congruence subgroup. By a *congruence subgroup* we understand a subgroup $\Gamma \subset \mathrm{Sp}(n, \mathbb{Z})$ which contains a principal congruence subgroup $\Gamma_n[q]$ (compare [FB], Definition VI.5.1). Generalizing [FB], Definition VI.5.5, we define a multiplier system as follows.

2.5 Definition. *Let $\Gamma \in \mathrm{Sp}(n, \mathbb{Z})$ be a congruence subgroup. A map*

$$v : \Gamma \longrightarrow \mathbb{C}$$

is called a multiplier system of weight $r/2$, $r \in \mathbb{Z}$, if the following conditions are satisfied:
a) *There exists a natural number l such that*

$$v(M)^l = 1 \text{ for all } M \in \Gamma.$$

b) *If we define*

$$j_r(M, Z) = v(M) \det(CZ + D)^{r/2} \quad (M \in \Gamma),$$

we have
  b1) $j_r(MN, z) = j_r(M, NZ)j_r(N, Z);$
  b2) $j_r(-E, Z) = 1, \text{ if } -E \in \Gamma.$

As in the case $n = 1$ ([FB], Remark VI.5.7), we can also define the *conjugate multiplier system*.

2.6 Lemma. *Let v be a multiplier system on the congruence group $\Gamma \subset \mathrm{Sp}(n, \mathbb{Z})$ and let $L \subset \mathrm{Sp}(n, \mathbb{Z})$. Then*

$$v^L(LML^{-1}) = w(L^{-1}, L)^r w(L, ML^{-1})^r w(M, L^{-1})^r v(M)$$

is a multiplier system on the conjugate group $L\Gamma L^{-1}$.

We call v^L the conjugate multiplier system.

Now we can introduce the notion of a modular form in analogy to the case $n = 1$ (compare [FB], Definition VI.5.8).

2.7 Definition.   *Let* $\Gamma \subset \mathrm{Sp}(n, \mathbb{Z})$ *be a congruence subgroup, let* $r \in \mathbb{Z}$ *be an integer, and let* v *be a multiplier system of weight* $r/2$ *on* Γ. *A modular form of weight* $r/2$ *with respect to the multiplier system* v *is a holomorphic function* $f : \mathbb{H}_n \to \mathbb{C}$ *with the following properties:*

1) $f(MZ) = v(M) \det(CZ + D)^{r/2} f(Z)$ *for all* $M \in \Gamma$.

2)   *For every* $M \in \mathrm{Sp}(n, \mathbb{Z})$, *the function*

$$(f|M)(Z) := \det(CZ + D)^{-r/2} f(Mz)$$

is bounded in domains of the kind

$$Y \geq Y_0 > 0 \quad (Y_0 \text{ arbitrary}).$$

The set of all modular forms for given Γ, v, r is a vector space. We denote this set by

$$[\Gamma, r/2, v].$$

If r is even and $v \equiv 1$ is the trivial multiplier system, we write simply

$$[\Gamma, r/2].$$

Exercises for Sect. VII.2

1. Using the Petersson notation

$$(f|M)(Z) = (f|_r M)(Z) = \sqrt{\det(CZ + D)}^{-r} f(Z),$$

    show that

$$f|MN = w^r(M, N)(f|M)|N.$$

2. Show that a system of lth roots of unity $\{v(M)\}_{M \in \Gamma}$ is a multiplier system iff there exists a (not necessarily continuous) non-vanishing function $f : \mathbb{H} \to \mathbb{C}$ which satisfies the transformation formula

$$f|M = v(M)f.$$

3. Let $L \in \Gamma_n$ and let $\Gamma \subset \Gamma_n$ be a congruence subgroup. Show that $\tilde{\Gamma} := L\Gamma L^{-1}$ is a congruence subgroup as well and that the map

$$f \longmapsto f|_r L^{-1}$$

    defines an isomorphism

$$[\Gamma, r/2, v] \xrightarrow{\sim} [\tilde{\Gamma}, r/2, \tilde{v}].$$

    Here $\tilde{v}$ denotes the multiplier system which is conjugate to v.

3. Koecher's Principle

In this section, we want to show that the condition of boundedness in the definition of modular forms (Definition 2.7) is automatically fulfilled in the case $n > 1$. This was proved in 1954 by Max Koecher [Koc]: A corresponding principle for Hilbert modular forms was apparently already known in 1928 by Fritz Götzky [Go].

We have to study periodic analytic functions

$$f : \mathbb{H}_n \longrightarrow \mathbb{C}, \quad f(Z + S) = f(Z), \quad S = S^t \text{ integral.}$$

As we know, these can be expended as Fourier series (Appendix to Sect. VI.8):

$$f(Z) = \sum a((n_{ij})) \exp\left(2\pi i \sum_{1 \le i \le j \le n} n_{ij} z_{ij}\right).$$

Here, there is an arbitrary integral linear combination of the N variables in the exponent. If we define the symmetric matrix T by

$$t_{ij} = t_{ji} = n_{ij} \text{ for } i > j \quad \text{and} \quad t_{ii} = 2n_{ii} \text{ for } i = j,$$

we can write this sum as

$$\frac{1}{2}\sigma(TZ) \quad (\sigma = \text{trace}).$$

Then the Fourier expansion has the form

$$\sum_T a(T)e^{\pi i \sigma(TZ)}.$$

Here, we have to sum over all symmetric integral matrices with even diagonal elements. Such matrices are called *even*. They can be characterized by the property that $T[g] = g^t T g$ is even for all integral columns g.

We recall the notation

$$\mathrm{SL}(n, \mathbb{Z})[q] = \mathrm{Kernel}(\mathrm{SL}(n, \mathbb{Z}) \longrightarrow \mathrm{SL}(n, \mathbb{Z}/q\mathbb{Z})).$$

3.1 Proposition. *Let $f : \mathbb{H} \to \mathbb{C}$ be a holomorphic function with a Fourier expansion*

$$f(Z) = \sum_{T=T^t \text{ even}} a(T)e^{\pi i \sigma(TZ)}.$$

Assume that there exists a natural number q with the property

$$a(T[U]) = a(T) \text{ for all } U \in \mathrm{SL}(n, \mathbb{Z})[q].$$

Under the assumption $n > 1$, we have

$$a(T) \ne 0 \Longrightarrow T \ge 0.$$

Here "$T \ge 0$" means that T is semipositive, i.e. $T[g] \ge 0$ for all $g \in \mathbb{R}^n$. By a continuity argument, it is sufficient to take $g \in \mathbb{Q}^n$ and, by homogeneity, even $g \in \mathbb{Z}^n$. In addition, one can assume that the components of g are coprime.

For the proof of Proposition 3.1, we need a well-known lemma of Gauss, as below.

3.2 Lemma.   *For every column $g \in \mathbb{Z}^n$, there exists a unimodular matrix $U \in \mathrm{GL}(n, \mathbb{Z})$ with the property*

$$Ug = \begin{pmatrix} a_1 \\ 0 \\ \vdots \\ 0 \end{pmatrix}.$$

Corollary. *Every vector $g \in \mathbb{Z}^n$ with coprime components occurs as the first column of a unimodular matrix $U \in \mathrm{GL}(n, \mathbb{Z})$:*

$$U = (g, *).$$

In the case $n > 1$, one can obtain $U \in \mathrm{SL}(n, \mathbb{Z})$.

Proof. We can assume that g is different from zero. In the first step, we find a matrix $U \in \mathrm{GL}(n, \mathbb{Z})$ (actually a permutation matrix) such that

$$Ug = \begin{pmatrix} a_1 \\ \vdots \\ a_n \end{pmatrix}, \quad a_1 \neq 0.$$

We choose one such matrix with minimal $|a_1|$. Next we find a triangular matrix $V \in \mathrm{GL}(n, \mathbb{Z})$ such that

$$VUg = \begin{pmatrix} a_1 \\ a_1 + x_2 a_1 \\ \vdots \\ a_n + x_n a_n \end{pmatrix},$$

with given $x_2, \ldots, x_n$. By means of the Euclidean algorithm, we can choose $x_2, \ldots, x_n$ in such a way that

$$|a_\nu + x_\nu a_1| < |a_1| \qquad (2 \leq \nu \leq n).$$

From the minimality, we get $a_\nu + x_\nu a_1 = 0$. □

Proof of Proposition 3.1. We give an indirect proof and assume that there exists a Fourier coefficient $a(T) \neq 0$ for a matrix T which is not semipositive. We choose a vector $g \in \mathbb{Z}^n$, $T[g] < 0$, with coprime components and complete it to a unimodular matrix $U = (g, *) \in \mathrm{SL}(n, \mathbb{Z})$. Then the first diagonal element of $\tilde{T} = T[U]$ is negative. Now we consider

$$U(x) := U \begin{pmatrix} 1 & x & 0 & \cdots & 0 \\ 0 & 1 & 0 & \cdots & 0 \\ 0 & 0 & 1 & \cdots & 0 \\ \vdots & \vdots & & \ddots & \vdots \\ 0 & 0 & 0 & \cdots & 1 \end{pmatrix}.$$

By assumption,
$$a(T[U(x)]) = a(T) \text{ if } x \equiv 0 \mod q.$$

Since any subseries of $f(Z)$ converges, the series
$$\sum_{T_1} e^{\pi i \sigma(T_1 Z)}$$

has to converge, where the sum is taken over all T_1 which can be written in the form $T_1 = T[U(x)]$. In particular,
$$e^{-\pi\sigma(T[U(x)])}, \quad x \equiv 0 \mod q,$$

has to be bounded. But we have
$$\sigma(T[U(x)]) = x^2 \tilde{t}_{11} + O(x) \to -\infty \text{ for } |x| \to \infty. \qquad \square$$

Fourier Expansion of a Modular Form

If we apply the transformation property of modular forms to translation matrices, we obtain
$$f(Z+S) = v \begin{pmatrix} E & S \\ 0 & E \end{pmatrix} \sqrt{0Z + E}^r f(Z)$$

for all $S \equiv 0 \mod q$ (for suitable q). We take $+1$ for the square root. Then v is a homomorphism of the additive group of all symmetric matrices $S \equiv 0 \mod q$ into a finite group of roots of unity. Hence there exists a multiple l of q such that all $S \equiv 0 \mod l$ are contained in the kernel. We obtain
$$f(Z+S) = f(Z) \text{ for all integral symmetric } S \equiv 0 \mod l.$$

As described, the function f can be expanded into a Fourier series. Using this fact together with Koecher's principle, we obtain the following result.

3.3 Proposition. *Let f be a modular form. There exists a natural number l such that f has an expansion of the form*
$$f(Z) = \sum \exp(\pi i \sigma(TZ)).$$

Here T runs through all symmetric matrices such that lT is integral and such that $T \geq 0$.

Exercises for Sect. VII.3

1. Construct a matrix in $\mathrm{SL}(3, \mathbb{Z})$ with first row $(2, 3, 5)$.

2. Show that the condition $a(T[U]) = a(T)$ in Proposition 3.1 can be weakened to
$$|a(T[U])| \leq \|U\|\, |a(T)|.$$
Here $\|U\|$ denotes some norm on the vector space of matrices.

3. Let $f : \mathbb{H}_n \to \mathcal{Z}_n$, $n > 1$, be a matrix-valued function with the properties
$$f(Z + S) = f(Z), \quad S \text{ integral}, \quad f(Z[U]) = f(Z)[U], \quad U \in \mathrm{SL}(n, \mathbb{Z}).$$
Show that f is bounded in domains of the kind $Y - \delta E \geq 0$ for $\delta > 0$.

4. Show that the Fourier expansion of a modular form is independent of the choice of l in Proposition 3.3.

4. Specialization of Modular Forms

Let $S = S^t = S^{(n)} > 0$ be a positive real matrix. For a point $z \in \mathbb{H}$ in the usual upper half-plane, the point Sz is contained in the half-plane $\mathbb{H}_n$ of degree n. Let $M \in \mathrm{SL}(2, \mathbb{R})$. We seek a symplectic matrix $M^S \in \mathrm{Sp}(n, \mathbb{R})$ which is compatible with the map $z \mapsto Sz$. This means that $M^S(Sz) = SM(z)$, or

$$(ASz + B)(CSz + D)^{-1} = S(az + b)(cz + d)^{-1}$$
$$\left(M = \begin{pmatrix} a & b \\ c & d \end{pmatrix}, \quad M^S = \begin{pmatrix} A & B \\ C & D \end{pmatrix} \right).$$

This equation is satisfied if we define

$$M^S = \begin{pmatrix} aE & bS \\ cS^{-1} & dE \end{pmatrix}.$$

An easy calculation shows that M^S is actually symplectic.

4.1 Lemma. *Let $S = S^{(n)} = S^t > 0$ be a positive real matrix. The map*

$$\mathrm{SL}(2, \mathbb{R}) \longrightarrow \mathrm{Sp}(n, \mathbb{R}),$$
$$M = \begin{pmatrix} a & b \\ c & d \end{pmatrix} \longmapsto M^S = \begin{pmatrix} aE & bS \\ cS^{-1} & dE \end{pmatrix},$$

is an injective group homomorphism. It is compatible with the embedding

$$\mathbb{H} \longrightarrow \mathbb{H}_n, \quad z \longmapsto Sz,$$

in the following sense:

$$S \cdot (Mz) = M^S(Sz).$$

Supplement. *If* $\Gamma \subset \mathrm{Sp}(n, \mathbb{Z})$ *is a congruence subgroup, then its inverse image*

$$\Gamma_0 := \{M \in \mathrm{SL}(2, \mathbb{Z}), \quad M^S \in \Gamma\}$$

is a congruence subgroup in $\mathrm{SL}(2, \mathbb{Z})$.

Proof. Excluding the supplement, the proof is given by straightforward computation. For the proof of the supplement, we choose q in such a way that $\Gamma \supset \Gamma_n[q]$. Then we determine a natural number Q such that Q is a multiple of q and such that QS and QS^{-1} are integral. We then have $\Gamma_0 \supset \Gamma_1[Q]$. □

We can verify the formula

$$\det(S^{-1}cSz + dE) = (cz + d)^n.$$

If v is a multiplier system of weight $r/2$, then

$$v_0(M) := v(M^S)$$

is a multiplier system on Γ_0. So, we obtain the following lemma.

4.2 Lemma. *If* $f \in [\Gamma, r/2, v]$, $\Gamma \subset \mathrm{Sp}(n, \mathbb{Z})$, *is a Siegel modular form of weight* $r/2$, *then*

$$f_0(z) := f(Sz)$$

is an elliptic modular form of weight $nr/2$,

$$f_0 \in \left[\Gamma_0, \frac{rn}{2}, v_0\right].$$

This lemma allows us to reduce some basis facts for Siegel modular forms to the case of elliptic modular forms.

4.3 Proposition. *A Siegel modular form of negative weight vanishes. Any modular form of weight zero is constant. (The constant can be different from zero only if the multiplier system is trivial.)*

Proof. The case $n = 1$ is known ([FB], Proposition VI.5.11). Now let f be a Siegel modular form of negative weight. Using the specialization of Lemma 4.2, we can show that $f(Sz) = 0$ for arbitrary rational $S > 0$ and then, by continuity, also for real $S > 0$. In particular, $f(iY) = 0$. Simple arguments from the complex analysis of one variable show that $f = 0$.

Now we assume that the weight of f is zero. Since we know that elliptic modular forms of weight zero are constant ([FB], Proposition VI.5.11), we obtain the result that $f(Sz)$ is a constant C_S (depending on S), first for all positive rational S and then, by continuity, for all positive real S. Obviously, $C_S = \lim_{y\to\infty} f(Sz)$ is the zero Fourier coefficient of f and hence $C = C_S$ is independent of S. Now $f(Z) - C$ vanishes on matrices of the form $Z = iY$ and hence is identically zero. $\qquad\qquad\qquad\qquad\qquad\qquad\qquad\qquad\qquad\qquad\qquad\qquad\square$

Exercises for Sect. VII.4

1. In the case $n > 1$, any modular form without zeros in $\mathbb{H}_n$ is constant. Is this true for $n = 1$?

2. Assume that it has been proved that every modular form of weight 0 is constant but not yet that modular forms of negative weight vanish. Show that it is not possible that there exist nonvanishing modular forms of positive weight and also of negative weight.

3. You may use the fact that

$$\vartheta(Z) = \sum_{g\in\mathbb{Z}^n} \exp(\pi i Z[g])$$

is a modular form (Proposition 7.8). What is its weight?

3. Again, you may use the fact that

$$\vartheta(Z) = \sum_{g\in\mathbb{Z}^n} \exp(\pi i Z[g])$$

is a modular form. Show that

$$\sum_{g\in\mathbb{Z}^n} (-1)^{g_1} \exp(\pi i Z[g])$$

is a modular form too.

5. Generators for Some Modular Groups

Recall that the group $SL(2, \mathbb{Z})$ is generated by the two matrices

$$\begin{pmatrix} 1 & 1 \\ 0 & 1 \end{pmatrix} \quad \text{and} \quad \begin{pmatrix} 0 & -1 \\ 1 & 0 \end{pmatrix}$$

(see [FB], Proposition VI.1.8). A variant states that it is generated by the two matrices

$$\begin{pmatrix} 1 & 1 \\ 0 & 1 \end{pmatrix} \quad \text{and} \quad \begin{pmatrix} 1 & 0 \\ 1 & 1 \end{pmatrix},$$

and hence by strict triangular matrices. For a proof, one can consider the formula

$$\begin{pmatrix} 1 & 0 \\ 1 & 1 \end{pmatrix} \begin{pmatrix} 1 & 1 \\ 0 & 1 \end{pmatrix}^{-1} \begin{pmatrix} 1 & 0 \\ 1 & 1 \end{pmatrix} = \begin{pmatrix} 0 & -1 \\ 1 & 0 \end{pmatrix}.$$

In [FB], we proved this by means of the fundamental domain of the modular group. Because of its importance, we give another proof here. It rests on the following statement.

5.1 Remark. *For each pair (a, b) of integers, there exists a matrix U in the group*

$$G = \left\langle \begin{pmatrix} 1 & 1 \\ 0 & 1 \end{pmatrix}, \begin{pmatrix} 1 & 0 \\ 1 & 1 \end{pmatrix} \right\rangle \subset SL(2, \mathbb{Z}),$$

which is generated by the two given matrices, such that

$$U \begin{pmatrix} a \\ b \end{pmatrix} = \begin{pmatrix} \alpha \\ 0 \end{pmatrix}.$$

The proof can be obtained by induction on $|a||b|$.
Beginning of the induction. $|a||b| = 0$. Then $a = 0$ or $b = 0$. Since the matrix $\begin{pmatrix} 0 & -1 \\ 1 & 0 \end{pmatrix}$ is contained in G, we can assume that $b = 0$.

Induction step. We now assume that $|a||b| > 0$. Multiplication by powers of the two generating matrices has the effect

$$\begin{pmatrix} a \\ b \end{pmatrix} \longmapsto \begin{pmatrix} a \\ b + xa \end{pmatrix} \quad \text{or} \quad \begin{pmatrix} a + yb \\ b \end{pmatrix},$$

respecitvely. By means of the Euclidean algorithm, we can make $|a||b|$ smaller. $\qquad \square$

Besides the ring $\mathbb{Z}$, we consider factor rings

$$R = \mathbb{Z}/q\mathbb{Z}, \qquad q \geq 0.$$

In the case $q = 0$ we obtain $\mathbb{Z}$, but in the case $q > 0$ we obtain a finite ring. When q is prime, then R is a finite field. It is clear (and follows from Remark 5.1) that Remark 5.1 holds for R instead of $\mathbb{Z}$.

5.2 Lemma.   *Let R be a factor ring of $\mathbb{Z}$ (and hence $R = \mathbb{Z}$ or $R = \mathbb{Z}/q\mathbb{Z}$, $q > 0$). The group $\mathrm{SL}(2, R)$ is generated by the two matrices*

$$\begin{pmatrix} 1 & 1 \\ 0 & 1 \end{pmatrix} \qquad \begin{pmatrix} 1 & 0 \\ 1 & 1 \end{pmatrix}.$$

Proof. By Remark 5.1, there exists for a given $U \in \mathrm{SL}(2, R)$ a matrix $V \in G$ with the property

$$UV = \begin{pmatrix} a & b \\ 0 & d \end{pmatrix} = \begin{pmatrix} a & 0 \\ 0 & a^{-1} \end{pmatrix} \begin{pmatrix} 1 & a^{-1}b \\ 0 & 1 \end{pmatrix}.$$

The prove now follows from the following formula.

5.3 Formula.   *We have*

$$\begin{pmatrix} a & 0 \\ 0 & a^{-1} \end{pmatrix} = \begin{pmatrix} 1 & a \\ 0 & 1 \end{pmatrix} \begin{pmatrix} 0 & 1 \\ -1 & 0 \end{pmatrix} \begin{pmatrix} 1 & a^{-1} \\ 0 & 1 \end{pmatrix} \begin{pmatrix} 0 & 1 \\ -1 & 0 \end{pmatrix} \begin{pmatrix} 1 & a \\ 0 & 1 \end{pmatrix} \begin{pmatrix} 0 & 1 \\ -1 & 0 \end{pmatrix}.$$

We define for each pair (μ, ν), $1 \leq \mu < \nu \leq n$, an embedding (= injective homomorphism)

$$\alpha_{\mu\nu} : \mathrm{SL}(2, R) \longrightarrow \mathrm{SL}(n, R),$$

$$\begin{pmatrix} a & b \\ c & d \end{pmatrix} \longmapsto \begin{pmatrix} 1 & & & & & & & \\ & \ddots & & & & & & \\ & & 1 & & & & & \\ & & & a & \cdots & b & & \\ & & & \vdots & \ddots & \vdots & & \\ & & & c & \cdots & d & & \\ & & & & & & 1 & \\ & & & & & & & \ddots \\ & & & & & & & & 1 \end{pmatrix} \begin{array}{l} \\ \\ \\ \leftarrow \mu\text{th row} \\ \\ \leftarrow \nu\text{th column} \\ \\ \\ \end{array}$$

The image is called an embedded $\mathrm{SL}(2, R)$. We denote by

$$G_n = \langle \alpha_{\mu\nu}(\mathrm{SL}(2, R), \quad 1 \leq \mu < \nu \leq n \rangle$$

the subgroup of $\mathrm{SL}(n, R)$ which is generated by all embedded $\mathrm{SL}(2, R)$.

We need the following seeming generalization of Gauss's lemma.

5.4 Lemma. *For every column $g \in R^n$, there exists $U \in G_n$ with*

$$Ug = \begin{pmatrix} a_1 \\ 0 \\ \vdots \\ 0 \end{pmatrix}.$$

Proof. By means of the matrix $\begin{pmatrix} 0 & -1 \\ 1 & 0 \end{pmatrix}$, we construct matrices $U \in G_n$ such that the transformation $g \mapsto Ug$ permutes two components of g up to a sign. Hence we can assume that $g_1 \neq 0$. Now we multiply g step by step by matrices from the images of $\alpha_{1n} \ldots, \alpha_{12}$ to obtain $g_n = \ldots = g_2 = 0$.

As an application, we prove the following proposition for factor rings R of $\mathbb{Z}$.

5.5 Proposition. *The group $\mathrm{SL}(n, R)$ $(n \geq 2)$ is generated by the embedded $\mathrm{SL}(2, R)$.*

Before giving the proof, we formulate two obvious consequences.

5.6 Corollary. *The group $\mathrm{SL}(n, R)$ is generated by (upper and lower) strict triangular matrices.*

5.7 Corollary. *The natural homomorphism*

$$\mathrm{SL}(n, \mathbb{Z}) \longrightarrow \mathrm{SL}(n, \mathbb{Z}/q\mathbb{Z})$$

is surjective.

Proof of Proposition 5.5. We use induction on n. The proposition is assumed to have been proved for $n - 1$ in place of n. By Lemma 5.4, there exists a matrix $V \in G_n$ with

$$VU = \begin{pmatrix} a_1 & * \ldots * \\ 0 & \\ \vdots & A \\ 0 & \end{pmatrix}.$$

We can achieve $a_1 = 1$. We then have $A \in \mathrm{SL}(n-1)$ and hence $A \in G_{n-1}$. We obtain

$$\begin{pmatrix} 1 & 0 \\ 0 & A \end{pmatrix} \in G_n$$

and

$$\begin{pmatrix} 1 & 0 \\ 0 & A \end{pmatrix}^{-1} VU = \begin{pmatrix} 1 & * \ldots * \\ 0 & \\ \vdots & E \\ 0 & \end{pmatrix}.$$

Such a matrix can easily be written as a product of embedded $\mathrm{SL}(2,R)$ matrices. $\square$

A third corollary refers to the group $\mathrm{GL}(n,R)$. This group is generated by $\mathrm{SL}(n,R)$ and diagonal matrices. Making use of the formula

$$\begin{pmatrix} 1 & 1 \\ 0 & 1 \end{pmatrix} = \begin{pmatrix} 1 & 1 \\ 1 & 0 \end{pmatrix} \begin{pmatrix} 0 & 1 \\ 1 & 0 \end{pmatrix},$$

we obtain the following result.

5.8 Lemma.   *The group $\mathrm{GL}(n,R)$ can be generated by symmetric matrices.*

There are similar results for the symplectic group . The main-result is the following.

5.9 Proposition.   *Let R be a factor ring of $\mathbb{Z}$. The group $\mathrm{Sp}(n,R)$ is generated by the matrices*

$$\begin{pmatrix} A & B \\ C & D \end{pmatrix}, \quad B = 0 \quad or \quad C = 0.$$

Corollary 1. *The group $\mathrm{Sp}(n,R)$ is generated by the matrices*

$$\begin{pmatrix} 0 & E \\ -E & 0 \end{pmatrix}, \quad \begin{pmatrix} E & S \\ 0 & E \end{pmatrix}.$$

Corollary 2. *The group $\mathrm{Sp}(n,R)$ is generated by the matrices*

$$\begin{pmatrix} E & S \\ 0 & E \end{pmatrix}, \quad \begin{pmatrix} E & 0 \\ S & E \end{pmatrix}.$$

Corollary 3. *The natural homomorphism*

$$\mathrm{Sp}(n,\mathbb{Z}) \longrightarrow \mathrm{Sp}(n,\mathbb{Z}/q\mathbb{Z})$$

is surjective.

For the proof, we have to use the formulae

$$\begin{pmatrix} A & B \\ 0 & D \end{pmatrix} = \begin{pmatrix} U^t & 0 \\ 0 & U^{-1} \end{pmatrix} \begin{pmatrix} E & S \\ 0 & E \end{pmatrix} \quad \text{with } U = A^t, \quad S = S^t = A^{-1}B.$$

Obviously, Formula 5.3 holds for the matrices E instead of 1 and U instead of a. We obtain the result that for symmetric U,

$$\begin{pmatrix} U^t & 0 \\ 0 & U^{-1} \end{pmatrix}$$

can be expressed by

$$\begin{pmatrix} E & S \\ 0 & E \end{pmatrix} \quad \text{and} \quad \begin{pmatrix} 0 & E \\ -E & 0 \end{pmatrix}.$$

Because of Lemma 5.8, this is true for arbitrary $U \in \mathrm{GL}(n,\mathbb{Z})$.

Proof of Proposition 5.9. We use induction on n and assume that the proposition has been proved for $n-1$ in place of n. The proof rests on a simple variant of Gauss's lemma.

5.10 Lemma. *For each vector* $g \in \mathbb{Z}^{2n}$, *there exists a matrix* $M \in \mathrm{Sp}(n, \mathbb{Z})$ *such that*

$$Mg = \begin{pmatrix} a_1 \\ 0 \\ \vdots \\ 0 \end{pmatrix}.$$

Furthermore, M can be found in the subgroup which is generated by the special matrices "$B = 0$ or $C = 0$".

The proof is similar to that of Gauss's lemma. Hence we shall keep it short. The formulae

$$\begin{pmatrix} U^t & 0 \\ 0 & U^{-1} \end{pmatrix} \begin{pmatrix} a \\ b \end{pmatrix} = \begin{pmatrix} U^t a \\ U^{-1} b \end{pmatrix}, \quad \begin{pmatrix} 0 & E \\ -E & 0 \end{pmatrix} \begin{pmatrix} a \\ b \end{pmatrix} = \begin{pmatrix} b \\ -a \end{pmatrix}$$

show that we can find an M such that the first component of Mg is different from 0. We choose M such that the first entry of Mg is different from 0 and has minimal modulus. If we replace M by

$$\begin{pmatrix} E & 0 \\ S & E \end{pmatrix} \begin{pmatrix} U^t & 0 \\ 0 & U^{-1} \end{pmatrix} M,$$

the claim can be obtained by means of the Euclidean algorithm as in the case of the linear group. $\qquad\square$

Now we can prove Proposition 5.9. Let $M \in \mathrm{Sp}(n, \mathbb{Z})$ be given. By the above lemma, there exists a matrix N in the group H_n which is generated by the special matrices, such that

$$NM = \begin{pmatrix} 1 & * & * & * \\ 0 & A_1 & * & B_1 \\ 0 & * & * & * \\ 0 & C_1 & * & D_1 \end{pmatrix}.$$

We can easily check that the matrix $\begin{pmatrix} A_1 & B_1 \\ C_1 & D_1 \end{pmatrix}$ is symplectic. By the induction hypothesis, this matrix is contained in H_{n-1}. We obtain the result that

$$\tilde{M}_1 = \begin{pmatrix} 1 & 0 & 0 & 0 \\ 0 & A_1 & 0 & B_1 \\ 0 & 0 & 0 & 0 \\ 0 & C_1 & 0 & D_1 \end{pmatrix}$$

is contained in H_n. A simple calculation shows that

$$\tilde{M}_1^{-1} NM = \begin{pmatrix} \tilde{A} & \tilde{B} \\ \tilde{C} & \tilde{D} \end{pmatrix}, \quad \tilde{A} = \begin{pmatrix} 1 & * \\ 0 & E \end{pmatrix}, \quad \tilde{C} = \begin{pmatrix} 0 & * \\ 0 & 0 \end{pmatrix}.$$

From the symplectic relations $\tilde{A}^t \tilde{C} = \tilde{C}^t \tilde{A}$, $\tilde{A} \tilde{D}^t = E$, we get $\tilde{C} = 0$ and hence $\tilde{M}_1^{-1} NM \in H_n$. This implies $M \in H_n$. $\qquad\square$

Congruence Subgroups of Level Two

It is difficult to find generators for arbitrary congruence subgroups. An exceptional case is the principal congruence subgroup of level two. In the first volume ([FB], Appendix to Sect. VI, Proposition A.6), we showed that the elliptic principal congruence subgroup of level two $SL(2, \mathbb{Z})[2] = \Gamma_1[2]$ can be generated by the matrices

$$\begin{pmatrix} 1 & 2 \\ 0 & 1 \end{pmatrix}, \quad \begin{pmatrix} 1 & 0 \\ 2 & 1 \end{pmatrix}, \quad \begin{pmatrix} -1 & 0 \\ 0 & -1 \end{pmatrix}.$$

The proof used properties of the fundamental domain. In the exercises for this section, we explain an algebraic proof.

The technique of embedded $SL(2)$ allows to generalize this result to arbitrary $SL(n)$. The restrictions of the embeddings $\alpha_{\mu\nu}$ define embeddings

$$\alpha_{\mu\nu} : SL(2, \mathbb{Z})[2] \longrightarrow SL(n, \mathbb{Z})[2].$$

We denote by

$$G_n[2] := \langle \alpha_{\mu\nu}(SL(2, \mathbb{Z})[2]), \quad 1 \leq \mu < \nu \leq n \rangle$$

the subgroup which is generated by their images.

5.11 Lemma. *Let $g \in \mathbb{Z}^n$ be a column whose first component g_1 is odd, and all other components of which are even. Then there exists*

$$U \in G_n[2] \text{ with } Ug = \begin{pmatrix} a_1 \\ 0 \\ \vdots \\ 0 \end{pmatrix}.$$

Proof. It is sufficient to prove this for $n = 2$, since then one can use embedded $SL(2)$ matrices to annul $g_n, \ldots, g_2$ successively. In the case $n = 2$, we can use $G_2[2] = SL(2, \mathbb{Z})[2]$. We can assume that that g_1 and g_2 are coprime. Hence they can be completed to a matrix from $SL(2, \mathbb{Z})$,

$$U = \begin{pmatrix} g_1 & g_3 \\ g_2 & g_4 \end{pmatrix} \in SL(2, \mathbb{Z}).$$

Because of the determinant condition, g_4 has to be odd. If g_3 is also odd, we make the replacement

$$U \longmapsto U \begin{pmatrix} 1 & 1 \\ 0 & 1 \end{pmatrix}.$$

After that, we can assume $U \in SL(2, \mathbb{Z})[2]$, and U^{-1} has the desired property.
□

Now the same proof as in the case of the full modular group can be used to obtain the following result.

5.12 Proposition.   *The group* SL$(n, \mathbb{Z})[2]$ *is generated by the embedded* SL$(2, \mathbb{Z})[2]$.

Corollary. *The group* SL$(n, \mathbb{Z})[2]$ *is generated by (upper and lower) triangular matrices.*

An analogous result in the case of the symplectic group states the following.

5.13 Proposition.   *The principal congruence subgroup of level two,* $\Gamma_n[2]$, *is generated by the special matrices*

$$ M = \begin{pmatrix} A & B \\ C & D \end{pmatrix}, \quad B = 0 \text{ or } C = 0. $$

The same proof as in the case of the full modular group works if we use the following variant of Gauss's lemma.

5.14 Lemma.   *Let* $H_n[2]$ *be the subgroup of* $\Gamma_n[2]$ *which is generated by the special matrices*

$$ M = \begin{pmatrix} A & B \\ C & D \end{pmatrix}, \quad B = 0 \text{ or } C = 0. $$

For each column $g \in \mathbb{Z}^{2n}$ *whose first column is odd but the other columns of which are even, there exists a matrix* $M \in H_n[2]$ *with*

$$ Mg = \begin{pmatrix} a_1 \\ 0 \\ \vdots \\ 0 \end{pmatrix}. $$

Proof. We decompose g:

$$ g = \begin{pmatrix} a \\ b \end{pmatrix}, \quad a \in \mathbb{Z}^n, \ b \in \mathbb{Z}^n. $$

First we find $U \in$ SL$(n, \mathbb{Z})[2]$ with the property

$$ Ua = \begin{pmatrix} a_1 \\ 0 \\ \vdots \\ 0 \end{pmatrix}. $$

Because

$$ \begin{pmatrix} U^t & 0 \\ 0 & U^{-1} \end{pmatrix} \in H_n[2], $$

we can assume that

$$a = \begin{pmatrix} a_1 \\ 0 \\ \vdots \\ 0 \end{pmatrix} \quad (a_1 \text{ odd}).$$

Now we use the embedded SL(2) matrix

$$\alpha_{1,n+1} \begin{pmatrix} a & b \\ c & d \end{pmatrix} = \begin{pmatrix} a & 0 & b & 0 \\ 0 & E^{(n-1)} & 0 & 0 \\ c & 0 & d & 0 \\ 0 & 0 & 0 & E^{(n-1)} \end{pmatrix}$$

to obtain the result that besides $a_2 = \ldots = a_n = 0$, also $b_1 = 0$.

Now we consider, for given $\nu \in \{2, \ldots, n\}$, a special symmetric matrix S. This matrix includes the values

$$s_{11} = s_{1\nu} = s_{\nu 1} = 1.$$

All other entries are 0. The transformations

$$\begin{pmatrix} a \\ b \end{pmatrix} \longmapsto \begin{pmatrix} E & 2S \\ 0 & E \end{pmatrix} \begin{pmatrix} a \\ b \end{pmatrix} \quad \text{and} \quad \begin{pmatrix} E & 0 \\ 2S & E \end{pmatrix} \begin{pmatrix} a \\ b \end{pmatrix}$$

have the effect

$$\begin{matrix} a_1 \mapsto a_1 + 2b_\nu, & & a_1 \mapsto a_1, \\ b_\nu \mapsto b_\nu, & \text{and} & b_\nu \mapsto b_\nu + 2a_1, \end{matrix}$$

respectively. All other components are unchanged. The same transformation is obtained from

$$\begin{pmatrix} a_1 \\ b_\nu \end{pmatrix} \longmapsto \begin{pmatrix} 1 & 2 \\ 0 & 1 \end{pmatrix} \begin{pmatrix} a_1 \\ b_\nu \end{pmatrix} \quad \text{and} \quad \begin{pmatrix} 1 & 0 \\ 2 & 1 \end{pmatrix} \begin{pmatrix} a_1 \\ b_\nu \end{pmatrix},$$

respectively. By means of Lemma 5.11 (applied in the case $n = 2$), we obtain $b = 0$. $\qquad\square$

Exercises for Sect. VII.5

1. Construct a matrix $M \in \Gamma_2$ with first column $(2, 3, 5, 7)$.

2. Construct a matrix from $\Gamma_2[2]$ with first row $(3, 10, 14, 22)$.

3. Write $\begin{pmatrix} 3 & 7 \\ 2 & 5 \end{pmatrix}$ as a product of symmetric integral matrices.

4. Show that the homomorphism
$$GL(n, \mathbb{Z}) \longrightarrow GL(n, \mathbb{Z}/q\mathbb{Z}), \quad q > 2,$$
   is not surjective.

6. Computation of Some Indices

We start with the computation of the order of $GL(n, k)$, where k is a field of p elements. The first column of A can be an arbitrary vector different from 0. There are $p^n - 1$ such vectors. The second column is not allowed to be a multiple of the first column. There are $p^n - p$ possibilities for it. The third column has to avoid the subspace which is generated by the first two columns. If the first two columns are given, there remain $p^n - p^2$ possibilities, and so on.

6.1 Remark. *Let p be a prime. We have*

$$\# GL(n, \mathbb{Z}/p\mathbb{Z}) = \prod_{\nu=0}^{n-1} (p^n - p^\nu).$$

The homomorphism

$$GL(n, \mathbb{Z}/p\mathbb{Z}) \longmapsto (\mathbb{Z}/p\mathbb{Z})^*$$

is surjective; its kernel is $SL(n, \mathbb{Z}/p\mathbb{Z})$. We obtain the following result.

6.2 Remark. *For prime numbers p, we have*

$$[SL(n, \mathbb{Z}) : SL(n, \mathbb{Z})[p]] = \# SL(n, \mathbb{Z}/p\mathbb{Z}) = \frac{1}{p-1} \prod_{\nu=0}^{n-1} (p^n - p^\nu).$$

Similar considerations apply for the symplectic group. We know that the symplectic group $Sp(n, \mathbb{Z}/p\mathbb{Z})$ acts transitively on $(\mathbb{Z}/p\mathbb{Z})^{2n} - \{0\}$. Let

$$P_n \subset Sp(n, \mathbb{Z}/p\mathbb{Z})$$

be the subgroup which stabilizes the first unit vector. We have

$$\# Sp(n, \mathbb{Z}/p\mathbb{Z}) = (p^{2n} - 1)\# P_n.$$

We know that

$$P_n \longrightarrow Sp(n - 1, \mathbb{Z}/p\mathbb{Z}),$$

$$\begin{pmatrix} 1 & & & \\ & A_1 & & B_1 \\ & & 1 & \\ & C_1 & & D_1 \end{pmatrix} \longmapsto \begin{pmatrix} A_1 & B_1 \\ C_1 & D_1 \end{pmatrix}$$

is a surjective homomorphism. We denote its kernel by K_n. We have

$$\# P_n = \# Sp(n - 1, \mathbb{Z}/p\mathbb{Z}) \cdot \# K_n.$$

The elements of K_n are of the form

$$\begin{pmatrix} A & B \\ 0 & D \end{pmatrix}, \quad A = \begin{pmatrix} 1 & * \\ 0 & A_1 \end{pmatrix}, \quad B = \begin{pmatrix} * & * \\ * & 0 \end{pmatrix},$$

where the symplectic relations $D = A^{t-1}$ and $AB^t = BA^t$ must hold. This shows that

$$\#K_n = p^{2n-1}.$$

Now we have proved the inductive formula

$$\# \operatorname{Sp}(n, \mathbb{Z}/p\mathbb{Z}) = p^{2n-1}(p^{2n} - 1)\# \operatorname{Sp}(n - 1, \mathbb{Z}/p\mathbb{Z}).$$

Induction on n gives the following lemma.

6.3 Lemma.   *Let p be a prime. We have*

$$[\Gamma_n : \Gamma_n[p]] = \# \operatorname{Sp}(n, \mathbb{Z}/p\mathbb{Z}) = p^{n(2n+1)} \prod_{\nu=1}^{n} \left(1 - \frac{1}{p^{2\nu}}\right).$$

Now we consider the case of a power of a prime $q = p^m$. Again we start with the case of the general linear group. The idea is to study the natural homomorphism

$$\operatorname{GL}(n, \mathbb{Z}/p^m\mathbb{Z}) \longrightarrow \operatorname{GL}(n, \mathbb{Z}/p^{m-1}\mathbb{Z}), \quad m > 1.$$

This consists of all matrices of the form

$$E + p^{m-1}A, \quad A \in (\mathbb{Z}/p^m\mathbb{Z})^{(n,n)}.$$

Because $m > 1$, we have, modulo p^m,

$$(E + p^{m-1}A)(E + p^{m-1}B) = E + p^{m-1}(A + B).$$

So, the kernel is abelian. We also observe that $E + p^m A$ is invertible for all $A \in \mathbb{Z}/p^m\mathbb{Z}$. The inverse is obtained by replacing A by $-A$. The condition

$$p^{m-1}a = p^{m-1}b \quad \text{for } a, b \in \mathbb{Z}/p^m\mathbb{Z}$$

means simply that the images of a and b in $\mathbb{Z}/p\mathbb{Z}$ agree. Hence the groups

$$p^{m-1}(\mathbb{Z}/p^m\mathbb{Z}) \text{ and } \mathbb{Z}/p\mathbb{Z}$$

are isomorphic. We obtain the following lemma.

6.4 Lemma. *Assume $m > 1$. The kernel of the natural homomorphism*

$$\mathrm{GL}(n, \mathbb{Z}/p^m\mathbb{Z}) \longrightarrow \mathrm{GL}(n, \mathbb{Z}/p^{m-1}\mathbb{Z})$$

and the additive group

$$(\mathbb{Z}/p\mathbb{Z}))^{(n,n)}$$

are isomorphic.

Supplement 1. *The kernel of the natural homomorphism*

$$\mathrm{SL}(n, \mathbb{Z}/p^m\mathbb{Z}) \longrightarrow \mathrm{SL}(n, \mathbb{Z}/p^{m-1}\mathbb{Z})$$

is isomorphic to the additive group

$$A \in (\mathbb{Z}/p^m\mathbb{Z})^{(n,n)} \text{ with } \mathrm{trace}(A) = 0.$$

Supplement 2. *The kernel of the natural homomorphism*

$$\mathrm{Sp}(n, \mathbb{Z}/p^m\mathbb{Z}) \longrightarrow \mathrm{Sp}(n, \mathbb{Z}/p^{m-1}\mathbb{Z})$$

is isomorphic to the set of all symmetric matrices

$$N = N^t \in \mathbb{Z}/p^m\mathbb{Z}^{(2n,2n)}.$$

Only the supplements have to be proved. We restrict ourselves to the second supplement. We have to consider, in $\mathbb{Z}/p^{m-1}\mathbb{Z}^{(2n,2n)}$, the equation

$$(E + p^{m-1}M)^t I(E + p^{m-1}M) = I.$$

This means

$$p^{m-1}M^t I = -p^{m-1}IM.$$

Because $I^t = -I$, it is equivalent to the symmetry of $N = IM$. $\qquad\square$

If $q = q_1 q_2$ is the product of two coprime natural numbers, then, by the Chinese remainder theorem, a number $a \in \mathbb{Z}$ is determined modulo q by its remainders modulo q_1 and q_2. This means that the natural homomorphism

$$\mathbb{Z}/q\mathbb{Z} \longrightarrow \mathbb{Z}/q_1\mathbb{Z} \times \mathbb{Z}/q_2\mathbb{Z}$$

is an isomorphism. This gives an isomorphism

$$\mathrm{GL}(n, \mathbb{Z}/q\mathbb{Z}) \xrightarrow{\sim} \mathrm{GL}(n, \mathbb{Z}/q_1\mathbb{Z}) \times \mathrm{GL}(n, \mathbb{Z}/q_2\mathbb{Z}),$$

and analogously for the special linear and the symplectic group. Now we obtain the general index formulae.

6.5 Proposition.   *We have*

a) $\left[\mathrm{SL}(n,\mathbb{Z}) : \mathrm{SL}(n,\mathbb{Z})[q] \right] = \#\,\mathrm{SL}(n,\mathbb{Z}/q\mathbb{Z}) = q^{n^2-1} \prod_{p|q} \prod_{\nu=2}^{n} \left(1 - \frac{1}{p^\nu} \right),$

b) $\qquad\qquad \left[\Gamma_n : \Gamma_n[q] \right] = \#\,\mathrm{Sp}(n,\mathbb{Z}/q\mathbb{Z}) = q^{n(2n+1)} \prod_{p|q} \prod_{\nu=1}^{n} \left(1 - \frac{1}{p^{2\nu}} \right).$

Exercises for Sect. VII.6

1. A congruence subgroup is defined by
    $$\Gamma_{n,0}[q] = \{ M \in \Gamma_n; \quad C \equiv 0 \bmod q \}.$$
    Compute its index in Γ_n.

2. A congruence subgroup is defined by
    $$\Gamma_{n,1}[q] = \{ M \in \Gamma_n; \quad C \equiv 0, \ A \equiv D \equiv E \bmod q \}.$$
    Compute its index in Γ_n.

3. Show that the index of $\Gamma_2[2]$ in the full modular group is 720. Is there a full permutation group S_n of the same order?

4. If you have a computer and, for example, the program GAP, show by computation that the groups S_6 and $\mathrm{Sp}(2,\mathbb{Z}/2\mathbb{Z})$ are isomorphic.

7. Theta series

The theta functions, as they occurred in the theory of abelian functions (Definition VI.8.1),

$$\vartheta \begin{bmatrix} a \\ b \end{bmatrix} (Z,z) := \sum_{g\in\mathbb{Z}^n} e^{\pi \mathrm{i} \{ Z[g+a]+2(g+a)^t(z+b) \}},$$

are functions of two variables Z, z. In Chap. VI, we studied them in terms of the variable z for fixed Z. Now we are interested in Z as a variable. It turns out that these series are now of special interest if z is rational, since then they turn out to be modular forms. The most interesting case is $z = 0$, and $\vartheta \begin{bmatrix} a \\ b \end{bmatrix} (Z, 0)$ is called a theta nullwert. We shall try to keep our investigations of theta nullwerte independent of Chap. VI.

The simplest "theta nullwert" in the case $n = 1$ was

$$\vartheta(z) = \sum_{n=-\infty}^{\infty} e^{\pi i n^2 z}.$$

Its obvious generalization to the case $n > 1$ is

$$\vartheta(Z) = \vartheta^{(n)}(Z) = \sum_{g \in \mathbb{Z}^n} e^{\pi i Z[g]}.$$

In the following, we use the notation

$$e(a) = e^{\pi i a}.$$

In the case $n = 1$, it was necessary to consider besides $\vartheta(Z)$ also the conjugate forms

$$\widetilde{\vartheta}(z) = \sum (-1)^n e(n^2 z),$$
$$\widetilde{\widetilde{\vartheta}}(z) = \sum e((n + 1/2)^2 z).$$

Using the notation

$$\vartheta \begin{bmatrix} a \\ b \end{bmatrix} (z) = \sum e((n + a/2)^2 z + b(n + a/2)),$$

we obtain

$$\vartheta = \vartheta \begin{bmatrix} 0 \\ 0 \end{bmatrix}, \quad \widetilde{\vartheta} = \vartheta \begin{bmatrix} 0 \\ 1 \end{bmatrix}, \quad \vartheta \begin{bmatrix} 1 \\ 0 \end{bmatrix}.$$

This suggests how the *satellites* of ϑ have to be generalized:

$$\vartheta \begin{bmatrix} a \\ b \end{bmatrix} (Z) = \sum_{g \in \mathbb{Z}^n} e(Z[g + a/2] + b^t(g + a/2)).$$

Here a, b could be arbitrary vectors from $\mathbb{C}^n$. But for us, only integral a, b are of interest. This is the reason for the deviation from the notation in Definition VI.8.1,

$$\vartheta \begin{bmatrix} a \\ b \end{bmatrix} (Z) = \vartheta \begin{bmatrix} a/2 \\ b/2 \end{bmatrix} (Z, 0).$$

7.1 Lemma.   *The series*

$$\vartheta \begin{bmatrix} a \\ b \end{bmatrix} (Z) = \sum_{g \in \mathbb{Z}^n} e\big(Z[g + a/2] + b^t(g + a/2)\big)$$

converges normally in $\mathbb{H}_n$ *for* $a, b \in \mathbb{Z}^n$ *and defines an analytic function there. Moreover, it converges uniformly in domains of the kind* $Y \geq \delta E$, $\delta > 0$, *and defines a bounded function there. In the case*

$$a^t b \equiv 1 \bmod 2,$$

it is identically zero. In all other cases, it does not vanish identically.

Proof. The statements about convergence are simple. In principle, they already have been proved in the first volume. We shall repeat the argument briefly. In the domain $Y \geq \delta E$, we can easily verify the estimate

$$\big|\big(Z[g + a/2] + b^t(g + a/2)\big)\big| \leq e^{-\varepsilon(g_1^2 + \cdots g_n^2)},$$

with a suitable positive number $\varepsilon = \varepsilon(\delta, a, b)$. The statements about the convergence follow from this.

We now study the vanishing of the theta series. For this, we rewrite these series as Fourier series. This is possible, since

$$Z[g] = g^t Z g = \sigma(gg^t Z).$$

We obtain

$$\vartheta \begin{bmatrix} a \\ b \end{bmatrix} = \sum_{T = T^t} a(T) e(\sigma(TZ)/4),$$

with

$$a(T) = \sum_{\substack{g \text{ integral} \\ (2g+a)(2g+b)^t = T}} e(b^t(g + a/2)).$$

If we replace g by $-g - a$, the sums change by a factor $(-1)^{a^t b}$. Hence $a(T) = 0$ for odd $a^t b$. We now compute $a(T)$ for a special T. This contains only entries $0, 1$, and it suffices $T \equiv aa^t \bmod 2$. The equation $(2g + a)(2g + b)^t = T$ has only two solutions, and they satisfy $g \equiv 0$ and $g \equiv -a \bmod 2$. We get $a(T) = \pm 2$ for even $a^t b$. This proves Lemma 7.1.

Up to a sign, the theta series depend on a, b only modulo 2, since it follows from

$$\tilde{a} \equiv a \bmod 2, \quad \tilde{b} \equiv b \bmod 2$$

that

$$\vartheta \begin{bmatrix} \tilde{a} \\ \tilde{b} \end{bmatrix} = (-1)^{a^t(\tilde{b}-b)/2} \vartheta \begin{bmatrix} a \\ b \end{bmatrix},$$

as one can easily verify. In this context, the pair

$$\mathfrak{m} := \begin{pmatrix} a \\ b \end{pmatrix} \in \mathbb{Z}^{2n}$$

is called a *theta characteristic*. This is frequently normalized in such a way that the components of $\mathfrak{m}$ are contained in $\{0, 1\}$. A theta characteristic is called *even* iff

$$a^t b \equiv 0 \bmod 2,$$

and odd otherwise. In the case $n = 1$, there are three even characteristics mod 2,

$$\mathfrak{m} = \begin{pmatrix} 0 \\ 0 \end{pmatrix}, \quad \begin{pmatrix} 0 \\ 1 \end{pmatrix}, \quad \begin{pmatrix} 1 \\ 0 \end{pmatrix}.$$

The following lemma can easily be shown by induction on n.

7.2 Lemma. *Modulo 2, there are*

$$2^{n-1}(2^n + 1)$$

even characteristics, and hence 3 in the case $n = 1$, 10 in the case $n = 2$, and 36 in the case $n = 3$.

We have to investigate the transformation behavior of these theta series under modular substitutions. This theta transformation formalism goes back to the nineteenth century.

7.3 Proposition. *Let $M \in \Gamma_n = \mathrm{Sp}(n, \mathbb{Z})$ be a modular substitution of degree n. For each characteristic*

$$\mathfrak{m} \in \{0, 1\}^{2n},$$

there exists a characteristic

$$M\{\mathfrak{m}\} \in \{0, 1\}^{2n}$$

and an eighth root of unity $v(M, \mathfrak{m})$ such that the transformation formula

$$\vartheta[M\{\mathfrak{m}\}](MZ) = v(M, \mathfrak{m}) \det(CZ + D)^{1/2} \vartheta[\mathfrak{m}](Z)$$

is valid.

Of course, $v(M, \mathfrak{m})$ depends on the choice of the square root of $\det(CZ + D)$.

It is sufficient to prove Proposition 7.3 for generators of the modular group. Hence Proposition 7.3 follows from the next statement.

7.4 Lemma. *We have*

1)
$$\vartheta\begin{bmatrix} a \\ b \end{bmatrix}(Z+S) = e^{\pi i S[a]/4}\vartheta\begin{bmatrix} a \\ b+Sa+S_0 \end{bmatrix}(Z),$$

where S_0 is the column built from the diagonal of S.

2)
$$\vartheta\begin{bmatrix} a \\ b \end{bmatrix}(Z[U]) = \vartheta\begin{bmatrix} Ua \\ U^{t-1}b \end{bmatrix}(Z).$$

3)
$$\vartheta\begin{bmatrix} a \\ b \end{bmatrix}(-Z^{-1}) = e^{\pi i a^t b/2}\vartheta\begin{bmatrix} b \\ -a \end{bmatrix}(Z).$$

Proof. The proof of the first two formulae is very simple:

1) We observe that

$$S\left[g+\frac{1}{2}a\right] = S[g] + (Sa)^t g + \frac{1}{4}S[a]$$

and note the congruence

$$S[g] = \sum_{i=1}^{n} s_{ij}g_i + 2\sum_{i<j} s_{ij}g_i g_j \equiv S_0^t g \bmod 2.$$

2) In the series of $\vartheta[m](Z[U])$, we perform the transformation $g \mapsto U^{-1}$ of the summation variable.

3) It is sufficient to prove the formula for purely imaginary Z, and hence for matrices of the form $Z = Sz$, S real, $z \in \mathbb{H}$. It is then a consequence of the theta transformation formula for theta series with respect to quadratic forms on the usual upper half-plane ([FB], Theorem VI.4.7). □

One can work out an explicit formula for $m \mapsto M\{m\}$.

7.5 Lemma. *Let* $m = \begin{pmatrix} a \\ b \end{pmatrix} \in \{0,1\}^{2n}$ *and* $M \in \mathrm{Sp}(n,\mathbb{Z})$. *If we define* $M\{m\} \in \{0,1\}^{2n}$ *by*

$$M\left\{\begin{matrix} a \\ b \end{matrix}\right\} \equiv M^{t-1}\begin{pmatrix} a \\ b \end{pmatrix} + \begin{pmatrix} (CD^t)_0 \\ (AB^t)_0 \end{pmatrix} \bmod 2,$$

then Proposition 7.3 holds.

For the generators of the modular group, this follows from Lemma 7.4. For this reason, it is sufficient to show the following.

7.6 Lemma. *If we define, for* $\mathfrak{m} \in (\mathbb{Z}/2\mathbb{Z})^{2n}$ *and* $M \in \mathrm{Sp}(n, \mathbb{Z})$,

$$M\{\mathfrak{m}\} = M\left\{\begin{matrix} a \\ b \end{matrix}\right\} \equiv M^{t-1}\begin{pmatrix} a \\ b \end{pmatrix} + \begin{pmatrix} (CD^t)_0 \\ (AB^t)_0 \end{pmatrix} \in (\mathbb{Z}/2\mathbb{Z})^{2n},$$

we have

$$(MN)\{\mathfrak{m}\} = M\{N\{\mathfrak{m}\}\}.$$

Proof. It is sufficient to assume that M is an arbitrary modular matrix and N a generator. For example, in the case $N = \begin{pmatrix} E & S \\ 0 & E \end{pmatrix}$, the statement is equivalent to

$$CS_0 \equiv (C^t CS)_0 \quad \text{and} \quad AS_0 = (A^t AS)_0.$$

This follows from $x^2 = x \pmod 2$. $\qquad\square$

7.7 Lemma. *The set*

$$\Gamma_{n,\vartheta} := \{M \in \Gamma_n, \quad CD^t \text{ and } AB^t \text{ have even diagonal}\}$$

is a congruence subgroup.

We call $\Gamma_{n,\vartheta}$ the *theta group of degree* n. It generalizes the theta group introduced in [FB], Appendix to Sect. VI.5.

Proof of Lemma 7.7. The set $\Gamma_{n,\vartheta}$ is characterized by the condition $M\{0\} = 0$. The claim follows from Lemma 7.6. Moreover, $\Gamma_{n,\vartheta} \supset \Gamma_n[2]$. $\qquad\square$

The next proposition now follows from Proposition 7.3.

7.8 Proposition. *The theta series*

$$\vartheta(Z) = \vartheta[0](Z) = \sum_{g \in \mathbb{Z}^n} e^{\pi i Z[g]}$$

is a modular form of weight $1/2$ *with respect to a certain multiplier system* v_ϑ *for the theta group* $\Gamma_{n,\vartheta}$.

The multiplier system

$$v_\vartheta : \Gamma_{n,\vartheta} \longrightarrow \mathbb{C}^{\bullet}$$

is called the *theta multiplier system*. It is the most important multiplier system of nonintegral weight.

7.9 Lemma. *Every even theta characteristic* $\mathfrak{m} \in \{0,1\}^{2n}$ *can be written in the form* $\mathfrak{m} = M\{0\}$. *The stabilizer of* $\mathfrak{m}$ *in the full modular group*

$$\Gamma_n(\mathfrak{m}) := \{M \in \Gamma_n, \quad M\{\mathfrak{m}\} = \mathfrak{m}\}$$

is conjugate to the theta group:

$$\Gamma(\mathfrak{m}) = M\Gamma_{n,\vartheta}M^{-1} \quad (M\{0\} = \mathfrak{m}).$$

We have

$$\vartheta[\mathfrak{m}] \in [\Gamma_n(\mathfrak{m}), 1/2, v^{\mathfrak{m}}].$$

Here $v^{\mathfrak{m}}$ *denotes the conjugate multiplier system of* v_ϑ.

Proof. Only the statement about transitivity remains to be proved. This statement says that for two

$$a, b \in \mathbb{Z}^n$$

with $a^t b \equiv 0 \bmod 2$, there exists a modular matrix $M \in \Gamma_n$ with the property

$$a \equiv (CD^t)_0, \quad b \equiv (AB^t)_0.$$

We argue by induction on n and decompose

$$a = \begin{pmatrix} a_1 \\ a_2 \end{pmatrix}, \ a_1 \in \mathbb{Z}, \ a_2 \in \mathbb{Z}^{n-1}, \quad \text{and correspondingly } b = \begin{pmatrix} b_1 \\ b_2 \end{pmatrix}.$$

In the case $a_1 b_1 \equiv 0$, the characteristic $\begin{pmatrix} a_2 \\ b_2 \end{pmatrix}$ is even. Using the induction hypothesis, we reduce the problem to the case $a_2 = b_2 = 0$. Now it is enough to apply a suitable embedded $\mathrm{SL}(2)$ matrix.

In the case $a_1 b_1 \equiv 1$, there must exist a further index ν such that $a_\nu b_\nu$ is odd. Now we can easily construct a translation matrix such that

$$\begin{pmatrix} E & S \\ 0 & E \end{pmatrix} \begin{Bmatrix} a \\ b \end{Bmatrix}$$

suffices for the assumption of the first case.   □

The product of all theta series is of special importance. The next proposition follows from Proposition 7.3.

7.10 Proposition. *The product*

$$\Delta^{(n)} = \prod \vartheta \begin{bmatrix} a \\ b \end{bmatrix}$$

of all theta series is a modular form of weight $2^{n-2}(2^n+1)$ *for the full modular group with respect to some multiplier system* $v^{(n)}$:

$$\Delta^{(n)} \in [\Gamma_n, 2^{n-2}(2^n+1), v^{(n)}].$$

In the case $n \geq 2$, *the weight is integral. Hence, in this case,* $v^{(n)}$ *is a character on* Γ_n.

It is possible to show that this character is trivial for $n \geq 3$.

Exercises for Sect. VII.7

1. Verify by means of generators that the character of $\Delta^{(3)}$ is trivial.

2. What is the connection between $\Delta^{(1)}$ and the discriminant Δ from the theory of elliptic functions?

3. Show that $\vartheta(Z)$ is positive for real imaginary Z and use Lemma 7.9 to prove that the theta series with even characteristics do not vanish.

8. Group-Theoretic Considerations

We consider the set of all left cosets

$$\Gamma_n / \Gamma_{n,\vartheta} = \{M\Gamma_{n,\vartheta}, \quad M \in \Gamma_n\}.$$

By the definition of $\Gamma_{n,\vartheta}$, the map

$$\Gamma / \Gamma_{n,\vartheta} \longrightarrow \left\{ \mathfrak{m} = \begin{pmatrix} a \\ b \end{pmatrix} \in \{0,1\}^n, \quad a^t b \text{ even} \right\},$$

$$M\Gamma_{n,\vartheta} \longrightarrow M\{0\},$$

is well defined and injective. By Lemma 7.9, it is also surjective. This shows the following.

8.1 Remark. *The theta group has the index $2^{n-1}(2^n + 1)$ in the full modular group.*

The theta group is not normal in Γ_n. On the contrary, the following is true.

8.2 Remark. *The $2^{n-1}(2^n + 1)$ conjugate groups*

$$\Gamma_n(\mathfrak{m}), \quad \mathfrak{m} \in \{0,1\} \text{ even,}$$

of the theta group are pairwise distinct.

Proof. It suffices to show that $\Gamma_n(0)$ and $\Gamma_n(\mathfrak{m})$ are different. Therefore we have to show that if $\mathfrak{m} \in \mathbb{Z}^{2n}$ and if

$$M\mathfrak{m} \equiv \mathfrak{m} \bmod 2 \text{ for all } M \in \Gamma_{n,\vartheta},$$

then $\mathfrak{m} \equiv 0 \bmod 2$. In the case $n = 1$, this is obvious; in the case $n > 1$, we consider embedded SL(2).   □

Some Exceptional Isomorphisms

We now investigate the group $\mathrm{Sp}(n, \mathbb{Z}/2\mathbb{Z})$ for small n. We make use of the action of $\mathrm{Sp}(n, \mathbb{Z}/2\mathbb{Z})$ on $(\mathbb{Z}/2\mathbb{Z})^{2n}$. An element $\mathfrak{m} = \left(\begin{smallmatrix} a \\ b \end{smallmatrix}\right)$ of $(\mathbb{Z}/2\mathbb{Z})^{2n}$ is said to be even, if $a^t b = 0$ (in $\mathbb{Z}/2\mathbb{Z}$).

Let M be an element which fixes all $\mathfrak{m} \in (\mathbb{Z}/2\mathbb{Z})^{2n}$. Then M fixes 0:

$$M\{0\} = M^{t^{-1}} \cdot \mathfrak{m}.$$

It follows that M is the unit matrix. But we have further results, as below.

8.3 Lemma.
1) If M is an element which fixes all even $\mathfrak{m} \in (\mathbb{Z}/2\mathbb{Z})^{2n}$, $M\{\mathfrak{m}\} = \mathfrak{m}$, then M is the unit matrix.

2) Let $n > 1$. If M is an element which fixes all odd $\mathfrak{m} \in (\mathbb{Z}/2\mathbb{Z})^{2n}$, then M is the unit matrix.

Proof. 1) If M fixes all even $\mathfrak{m}$, then M fixes 0. Because $M\{0\} = M^{t^{-1}} \cdot \mathfrak{m}$, the sum of two even characteristics is fixed. But we can easily show that each odd element of $(\mathbb{Z}/2\mathbb{Z})^{2n}$ is the sum of two even elements. In the case $n = 1$, this follows from the formula

$$\begin{pmatrix} 1 \\ 1 \end{pmatrix} = \begin{pmatrix} 1 \\ 0 \end{pmatrix} + \begin{pmatrix} 0 \\ 1 \end{pmatrix}.$$

The general case can be reduced to this case.

2) So, let $n > 1$. There then exist two odd $\mathfrak{m}$, $\mathfrak{n}$ such that $\mathfrak{m} + \mathfrak{n}$ is odd. The general rule

$$M\{\mathfrak{m} + \mathfrak{n}\} = \mathfrak{m} + \mathfrak{n} + M\{0\},$$

together with the assumption

$$M\{\mathfrak{m}\} = \mathfrak{m}, \quad M\{\mathfrak{n}\} = \mathfrak{m}, \quad M\{\mathfrak{m} + \mathfrak{n}\} = \mathfrak{m} + \mathfrak{n},$$

shows that $M\{0\} = 0$. Again it follows that the sum of two odd elements is fixed. But in the case $n > 1$, each element of $(\mathbb{Z}/2\mathbb{Z})^{2n}$ is the sum of two odd characteristics.

An immediate consequence of the first part of Lemma 8.3 is the following.

8.4 Proposition. *The intersection of the conjugates of the theta group is the principal congruence subgroup of level two.*

There is an other way to look at Lemma 8.3. We order the even and odd elements of $(\mathbb{Z}/2\mathbb{Z})^{2n}$. Their numbers are denoted by $g = g(n)$ and $u = u(n)$, respectively. The operation $\mathfrak{m} \mapsto M\{\mathfrak{m}\}$ permutes these elements and can now be considered as a permutation of the digits $1, \ldots, g$ or $1, \ldots, u$, respectively. This gives homomorphisms

$$\mathrm{Sp}(n, (\mathbb{Z}/2\mathbb{Z})) \longrightarrow S_g \quad \text{and} \quad S_u.$$

It follows from Lemma 8.3 that the first of these homomorphisms is always injective, and that the second is injective in the case $n > 1$. We are interested in the cases where such a homomorphism is an isomorphism. It turns out that this happens in exactly two cases.

An injective homomorphism of finite groups is an isomorphism iff the orders agree. Hence we compare the orders.

1) We have $\# \mathrm{SL}(2, (\mathbb{Z}/2\mathbb{Z})) = 6$. On the other hand, $g(1) = 3$ and $\#S_3 = 6$.

2) By our index formulae, $\# \mathrm{Sp}(2, (\mathbb{Z}/2\mathbb{Z})) = 720$. On the other hand, $u(2) = 6$ and $\#S_6 = 720$.

We obtain the following result.

8.5 Proposition.
1) *The group* $\# \mathrm{SL}(2, (\mathbb{Z}/2\mathbb{Z}))$ *is isomorphic to* S_3.
2) *The group* $\# \mathrm{Sp}(2, (\mathbb{Z}/2\mathbb{Z}))$ *is isomorphic* S_6.

The isomorphisms are realized through the action of these groups on the even elements of $(\mathbb{Z}/2\mathbb{Z})^2$ *and the odd elements of* $(\mathbb{Z}/2\mathbb{Z})^4$, *respectively.*

Exercises for Sect. VII.8

1. Show that in the case $n \leq 2$, the modular group Γ_n admits a subgroup of index two.

2. Show that in the case $n \geq 3$, the images of Γ_n in S_g and S_u are contained in the alternating group.

3. Find an element in Γ_2 such that its image in S_6 is a transposition.

9. Igusa's Congruence Subgroups

Igusa's congruence subgroup $\Gamma_n[q, 2q]$ $(q \in \mathbb{N})$ is a generalization of the theta group (compare [FB], Sect. VI.3).

9.1 Remark. *The set*

$$\Gamma_n[q, 2q] := \left\{ M \in \Gamma_n[q], \quad \frac{1}{q}(CD^t) \equiv \frac{1}{q}(AB^t) \equiv 0 \mod 2 \right\}$$

is a congruence subgroup.

Because

$$\Gamma_n[q] \supset \Gamma_n[q, 2q] \supset \Gamma_n[2q],$$

it is sufficient to show that $\Gamma_n[q, 2q]$ is group. For odd q, we obviously have

$$\Gamma_n[q, 2q] = \Gamma_n[q] \cap \Gamma_{n, \vartheta}.$$

Hence we can restrict ourselves to the even case. In this case, we have

$$q^2 \equiv 0 \mod 2q,$$

which will be used frequently in the following. If we write a matrix $M \in \Gamma_n[q]$ in the form

$$M = \begin{pmatrix} A & B \\ C & D \end{pmatrix} = \begin{pmatrix} E + q\tilde{A} & q\tilde{B} \\ q\tilde{C} & E + q\tilde{D} \end{pmatrix},$$

we obtain

$$\frac{1}{q} AB^t \equiv \tilde{B}^t \mod 2q.$$

Hence the condition which defines $\Gamma_n[q, 2q]$ inside $\Gamma_n[q]$ is simply

$$\tilde{B}_0 \equiv \tilde{D}_0 \mod 2.$$

Now Remark 9.1 is consequence of the following statement.

9.2 Remark. *Assume that q is even. The map*

$$\eta : \Gamma_n[q] \longrightarrow (\mathbb{Z}/2\mathbb{Z})^{2n},$$

$$\begin{pmatrix} A & B \\ C & D \end{pmatrix} \longmapsto \frac{1}{q} \begin{pmatrix} B_0 \\ C_0 \end{pmatrix},$$

is a surjective homomorphism. Its kernel is $\Gamma_n[q, 2q]$.

Corollary. *We have the index formula*

$$[\Gamma_n[q] : \Gamma_n[q, 2q]] = 2^{2n}.$$

The homomorphy property $\eta(MN) = \eta(M) + \eta(N)$ follows easily from $q^2 \equiv 0 \mod 2q$. For the surjectivity, we use special matrices where $A = D = E$, $B = 0$, or $C = 0$. $\square$

We know that the theta group $\Gamma_{n, \vartheta} = \Gamma[1, 2]$ is not normal in Γ_n. But, remarkably, we can make the following statement.

9.3 Remark. *Let q be even. The group $\Gamma_n[q, 2q]$ is a normal subgroup of the full modular group.*

For the proof, it is sufficient to show that $M\Gamma_n[q, 2q]M^{-1} \subset \Gamma_n[q, 2q]$ for the generators of the modular group. The simple computation will be skipped here. □

Because of Remark 9.3, the group $\Gamma_n[2q, 4q]$ is normal in $\Gamma_n[q]$.

9.4 Lemma. *The group*

$$\Gamma_n[2]/\Gamma_n[4, 8]$$

is abelian.

Proof. We have to show that the commutator of two elements of $\Gamma_n[2]$ is contained in $\Gamma_n[4, 8]$. It is enough to verify this for the generators, and easy to do so. □

We want to determine the group which occurs in Lemma 9.4 when we take the quotient by the subgroup $\pm E$, since we are interested only in the mapping groups. Hence we consider the group

$$\widetilde{\Gamma}_n[4, 8] = \Gamma_n[4, 8] \cup (-\Gamma_n[4, 8]),$$

which also is normal in $\Gamma_n[2]$, and consider

$$\mathcal{G}_n := \Gamma_n[2]/\widetilde{\Gamma}_n[4, 8].$$

In the first volume ([FB], Lemma VI.6.2), we have shown in the case $n = 1$ that

$$\mathcal{G}_1 = \mathbb{Z}/4\mathbb{Z} \times \mathbb{Z}/4\mathbb{Z}.$$

We have to generalize this result. We make use of our knowledge of the generators of the principal congruence subgroup of level two,

$$\begin{pmatrix} E & S \\ 0 & E \end{pmatrix}, \quad \begin{pmatrix} E & 0 \\ S & E \end{pmatrix}, \quad \begin{pmatrix} U^t & 0 \\ 0 & U^{-1} \end{pmatrix}.$$

For the sake of simplicity, we restrict ourselves to the case $n = 2$, since we shall not use the case $n > 2$.

The matrices S which we have to use are, in the case $n = 2$, only

$$S = \begin{pmatrix} 2 & 0 \\ 0 & 0 \end{pmatrix}, \quad \begin{pmatrix} 0 & 0 \\ 0 & 2 \end{pmatrix}, \quad \begin{pmatrix} 0 & 2 \\ 2 & 0 \end{pmatrix},$$

and, since we know the generators of $\mathrm{SL}(2, \mathbb{Z})[2]$ and hence of $\mathrm{GL}(2, \mathbb{Z})[2]$, we need the matrices U only in the cases

$$U = \begin{pmatrix} 1 & 2 \\ 0 & 1 \end{pmatrix}, \quad \begin{pmatrix} 1 & 0 \\ 2 & 1 \end{pmatrix}, \quad \begin{pmatrix} -1 & 0 \\ 0 & -1 \end{pmatrix}, \quad \text{and} \quad \begin{pmatrix} 1 & 0 \\ 0 & -1 \end{pmatrix}.$$

Hence we obtain $2 \cdot 3 + 4 = 10$ generators of $\Gamma_2[2]$. Since we factor out $\{\pm E^{(2n)}\}$, the negative unit matrix can be canceled. So,

$$\mathcal{G}_2 = \Gamma_2[2]/\widetilde{\Gamma}_2[4,8]$$

can be generated by nine elements. Since $\mathcal{G}_2$ is abelian, this means that we get a *surjective homomorphism*

$$\mathbb{Z}^9 \longrightarrow \mathcal{G}_2.$$

In the following, we denote the generators by

$$T_1, T_2, \widetilde{T}_1, \widetilde{T}_2, T_3, \widetilde{T}_3, P_1, P_2, P_3,$$

where

$$T_\nu = \begin{pmatrix} E & S_\nu \\ 0 & E \end{pmatrix}, \quad \widetilde{T}_\nu = \begin{pmatrix} E & 0 \\ S_\nu & E \end{pmatrix},$$

$$S_1 = \begin{pmatrix} 2 & 0 \\ 0 & 0 \end{pmatrix}, \quad S_2 = \begin{pmatrix} 0 & 0 \\ 0 & 2 \end{pmatrix}, \quad S_3 = \begin{pmatrix} 0 & 0 \\ 2 & 0 \end{pmatrix}$$

and

$$P_\nu = \begin{pmatrix} U_\nu^t & 0 \\ 0 & U_\nu^{-1} \end{pmatrix},$$

$$U_1 = \begin{pmatrix} 1 & 2 \\ 0 & 1 \end{pmatrix}, \quad U_2 = \begin{pmatrix} 1 & 0 \\ 2 & 1 \end{pmatrix}, \quad U_3 = \begin{pmatrix} 1 & 0 \\ 0 & -1 \end{pmatrix}.$$

The homomorphism is given by

$$(a_1, b_1, a_2, b_2, a_3, b_3, c_1, c_2, c_3) \longmapsto T_1^{a_1} \widetilde{T}_1^{b_1} T_2^{a_2} \widetilde{T}_2^{b_2} T_3^{a_3} \widetilde{T}_3^{b_3} P_1^{c_1} P_2^{c_2} P_3^{c_3}.$$

The images of the first four matrices have order 4; the other five have order 2. So we obtain a surjective homomorphism

$$(\mathbb{Z}/4\mathbb{Z})^4 \times (\mathbb{Z}/2\mathbb{Z})^5 \longrightarrow \Gamma_2[2] \, \widetilde{\Gamma}_2[4,8].$$

This turns out to be an isomorphism.

9.5 Lemma. *The map*

$$(a_1, b_1, a_2, b_2, a_3, b_3, c_1, c_2, c_3) \longmapsto T_1^{a_1} \widetilde{T}_1^{b_1} T_2^{a_2} \widetilde{T}_2^{b_2} T_3^{a_3} \widetilde{T}_3^{b_3} P_1^{c_1} P_2^{c_2} P_3^{c_3}$$

$(\mathrm{mod}\ \widetilde{\Gamma}_2[4,8])$ *induces an isomorphism*

$$(\mathbb{Z}/4\mathbb{Z})^4 \times (\mathbb{Z}/2\mathbb{Z})^5 \longrightarrow \Gamma_2[2]/\widetilde{\Gamma}_2[4,8].$$

For the proof, it is sufficient to compare the orders of the two groups. The index formulae, in particular, that in Remark 9.2 show that they are equal.   □

Decomposition into Eigenspaces

Let $\Gamma' \subset \Gamma$ be congruence subgroups and let Γ' be a normal subgroup with an *abelian* factor group $G := \Gamma/\Gamma'$. Assume that a multiplier system v of weight $r/2$ on the larger group Γ is given. Then the map

$$f \longmapsto v(M)^{-r} \det(CZ + D)^{-r} f(MZ)$$

associates each $M \in \Gamma$ with an endomorphism of $[\Gamma', r/2, v]$. This linear map depends only on the image of M in G. So, it defines an action of G on this space. Since G is abelian, we can decompose the space into eigenspaces (compare [FB], Remark VI.6.4). In the following, we identify characters of G with characters on Γ which are trivial on Γ'. We obtain

$$[\Gamma', r/2, v] = \sum_{\chi} [\Gamma, r/2, v\chi],$$

where χ runs through all characters of G. We shall use this decomposition in the special case $\Gamma' = \Gamma_2[2]$ and $\Gamma = \Gamma_2[4, 8]$. As in [FB], Sect. VI.6, we have the following lemma.

9.6 Lemma. *We have*

$$[\Gamma_2[4, 8], r/2, v_\vartheta^r] = \sum_{\chi} [\Gamma[2], r/2, v_\vartheta^r \chi],$$

where χ runs through all characters of $\mathcal{G}_2 = \Gamma[2]/\widetilde{\Gamma}[4, 8]$. The characters χ are determined by their values on

$$T_1, T_2, \widetilde{T}_1, \widetilde{T}_2, T_3, \widetilde{T}_3, P_1, P_2, P_3.$$

On the first four matrices, they can be arbitrary fourth roots of unity, and on the rest they can be ± 1. So, there are 2048 such characters.

The group $\Gamma_2[4, 8]$ is generated by the commutators of $\Gamma_2[2]$, the fourth powers of the first four, and the squares of the remaining five. The multiplier systems of two $\vartheta[\mathbf{m}]$ differ only by a character. This character is trivial on all commutators. One can easily check that it is trivial on the other nine generators. This gives us the following result (compare [FB], Lemma VI.6.2):

9.7 Proposition. *The multiplier systems of the ten theta series agree on $\Gamma_2[4, 8]$. In particular, they are contained in eigenspaces of the decomposition in Lemma 9.6.*

Exercises for Sect. VII.9

1. Determine the index of $\Gamma_n[q, 2q]$ in $\Gamma_n[q]$ and in Γ_n.

2. Show that in the case $r = 1$, at least ten of the eigenspaces in Lemma 9.6 are different from zero. (We will see later that there are exactly ten.)

3. To illustrate the complexity of congruence subgroups, we ask: How many groups are between $\tilde{\Gamma}[4, 8]$ and $\Gamma[2]$?

10. The Fundamental Domain of the Modular Group of Degree Two

The fundamental domain of the Siegel modular group was constructed by Siegel in his original paper of 1935. The essential tool was *Minkowski's reduction theory* for the group $GL(n, \mathbb{Z})$. In the case $n = 2$, this is essentially the elliptic modular group, and Minkowski's reduction theory is equivalent to the construction of the fundamental domain of the elliptic modular group. So, in the case $n = 2$, the theory becomes very simple. We restrict ourselves to this case.

Let

$$\mathcal{R}_2 = \left\{ Y = \begin{pmatrix} y_0 & y_1 \\ y_1 & y_2 \end{pmatrix}, \quad 0 \leq 2y_1 \leq y_0 \leq y_2, \ 0 < y_0 \right\}.$$

10.1 Remark. *Each matrix from $\mathcal{R}_2$ is positive definite. In $\mathcal{R}_2$, the inequalities*

$$\det Y \leq y_0 y_2 \leq \frac{4}{3} \det Y$$

hold.

The proof is simple and will be omitted. The following lemma is a little more difficult.

10.2 Lemma. *For each positive 2×2 matrix $Y \in \mathcal{P}_2$, there exists a unimodular matrix*

$$U \in GL(2, \mathbb{Z})$$

with the property

$$Y[U] \in \mathcal{R}_2.$$

Proof. The operation $Y \mapsto Y[U]$ is compatible with the substitution $Y \mapsto tY$, $t > 0$. For this reason, we can restrict ourselves to matrices of determinant one,

$$\mathcal{P}_n(1) := \{Y \in \mathcal{P}_2, \quad \det Y = 1\}.$$

For the proof, we make use of the map

$$\varphi : \mathbb{H} \longrightarrow \mathcal{P}_2(1),$$

$$z \longmapsto S := \begin{pmatrix} 1 & 0 \\ x & 1 \end{pmatrix} \begin{pmatrix} y^{-1} & 0 \\ 0 & y \end{pmatrix} \begin{pmatrix} 1 & x \\ 0 & 1 \end{pmatrix}.$$

We can easily verify the following three simple facts:
a) The map φ is bijective.
b) $\varphi(Mz) = S[M]$ for $M \in \mathrm{SL}(2, \mathbb{Z})$.
c) We have

$$\varphi(\mathcal{F}) = \{Y \in \mathcal{P}_2(1), \quad 0 \le |2y_1| \le y_0 \le y_2, \ 0 < y_0\}.$$

Here $\mathcal{F}$ is the fundamental domain of the elliptic modular group, ([FB], Proposition V.8.7).

So, each matrix from $\mathcal{P}_2(1)$ can be transformed into $\varphi(\mathcal{F})$ by means of a matrix $\mathrm{SL}(2, \mathbb{Z})$. Making use of the fact that the diagonal matrix with entries 1 and -1 is contained in $\mathrm{GL}(2, \mathbb{Z})$, we can enforce the condition $y_1 \ge 0$.   □

By the *height* of a point $Z \in \mathbb{H}_n$, we understand the positive number

$$h(Z) = \det Y.$$

10.3 Lemma. *Assume that a point $Z \in \mathbb{H}_2$ and a positive number $\varepsilon > 0$ are given. There are only finitely many numbers h_0 with the properties*
a) $h_0 \ge \varepsilon$;
b) $h_0 = h(MZ)$ for an $M \in \Gamma_2$.

Proof.[*]) Let $Z^* = MZ$, $M \in \Gamma_n$, $h_0 = h(Z^*) \ge \varepsilon$. The height is invariant under unimodular transformations $Z^* \mapsto Z^*[U]$. Hence we can assume

$$Y^{*^{-1}} \in \mathcal{R}_2.$$

If we denote the diagonal elements of $Y^{*^{-1}}$ by r_1, r_2, we obtain from Remark 10.1

$$r_1 r_2 \frac{4}{3} \det(Y^*)^{-1} \le \frac{4}{3}\varepsilon^{-1}.$$

The formula

$$Y^{*^{-1}} = Y^{-1}[(CX + D)^t] + Y[C^t]$$

---

*) A proof for arbitrary n can be found in [Fr1].

shows that

$$r_k = [Xc_k^t + d_k^t] + Y[c_k^t] \quad (k = 1, 2),$$

where c_k, d_k denote the kth rows of C, D. The rows c_k, d_k cannot both vanish. Therefore r_k has a lower bound depending on Y. Since the product of the r_k is bounded from above, both have a lower bound (depending only on Y and ε). This shows that the vectors c_k, d_k belong to a finite set.   □

10.4 Proposition. *The subset* $\mathfrak{M} \subset \mathbb{H}_2$ *defined by*

$$0 \le 2y_1 \le y_0 \le y_2, \qquad \frac{\sqrt{2}}{2} \le y_0$$

is a fundamental set of the modular group of degree two. This means that for each $Z \in \mathbb{H}_2$ *there exists* $M \in \Gamma_2$ *with* $MZ \in \mathfrak{M}$.

Proof. Let $Z \in \mathbb{H}_2$ be an arbitrary point. Because of Lemma 10.3, the orbit $\{MZ, \ M \in \Gamma_2\}$ contains a point Z_0 with maximal height. This point satisfies

$$|\det(CZ_0 + D)|^{-2} h(Z_0) = h(MZ) \le h(Z_0).$$

This means

$$|\det(CZ_0 + D)| \ge 1.$$

Since the height remains unchanged under unimodular transformations, we can assume that $Y_0 \in \mathcal{R}_2$. Now we use the condition $|\det(CZ_0 + D)| \ge 1$ for the matrices

$$M = \begin{pmatrix} A & B \\ C & D \end{pmatrix}, \ A = \begin{pmatrix} a & 0 \\ 0 & 1 \end{pmatrix}, \ B = \begin{pmatrix} b & 0 \\ 0 & 0 \end{pmatrix}, C = \begin{pmatrix} c & 0 \\ 0 & 0 \end{pmatrix}, \ D = \begin{pmatrix} d & 0 \\ 0 & 1 \end{pmatrix}.$$

We obtain the result that the first diagonal element z of Z_0 satisfies the inequality

$$|cz + d| \ge 1 \text{ for all } \begin{pmatrix} a & b \\ c & d \end{pmatrix} \in \mathrm{SL}(2, \mathbb{Z}).$$

This gives $y \ge \sqrt{3}/2$. The same argument works for the second diagonal element.   □

Exercises for Sect. VII.10

1. Show that, in the case $n \le 2$, there exists a number $\delta_n > 0$ such that the set defined by $Y - \delta_n E \ge 0$ is a fundamental set of the modular group (This true for arbitrary n. But the general proof needs the complete Minkowski reduction theory; see for example, [Fr1].)

2. Show that two different points of the form iyE for sufficiently large y can never be equivalent. Show that it follows that there is no compact fundamental set for Γ_n.

3. Let dv be the Euclidean volume element on $\mathbb{H}_n$ and let

$$dw = \frac{dv}{\det Y^{n+1}}.$$

Show that dw is invariant under $\mathrm{Sp}(n, \mathbb{R})$, i.e.

$$\int_{\mathbb{H}_n} f(MZ)\, dw = \int_{\mathbb{H}_n} f(Z)\, dw,$$

for example for continuous functions with compact support.

4. The set

$$\{Z \in \mathbb{H}_2;\ |x_0|, |x_1|, |x_2| \leq 1/2,\ 0 \leq 2y_1 \leq y_0 \leq y_2,\ \frac{\sqrt{3}}{2} \leq y_0\}$$

is a fundamental set of Γ_2. Show that its volume with respect to dw is finite.

Hint. Use Remark 10.1. Integration over the x-coordinates is harmless. After that, integrate over y_1.

11. The Zeros of the Theta Series of Degree two

In the first volume, we saw that in the case $n = 1$ the theta series $\vartheta[m]$ have no zeros in the upper half-plane ([FB], Lemma VI.6.6). This is false for $n > 1$. But in the case $n = 2$, the zeros can be described by simple equations. This has been done by Igusa [Ig3]. A short, elementary proof can be found in [Fr3]; see also [Fr1]. This proof will be reproduced here.

11.1 Lemma. *Let*

$$a = \begin{pmatrix} a_1 \\ a_2 \end{pmatrix}, \quad b = \begin{pmatrix} b_1 \\ b_2 \end{pmatrix}, \quad a_1, b_1 \in \mathbb{Z}^{n_1},\ a_2, b_2 \in \mathbb{Z}^{n_2}.$$

Then

$$\vartheta\begin{bmatrix} a \\ b \end{bmatrix}\begin{pmatrix} Z_1^{(n_1)} & 0 \\ 0 & Z_2^{(n_2)} \end{pmatrix} = \vartheta\begin{bmatrix} a_1 \\ b_1 \end{bmatrix}\vartheta\begin{bmatrix} a_2 \\ b_2 \end{bmatrix}.$$

Proof. The proof follows easily from the formula

$$\begin{pmatrix} Z_1 & 0 \\ 0 & Z_2 \end{pmatrix} \begin{bmatrix} g_1 \\ g_2 \end{bmatrix} = Z_1[g_1] + Z_2[g_2]$$

together with the Cauchy multiplication theorem for infinite series. □

The expression in Lemma 11.1 vanishes if one of the two characteristics is odd. For example,

$$\vartheta \begin{bmatrix} 1 \\ 1 \\ 1 \\ 1 \end{bmatrix} \begin{pmatrix} z_0 & 0 \\ 0 & z_2 \end{pmatrix} = 0.$$

In this section, we shall show that in some sense this describes all zeros of theta series of degree two. The ten even characteristics are

$$\begin{pmatrix} a \\ b \end{pmatrix} \in \left\{ \begin{array}{cccccccccc} 0 & 0 & 0 & 0 & 1 & 0 & 1 & 0 & 1 & 1 \\ 0 & 0 & 0 & 0 & 0 & 1 & 0 & 1 & 1 & 1 \\ 0 & 1 & 0 & 1 & 0 & 0 & 0 & 1 & 0 & 1 \\ 0 & 0 & 1 & 1 & 0 & 0 & 1 & 0 & 0 & 1 \end{array} \right\}.$$

We now determine the zeros of the ten theta series on the fundamental set $\mathfrak{M}$ $(0 \le 2y_1 \le y_0 \le y_2, \ \sqrt{3}/2 \le y_0)$.

11.2 Lemma. 1) *The eight theta series*

$$\vartheta \begin{bmatrix} a \\ b \end{bmatrix}, \quad a \neq \begin{bmatrix} 1 \\ 1 \end{bmatrix},$$

have no zeros in $\mathfrak{M}$.

2) *The two functions*

$$\frac{\vartheta \begin{bmatrix} 1 \\ 1 \\ 1 \\ 1 \end{bmatrix}}{e^{\pi i z_1} - 1} \quad and \quad \frac{\vartheta \begin{bmatrix} 1 \\ 1 \\ 0 \\ 0 \end{bmatrix}}{e^{\pi i z_1} + 1}$$

are analytic in $\mathbb{H}_2$ *and have no zeros in* $\mathfrak{M}$.

The proof rests on elementary estimates:

1) Let $a = 0$. We extract from the theta series the constant term $(g = 0)$, and estimate the rest by means of the series of the absolute values

$$\left| \vartheta \begin{bmatrix} 0 \\ b \end{bmatrix} - 1 \right| \le \sum_{g \neq 0} e^{-\pi Y[g]}.$$

Then we extract the terms for $g_1^2 + g_2^2 = 1$ from the sums to obtain

$$\sum_{g \neq 0} e^{-\pi Y[g]} \leq 4e^{-(\pi/2)\sqrt{3}} + \sum_{g_1^2+g_2^2} e^{-(\pi/4)(g_1^2+g_2^2)} < 1.$$

A numerical computation shows that the expression on the right-hand side is smaller than one. This shows that

$$\vartheta \begin{bmatrix} 0 \\ b \end{bmatrix} (Z) \neq 0 \quad \text{for } Z \in \mathfrak{M}.$$

2) Let $a = \begin{pmatrix} 1 \\ 0 \end{pmatrix}$. First we divide the theta series by $e^{\pi i z/4}$, and then we extract the constant terms which belong to $g = \begin{pmatrix} 0 \\ 0 \end{pmatrix}, \begin{pmatrix} -1 \\ 0 \end{pmatrix}$. The rest is estimated by means of the series of the absolute values,

$$\left| \vartheta \begin{bmatrix} \begin{pmatrix} 1 \\ 1 \\ 0 \end{pmatrix} \end{bmatrix} \begin{pmatrix} 0 \\ * \end{pmatrix} [Z] e^{-\pi i z_0} - 2 \right| \leq \sum_{g \neq \begin{pmatrix} 0 \\ 0 \end{pmatrix}, \begin{pmatrix} -1 \\ 0 \end{pmatrix}} e^{-\pi \{y_0 g_1(g_1+1)+y_1(2g_1+1)g_2+y_2 g_2^2\}}.$$

By means of the identity

$$(2g_1 + 1)g_2 = (g_1 + g_2 + 1)(g_1 + g_2) - g_1(g_1 + 1) - g_2^2,$$

we see that

$$y_0 g_1(g_1 + 1) + y_1(2g_1 + 1)g_2 + y_2 g_2^2$$
$$\geq (y_0 - y_1)g_1(g_1 + 1) + (y_2 - y_1)g_2^2 \geq \frac{1}{4}\sqrt{3}[g_1(g_1 + 1) + g_2^2].$$

As in the first case, we now obtain

$$\vartheta \begin{bmatrix} a \\ b \end{bmatrix} \neq 0 \text{ in } \mathfrak{M} \text{ for } a = \begin{pmatrix} 1 \\ 0 \end{pmatrix} \text{ and similarly for } a = \begin{pmatrix} 1 \\ 1 \end{pmatrix}.$$

3) It remains to consider $a = \begin{pmatrix} 1 \\ 1 \end{pmatrix}$ and $a = \varepsilon \begin{pmatrix} 1 \\ 1 \end{pmatrix}$ with $\varepsilon = 0$ or 1. A simple conversion of the theta series shows that

$$e^{-\frac{1}{4}\pi i Z[a] + \pi i z_1} \vartheta \begin{bmatrix} a \\ b \end{bmatrix} (Z)$$

$$= 2 \sum_{g_1, g_2 \geq 0} (-1)^{\varepsilon(g_1+g_2)} e^{\pi i g_1(g_1+1)(z_0-z_1)+\pi i g_2(g_2+1)(z_2-z_1)}$$

$$\cdot \left\{ e^{\pi i(g_1+g_2+1)^2 z_1} + (-1)^{\varepsilon} e^{\pi i(g_2-g_1)^2 z_1} \right\}.$$

The expression in the curly brackets can be divided by $2(1+(-1)^\varepsilon e^{\pi i z_1})$. After this, we bring the term for $g_1 = g_2 = 0$ to the left-hand side and estimate the rest by means of the series of the absolute values:

$$\left| \frac{\vartheta \begin{bmatrix} a \\ b \end{bmatrix}(Z)}{2e^{\pi i Z[a]/4}(1+(-1)^\varepsilon e^{\pi i z_1})} - (-1)^r \right|$$

$$\leq -1 + \sum_{g_1, g_2 \geq 0} e^{-\pi g_1(g_1+1)(y_0-y_2)-\pi g_2(g_2+1)(y_2-y_1)}$$

$$\cdot \left\{ \sum_{n=0}^{(2g_1+1)(2g_2+1)-1} e^{-\pi n y_1} \right\}.$$

Here we have used

$$e^{\pi i (g_1+g_2+1)^2} + (-1)^\varepsilon e^{\pi i (g_1-g_2)^2 z_1}$$

$$= e^{\pi i (g_1-g_2)^2 z_1}(1+(-1)^\varepsilon e^{\pi i z_1})(-1)^\varepsilon \cdot \sum_{n=0}^{(2g_1+1)(2g_2+1)-1} (-1)^{n(\varepsilon-1)} e^{\pi i n z_1}.$$

For Z in $\mathfrak{M}$, we immediately obtain

$$\left| \frac{\vartheta \begin{bmatrix} a \\ b \end{bmatrix}(Z)}{2e^{\pi i Z[a]/4}(1+(-1)^\varepsilon e^{\pi i z_1})} - (-1)^r \right| \leq -1 + \left(\sum e^{-(\pi \sqrt{3}/4)n(n+1)}(2n+1)^2 \right).$$

Since the expression on the right hand side is < 1, we obtain that

$$\frac{\vartheta \begin{bmatrix} a \\ b \end{bmatrix}}{1+(-1)^\varepsilon e^{\pi i z_1}} \quad \text{in the case } a = \begin{pmatrix} 1 \\ 1 \end{pmatrix}, \ b = \varepsilon \begin{pmatrix} 1 \\ 1 \end{pmatrix}$$

has no zero in $\mathfrak{M}$.   □

We now determine all symplectic substitutions which transform the diagonal into itself. For this, we recall the embedding

$$\mathrm{SL}(2,\mathbb{R}) \times \mathrm{SL}(2,\mathbb{R}) \longrightarrow \mathrm{Sp}(2,\mathbb{R}),$$

$$\begin{pmatrix} a & b \\ c & d \end{pmatrix}, \begin{pmatrix} \alpha & \beta \\ \gamma & \delta \end{pmatrix} \longmapsto \begin{pmatrix} a & & b & \\ 0 & \alpha & & \beta \\ c & & d & \\ & \gamma & & \delta \end{pmatrix}.$$

The image acts on the diagonal componentwise. There is another substitution which transforms the diagonal into itself, namely the unimodular transformation

$$\begin{pmatrix} U^t & 0 \\ 0 & U^{-1} \end{pmatrix}, \quad U = \begin{pmatrix} 0 & 1 \\ 1 & 0 \end{pmatrix}.$$

One can easily verify the following statement.

11.3 Remark. *The set $\mathcal{N}$ of all matrices of the form*

$$\begin{pmatrix} a & & b & \\ 0 & \alpha & & \beta \\ c & & d & \\ & \delta & & \Gamma \end{pmatrix} \quad and \quad \begin{pmatrix} 0 & 1 & & \\ 1 & 0 & & \\ & & 0 & 1 \\ & & 1 & 0 \end{pmatrix} \begin{pmatrix} a & & b & \\ 0 & \alpha & & \beta \\ c & & d & \\ & \delta & & \Gamma \end{pmatrix}$$

is a subgroup of $\mathrm{Sp}(2, \mathbb{R})$ which transforms the diagonal into itself.

We denote by

$$\mathcal{N}(\mathbb{Z}) = \mathcal{N} \cap \mathrm{Sp}(2, \mathbb{Z})$$

the subgroup of integral matrices. It contains a subgroup which is isomorphic to $\mathrm{SL}(2, \mathbb{Z}) \times \mathrm{SL}(2, \mathbb{Z})$. Obviously, $\mathcal{N}(\mathbb{Z})$ fixes the characteristic

$$\mathfrak{m} = \begin{pmatrix} 1 \\ 1 \\ 1 \\ 1 \end{pmatrix}.$$

Hence we have

$$\mathcal{N}(\mathbb{Z}) \subset \Gamma_2(\mathfrak{m}).$$

The group $\Gamma_2[2]$ is also contained in $\Gamma_2(\mathfrak{m})$. We shall show that $\Gamma_2(\mathfrak{m})$ is generated by both subgroups.

11.4 Lemma. *The homomorphism*

$$\mathcal{N}(\mathbb{Z})/(\mathcal{N}(\mathbb{Z}) \cap \Gamma_2[2]) \longrightarrow \Gamma_2[\mathfrak{m}]/\Gamma_2[2],$$

which is induced by the inclusion

$$\Gamma_2[2] \longrightarrow \Gamma_2(\mathfrak{m}),$$

is an isomorphism.

Proof. It suffices to show that the orders of the groups agree. On one side, we have

$$\#(\Gamma_2(\mathfrak{m})/\Gamma_2[2]) = [\Gamma_2(\mathfrak{m}) : \Gamma_2[2]] = \frac{[\Gamma_2 : \Gamma_2[2]]}{[\Gamma_2 : \Gamma_2(\mathfrak{m})]} = \frac{[\Gamma_2 : \Gamma_2[2]]}{[\Gamma_2 : \Gamma_{2,\vartheta}]} = \frac{720}{10}$$

and, on the other side, $\mathcal{N}(\mathbb{Z})/(\mathcal{N}(\mathbb{Z}) \cap \Gamma_2[2])$ is an extension of index two of $\mathrm{SL}(2, \mathbb{Z}/2\mathbb{Z}) \times \mathrm{SL}(2, \mathbb{Z}/2\mathbb{Z})$. Hence the order of both groups is 72. □

11.5 Proposition.   *Let $f \in [\Gamma_2[2], r/2, v]$ be a modular form with respect to the principal group of level two which vanishes on the diagonal. Then f is divisible by $\vartheta[\mathrm{m}]$, and we have*

$$\frac{f}{\vartheta[\mathrm{m}]} \in \left[\Gamma_2[2], \; \frac{r-1}{2}, \; \frac{v}{v^{\mathrm{m}}}\right], \qquad \mathrm{m} = \begin{pmatrix} 1 \\ 1 \\ 1 \\ 1 \end{pmatrix}.$$

Proof. Since $\mathfrak{M}$ is a fundamental set of the full modular group, it is sufficient to show the following:

For an arbitrary modular substitution $M \in \mathrm{Sp}(2, \mathbb{Z})$, the function

$$f(MZ)/\vartheta[\mathrm{m}](MZ)$$

is analytic in an open neighborhood of $M(\mathfrak{M})$.

We know that $\vartheta[\mathrm{m}](MZ)$ equals $\vartheta[M\{\mathrm{m}\}](Z)$ up to a factor without zeros. The first part of Lemma 11.2 shows that it is sufficient to consider the cases

$$M\{\mathrm{m}\} = \begin{pmatrix} 1 \\ 1 \\ 1 \\ 1 \end{pmatrix} \quad \text{and} \quad \begin{pmatrix} 1 \\ 1 \\ 0 \\ 0 \end{pmatrix}.$$

First case. $M\{\mathrm{m}\} = \mathrm{m}$. Then $M \in \Gamma_2(\mathrm{m})$. Because of Lemma 11.4, we can assume that $M \in \mathcal{N}(\mathbb{Z})$. Like f, the modular form $f(MZ) \det(CZ + D)^{-r/2}$ is a modular form of level two which vanishes on the diagonal. Hence it is sufficient to show that f is divisible by $\vartheta[\mathrm{m}]$ in a full open neighborhood of $\mathfrak{M}$. By Lemma 11.2, this means that $f(Z)/(e^{\pi i z_1} - 1)$ is analytic or that the functions $f(Z)/(z_1 - 2k)$ are analytic for all $k \in \mathbb{Z}$. This follows from the fact that f vanishes on the diagonal and hence on all $z_1 = 2k$.

Second case. $M\{\mathrm{m}\} = \begin{pmatrix} 1 \\ 1 \\ 0 \\ 0 \end{pmatrix}.$

As in the first case, it is sufficient to consider a special M. We take

$$M = \begin{pmatrix} E & S \\ 0 & E \end{pmatrix} \text{ with } S = \begin{pmatrix} 0 & 1 \\ 1 & 0 \end{pmatrix}.$$

So, it suffices to show that

$$\frac{f(Z + S)}{\left(e^{\pi i z_1} \pm 1\right)}$$

is analytic in $\mathbb{H}_2$. This follows from the fact that $f(Z + S)$ vanishes on $z_1 = 2k + 1$, $k \in \mathbb{Z}$. $\qquad \square$

Exercises for Sect. VII.11

1. Let $\alpha, \beta, \gamma, \delta, \epsilon$ be five integers with the property
    $$\alpha\delta + \beta\gamma - \epsilon^2 = -1.$$
    We consider the sets
    $$\mathfrak{N}(\alpha, \beta, \gamma, \delta, \epsilon) = \{Z \in \mathbb{H}_2; \quad \alpha z_0 + \beta z_1 + \gamma z_2 + \delta(z_1^2 - z_0 z_2) + \epsilon\}$$
    and their union
    $$\mathfrak{N} = \bigcup_{\beta^2 - 4\alpha\gamma - 4\delta\varepsilon = 1} \mathfrak{N}(\alpha, \beta, \gamma, \delta, \epsilon).$$
    Show that a modular substitution $M \in \Gamma_2$ permutes the sets $\mathfrak{N}(\alpha, \beta, \gamma, \delta, \epsilon)$. As a consequence, the modular group acts on $\mathfrak{N}$.

2. The diagonal is contained in $\mathfrak{N}$ (see the previous exercise). Deduce from Lemma 11.2 that all zeros of the ten theta series are contained in $\mathfrak{N}$.

3. This exercise is not quite so simple: Show that the modular group Γ_2 permutes the sets $\mathfrak{N}(\alpha, \beta, \gamma, \delta, \epsilon)$ transitively. Hence $\mathfrak{N}$ is the precise zero set of the function $\Delta^{(2)}$ (the product of the ten theta series).

12. A Ring of Modular Forms

In this section, we shall give a proof of Igusa's beautiful structure theorem about the ring of modular forms on the group $\Gamma_2[4, 8]$. Igusa's proof can be found in his paper [Ig3] of 1964, which we quoted at the beginning of the last section. A completely different and much more elementary proof was given by A. Lober in his Heidelberg Diplomarbeit; a published version can be found in [Lo].

We study modular forms for the principal congruence subgroup of level two,

$$f \in [\Gamma_2[2], r/2, v].$$

We write the multiplier system in the form

$$v = \chi v_\vartheta^r,$$

with a *character* χ on $\Gamma_2[2]$. Now we use the fact that the substitution

$$P_3 = \begin{pmatrix} 1 & & & \\ & -1 & & \\ & & 1 & \\ & & & -1 \end{pmatrix}$$

is contained in $\Gamma_2[2]$. It acts as

$$P_3 \begin{pmatrix} z_0 & z_1 \\ z_1 & z_2 \end{pmatrix} = \begin{pmatrix} z_0 & -z_1 \\ -z_1 & z_2 \end{pmatrix}.$$

Since the theta series $\vartheta[0]$ is invariant under P_3, we have

$$\det \begin{pmatrix} 1 & 0 \\ 0 & -1 \end{pmatrix}^{1/2} v_\vartheta(P_3) = +1.$$

So, we obtain

$$f \begin{pmatrix} z_0 & z_1 \\ z_1 & z_2 \end{pmatrix} = \chi(P_3) f \begin{pmatrix} z_0 & -z_1 \\ -z_1 & z_2 \end{pmatrix}.$$

If $\chi(P_3)$ is different from 1, then f vanishes *by force* on the diagonal.

12.1   Definition.   *The diagonal $\mathcal{D}$ is an enforced zero for the space* $[\Gamma_2[2], r/2, v]$ *if*

$$\chi(P_3) \neq 1 \quad (v = \chi v_\vartheta^r).$$

By means of Koecher's principle, we can now deduce the following from Proposition 11.5.

12.2 Remark. *If the diagonal is an enforced zero for $[\Gamma_2[2], r/2, v]$, then the map*

$$\left[\Gamma_2[2], \frac{r-1}{2}, \frac{v}{v_\mathfrak{m}}\right] \longrightarrow [\Gamma_2[2], r/2, v], \qquad \mathfrak{m} = \begin{pmatrix} 1 \\ 1 \\ 1 \\ 1 \end{pmatrix},$$

$$f \longmapsto f \cdot \vartheta_\mathfrak{m},$$

is an isomorphism.

Example. If the diagonal is an enforced zero for the space $[\Gamma_2[2], 1/2, v]$ and if v is different from $v_\mathfrak{m}$, then this space is the zero space, since any modular form of weight 0 for a nontrivial multiplier system does not vanish. This argument shows that the space $[\Gamma_2[2], 1/2, v]$, $v = v_\vartheta \chi$, vanishes for at least 1023 of the 2048 characters.

   If $f \in [\Gamma_2[2], r/2, v]$ does not vanish on the diagonal, we can consider one of the conjugate forms

$$(f|M)(Z) = \det(CZ + D)^{-r/2} f(MZ)$$

instead of f. The form belongs to a conjugate multiplier system

$$f|M \in [\Gamma_2[2], r/2, v^M].$$

It may happen that the diagonal is an enforced zero for this space. In this case we can divide f by

$$\vartheta[\mathfrak{m}], \quad \mathfrak{m} = M \begin{Bmatrix} 1 \\ 1 \\ 1 \\ 1 \end{Bmatrix}.$$

These considerations lead to the following definition.

12.3 Definition. *The space* $[\Gamma_2[2], r/2, v]$ *has an enforced zero if there exists a matrix* $M \in \mathrm{Sp}(2, \mathbb{Z})$ *such that the diagonal is an enforced zero for* $[\Gamma_2[2], r/2, v^M]$.

These considerations show the following.

12.4 Proposition. *If the space*

$$[\Gamma_2[2], r/2, v]$$

has an enforced zero, then there exists a characteristic $\mathfrak{m}$ *such that the map*

$$\left[\Gamma_2[2], \frac{r-1}{2}, \frac{v}{v_\mathfrak{m}}\right] \longrightarrow \left[\Gamma_2[2], \frac{r}{2}, v\right],$$

$$f \longmapsto \vartheta[\mathfrak{m}],$$

is an isomorphism.

How can one decide whether a space $[\Gamma_2[2], r/2, v]$ has an enforced zero? From now on, we consider only multiplier systems of the form $v = v_\vartheta^r \chi$, where χ is one of the 2048 characters. As we know, each conjugate multiplier system v^M is again of this form, i.e.

$$v^M = v_\vartheta^r \tilde{\chi}.$$

The character $\tilde{\chi}$ depends on M and r, but on r only mod 4. We write

$$\tilde{\chi} = \chi^{(M,r)}.$$

12.5 Remark. *Consider* $M \in \mathrm{Sp}(2, \mathbb{Z})$ *and* $r \in \mathbb{Z}$. *By means of*

$$\chi^{(M,r)} = \frac{(v_\vartheta^r \chi)^M}{v_\vartheta{}^r},$$

we obtain a permutation of the 2048 *characters. This map depends only on the coset* $\Gamma_2[2]M$ *and on* r *mod 4.*

We now formulate some simple rules which rest on the formula

$$(v_1 v_2)^M = v_1^M v_2^M,$$

which is valid for two multiplier systems v_1, v_2. If v is a multiplier system of even weight (a character), we have

$$v^M(N) = v(MNM^{-1}).$$

12.6 Remark. *We have*

$$\chi^{(M,r)} = \frac{(v_\vartheta^r)^M}{v_\vartheta{}^r} \chi^M.$$

Here

$$\chi^M(N) := \chi(MNM^{-1}).$$

By the way, $\chi^M = \chi^{(M,0)}$. From Remark 12.6, we can conclude the following.

12.7 Remark.   1) *If r is even and χ is the square of one of the 2048 characters, then $\chi^{(M,r)}$ is the square of one these characters too.*

2) *If $M \in \Gamma_{n,\vartheta}$ is an element of the theta group, then*

$$\chi^{(M,r)} = \chi^{(M,0)} = \chi^M$$

for all r.

Character squares will play an important role in what follows.

We recall that $[\Gamma_2[2], r/2, v]$ has an enforced zero if

$$\chi^{(M,r)}(P_3) \neq 1$$

for some M. Because $P_3^2 = E$, characters can have only the values ± 1 on P_3. Hence character squares have the value 1. If r is even and χ is a character square, then $\chi^{(M,r)}$ is a character square by Remark 12.7. Hence we have *no enforced zero* in this case. Fortunately, in all other cases we have an enforced zero.

12.8 Proposition.   *The space $[\Gamma_2[2], r/2, v_\vartheta^r \chi]$ has an enforced zero, if one of the following conditions is satisfied:*

1) *r is odd.*

2) *r is even and χ is not a character square.*

Proof. We have use the formula

$$\chi^{(M,r)}(P_3) = \left(\frac{v_\vartheta^M(P_3)}{v_\vartheta(P_3)} \right)^r \chi(MP_3M^{-1}).$$

For this, we need some information on

$$\varepsilon(M) := v_\vartheta^M(P_3)/v_\vartheta(P_3).$$

We shall extract this information only from the fact that v_ϑ^M is the multiplier system of a well-known modular form, namely the theta series $\vartheta[\mathfrak{m}]$ with $\mathfrak{m} = M\{0\}$. More exactly,

$$\frac{\vartheta[\mathfrak{m}](P_3 Z)}{\vartheta(P_3 Z)} = \varepsilon(M) \frac{\vartheta[\mathfrak{m}](Z)}{\vartheta(Z)}.$$

We know how theta series transform under P_3. The transformation formula shows that $\varepsilon(M)$ takes only the values ± 1 and that both values actually occur.

For the proof of Proposition 12.8, we have to distinguish between two cases:

First case. χ is a character square. Then, by assumption, r is odd and

$$\chi^{(M,r)}(P_3) = \varepsilon(M)^r$$

takes both values $\pm$, in particular -1.

Second case. χ is not a character square. In this case, we construct M from the theta group, such that $\chi^{(M,r)}(P_3) = -1$. For M in the theta group, we have $v_\vartheta^M = \vartheta$ and hence

$$\chi^{(M,r)}(P_3) = \chi(MP_3M^{-1}).$$

We have to exhibit an element M of the theta group such that this expression is -1. What does it mean that χ is not a character square? Character squares are 1 on squares of elements and, conversely, each element with this property is a character square. In our case, this follows easily from the isomorphy of $\mathcal{G}_2$ and its character group with $(\mathbb{Z}4\mathbb{Z})^4 \times (\mathbb{Z}/2\mathbb{Z})^5$. Since χ is assumed not to be a character square, we must have

$$\chi(N) \neq 1 \quad \text{for some } M \in \{T_1^2, T_2^2, \tilde{T}_1^2, \tilde{T}_2^2, T_3, \tilde{T}_3, P_1, P_2, P_3\},$$

since these elements generate the subgroup consisting of elements of order ≤ 2. So we have to show that each of them can be written modulo $\tilde{\Gamma}[4,8]$ in the form MP_3M^{-1}, with some element M of the theta group. This can easily be verified. $\qquad\square$

Now we investigate the case where r is even and χ is a character square. For this, we recall the embedding

$$\Gamma_1[2] \times \Gamma_1[2] \longrightarrow \Gamma_2[2],$$

$$\begin{pmatrix} a & b \\ c & d \end{pmatrix}, \begin{pmatrix} \alpha & \beta \\ \gamma & \delta \end{pmatrix} \longmapsto \begin{pmatrix} a & & b & \\ & \alpha & & \beta \\ c & & d & \\ & \delta & & \gamma \end{pmatrix}.$$

This induces an embedding

$$\mathcal{G}_1 \times \mathcal{G}_1 \longrightarrow \mathcal{G}_2 \quad (\mathcal{G}_n = \Gamma_n[2]/\tilde{\Gamma}_n[4,8]).$$

The "restriction" of a character χ on $\mathcal{G}_2$ is a character on $\mathcal{G}_1 \times \mathcal{G}_1$. This corresponds to a pair of characters (χ_1, χ_2) of $\mathcal{G}_1$. We write

$$(\chi_1, \chi_2) = \chi|\mathcal{G}_1 \times \mathcal{G}_1.$$

We have

$$\chi\begin{pmatrix} a & b \\ c & d \end{pmatrix}, \begin{pmatrix} \alpha & \beta \\ \gamma & \delta \end{pmatrix} \longmapsto \begin{pmatrix} a & & b & \\ 0 & \alpha & & \beta \\ c & & d & \\ & \delta & & \gamma \end{pmatrix} = \chi_1\begin{pmatrix} a & b \\ c & d \end{pmatrix} \cdot \chi_2\begin{pmatrix} \alpha & \beta \\ \gamma & \delta \end{pmatrix}.$$

The map $\chi \mapsto (\chi_1, \chi_2)$ cannot be injective. But we have the following fact.

12.9 Lemma. *The map*

$$\chi \mapsto (\chi_1, \chi_2) = \chi | \mathcal{G}_1 \times \mathcal{G}_1$$

is a bijection between the set of all character squares of $\mathcal{G}_2$ and the set of pairs of character squares of $\mathcal{G}_1$.

The proof follows from the explicit structure: we shall give only a hint. The group of characters of $\mathcal{G}_2$ is isomorphic to $(\mathbb{Z}/4\mathbb{Z})^4 \times (\mathbb{Z}/2\mathbb{Z})^5$, and hence the group of character squares is isomorphic to $(\mathbb{Z}/2\mathbb{Z})^4$. On the other hand, the group of characters of $\mathcal{G}_1$ is isomorphic to $(\mathbb{Z}/4\mathbb{Z})^2$. The group of its squares is isomorphic to $(\mathbb{Z}/2\mathbb{Z})^2$. $\square$

12.10 lemma. *Let χ be a character of $\mathcal{G}_2$ and let*

$$f \in [\Gamma_2[2], r/2, v_\vartheta^r \chi].$$

Then $f\left(\begin{smallmatrix} z_0 & 0 \\ 0 & z_2 \end{smallmatrix}\right)$ is a finite sum of functions

$$f_1(z_0) f_2(z_2) \text{ with } f_\nu \in [\Gamma_1[2], r/2, v_\vartheta^r \chi_\nu] \quad (\nu = 1, 2).$$

Here, $(\chi_1, \chi_2) = \chi | \mathcal{G}_1 \times \mathcal{G}_1$.

Proof. We use the transformation behavior of f under the elements of the image of $\Gamma_1[2] \times \Gamma_1[2] \hookrightarrow \Gamma_2[2]$. We can see that $f\left(\begin{smallmatrix} z_0 & 0 \\ 0 & z_2 \end{smallmatrix}\right)$, for fixed z_0 as function of z_2, is contained in $[\Gamma_1[2], r/2, v_\vartheta^r \chi_2]$, and conversely. We choose a basis $g_1, \ldots, g_m$ of $[\Gamma_1[2], r/2, v_\vartheta^r \chi_2]$. We have

$$f \left(\begin{matrix} z_0 & 0 \\ 0 & z_2 \end{matrix} \right) = \sum h_i(z_0) g_i(z_2)$$

with certain functions $h_i \in [\Gamma_1[2], r/2, v_\vartheta^r \chi_1]$. $\square$

Now we use the fact that the structure theorem is known in the case $n = 1$. The functions $f_\nu \in [\Gamma_1[2], r/2, v_\vartheta^r \chi_\nu]$ are linear combinations of monomials

$$\vartheta \begin{bmatrix} 0 \\ 0 \end{bmatrix}^{r_1} \vartheta \begin{bmatrix} 0 \\ 1 \end{bmatrix}^{r_2} \vartheta \begin{bmatrix} 1 \\ 0 \end{bmatrix}^{r_3}, \quad r_1 + r_2 + r_3 = r,$$

where the characters have to satisfy the condition

$$\chi_\nu \begin{pmatrix} 1 & 2 \\ 0 & 1 \end{pmatrix} = i^{r_2}, \quad \chi_\nu \begin{pmatrix} 1 & 0 \\ 2 & 1 \end{pmatrix} = i^{r_3}.$$

This formula shows that if χ is a character square, then r_2 and r_3 both have to be even. Also, if r is even, then r_3 has to be even. Now we make use of the formula

$$\vartheta \begin{bmatrix} a_1 \\ a_2 \\ b_1 \\ b_2 \end{bmatrix} \left(\begin{matrix} z_0 & 0 \\ 0 & z_2 \end{matrix} \right) = \vartheta \begin{bmatrix} a_1 \\ b_1 \end{bmatrix} \vartheta \begin{bmatrix} a_2 \\ b_2 \end{bmatrix}$$

to obtain the following lemma.

12.11 Lemma. *Let χ be a character square of $\mathcal{G}_2$ and*

$$[\Gamma_2[2], r/2, v_\vartheta^r \chi], \quad r \text{ even.}$$

Then there exists a linear combination g of monomials in the squares of theta series $\vartheta[\mathbf{m}]$ with the following properties:

1) $\quad f\begin{pmatrix} z_0 & 0 \\ 0 & z_2 \end{pmatrix} = g\begin{pmatrix} z_0 & 0 \\ 0 & z_2 \end{pmatrix}.$

2) *If $\tilde{\chi}$ is the character which belongs to this monomial, then*

$$\tilde{\chi}|\mathcal{G}_1 \times \mathcal{G}_1 = \chi|\mathcal{G}_1 \times \mathcal{G}_1.$$

For trivial reason, the characters corresponding to theta squares are character squares. We obtain the following result from Lemma 12.9.

Supplement. *We have $\tilde{\chi} = \chi$.* So, we have proved the following proposition.

12.12 Proposition. *Let χ be a character square of $\mathcal{G}_2$ and*

$$[\Gamma_2[2], r/2, v_\vartheta^r \chi], \quad r \text{ even.}$$

Then there exists a linear combination of monomials in the theta squares which are all contained in the space $[\Gamma_2[2], r/2, v_\vartheta^r \chi]$ and are such that

$$f\begin{pmatrix} z_0 & 0 \\ 0 & z_2 \end{pmatrix} = g\begin{pmatrix} z_0 & 0 \\ 0 & z_2 \end{pmatrix}.$$

Induction on r now gives our main result:

12.13 Theorem. *The space*

$$[\Gamma_2[4,8], r/2, v_\vartheta^r]$$

is generated by the monomials of degree r in the ten theta series.

Exercises for Sect. VII.12

1. Show that the ten theta series are linearly independent.

2. Show that the 55 functions $\vartheta[\mathbf{m}]\vartheta[\mathbf{n}]$ are linearly independent.

3. The function $\vartheta(2Z)^2$ (observe the factor 2) is a modular form of weight one with respect to $\Gamma_2[2]$. Verify this by means of generators.

4. Show that the multiplier system of $\vartheta(2Z)^2$ is trivial on $\Gamma_2[4,8]$.

5. By the structure theorem, $\vartheta(2Z)^2$ must be expressible as a linear combination of the $\vartheta[\mathfrak{m}]\vartheta[\mathfrak{n}]$. Compute this combination explicitly.

    Hint. It is enough to take the ten squares $\vartheta[\mathfrak{m}]^2$.

VIII. Appendix: Algebraic Tools

1. Divisibility

Here, we collect together the basic notions of divisibility. In the following, R is an *integral domain*, i.e. an *associative and commutative ring with unity* $1 \neq 0$ which is free of zero divisors:

$$a \cdot b = 0 \quad \Longrightarrow \quad a = 0 \text{ or } b = 0 \quad (a, b \in R).$$

Fields are examples of integral domains. Every integral domain is a subring of a field K. One can arrange that K consists of all fractions a/b with $a, b \in R$, $b \neq 0$. This field is unique up to isomorphism and is called the quotient field of R. The construction of the quotient field is standard. We define a/b to be the equivalence class of (a, b) with respect to an obvious equivalence relation.

The *polynomial ring* in n variables $R[X_1, \ldots, X_n]$ was introduced in Sect. V.3. It consists of all formal finite sums

$$\sum_{0 \leq \nu_1, \ldots, \nu_n} a_{\nu_1 \ldots \nu_n} X_1^{\nu_1} \cdots X_n^{\nu_n}, \quad a_{\nu_1 \ldots \nu_n} \in R,$$

with the usual rules for addition and multiplication. The base ring R is embedded into the polynomial ring. The element $a \in R$ is identified with the polynomial whose zero coefficient equals a and whose other coefficients vanish.

The case $n = 1$ is of special importance. By the degree d of an element $P \in R[X]$ which is different from zero, we understand the index of the highest nonzero coefficient of P in the representation

$$P = a_d X^d + \ldots + a_1 X + a_0, \quad a_i \in R \text{ for } 1 \leq i \leq d - 1, \quad a_d \neq 0.$$

Additionally, we define

$$\deg(0) := -\infty.$$

If R is an integral domain, which we will assume from now on, we have

$$\deg(PQ) = \deg(P) + \deg(Q).$$

Hence $R[X]$ is also an integral domain. By induction on n, using the isomorphism

$$(R[X_1, \ldots, X_{n-1}]) [X_n] \cong R[X_1, \ldots, X_n],$$

we can prove that the polynomial ring in arbitrarily many variables is an integral domain. An element ε is called a *unit* in R if the equation

$$\varepsilon x = 1$$

can be solved in R, i.e. if ε^{-1} exists in R. The set R^* of all units is a group under multiplication.

Examples.

1) $\mathbb{Z}^* = \{\pm 1\}$.
2) If K is a field, then $K^* = K - \{0\}$.
3) Let $R[X_1, \ldots, X_n]$ be the polynomial ring in n variables. Then

$$R[X_1, \ldots, X_n]^* = R^*.$$

4) Let $R = \mathbb{C}\{z_1, \ldots, z_n\}$ be the ring of convergent power series in n variables. Then

$$R^* = \{P \in \mathbb{C}\{z_1, \ldots, z_n\}; \quad P(0) \neq 0\}.$$

($P(0)$ is the constant term of the power series.)

One can make use of the fact that a convergent power series with nonzero constant term defines an analytic function f without zeros in a small open neighborhood of 0. The function $1/f$ is also analytic and can be expanded into a power series.

We recall the notion of a *greatest common divisor* $\gcd(r, s)$ of two elements r, s of an integral domain R. We assume that r and s are not both 0. An element $d \in R$ is called the *greatest common divisor* of r and s,

$$d = \gcd(r, s),$$

if r and s are both divisible by d and if any element s which divides r and s is also a divisor of d:

(a) $d|r$ and $d|s$;

(b) $t|r$ and $t|s$ $\Longrightarrow$ $d|t$.

It is easy to see that the greatest common divisor, if it exists, is uniquely determined up to a unit.

1.1 Definition. *An element $a \in R - R^*$ is called*

a) *indecomposable if*

$$a = bc \quad \Longrightarrow \quad b \text{ or } c \text{ is a unit};$$

b) *a prime element if*

$$a|bc \quad \Longrightarrow \quad a|b \text{ or } a|c$$

($a|b$ means that the equation $b = ax$ is solvable in R).

Of course, prime elements are indecomposable, but the converse is usually false.

Example. Let $R = \mathbb{C}[X]$ be the polynomial ring in one variable over $\mathbb{C}$ and let R_0 be the subring of all polynomials without a linear term. The element X^3 is indecomposable in R_0 but not a prime: $X^3 | X^2 \cdot X^4$.

2. Factorial Rings (UFD rings)

2.1 Definition. *The integral domain R is called a **factorial** ring or **UFD** ring if the following two conditions are satisfied:*

1) *Each element $a \in R - R^*$ can be written as a product of finitely many indecomposable elements.*

2) *Each indecomposable element is prime.*

In factorial rings, the decomposition into primes is unique in the following sense. Let

$$a = u_1 \cdot \ldots \cdot u_n = v_1 \cdot \ldots \cdot v_m$$

be two decompositions of $a \in R - R^*$ into primes. Then we have:

a) $m = n$.

b) There exists a permutation σ of the digits $1, \ldots, n$ such that

$$u_\nu = \varepsilon_\nu v_{\sigma(\nu)}, \quad \varepsilon_\nu \in R^* \quad \text{for } 1 \le \nu \le n.$$

It is easy to prove this by induction.

Examples of factorial rings.

1) Every field is factorial.
2) $\mathbb{Z}$ is factorial.
3) By an important theorem of Gauss, the polynomial ring $R[X_1, \ldots, X_n]$ over a factorial ring is factorial too.

2.2 Theorem (Gauss). *The polynomial ring $R[X]$ over a factorial ring is factorial too.*

Because of the great importance of this theorem, we shall give a proof here. It rests on the Euclidean algorithm for polynomials:

Let $Q \in R[X]$ be a normalized polynomial, i.e. the highest coefficient of Q is 1. Then any polynomial $P \in R[X]$ admits a unique decomposition

$$P = AQ + B, \quad \deg B < \deg Q \qquad \text{(division with remainder)}.$$

A simple proof can be given by means of induction on the degree of P and will be skipped. Of course, "division with remainder" holds already if the highest coefficient of Q is a unit (not necessarily 1). Since, in a field, any nonzero element is a unit, we can divide with remainder through any nonzero polynomial. This has the effect that divisibility in a polynomial ring of one variable over a field is very simple. Hence we shall now prove Proposition 2.2, in an initial step, for a field K instead of R.

2.3 Lemma. *In the polynomial ring of one variable over a field K, there always exists the **greatest common divisor** of two elements P, Q. This greatest common divisor can be obtained by combining P and Q, i.e. it is of the form*

$$D = RP + SQ \qquad (R, S \in K[X]).$$

Sketch of the proof. We consider the set

$$\mathfrak{a} = \{ RP + SQ \mid R, S \in K[X] \}$$

and choose from it an element $D \neq 0$ of minimal degree. By means of division with remainder, we can show that

$$\mathfrak{a} = D \cdot K[X].$$

It is easy to see that D is the greatest common divisor of P and Q.   □

2.4 Corollary. *The polynomial ring in one variable over a field is factorial.*

Proof. We restrict ourselves to the essential part: each indecomposable element is prime. So, let P be indecomposable. But we assume that

$$P \mid RS, \quad P \nmid R, \quad P \nmid S.$$

From Lemma 2.3, we obtain the result that the equations

$$PX + RY = 1 \quad \text{and} \quad P\tilde{X} + S\tilde{Y} = 1$$

are solvable. Taking the product of the two equations, we obtain $P\mid 1$. This contradicts $P \nmid R$.   □

    Two elements of an integral domain are called *coprime* if the only common divisors are units.

2.5 Corollary. *Let $K \subset L$ be a subfield L and let P, Q be two polynomials from $K[X]$. The two polynomials are coprime in $K[X]$ iff they are coprime in the larger ring $L[X]$.*

The prove follows from Lemma 2.3 and the simple fact that

$$P \neq 0 \in K[X], \ Q \in K[X], \text{ and } Q/P \in L[X] \implies Q/P \in K[X].$$

Now we consider, instead of a field, an arbitrary factorial ring R. The prime factorization shows that in a factorial domain, the greatest common divisor always exists. Slightly more generally, we can define the greatest common divisor of n elements which are not zero and show that it exists in factorial rings. It is unique up to a unit.

2.6 Definition. *The **content** of a polynomial $P \in R[X]$ which is different from zero over a factorial ring R is the greatest common divisor of all coefficients of P:*

$$\mathcal{I}(P) := \gcd(a_0, \ldots, a_n) \text{ for } P = a_n X^n + \ldots + a_0.$$

Of course, the content is determined only up to a unit.

2.7 Lemma. *Let P, Q be two polynomials (in one variable) over a factorial ring R; then*

$$\mathcal{I}(PQ) \sim \mathcal{I}(P) \cdot \mathcal{I}(Q).$$

Here the symbol "$\sim$" means "equal up to a unit". Two elements which differ by a unit are also called *associated*.

We leave the proof of this lemma to the reader as a not quite simple exercise. All that one has to show is

$$\mathcal{I}(P) \sim 1, \quad \mathcal{I}(Q) \sim 1 \quad \implies \quad \mathcal{I}(PQ) \sim 1.$$

Lemma 2.7 is a tool to compare divisibility in $R[X]$ and in $K[X]$, where K denotes the field of fractions of R.

A polynomial $P \neq 0 \in R[X]$ is called *primitive* if its content is a unit. For example, normalized polynomials are primitive.

2.8 Lemma. *Let R be a factorial ring with a field of fractions K, and let $P \in R[X]$ be a primitive polynomial. Let $Q \in R[X]$ be a further polynomial over R. We then have*

$$P|Q \text{ in } R[X] \quad \Longleftrightarrow \quad P|Q \text{ in } K[X].$$

This lemma is an easy corollary of Lemma 2.7 if we take into account the fact that every polynomial over K can be transformed, by multiplication by a suitable nonzero element of R, into a polynomial over R. □

We have now collected together everything that we need to prove Gauss's theorem. All that we have to show is:

Every irreducible (= indecompasable) polynomial $P \neq 0 \in R[X]$ is prime.

First case. $\deg P = 0$, i.e. $P \in R$. Obviously, an element $r \in R$ is indecomposable (or prime) in R if it is so in $R[X]$. Hence the claim follows from the fact that R is factorial.

Second case. $\deg P > 0$. Since P is indecomposable, P must be primitive. We assume

$$P|QS \text{ in } R[X] \qquad (Q, S \in R[X]).$$

We then obtain

$$P|QS \text{ in } K[X],$$

and hence

$$P|Q \quad \text{or} \quad P|S \text{ in } K[X]$$

(because of Corollary 2.4). The claim follows from Lemma 2.8. □

3. The Discriminant

Let K be a field and let

$$P = X^n + a_{n-1}X^{n-1} + \ldots + a_0$$

be a normalized polynomial over K. It is well known that there exists a splitting field for P. This is a field L which contains K as a subfield and in which P decomposes into a product of linear factors,

$$P = (X - \alpha_1) \cdot \ldots \cdot (X - \alpha_n).$$

(The elements $\alpha_1, \ldots, \alpha_n$ are the zeros of P.) The *discriminant* of P is defined as

$$\Delta := \Delta_P := \prod_{i<j}(\alpha_j - \alpha_i)^2.$$

3.1 Remark. *For each natural number n, there exists a unique "universal polynomial"*

$$\Delta^{(n)} \in \mathbb{Z}[X_1, \ldots, X_n]$$

in n variables over $\mathbb{Z}$, such that for each normalized polynomial

$$P = X^n + a_{n-1}X^{n-1} + \ldots + a_0$$

over an arbitrary field the relation

$$\Delta_P = \Delta^{(n)}(a_0, \ldots, a_{n-1})$$

holds.

We should mention here that the elements a of an abelian group can be multiplied by elements of $\mathbb{Z}$:

$$na := \begin{cases} \overbrace{a + \ldots + a}^{n \text{ times}} & \text{if } n > 0, \\ -(-n)a & \text{if } n < 0, \\ 0 & \text{if } n = 0. \end{cases}$$

As a consequence, in a polynomial over $\mathbb{Z}$, we can substitute the variables by elements of an arbitrary commutative ring with unity (and obtain an element of this ring again).

We now indicate the *proof* of Remark 3.1. As is well known (and trivial), the coefficients a_i of P are, up to a sign, the elementary symmetric polynomials in the roots $\alpha_1, \ldots, \alpha_n$. The expression

$$\prod_{i<j}(\alpha_j - \alpha_i)^2$$

is a symmetric polynomial in $\alpha_1, \ldots, \alpha_n$. Now the claim follows from a well-known result about *elementary symmetric polynomials,* which states that every symmetric polynomial with integral coefficients can be written as a polynomial with integral coefficients in the elementary symmetric polynomials. □

By means of the discriminant, we can characterize the square-free normalized polynomials in one variable over a field or, more generally, over a factorial ring. Here an element $r \in R$ is called square-free, if

$$x^2 | r \quad \Longrightarrow \quad x \text{ is a unit.}$$

3.2 Proposition. *Let $P \in R[X]$ be a normalized polynomial over a factorial ring R. Then the following two statements are equivalent:*
1) *P is square-free in $R[X]$.*
2) *$\Delta_P \neq 0$.*

Proof. Let K be the quotient field of R. It follows from Remark 3.1 that *P is square-free in $R[X]$ iff P is square-free in $K[X]$.* Hence we can assume that

$$R = K \quad \text{(field)}.$$

For the proof of Proposition 3.2, we need the *derivative* of a polynomial

$$P = a_n X^n + \ldots + a_0 \in K[X].$$

This is defined formally by

$$P' = n a_n X^{n-1} + \ldots + a_1.$$

We can verify the usual rules

$$(P + Q)' = P' + Q',$$
$$(P \cdot Q)' = P' \cdot Q + P \cdot Q'.$$

Now we show that *a polynomial over a field K (of characteristic zero) is square-free if and only if P and P' are coprime.*
Proof. 1) Assume $P = S^2 Q$. Then S is a common divisor of P and P'.
2) Now assume, conversely, that S is a common divisor of P and P'. We can assume that S is a prime element. We define Q by

$$P = SQ.$$

Then

$$P' = S'Q + SQ'.$$

It follows from $S|P'$ that $S|S'Q$ and hence $S|Q$. This shows that P is not square-free. □

For the proof of Proposition 3.2, we can replace K by a splitting field L of P. (The statement "$\gcd(P, P') = 1$" does not change if we replace K by L.) So we can assume that P decomposes,

$$P = (X - \alpha_1) \ldots (X - \alpha_n).$$

But then Proposition 3.2 is trivial, since P is square-free iff it has no multiple roots, and this means that the discriminant is different from zero. □

4. Algebraic Function Fields

Let K be a subfield of the field Ω. For a subset $\mathcal{M} \subset \Omega$, we use the following notations:

a) $K\langle\mathcal{M}\rangle$ is the smallest K-subvector space of Ω which contains $\mathcal{M}$.
b) $K[\mathcal{M}]$ is the smallest subring of Ω which contains K and $\mathcal{M}$.
c) $K(\mathcal{M})$ is the smallest subfield of Ω which contains K and $\mathcal{M}$.

We are mainly interested in the case where

$$\mathcal{M} = \{x_1, \ldots, x_m\}$$

is a finite set. Then

a) $K\langle x_1, \ldots, x_m\rangle = \sum_{i=1}^{m} Kx_i;$
b) $K[x_1, \ldots, x_m] = \{P(x_1, \ldots, x_m); \quad P \in K[X_1, \ldots, X_m]\};$
c) $K(x_1, \ldots, x_m) = \{a/b; \quad a, b \in K[x_1, \ldots, x_m], \ b \neq 0\}.$

(This field is naturally isomorphic to the field of fractions of $K[x_1, \ldots, x_m]$.)

We call Ω *finitely generated as a ring* over K if

$$\Omega = K[x_1, \ldots, x_m],$$

and *finitely generated as a field* over K if

$$\Omega = K(x_1, \ldots, x_m)$$

for suitable elements $x_1, \ldots, x_m$.

Definition. The field Ω is called an *algebraic function field over K* if it is finitely generated as a field over K.

The Transcendental Degree.

Let $x_1, \ldots, x_m$ be elements of Ω. There is a natural homomorphism of the polynomial ring in m variables into Ω,

$$K[X_1, \ldots, X_m] \longrightarrow \quad \Omega, \quad P \longmapsto P(x_1, \ldots, x_m).$$

Its image is $K[x_1, \ldots, x_m]$. The elements $x_1, \ldots, x_m$ are called *algebraically independent* if this homomorphism is injective. Then $K[x_1, \ldots, x_m]$ is isomorphic to the polynomial ring. The field $K(x_1, \ldots, x_m)$ is then isomorphic to the *field of rational functions* (This is the quotient field of the polynomial ring.)

By the way, all of the homomorphisms which we consider here keep K elementwise fixed.

4.1 Lemma. $n + 1$ *polynomials in n variables over a field K are always algebraically dependent (which means not algebraically independent).*

This lemma is not deep, but it is not trivial. We skip a proof of it.   □

The next statement follows from 4.1.

4.2 Lemma. *If a field Ω can be generated over K by m elements, i.e. $\Omega = K(x_1, \ldots, x_m)$, then $m + 1$ elements of Ω will be algebraically dependent over K.*

As a consequence, in an algebraic function field there exists a maximal number of algebraically independent elements. This maximal number is called the *transcendental degree of the function field,*

$$n = \mathrm{tr}(\Omega/K).$$

So, we can say that in an algebraic function field of transcendental degree n, there exist n algebraically independent elements. $n + 1$ elements are algebraically dependent. A system of n algebraically independent elements is called a *transcendental basis* of Ω/K.

Not quite obvious (but not deep) is the following fact.

4.3 Lemma. *Let*

$$\Omega = K(x_1, \ldots, x_m)$$

be an algebraic function field. Any maximal algebraically independent subsystem of $x_1, \ldots, x_m$ is a transcendental basis.

Algebraic Extensions

An element $x \in \Omega$ is called *algebraic* over K if there exists an equation of the kind

$$x^n + a_{n-1}x^{n-1} + \ldots + a_0 = 0, \quad a_\nu \in K, \quad 0 \le \nu < n.$$

A field extension Ω/K is called algebraic if each element is algebraic over K. It is called a *finite* (or finite algebraic) extension if, in addition, Ω is finitely generated over K. This means that a finite algebraic extension is the same as an algebraic function field of transcendental degree 0.

4.4 Remark. *The following two statements for a field extension $\Omega \supset K$ are equivalent:*

a) *Ω is a finite algebraic extension over K;*
b) *Ω is a finite-dimensional K-vector space.*

Proof.

a) $\Rightarrow$ b). Let $\Omega = K(x_1,\dots,x_m)$. The elements x_ν $(1 \le \nu \le m)$ satisfy polynomial equations of some degree $\le N$. It is easy to show that the K-vector space Ω is generated by the elements

$$x_1^{\nu_1}\dots x_m^{\nu_m}, \quad 0 \le \nu_1,\dots,\nu_m < N.$$

b) $\Rightarrow$ a). Let $x \in \Omega$. The powers

$$1, x, x^2,\dots,x^m$$

are linearly dependent if m is large enough. $\qquad\square$

We call

$$[\Omega : K] := \dim_K \Omega$$

the *degree* of the finite algebraic extension. If it is monogenous, i.e. $\Omega = K[x]$ for a suitable $x \in \Omega$, then $[\Omega : K]$ is the minimal degree of a nonzero polynomial with root x.

The characterization b) shows the following.

4.5 Remark. *Let $K \subset L$ and $L \subset \Omega$ be two finite algebraic extensions; then $K \subset \Omega$ is also a finite algebraic extension and we have*

$$[\Omega : L][L : K] = [\Omega : K].$$

The following theorem is an important result of elementary algebra.

4.6 Theorem (theorem of the primitive element). *Let $\Omega \supset K$ be a finite algebraic extension, and let K (and then also Ω) be of characteristic 0 ($n \cdot 1 \ne 0$ for $n \in \mathbb{N}$). Then this extension is monogenous:*

$$\Omega = K[x], \quad x \in \Omega \text{ suitable.}$$

An important consequence of the theorem of primitive element states the following.

4.7 Proposition. *Let $\Omega \supset K$ be an algebraic field extension of characteristic 0. Assume that there exists a number $N \in \mathbb{N}$ such that each $x \in \Omega$ satisfies a polynomial equation of degree $\le N$. Then Ω/K is a finite algebraic extension.*

If Ω/K is an algebraic function field and $x_1,\dots,x_n$ a transcendental basis, then

$$\Omega \supset K(x_1,\dots,x_n)$$

is a finite algebraic extension. Using the theorem of the primitive element, we obtain the following result.

4.8 Proposition. *An algebraic function field of transcendental degree n over a field K of characteristic 0 can be generated by $n+1$ elements.*

Final Remarks

Let $\Omega = K(x_0,\ldots,x_n)$ be a function field of transcendental degree n; then there exists an irreducible polynomial P (= prime element) in the polynomial ring of $n+1$ variables with the property

$$P(x_0,\ldots,x_n) = 0.$$

This polynomial essentially determines the field extension, since Ω is isomorphic to the quotient field of the factor ring

$$K[X_0,\ldots,X_n]/(P).$$

But P depends on the choice of the generating system.

Even in the case $n=1$ and $K=\mathbb{C}$ it is a complicated problem to decide when two polynomials lead to the same function field. For each function field of transcendental degree one, there exists a compact Riemann surface X and an isomorphism

$$K \xrightarrow{\sim} \mathcal{M}(X) = \text{field of meromorphic fucntions on } X,$$

which fixes $\mathbb{C}$ elementwise. Two function fields of transcendental degree one over $\mathbb{C}$ are isomorphic if and only if the corresponding Riemann surfaces are biholomorphically equivalent. Torelli's theorem says that this the case if and only if the corresponding period lattices are equivalent (in a sense which has to be made precise). This, historically, was the starting point for the interest in abelian functions and higher modular functions.

References

[Ac] Accola, R.D.M.: *Riemann Surfaces, Theta Functions, and Abelian Automorphism Groups*, Lecture Notes in Mathematics, No. 483. Springer, Berlin, Heidelberg, 1975

[AS] Andreotti, A., Stoll, W.: *Analytic and Algebraic Dependence of Meromorphic Functions*, Lecture Notes in Mathematics, No. 234. Springer, Berlin, Heidelberg, 1971

[An] Andrianov, A.N.: *Quadratic Forms and Hecke Operators*, Grundlehren der mathematischen Wissenschaften in Einzeldarstellungen. Bd. 286. Springer, Berlin, Heidelberg, 1987

[Bel] Bellman, R.E.: *A Brief Introduction to Theta Functions*, Holt, Rinehart and Winston, New York, 1961

[Ber] Bers, L.: *Introduction to Several Complex Variables*, Courant Institute of Mathematical Sciences, New York University, 1964

[CL] Chenkin, G.M., Leiterer, J.: *Theory of Functions on Complex Manifolds*, Monographs in Mathematics, No. 79. Birkhäuser, Basel, Stuttgart, 1984

[Ca] Cartan, H.: *Elementary Theory of Analytic Functions of one or Several Complex Variables*, Hermann, Paris, 1963

[Co] Coble, A.B.: *Algebraic Geometry and Theta Functions*, American Mathematical Society, New York, 1929

[Co] Conforto, F.: *Abel'sche Funktionen und algebraische Geometrie*, Grundlehren der mathematischen Wissenschaften in Einzeldarstellungen, Nr. 84. Springer, Berlin, 1956

[Du] Duma, A.: *Holomorphe Differentiale höherer Ordnung auf kompakten Riemann'schen Flächen*, Schriftenreihe des Mathematischen Instituts der Universität Münster, Ser. 2, Nr.14. Münster, 1978

[Fa] Fay, J.D.: *Theta Functions on Riemann Surfaces*, Lecture Notes in Mathematics, No. 352. Springer, Berlin, Heidelberg, 1973

[FG] Fischer, G.: *Complex Analytic Geometry*, Lecture Notes in Mathematics, No. 538. Springer Berlin, Heidelberg, 1976.

[Fo1] Forster, O.: *Lectures on Riemann Surfaces*, Graduate Texts in Mathematics, No. 81. Springer, New York, Heidelberg, 1999

[Fo2] Forster, O.: *Riemannsche Flächen*, Heidelberger Taschenbücher, Nr. 184. Springer, Berlin, Heidelberg, 1977

[FB] Freitag, E., Busam, R.: *Complex Analysis*, 2nd edn. Springer, Berlin, Heidelberg, 2008

[Fr1] Freitag, E.: *Siegelsche Modulfunktionen*, Grundlehren der mathematischen Wissenschaften, Bd. 254. Springer, Berlin, Heidelberg, 1983

[Fr2] Freitag, E.: *Singular Modular Forms and Theta Relations*, Lecture Notes in Mathematics, No. 1487. Springer, Berlin, Heidelberg, 1991

[Fr3] Freitag, E.: *Zur Theorie der Modulformen zweiten Grades*, Nachr. Akad. Wiss. Göttingen, 1965

[Fu1] Fuks, B.A.: *Special Chapters in the Theory of Analytic Functions of Several Complex Variables*, Translations of Mathematical Monographs, No 14. Providence, RI, 1965.

[Fu2] Fuks, B.A.: *Theory of Analytic Functions of Several Complex Variables*, Translations of Mathematical Monographs, No. 8. Providence, RI, 1963.

[GF] Grauert, H., Fritzsche, K.: *Several Complex Variables*, Graduate Texts in Mathematics, No. 38. Springer, New York, Heidelberg, 1976

[Go] Götzky, F.: *Über eine zahlentheoretische Anwendung von Modulfunktionen zweier Veränderlicher*, Math. Ann. **100**, 411–437 (1928)

[GRe] Grauert, H. Remmert, R.: *Theorie der Steinschen Räume*, Grundlehren der mathematischen Wissenschaften in Einzeldarstellungen, Nr. 227. Springer, Berlin, Heidelberg, 1977

[GRo] Gunning, R.C., Rossi, H.: *Analytic Functions of Several Complex Variables*, Prentice-Hall, Englewood Cliffs, NJ, 1965

[Gr] Grauert, H.: *Einführung in die Funktionentheorie mehrerer Veränderlicher*, Springer, Berlin, Heidelberg, 1974

[Gu1] Gunning, R.C.: *Riemann Surfaces and Generalized Theta Functions*, Ergebnisse der Mathematik und ihrer Grenzgebiete, Bd. 91. Springer, Berlin, Heidelberg, 1976

[Gu2] Gunning, R.C.: *Vorlesungen über Riemann'sche Flächen*, B.I.-Hochschultaschenbücher, Nr. 837. Bibliographisches Institut, Mannheim, 1972

[Hö] Hörmander, L.: *An Introduction to Complex Analysis in Several Variables*, North-Holland Mathematical Library, No. 7. Van Nostrand, Princeton, NJ, 1966

[Jo] Jost, J.: *Compact Riemann Surfaces: An Introduction to Contemporary Mathematics*, Springer, Berlin, Heidelberg, 1997.

[Ig1] Igusa, J,I.: *On the graded ring of theta constants*, Am. J. Math. **86**, 219–246 (1964)

[Ig2] Igusa, J,I.: *On the graded ring of theta constants II*, Am. J. Math. **88**, 221–236 (1966)

[Ig3] Igusa, J,I.: *On Siegel's modular forms of Genus II*, Am. J. Math. **89**, 817–855 (1964).

[Ig4] Igusa, J,I.: *Theta Functions*, Grundlehren der mathematischen Wissenschaften in Einzeldarstellungen, Bd. 194. Springer, Berlin, Heidelberg, 1972

[Ke] Kempf, G.R.: *Complex Abelian Varieties and Theta Functions*, Springer, Berlin, Heidelberg, 1991

[Ki] Kitaoka, Y.: *Lectures on Siegel Modular Forms and Representation by Quadratic Forms*, Springer, Berlin, Heidelberg, 1986

[Kl] Klein, F.: *Riemann'sche Flächen*, Teubner-Archiv zur Mathematik, Bd. 5. Teubner, Leipzig, 1986

[Kl] Klingen, H.: *Introductory Lectures on Siegel Modular Forms*, Cambridge Studies in Advanced Mathematics, No. 20, Cambridge University Press, Cambridge, 1990

[Ko] Kodaira, K.: *Introduction to Complex Analysis*, Cambridge University Press, Cambridge, 1984

[Koc] Koecher, M.: *Zur Theorie der Modulformen n-ten Grades I*, Math. Zeitschrift **59**, 455–466 (1954)

[Koe] Koebe, P.: *Über die Uniformisierung beliebiger analytischer Kurven*, Nachr. Akad. Wiss. Göttingen, 197–210 (1907)

[La] Lamotke, K.: *Riemann'sche Flächen*, Springer, Berlin, Heidelberg, 2005

[Le] Lelong, P.: *Entire Functions of Several Complex Variables*, Grundlehren der mathematischen Wissenschaften in Einzeldarstellungen, Bd. 282. Springer, Berlin Heidelberg, 1986

[Lo] Lober, A.: *Ein Satz von Igusa über einen Ring von Modulformen zweiten Grades und halbzahligen Gewichts*, Abh. Math. Sem. Univ. Hamburg **65**, 155–163 (1995)

[Ma1] Maaß, H.: *Lectures on Siegel's Modular Functions*, Tata Institute of Fundamental Research, Bombay, 1955

[Ma2] Maaß, H.: *Siegel's modular Forms and Dirichlet Series*, Lecture Notes in Mathematics, Bd. 216, Springer, Berlin, Heidelberg, 1971

[Mal] Malgrange, B.: *Lectures on the Theory of Functions of Several Complex Variables*, Lectures on Mathematics and Physics: Mathematics, No. 13. Tata Institute of Fundamental Research, Bombay, 1965

References 497

[Mu] Mumford, D.: *Tata Lectures on Theta,* 3 vols. Birkhäuser, Boston, 1983, 1984, 1987

[Na2] Narasimhan, R.: *Introduction to the Theory of Analytic Spaces,* Lecture Notes in Mathematics, No. 25. Springer, Berlin, Heidelberg, 1966

[Na3] Narasimhan, R.: *Several Complex Variables,* Chicago Lectures in Mathematics, University of Chicago Press, Chicago, 1971.

[Nac] Nachbin, L.: *Holomorphic Functions, Domains of Holomorphy and Local Properties,* North-Holland Mathematics Studies, No. 1, North-Holland, Amsterdam, 1970.

[Ne] Nevanlinna, R.H.: *Uniformisierung,* 2nd edn. Grundlehren der mathematischen Wissenschaften in Einzeldarstellungen, Nr. 64. Springer, Berlin, Heidelberg, 1967

[Os] Osgood, W.F.: *Lehrbuch der Funktionentheorie,* Chelsea, New York, 1912

[Pf] Pfluger, A.: *Theorie der Riemann'schen Flächen,* Grundlehren der mathematischen Wissenschaften in Einzeldarstellungen, Nr. 89, Springer, Berlin, Heidelberg, 1957

[Po] Poincaré, H.: *Sur l'Uniformisation des Fonctions Analytiques,* Acta Math. **31**, 1–64 (1907)

[Pol] Polishchuk, A.: *Abelian Varieties, Theta Functions and the Fourier Transform,* Cambridge Tracts in Mathematics, No. 153. Cambridge University Press, Cambridge, 2003

[RF] Rauch, H.E., Farkas, H.M.: *Theta Functions with Applications to Riemann Surfaces,* Williams & Wilkins, Baltimore, MD, 1974

[RK] Rothstein, W., Kopfermann, K.: *Funktionentheorie mehrerer komplexer Veränderlicher,* Bibliogr. Inst., Mannheim 1982

[RL] Rauch, H.E., Lebowitz, A.: *Elliptic functions, theta functions, and Riemann surfaces,* Williams & Wilkins, Baltimore, Md. 1973

[Ru1] Rudin, W.: *Function Theory in Polydisks,* Mathematics Lecture Note Series, Benjamin, New York, 1969

[Si1] Siegel, C.L.: *Topics in Complex Function Theory,* vols. 1–3. Intersience Tracts in Pure and Applied Mathematics, No. 25. Wiley-Interscience, New York, 1969, 1971, 1973

[Si2] Siegel, C.L.: *Gesammelte Abhandlungen,* ed. K. Chandrasekharan. Springer, Berlin, Heidelberg, 1966

[Sp] Springer, G.: *Introduction to Riemann Surfaces,* Addison-Wesley, Reading, MA, 1957

[St] Strebel, K.: *Vorlesungen über Riemann'sche Flächen,* Studia mathematics, Skript Nr. 5. Vandenhoeck & Ruprecht, Göttingen, 1980

[Ti] Thimm, W.: *Vorlesung über Funktionentheorie mehrerer Veränderlicher,* Ausarbeitungen mathematischer und physikalischer Vorlesungen, Nr. 25, Aschendorff, Münster, 1961

[Vl] Vladimirov, V.S.: *Methods of the Theory of Functions of Many Complex Variables,* MIT Press, Cambridge, MA, 1966

[We] Weyl, H.: *Die Idee der Riemannschen Fläche,* herausgegeben von Reinhold Remmert , Teubner-Archiv zur Mathematik, Supplement 5, Teubner Stuttgart, Leipzig, 1997.

Index